Wearable Robots

Wearable Robots: Biomechatronic Exoskeletons

Edited by
José L. Pons
CSIC, Madrid, Spain

John Wiley & Sons, Ltd

Telephone (+44) 1243 779777

Email (for orders and customer service enquiries): cs-books@wiley.co.uk
Visit our Home Page on www.wileyeurope.com or www.wiley.com

Other Wiley Editorial Offices

John Wiley & Sons Inc., 111 River Street, Hoboken, NJ 07030, USA

Jossey-Bass, 989 Market Street, San Francisco, CA 94103-1741, USA

Wiley-VCH Verlag GmbH, Boschstr. 12, D-69469 Weinheim, Germany

John Wiley & Sons Australia Ltd, 42 McDougall Street, Milton, Queensland 4064, Australia

John Wiley & Sons (Asia) Pte Ltd, 2 Clementi Loop #02-01, Jin Xing Distripark, Singapore 129809

John Wiley & Sons Canada Ltd, 6045 Freemont Blvd, Mississauga, Ontario, L5R 4J3, Canada

Wiley also publishes its books in a variety of electronic formats. Some content that appears in print may not be available in electronic books.

British Library Cataloguing in Publication Data

A catalogue record for this book is available from the British Library

ISBN 978-0-470-51294-4 (HB)

Typeset in 9/11pt Times by Laserwords Private Limited, Chennai, India

Arms
manipulators,
prostheses,
assistive
robots,
orthoses...

Contents

Foreword

Being a multidisciplinary area involving subjects such as mechanics, electronics and computing, the evolution and spread of robotics to different application sectors still requires intense interaction with other fields of science and technology. This applies equally when dealing with wearable robots, meaning robotic systems that a person wears to enhance his/her capabilities in some way. Since the first wearable robots, conceived in the early 1990s as amplifiers of human force or reach, progress in all robotics-related areas has been moving in the direction of a symbiosis between humans and robots as a means of enhancing human abilities in the fields of perception, manipulation, walking and so on.

Although the number of books available on robotics is huge, the existing literature in specific fields of robotic application is not so extensive; moreover, it appears that there is no book conceived as a compendium of all the subject matter involved in such specific emerging areas. The present book is intended to fill the gap in the field of wearable robots – an emerging sector that constitutes a step forward in robotic systems, which rely on the fact of having a human in the loop. That progress in the field is continuously expanding is evident from the number of publications on advances in research and development, new prototypes and even commercial products. Therefore, a book that brings together all the different subject matter encompassed by this discipline will assuredly be of valuable assistance in gaining an appreciation of the wide range of knowledge required; furthermore, by identifying the main concepts involved in dealing with such robots, it can be of help to new researchers wishing to enter the field.

As this book shows, in the field of wearable robots human/robot interaction is a key issue, from a physical or a cognitive point of view, or from both. Therefore, besides a solid knowledge of robotic techniques, research and development in this area also requires some background in anatomical behaviour of the human body and in the human neurological and cognitive systems. In this context, bioinspired or biomimetic design is of special importance for purposes of reproducing human functions or copying human actions respectively. Wearable robots must be designed to cope with specific working conditions, such as the need to accommodate a nonfixed structure, i.e. the human body; to be compliant, light and intrinsically safe enough to be worn by a user; or to be equipped with the requisite interfaces to enable easy intuitive control by a human.

Within this context, before going on to deal with exoskeletons – in the form of upper or lower limbs, or the trunk – as orthotic/prosthetic elements, the book looks at bioinspired and biomimetic systems, describing the human neuromotor system, the body kinematics and dynamics, and the human–machine interface requirements. The biologically inspired design of wearable robots requires a study of computational counterparts, such as genetic algorithms, as well as other technical issues like lightness of components, power efficiency, and general technological aspects of the elements involved in the design. On the subject of design of robot architectures for wearable robots, the book presents a preliminary study of human biomechanics and human mobility modelling. Special emphasis is placed on the analysis of potential human–machine interfaces for such robots, distinguishing between cognitive and physical interaction, which require quite different technologies: in the former case these have more to do with medical and biological aspects such as EEG and EMG signals, while in the latter case there is more reliance on engineering. Given the large number of sensors and actuators embedded in wearable robots, and also robot design requirements, communication networks are a key issue, which is dealt with by analysing the various existing techniques, naturally with particular attention to the performance of wireless technology.

With so broad a scope, the book will be of interest to students and researchers having some background in robotics and an interest or some experience in rehabilitation robots and assistive technology. It is also intended to provide basic educational material with which to introduce medical personnel or other specialists to the capabilities of such robotic systems. Rather than being a collection of materials, the book is carefully structured in such a way that the consecutive chapters allow the reader to perceive the context and requirements and gain an idea of the current solutions and future trends in this exciting field.

Alicia Casals
Professor, UPC

Preface

This book is the result of several years of research and work by the Bioengineering Group (CSIC) on the use of Robotics to assist handicapped people. The aim of the book is to provide a comprehensive discussion of the field of Wearable Robotics. Rehabilitation, Assistance and Functional Compensation are not the only fields of application for Wearable Robotics, but they may be regarded as paradigmatic scenarios for robots of this kind. The book covers most of the scientific topics relating to Wearable Robotics, with particular focus on bioinspiration, biomechatronic design, cognitive and physical human–robot interaction, wearable robot technologies (including communication networks), kinematics, dynamics and control. The book was enriched by the contribution of outstanding scientists and experts in the different topics addressed here. I would like to thank them all.

This book could not have been written without help and contributions from many people. I wish to express my gratitude to M. Wisse for his contributions to Chapter 2, particularly in all those aspects relating to the bioinspired design of robots, and to A. Schiele, also of Delft University of Technology (The Netherlands), for his contributions to Chapters 3 and 5; his comments in the field of kinematics, ergonomics and human–robot physical interaction are particularly interesting.

Many research groups worldwide have contributed by means of case studies. J.M. Belda-Lois, R. Poveda, R. Barberà, J.M. Baydal-Bertomeu, D. Garrido, F. Moll, M.J. Vivas and J.M. Prat, of the Instituto de Biomecánica de Valencia (Spain), provided valuable contributions in the fields of biomechanics, bioinspired design of exoskeletons and kinematic compatibility, as well as microclimate sensing, comfort and ergonomics in orthotics in Chapters 3, 5 and 6.

J.M. Carmena, of the Department of Electrical Engineering and Computer Sciences, Helen-Wills Neuroscience Institute, University of California (USA), contributed to Chapter 4 with new concepts for the cortical control of robots. In the same field but with the help of surface EEG, T.F. Bastos-Filho, M. Sarcinelli-Filho, A. Ferreira, W.C. Celeste, R.L. Silva, V.R. Martins, D.C. Cavalieri, P.N.S. Filgueira and I.B. Arantes, of the Federal University of Espirito Santo (Brazil), provided a discussion of brain-controlled robots and introduced some preliminary results with healthy users as a first step towards clinical validation of these technologies.

The book also reflects Italy's place at the forefront of Robotics research. Several groups contributed to this book. L. Beccai, S. Micera, C. Cipriani, J. Carpaneto, M.C. Carrozza, S. Roccella, E. Cattin, N. Vitiello and F. Vecchi, of the ARTS Lab, Scuola Superiore Sant'Anna, Pisa (Italy), enriched it with contributions in the field of bioinspired and biomechatronic design of wearable robots, in particular in upper limb exoskeletons for neuromotor research and in novel neuroprosthetic control of upper limb robotic prostheses. I would like to thank in particular M.C. Carrozza and Prof. P. Dario for their support. E. Farella and L. Benini, of the Department of Electronics, Computer Science and Systems, University of Bologna (Italy), contributed to the area of wireless sensor networks and the implementation of the posture and gesture interaction scheme. Finally, N.G. Tsagarakis and D.G. Caldwell, of the Italian Institute of Technology, in cooperation with S. Kousidou, of the Centre of Robotics and Automation, University of Salford (UK), contributed to the field of upper limb exoskeletons in those aspects relating to soft arm design and control.

There are also five additional contributions by groups from Finland, the USA, Iceland and Japan. J. Vanhala, of the Tampere University of Technology, contributed a discussion on wearable technologies with applications both to wearable robots and to smart textiles. J.C. Perry and J. Rosen, of the Department of Electrical Engineering, University of Washington (USA), provided a thorough discussion of upper limb exoskeletons with particular emphasis on kinematic compatibility between the human limb and the robot kinematics, from the special perspective of fitting into activities of daily living. D.P. Ferris, of the Division of Kinesiology, Department of Biomedical Engineering and Department of Physical Medicine and Rehabilitation, The University of Michigan (USA), presented a discussion on the application of pneumatic actuators to lower limb orthoses. K. De Roy, of Össur (Iceland), contributed a discussion on walking dynamics under normal, impaired and restored conditions following the fitting of robotic lower limb prostheses. I would like to thank F. Thorsteinsson for supporting this project and for our collaboration during the last few years. Finally, a full–body exoskeleton with pneumatic actuation is presented by K. Yamamoto, of the Kanagawa Institute of Technology (Japan).

Most of the work presented in this book has been developed in the framework of four European projects. Therefore, I would like to acknowledge the European Commission for the partial funding of this work under the following contracts:

- MANUS – modular anthropomorphous user-adaptable hand prosthesis with enhanced mobility and force feedback (EU Telematics DE-4205).
- DRIFTS – dynamically responsive intervention for tremor suppression (EU Quality of Life QLRT-2001-00536).
- GAIT – intelligent knee and ankle orthosis for biomechanical evaluation and functional compensation of joint disorders (UE IST IST-2001-37751).
- ESBiRRo – biomimetic actuation, sensing and control technology for limit cycle bipedal walkers (UE FP6-2005-IST-61-045301-STP).

In writing this book I have received the unstinting support of my colleagues in the Bioengineering Group. Professor R. Ceres and Dr L. Calderón contributed to the introduction to *Wearable Robotics* and to the concluding remarks and the outlook. Dr E. Rocon and A.F. Ruiz have been behind the contributions on physical human–robot interaction and on upper limb wearable robots, and R. Raya cooperated with them on the interaction between humans and robots. Dr A. Forner-Cordero contributed in those topics relating to the biological basis and in the biomechanical foundations for the design of wearable robots. In this particular regard, E. Turowska provided input on the kinematic analysis of both robot and human limbs.

The analysis of the cognitive interaction between humans and robots comes from L. Bueno, F. Brunetti and A. Frizera. L. Bueno contributed, in cooperation with J.C. Moreno, to the discussion on wearable robot technologies. In addition, J.C. Moreno provided the discussions on lower limb wearable robots. The main contribution from F. Brunetti was in the area of communication networks for wearable robots and wearable technologies.

Finally, I would like to thank all my colleagues in the Bioengineering Group, CSIC, in particular Luis and Lola, my family and my parents to whom I owe everything, and to God.

José L. Pons
Research Scientist, CSIC

List of Contributors

F. Brunetti
Bioengineering Group,
Instituto de Automática Industrial,
CSIC, Madrid, Spain

L. Bueno
Bioengineering Group,
Instituto de Automática Industrial,
CSIC, Madrid, Spain

L. Calderón
Bioengineering Group,
Instituto de Automática Industrial,
CSIC, Madrid, Spain

R. Ceres
Bioengineering Group,
Instituto de Automática Industrial,
CSIC, Madrid, Spain

A. Forner-Cordero
Bioengineering Group,
Instituto de Automática Industrial,
CSIC, Madrid, Spain

A. Frizera
Bioengineering Group,
Instituto de Automática Industrial,
CSIC, Madrid, Spain

J. C. Moreno
Bioengineering Group,
Instituto de Automática Industrial,
CSIC, Madrid, Spain

J. L. Pons
Bioengineering Group,
Instituto de Automática Industrial,
CSIC, Madrid, Spain

R. Raya
Bioengineering Group,
Instituto de Automática Industrial,
CSIC, Madrid, Spain

E. Rocon
Bioengineering Group,
Instituto de Automática Industrial,
CSIC, Madrid, Spain

A. F. Ruiz
Bioengineering Group,
Instituto de Automática Industrial,
CSIC, Madrid, Spain

A. Schiele
Mechanical Engineering Department,
Automation & Robotics Section,
ESA, European Space Agency (ESA),
Noordwijk, The Netherlands
Mechanical Engineering Faculty,
Biomechanical Engineering
Department, DUT,
Delft University of Technology (DUT),
Delft, The Netherlands

E. A. Turowska
Bioengineering Group,
Instituto de Automática Industrial,
CSIC, Madrid, Spain

M. Wisse
Mechanical Engineering Faculty,
Biomechanical Engineering Department,
DUT, Delft University of
Technology (DUT),
Delft, The Netherlands

List of Contributors

1

Introduction to wearable robotics

J. L. Pons, R. Ceres and L. Calderón

Bioengineering Group, Instituto de Automática Industrial, CSIC, Madrid, Spain

1.1 WEARABLE ROBOTS AND EXOSKELETONS

The history of robotics is one of ever closer interaction with the human actor. Originally, robots were only intended for use in industrial environments to replace humans in tedious and repetitive tasks and tasks requiring precision, but the current scenario is one of transition towards increasing interaction with the human operator. This means that interaction with humans is expanding from a mere exchange of information (in teleoperation tasks) and service robotics to a close interaction involving physical and cognitive modalities.

It is in this context that the concept of *Wearable Robots* (WRs) has emerged. Wearable robots are person-oriented robots. They can be defined as those worn by human operators, whether to supplement the function of a limb or to replace it completely. Wearable robots may operate alongside human limbs, as in the case of orthotic robots or exoskeletons, or they may substitute for missing limbs, for instance following an amputation. Wearability does not necessarily imply that the robot is ambulatory, portable or autonomous. Where wearable robots are nonambulatory, this is in most instances a consequence of the lack of enabling technologies, in particular actuators and energy sources.

A wearable robot can be seen as a technology that extends, complements, substitutes or enhances human function and capability or empowers or replaces (a part of) the human limb where it is worn. A possible classification of wearable robots takes into account the function they perform in cooperation with the human actor. Thus, the following are instances of wearable robots:

- *Empowering robotic exoskeletons*. These were originally called *extenders* (Kazerooni, 1990) and were defined as a class of robots that extends the strength of the human hand beyond its natural ability while maintaining human control of the robot. A specific and singular aspect of extenders is that the exoskeleton structure maps on to the human actor's anatomy. Where the extension of the ability of the human operator's upper limb is more to do with reach than power, master–slave robot configurations occur, generally in teleoperation scenarios.

Wearable Robots: Biomechatronic Exoskeletons Edited by José L. Pons

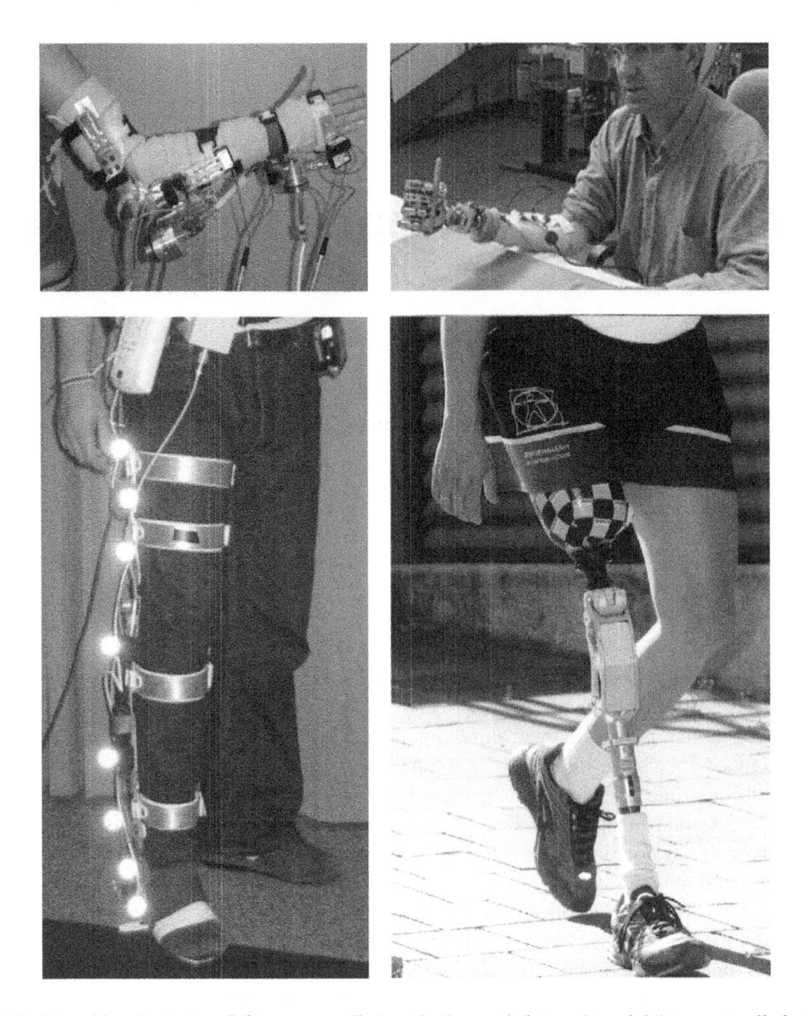

Figure 1.1 Wearable robots: (top left) an upper limb orthotic exoskeleton; (top right) an upper limb prosthetic robot; (bottom left) a lower limb orthotic exoskeleton; (bottom right) a lower limb prosthetic robot

- *Orthotic robots*. An orthosis is a mechanical structure that maps on to the anatomy of the human limb. Its purpose is to restore lost or weak functions, e.g. following a disease or a neurological condition, to their natural levels. The robotic counterparts of orthoses are robotic exoskeletons. In this case, the function of the exoskeleton is to complement the ability of the human limb and restore the handicapped function (see Figure 1.1).
- *Prosthetic robots*. A prosthesis is an electromechanical device that substitutes for lost limbs after amputation. The robotic counterparts of prostheses take the form of electromechanical wearable robotic limbs and make it possible to replace the lost limb function in a way that is closer to the natural human function. This is achieved by intelligent use of robotics technologies in terms of human–robot interaction (comprising sensing and control) and actuation (see Figure 1.1).

1.1.1 Dual human–robot interaction in wearable robotics

The key distinctive aspect in wearable robots is their intrinsic dual cognitive and physical interaction with humans. On the one hand, the key role of a robot in a *physical human–robot interaction* (pHRI) is the generation of supplementary forces to empower and overcome human physical limits (Alami *et al.*, 2006), be they natural or the result of a disease or trauma. This involves a net flux of power between both actors. On the other hand, one of the crucial roles of a *cognitive human–robot interaction* (cHRI) is to make the human aware of the possibilities of the robot while allowing him to maintain control of the robot at all times. Here, the term *cognitive* alludes to the close relationship between cognition – as the process comprising high-level functions carried out by the human brain, including comprehension and use of speech, visual perception and construction, the ability to calculate, attention (information processing), memory and executive functions such as planning, problem-solving, self-monitoring and perception – and motor control.

Both pHRI and cHRI are supported by a *human–robot interface* (HRi). An interface is a hardware and software link that connects two dissimilar systems, e.g. robot and human. Two devices are said to be interfaced when their operations are linked informationally, mechanically or electronically. In the context of wearable robotics, the interface is the link that supports interaction – the interaction between robot and human through control of the flow of information or power.

In wearable robotics, a *cognitive human–robot interface* (cHRi) is explicitly developed to support the flow of information in the cognitive interaction (possibly two-way) between the robot and the human. Information is the result of processing, manipulating and organizing of data, and so the cHRi in the human-robot direction is based on data acquired by a set of sensors to measure bioelectrical and biomechanical variables. Likewise, the cHRi in the robot–human direction may be based on biomechanical variables, a subset of bioelectrical variables, e.g. electroneurography (ENG), and modalities of natural perception, e.g. visual and auditory.

Similarly, a *physical human–robot interface* (pHRi) is explicitly developed to support the flow of power between the two actors. The pHRi is based on a set of actuators and a rigid structure that is used to transmit forces to the human musculoskeletal system. The close physical interaction through this interface imposes strict requirements on wearable robots as regards safety and dependability.

Cognitive and physical interactions are not independent. On the one hand, a perceptual cognitive process in the human can be triggered by physical interaction with the robot. One example is a wearable robot physically interacting with an operator to render haptic information on a virtual or remote object, so that the operator can feel the object (soft or rigid) (see Figure 1.2).

On the other hand, the cognitive interaction can be used to modify the physical interaction between human and robot, for instance to alter the compliance of an exoskeleton. One example is tremor suppression based on exoskeleton–human interaction: the onset of a tremor can be inferred from the biomechanical data of limb motion (cognitive process); this is used to modify the biomechanical characteristics of the human limb (damping and apparent inertia), which in turn leads to tremor reduction.

In this context, the cognitive interaction resulting from a human–robot (H–R) physical interaction can be either *conscious* or *involuntary*. The previous example of haptic rendering by means of wearable robots is a good example of conscious perceptual cognitive interaction. Involuntary cognitive interaction is produced by low-level, reflex-like mechanisms on either side of the human–robot interface. This is exemplified by a more subtle case of physically triggered human involuntary cognitive processes experienced in exoskeletons used to suppress tremor of the human upper limb. It has been shown (Manto *et al.*, 2007) that the modification of biomechanical characteristics of the human musculoskeletal system around a joint, e.g. the wrist, triggers a modification of human motor control processes that results in migration of tremor to adjacent joints, e.g. the elbow.

Involuntary cognitive interactions between robot and human can of course be nested at different levels. In the previous example of tremor reduction by means of exoskeletons, it was found that

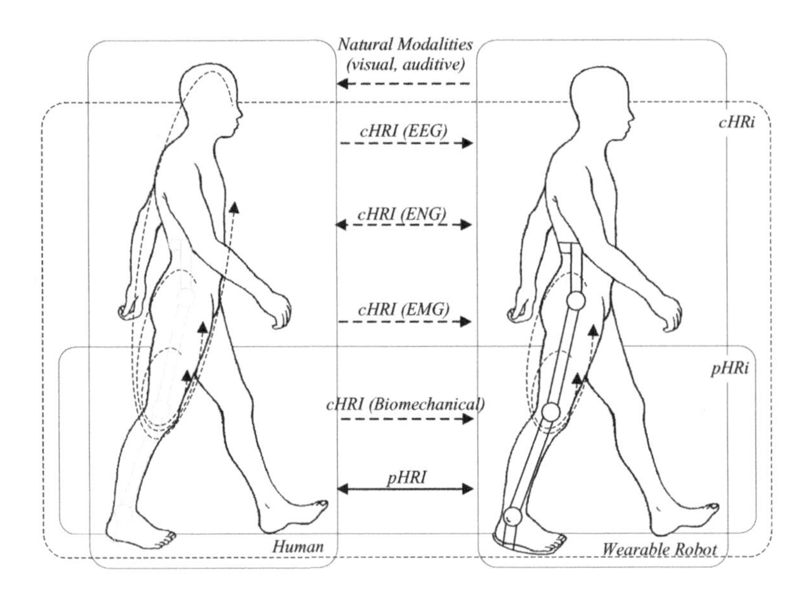

Figure 1.2 Schematic representation of dual cognitive and physical interaction in wearable robots

visual feedback of tremor reduction to the user – i.e. the use of natural perceptual visual information – triggers human motor control mechanisms that further reduce tremor. These human motor control mechanisms operate on the human side of the interface and are superimposed on the tremor migration mechanisms of the previous example; they are triggered by the pHRI and the cHRI though natural modes of perception (vision) and involve different motor control levels.

1.1.2 A historical note

Of the different wearable robots, exoskeletons are the ones in which the cognitive (information) and physical (power) interactions with the human operator are most intense. Scientific and technological work on exoskeletons began in the early 1960s. The US Department of Defense became interested in developing the concept of a powered 'suit of armor'. At the same time, at Cornel Aeronautical Laboratories work started to develop the concept of man–amplifiers – manipulators to enhance the strength of a human operator. The existing technological limitations on development of the concept were established in 1962; these related to servos, sensors and mechanical structure and design. Later on, in 1964, the hydraulic actuator technology was identified as an additional limiting factor.

General Electric Co. further developed the concept of human–amplifiers through the *Hardiman project* from 1966 to 1971. The Hardiman concept was more of a robotic master–slave configuration in which two overlapping exoskeletons were implemented. The inner one was set to follow human motion while the outer one implemented a hydraulically powered version of the motion performed by the inner exoskeleton. The concept of extenders versus master/slave robots as systems exhibiting genuine information and power transmission between the two actors was coined in 1990 (Kazerooni, 1990).

Efforts in the defence and military arena have continued up to the present, chiefly promoted by the US Defense Advanced Research Projects Agency (DARPA). Additional details on this can be found in Section 1.4.

Rehabilitation and functional compensation exoskeletons are another classic field of application for wearable robotics. Passive orthotic or prosthetic devices do not fall within the scope of this book, but they may be regarded as the forebears of current rehabilitation exoskeletons. More than a century ago, Prof. H. Wangenstein proposed the concept of a mobility assistant for scientists bereft of the use of their legs:

> This amazing feat shall revolutionize the way in which paraplegic Scientists continue their honorable work in the advancement of Science! Even in this modern day and age, some injuries cannot be healed. Even with all the Science at our command, some of our learned brethren today are without the use of their legs. This Device will change all that. From an ordinary-appearing wheelchair, the Pneumatic Bodyframe will transform into a light exoskeleton which will allow the Scientist to walk about normally. Even running and jumping are not beyond its capabilities, all controlled by the power of the user's mind. The user simply seats himself in the chair, fits the restraining belts around his chest, waist, thighs and calves, fastens the Neuro-Impulse Recognition Electrodes (N.I.R.E.) to his temples, and is ready to go!

The concept introduced by Prof. Wangenstein in 1883 contains the main features of current state-of-the-art wearable robotic exoskeletons: a pneumatically actuated body frame (in the form of a light exoskeleton), mapping on to the human lower limb, in which a cHRI is established by means of brain activity electrodes (known as NIRE).

Among the spinoff applications of robotic extenders are robotic upper limb orthoses (Rabischong, 1982). Although studies on active controlled orthoses date back to the mid 1950s (Battyke, Nightingale and Whilles, 1956), the first active implementations of powered orthoses were the work of Rahman *et al.* (2000). This functional upper limb orthosis was conceived for people with limited strength in their arms.

1.1.3 Exoskeletons: an instance of wearable robots

The exoskeleton is a species of wearable robot. The distinctive, specific and singular aspect of exoskeletons is that the exoskeleton's kinematic chain maps on to the human limb anatomy. There is a one-to-one correspondence between human anatomical joints and the robot's joints or sets of joints. This kinematic compliance is a key aspect in achieving ergonomic human–robot interfaces, as further illustrated in Chapters 3 and 5.

In exoskeletons, there is an effective transfer of power between the human and the robot. Humans and exoskeletons are in close physical interaction. This is the reverse of master–slave configurations, where there is no physical contact between the slave and the human operator, which are remote from one another. However, in some instances of teleoperation, an upper limb exoskeleton can be used as the interface between the human and the remote robot. According to this concept, the exoskeleton can be used as an input device (by establishing a pose correspondence between the human and the slave or remote manipulator), as a force feedback device (by providing haptic interaction between the slave robot and its environment), or both.

The interaction between the exoskeleton and the human limb can be achieved through *internal force* or *external force* systems. Which of these force interaction concepts is chosen depends chiefly on the application. On the one hand, empowering exoskeletons must be based on the concept of external force systems; empowering exoskeletons are used to multiply the force that a human wearer can withstand, and therefore the force that the environment exerts on the exoskeleton must be grounded: i.e. in external force systems the exoskeleton's mechanical structure acts as a load-carrying device and only a small part of the force is exerted on the wearer. The power is transmitted to an external base, be it fixed or portable with the operator. The only power transmission is between the human limbs and the robot as a means of implementing control inputs and/or force feedback. This concept is illustrated in Figure 1.3 (right).

Figure 1.3 Schematic representation of internal force (left) and external force (right) exoskeletal systems

On the other hand, orthotic exoskeletons, i.e. exoskeletons for functional compensation of human limbs, work on the internal force principle. In this instance of a wearable robot, the force and power are transmitted by means of the exoskeleton between segments of the human limb. Orthotic exoskeletons are applicable whenever there is weakness or loss of human limb function. In such a scenario, the exoskeleton complements or replaces the function of the human musculoskeletal system. In internal force exoskeletons, the force is nongrounded; force is applied only between the exoskeleton and the limb. The concept of internal force exoskeletons is illustrated in Figure 1.3 (left).

Superimposing a robot on a human limb, as in the case of exoskeletons, is a difficult problem. Ideally, the human must feel no restriction to his/her natural motion patterns. Therefore, kinematics plays a key role in wearable exoskeletons: if robots and humans are not kinematically compliant, a source of nonergonomic interaction forces appears. This is comprehensively addressed in Sections 3.4 and 5.2. The former analyses the kinematics of interacting human–robot systems. The latter theoretically analyses the forces resulting from kinematically noncompliant human–robot systems; this theoretical analysis is then quantified in Case Study 5.5.

Kinematic compatibility is of paramount importance in robotic exoskeletons working on the principle of internal forces. The typical misalignment between exoskeleton and anatomical joints results in uncomfortable interaction forces where both systems are attached to each other. Given the complex kinematics of most human anatomical joints, this problem is hard to avoid. The issue of compliant kinematics calls for bioinspired design of wearable robots and imposes a strong need for control of the human–robot physical interaction.

Exoskeletons are also characterized by a close cognitive interaction with the wearer. This cHRI is in most instances supported by the physical interface. By means of this cognitive interaction, the human commands and controls the robot, and in turn the robot includes the human in the control loop and provides information on the tasks, either by means of a force reflexion mechanism or of some other kind of information.

1.2 THE ROLE OF BIOINSPIRATION AND BIOMECHATRONICS IN WEARABLE ROBOTS

It is widely recognized that evolutionary biological processes lead to efficient behavioural and motor mechanisms. Evolution in biology involves all aspects and functions of creatures, from perception to actuation–locomotion, in particular gait, and manipulation–through efficient organization of motor control. Evolution is a process whereby functional aspects of living creatures are optimized. This optimization process seeks the maximization of certain objective functions, e.g. manipulative dexterity in human hands and efficiency in terms of energy balance in performing a certain function. Chapter 2 of this book analyses the basis for bioinspiration and biomimetism in the design of wearable robots.

Neurobiology plays a crucial role in hypothesizing engineering-inspired biological models. For example, some biological models explain how energetically efficient locomotion and gait speed modulation of six-legged insects can be achieved through frequency and stride length modification resulting in effective speed change. Engineering in turn plays a crucial role in validating neurobiological models by looking at how artificial systems reproduce and explain biological behaviour and performance. For instance, parallax motion in insects is validated by means of Dro-o-boT, a robot whose motion proved identical to that of insects when programmed following the principle of parallax motion (Abbott, 2007).

It is clear that the design of wearable robots can benefit from biological models in a number of aspects like control, sensing and actuation. Likewise, wearable robots can be used to understand and formalize models of biological motor control in humans. This concurrent view calls for a multidisciplinary approach to wearable robot development, which is where the concept of *biomechatronics* comes in.

The term *mechatronics* was coined in Japan in the mid 1970s and has been defined as the engineering discipline dealing with the study, analysis, design and implementation of hybrid systems comprising mechanical, electrical and control (intelligence) components or subsystems (Pons, 2005). Mechatronic systems closely linked to biological systems have been referred to as *biocybernetic systems* in the context of electromyography (EMG) control of the full-body HAL-5 exoskeleton wearable robot system (see Case Study 9.4). The concept of *biomechatronics* is not limited to biocybernetic systems.

Biomechatronics can be analysed by analogy to biological systems integrating a musculoskeletal apparatus with a nervous system (Dario *et al.*, 2005). Following this analogy (see Figure 1.4), biomechatronic systems integrate mechanisms, embedded control and human–machine interaction (HMI), sensors, actuators and energy supply in such a way that each of these components, and the whole mechatronic system, is inspired by biological models. This book stresses the biomechatronic conception of wearable robots:

- Bioinspiration is analysed in Chapter 2. This chapter explains the essentials of the design of wearable robots based on biological models.
- Mechanisms (in the context of wearable robots) are analysed in Chapter 3. This chapter addresses the particular kinematic and dynamic considerations of mapping robots on to human limb anatomy.
- HMI in the context of wearable robots, i.e. human–robot interaction, is analysed in Chapters 4 and 5. The former focuses on the cognitive aspects of this interaction while the latter addresses the physical interaction.
- Sensors, actuators and energy supply–i.e. technologies enabling the implementation of wearable robots–are analysed in Chapter 6. In many instances, sensors, actuators and control components are included in the wearable robot structure as nodes of a communication network. Networks for WRs are analysed in Chapter 7.

Biomechatronics may in a sense be viewed as a scientific and engineering discipline whose goal is to explain biological behaviour by means of artificial models, e.g. the system's components: sensors, actuators, control etc. This is consistent with the dual role of bioinspiration: firstly, to gain insight by observing biological models and, secondly, to explain biological function by means of engineering models.

Biomechatronics may be regarded as an extension of mechatronics. The scope of biomechatronics is broader in three distinctive aspects: firstly, biomechatronics intrinsically includes bioinspiration

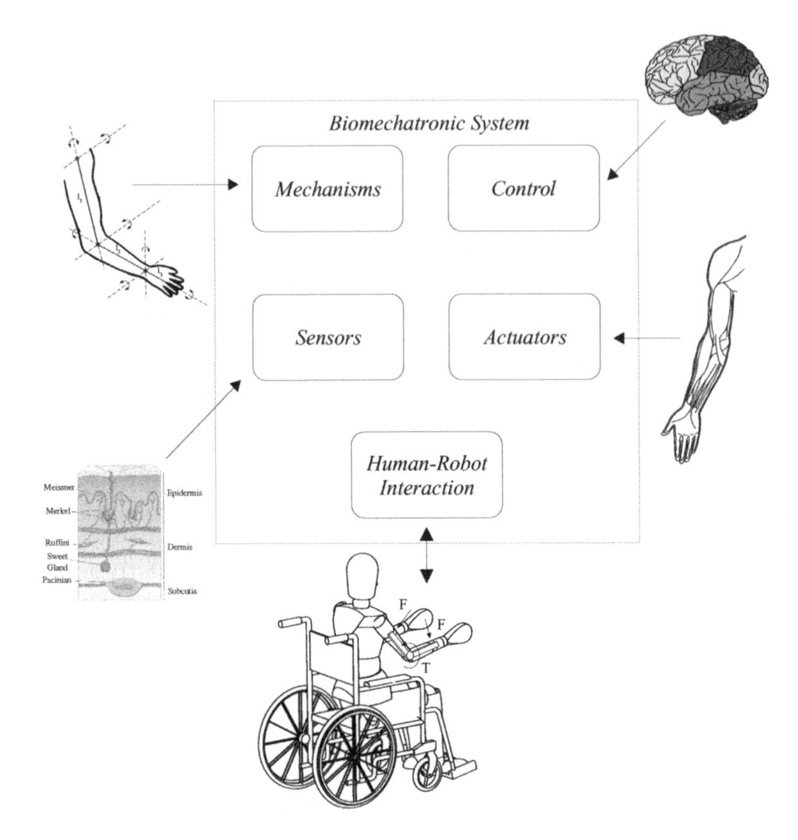

Figure 1.4 Components in a biomechatronic system

in the development of mechatronic systems, e.g. the development of bioinspired mechatronic components (control architectures, actuators, etc.); secondly, biomechatronics deals with mechatronic systems in close interaction with biological systems, e.g. a wearable robot interacting cognitively and physically with a human; and, finally, biomechatronics commonly adopts biologically inspired design and optimization procedures in the development of mechatronic systems, e.g. the adoption of genetic algorithms in the optimization of mechatronic components or systems. These three salient aspects of biomechatronics are further illustrated in the following paragraphs.

1.2.1 Bioinspiration in the design of biomechatronic wearable robots

Bioinspiration has been extensively adopted in the development of wearable robots. This includes the development of the complete robot system and its components. Bioinspiration in the context of actuator design has been studied in detail elsewhere (Pons, 2005). Here, a few examples are cited in the context of wearable robots, which are further detailed in case studies throughout this book.

Bioinspired actuators have also been developed in the context of wearable robots. A bioinspired knee actuator for a lower limb exoskeleton is analysed in Case Study 6.7. This shows that due to power and torque requirements in human gait, no state-of-the-art actuator technology can be applied to compensate quadriceps weakness during gait. It can be shown that the mechanical equivalent of the quadriceps muscle during the stance phase is a rigid spring–damper configuration and the mechanical equivalent of the quadriceps muscle during the swing phase is a soft spring–damper configuration. With this in mind, a passive knee actuator can be developed by switching between these mechanical configurations as a function of gait phase.

An example of bioinspired hierarchical motor control of manipulation and grasping is described in detail in Case Study 2.6. Here, the grasping control strategy is split into high-level grasp preshaping command primitives and the low-level grasping reflex. The high-level controller must set a reference for position and stiffness for each low-level controller so that a particular type of grasp can be implemented, e.g. lateral, precision or power. The low-level controllers implement reflexes to counteract slippage and other perturbations.

Finally, the HAL-5 full-body exoskeleton is introduced here as an example of biomimetism in robot design (see Case Study 9.4). The controller of the HAL-5 exoskeleton implements a dual control scheme. A first EMG control algorithm commands the system on the basis of the human electromyographic activity. A second control loop works on stored walking patterns for the human operator so that smooth, synchronized motion is achieved. In this way, the first time the exoskeleton is worn, HAL-5 stores gait patterns that are then mimicked during operation.

1.2.2 Biomechatronic systems in close interaction with biological systems

Any of the various different wearable robots described throughout this book would be a good example of a biomechatronic system in very close cognitive and physical interaction with a human. However, cortical control of robots (see Case Study 4.7) and the neural interface to control the CyberHand system (see Case Study 8.2) have been chosen to illustrate the intrinsically close interaction between biomechatronic systems and biological creatures.

Cortical control of robots constitutes a step forward in research into *brain–machine interfaces* (BMIs). It involves an intimate interaction between the biomechatronic system and the operator, a primate in this instance. The case of the CyberHand upper limb robotic prosthesis describes the development of a natural interface between the robot and the amputee. Classical prosthesis interfaces, e.g. EMG based, are intrinsically unidirectional as they only allow command generation (a cognitive interaction in the direction from human to prosthesis) and lack force or position – i.e. proprioceptive – feedback (a cognitive interaction in the direction from the robot to the human). The CyberHand system proposes a neural interface at the level of the *peripheral nervous system* (PNS) that allows both acquisition of neural information to command the robot and stimulation of the PNS in order to provide position and force feedback to the amputee.

1.2.3 Biologically inspired design and optimization procedures

Bioinspiration can be found both in the design of compliant kinematics for wearable robots and in the optimization method used in the design process, e.g. evolutionary optimization based on genetic algorithms. Case Study 3.5 illustrates the use of biologically inspired processes in the design of a knee joint for a lower limb exoskeleton. The study first introduces models for the kinematics of the anatomical joint and then goes on to examine a genetic algorithm approach to optimizing the design of a four-bar mechanism for the robot's joint.

1.3 TECHNOLOGIES INVOLVED IN ROBOTIC EXOSKELETONS

In most instances technologies are the limiting factor in developing novel robots. This is also true of wearable robots. Wearable robots are in many cases related to portable and ambulatory applications;

however, only a few examples of fully portable wearable robots can be found in the literature, one reason being a lack of enabling technologies.

Ambulatory scenarios require compact, miniaturized, energetically efficient technologies, e.g. control, sensors, actuators. Researchers have been looking to nature as a source of inspiration in the design of such efficient technologies, as discussed in detail in Chapter 2. All technologies involved in robotics need further development, but actuators and power sources are the ones that probably most limit wearability and portability at the present time.

As noted in Pons (2005), miniaturization is unlikely to be one of the main avenues of research in the particular case of actuator technologies. While miniaturization is a logical trend in sensor technologies (since sensors act as transducers between energy domains, ideally without influencing the physical phenomena they measure), actuators are designed to impose a mechanical state on the plant they drive without being influenced by perturbations. Therefore, power delivery is the key performance indicator, in most instances at the expense of miniaturization.

Portability is one important aspect of wearable robotics, but as has already been mentioned, the distinctive characteristic of wearable robots is dual cognitive and physical interaction with the human wearer. This immediately raises dependability and safety issues in robotics. Dependability and safety ultimately have a close bearing on control, sensor and actuator technologies, which interact directly with the human, and this again calls for the biomechatronic approach described in Section 1.2.

Chapters 3 to 6 of this book deal with the topics of wearability, portability, dependability and safety. Safety and dependability in human–robot interaction is addressed from the standpoint of mechanisms that map on to the human anatomy in Chapter 3. Here, the issue of the kinematically compliant design of WRs assures ergonomic physical interaction with the wearer. Dependability/safety in both cognitive and physical interaction with a robot are addressed in relation to control technologies in Chapters 4 and 5 respectively. Finally, the technologies (sensors, actuators and power sources) involved in the design of portable, dependable and safe wearable robots are discussed in Chapter 6.

1.4 A CLASSIFICATION OF WEARABLE EXOSKELETONS: APPLICATION DOMAINS

Wearable robots may be classified according to numerous different criteria. A division into orthotic and prosthetic robots was introduced in Section 1.1. According to this classification, orthotic wearable robots, e.g. exoskeletons, are those that operate mechanically parallel to the human body, whereas prosthetic wearable robots operate mechanically in series with the human body and their chief function is to substitute for lost body limbs, e.g. following an amputation.

Another previous section introduced wearable robots that interact with humans according to internal force or external force principles. This can likewise be considered a classification criterion. On the one hand, in internal force WRs, force and torque are applied only between the human and the robot, e.g. the robot can exert forces between consecutive segments in the human limb's kinematic chain. This concept of internal forces is mostly applied in the development of orthotic robots in the application domain of rehabilitation or in exoskeletons in master–slave teleoperation configurations. On the other hand, external force WRs are mainly used in empowering applications, whenever the role of the wearable robot is to ground a large proportion of the stress imposed on the human–robot system by the environment.

Wearable exoskeletons can also be classified according to the human limb on to which the robot's kinematic chain maps. Thus, robotic exoskeletons can be classified as upper limb (either including or excluding the hand), lower limb and full-body exoskeletons. This is the classification adopted in this book to present state-of-the-art worldwide wearable robot projects in Chapters 8 and 9. In general, the performance and design criteria in WRs differ considerably depending on the limb of interest. The main function of the upper limbs is manipulation; therefore, the kinematic chain consisting in the shoulder, elbow and wrist articulations together with the upper arm, forearm and hand segments has considerable mobility in order to provide a high degree of dexterity during manipulation. This imposes strict requirements in terms of kinematic compatibility between robot and human. In general, upper limb exoskeletons are required to provide less force and torque than lower limb exoskeletons.

The lower limb is generally less complex than the upper limb in terms of kinematics. The main function of human lower limbs is to provide support, stability and mobility (locomotion). Wearable lower limb exoskeletons to assist human gait constitute a paradigm of very close human–robot interaction. Human gait may be viewed as a cyclic process comprising a stance phase and a swing phase. This makes for easier implementation of cHRI schemes, but force and torque requirements for lower limb exoskeletons are very high due to weight support and stabilization demands.

Finally, a classification is possible according to the application domain. As in the case of the classification based on the human limb, here different application domains produce diverse robot design criteria. *Service robotics* is a vast and growing research domain. Service robotics include robots for rebuilding nuclear power plants, caring for the elderly, keeping watch in museums, exploring other planets or cleaning aircraft. Service robots may thus be seen as an intermediate stage in the evolution from industrial robots to personal robots, one instance of which are wearable robots.

Service robots are mobile, manipulative and interact with human beings or autonomously perform tasks to unburden the human being. A recent study by the International Federation of Robotics (IFR) and the European Commission envisaged high growth in cumulative sales of service robots for the period 2004–2007. Most of the wearable robots introduced in this book may be considered instances of service robots, as they are personal robots delivering services to the wearer.

Rehabilitation is a key application domain for the development of wearable robots. It is in Japan, with almost half of the world's nearly one million industrial robots, that adoption of exoskeletons is likely to take place first. Rapid population ageing – a common trend in most Western countries – has created a shortage of caregivers, to which a possible response is robotic-aided personal autonomy, including mobility, social and physical interactions, among others. Several of the WRs described in this book are clear examples of this trend. In particular, Case Study 9.5 shows the application of an empowering full-body exoskeleton to assisting caregivers in domestic tasks with impaired elderly individuals, and Case Studies 9.1 and 9.2 show the application of lower limb exoskeletons to assisting people whose locomotion is impaired by neurological disorders.

Space applications–in this case entailing teleoperation–are also an interesting domain for wearable robots. The EUROBOT exoskeleton is introduced in Case Study 8.3. It has been developed to assist crew during maintenance on the ISS and will support future manned or unmanned exploration missions to other celestial bodies in the solar system such as the Moon or Mars. When using force-feedback devices inside a low-gravity (μ-G) environment, any force or torque fed back to the user must be counteracted by the user's body. If it is not, they will push the operator away rather than helping him/her to interpret correctly the contact situation of the remotely located robot in relation to its environment. This particular characteristic of space applications makes the design of wearable robots in this domain a critical issue, especially in terms of kinematics.

Defence, homeland security and the military are logical application domains for wearable robots. One example of this is the *Exoskeletons for Human Performance Augmentation programme* set up by DARPA. The programme focuses on development of a fast-moving, heavily armoured, high-powered lower and upper body system that harnesses a number of technological innovations. These include, firstly, a combustion-based driver to support advanced hydraulic actuators that produce robotic limb movements with very high strength, speed, bandwidth and efficiency; and, secondly, a control system that allows the operator to move naturally, unencumbered and without additional fatigue while the exoskeleton carries the payload.

1.5 SCOPE OF THE BOOK

The book is organized in six thematic chapters (Chapters 2 to 7) dealing with all aspects relevant to the biomechatronic design and control of wearable robots in close cooperation with human actors. Following these chapters, a collection of case studies addressing outstanding research projects on wearable robots are presented in Chapters 8 and 9. At the end of the book, Chapter 10 summarizes the most salient topics discussed in the book and briefly presents the outlook for future development and likely avenues of research. Figure 1.5 illustrates the structure of the book and how the chapters relate to one another. The following paragraphs briefly summarize the scope of the book and of each chapter.

Chapter 2 presents the framework for *bioinspiration* and the biological basis for the design of wearable robots. Since they are intended to be worn by humans, an efficient wearable robot design must incorporate some basic knowledge of the biological system that will interact physically and cognitively with the robot. Biological systems are also a source of inspiration in robotics design: biological systems are able to deal with unpredictable situations, adapt and learn, and therefore it is desirable to build robots with comparable levels of ability. This chapter outlines the bases of

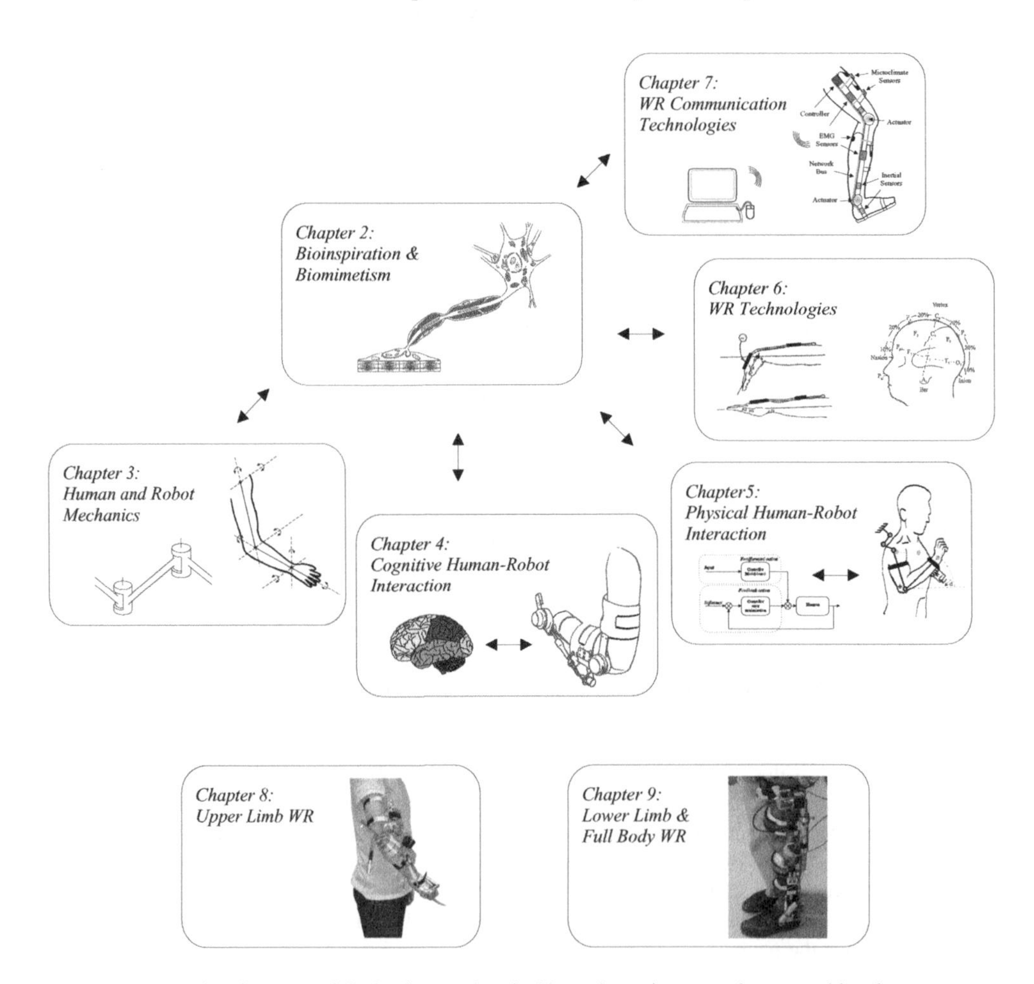

Figure 1.5 Structure of the book, stressing the biomechatronic approach to wearable robots

biological designs. The principles underlying biological systems are presented first, with emphasis on those relating to biomechanics and motor control. Also, energy efficiency in human gait is illustrated with a case study explaining limit cycle biped robots and how they imitate human gait. Another example of biological design procedures is evolution, which inspired the development of genetic algorithms. The application of these techniques to the design of a wearable robot is illustrated in a case study in the next chapter, which explains the design of a four-bar knee joint that mimics the human knee.

The chapter also presents an overview of the neuromuscular control system, including the concepts of internal models and explains the hierarchy in the neuromuscular control. This concept is further illustrated in two case studies, each presenting a wearable robotic design: the MANUS grasping control and the ESBiRRo gait and perturbation recovery control. Finally, the chapter introduces a new distinction between two levels in bioinspired engineering design: *biomimetism*, understood as replication of the external behaviour, and *bioimitation*, which replicates the dynamics of the system and requires an in-depth understanding of the biological phenomena.

The methods used to analyse the kinematics and dynamics of both robots and humans are introduced in Chapter 3, which highlights the parallel between the techniques used in robotics and the ones used in biomechanics, for the methods, given certain reasonable assumptions (such as modelling the human body as rigid links), are equivalent. The position and orientation of the robot is described using the *Denavit–Hartenberg* (D–H) convention algorithm. The dynamics are explained through a description of the equations of motion. Following a review of the methodology used to characterize the motions and forces in robotics, the same problem is addressed in humans. The biomechanics of the upper and lower limbs are briefly described in sufficient detail to derive the Denavit–Hartenberg parameters for the upper and lower limbs.

The first case study introduces a biomimetic knee joint that exactly follows the movement of the centre of rotation of the human knee. A second case study presents an application of the Denavit–Hartenberg parameters in the design of an exoskeleton for the upper limb. The dynamic analysis of human motion is introduced, with an overview of the different levels of biomechanical modelling. One example of dynamics calculation is presented in a case study that estimates tremor power with an upper limb exoskeleton. A major issue in wearable robotics is the *kinematic redundancy* that occurs when more degrees of freedom (DoFs) are available than are required to perform a given task. This problem is analysed in a section discussing the theoretical basis of kinematic redundancy in the design of a wearable robot for the upper limb.

Chapter 4 addresses cognitive human–robot interaction principles based on electroencephalography (EEG), EMG and biomechanical human information. It describes the basis of the bioelectrical phenomena and presents the algorithms for retrieving features, detecting events and classifying patterns of the different signals that can be used to command wearable robots. The chapter discusses separately cHRI technologies based on bioelectrical activity (EEG, EMG) and those based on biomechanical activity. The former pertain to higher levels in the human's cognitive process, and therefore the information is closer to the planning stage of the motor process. The latter in based on the acquisition of biomechanical data, i.e. once motor processes have produced effective limb motion. Therefore, cHRI based on bioelectrical activity is closer to human intentions, while cHRI based on biomechanical activity relates more to the early stages of human motion.

Classic and new approaches are exemplified in different case studies, including exoskeletons controlled by biomechanical means and electrodes implanted on the motor cortex. A first case study illustrates a lower limb exoskeleton in which stance and swing phase detection in a gait cycle is used to control the actuator system for stance stabilization in people suffering from quadriceps weakness. The cHRI is based on a finite state machine (FSM) in which state transition is triggered by rule-based classification of biomechanical (both kinematic and kinetic) data. A second case study presents the cortical control of neuroprosthetic devices. This case study is included to illustrate avenues of research in the field of new BMIs. Finally, a cHRI based on gesture and posture recognition is presented. Although not entirely fitting in with classical cHRI schemes in wearable robots, this case study presents an alternative interaction scheme imported from the field of ambient intelligence (AmI) with possible spinoff applications in wearable robotics.

Chapter 5 focuses on the dynamics of the man–machine physical interaction. Physical human–robot interface issues are examined, especially the mechanics of muscle and soft tissues. The chapter also reviews technologies aimed at enabling interaction within a maximum natural limb workspace and avoiding the creation of residual forces in the joints in the event of misalignments between the robot and human limb. Moreover, the application of controlled forces between the human and the robot requires the development of advanced control strategies, an issue that is analysed in detail. In particular, Chapter 5 describes the behaviour of both actors, human and robot, during the interaction. While the human is modelled as a variable impedance, the robot implements a variety of pHRI control strategies. A specific section is devoted to analysis of the human–robot control closed loop. Physically triggered cognitive processes are described and illustrated by means of an upper limb exoskeleton interacting with the human operator in a tremor suppression control strategy.

In brief, the chapter discusses the following key design aspects of WRs at some length: the kinematic compatibility between human and robot, the application of loads on humans and control strategies for better human physical interaction. Finally, four case studies are presented in order to illustrate the different aspects dealt with in this chapter. The first study illustrates the quantification of interaction forces between robot and wearer in nonergonomic pHR interfaces. In these interfaces, robot and human kinematic configurations are not perfectly aligned, so that interaction forces occur at the human–robot supports.

Human–robot interaction is in most cases accomplished by means of loading through the soft tissues of the limb, and so the second case study analyses the application of load through soft tissue and describes human tolerance of pressure and shear forces, dealing with both the upper and the lower limb. The third case study in the chapter illustrates control of the combined mechanical impedance of human–robot joints for the particular purpose of suppressing tremor by means of limb loading. Finally, the last case study describes control of lower limb impedance during stance as a means for compensating weak quadriceps during gait.

Chapter 6 reviews the key technologies for the development of biomechatronic exoskeletons, which are classified into sensor, actuator and battery technologies. The most suitable technologies for motion, force and pressure sensing are presented, as are a number of methods being investigated for measurement of biological muscular and brain activity signals to control and provide feedback to a WR. The main technologies for monitoring biological muscular and brain activity are presented, along with key issues regarding implementation in wearable applications. The review of actuator technologies in Section 6.3 focuses on their principles, practical availability and limitations with respect to application in lower, upper and full-body exoskeletons. The most suitable portable energy storage technologies for WR technologies are analysed and compared in Section 6.4, with a description of the current trends. Case studies are presented dealing with sensing of microclimate conditions in a human–robot interface, fusion of inertial sensor data in a controllable leg exoskeleton and a biologically based design of a knee actuator system.

WRs also have computer networks embedded. Chapter 7 presents the advantages of networks in this domain, setting out the principles for WR networks, including parameters, profiles, topologies, architectures and protocols. There is a brief description of wired and wireless candidate technologies. Finally, case studies illustrate the most innovative approaches in the field to implementation of a networked architecture. A practical example of smart textiles as a WR supporting technology is also included.

It has been seen that each thematic chapter includes illustrative examples of the topics addressed in the chapter, in the form of case studies. Case studies have been selected to provide examples of the most salient features in each chapter. In addition, there is a collection of case studies in which worldwide research projects relevant to upper limb (Chapter 8), lower limb (Chapter 9) and full-body wearable robots (Chapter 9) are discussed.

Chapter 8 is devoted to upper limb wearable robots. The upper limb is very important because it is responsible for cognition-driven, expression-driven and manipulation activities. In addition, it intervenes in the exploration of the environment and in all reflex motor acts. Therefore, dexterity of upper limb wearable robots is a major requirement. This chapter includes case studies illustrating upper limb wearable robots in the domains of rehabilitation and functional compensation of neurological disorders (the exoskeleton presented in Case Study 8.1), rehabilitation of upper limb amputees (CyberHand, the wearable upper limb robotic prosthesis presented in Case Study 8.2), human–machine interfaces in teleoperation activities for space applications (the EXARM exoskeleton introduced in Case Study 8.3), research in neuroscience and robotics (NEUROBOTICS, the upper limb exoskeleton introduced in Case Study 8.4), and physiotherapy and training by means of a soft-actuated upper limb wearable robot in Case Study 8.6.

Salient examples of lower limb and full-body wearable robots are discussed in Chapter 9. The main function of the lower limb is weight-bearing and locomotion, and therefore it must exhibit stability and high force and torque delivery. These specific functions impose strict technological requirements on lower limb and full-body wearable robots. A collection of representative research projects worldwide is presented in this chapter. Case Study 9.1 illustrates design and control aspects of the GAIT-ESBiRRo lower limb wearable exoskeleton in the arena of functional compensation of neurological disorders, which result in unstable gait. A similar application scenario, in this case concerning assistance of ankle motion during human gait by means of artificial pneumatic muscles, is presented in Case Study 9.2. Lower limb wearable robotic prostheses are analysed in Case Study 9.3; there, the focus is on functional analysis of the prosthetic leg in a comparison between normal walking, impaired walking and restored walking dynamics.

Next, the hybrid assistive limb (HAL) wearable full-body exoskeleton is introduced in Case Study 9.4, from the standpoint of control and cognitive HRI aspects. The analysis of full-body robotic suits is supplemented by a reference to the wearable exoskeleton developed at the Kanagawa Institute of Technology in Case Study 9.5. This study describes all the exoskeleton components in detail, with the focus on pneumatic actuator units and muscle hardness sensors in charge of controlling assistance to the wearer. The chapter finishes with a study devoted to cognitive HRI based on EEG activity (Case Study 9.6). The robot presented in this section cannot be considered a wearable robot as conceived in this book; nonetheless, cognitive interaction with a robotic wheelchair by means of EEG-based commands provides an illustrative example of novel cHRI approaches.

REFERENCES

Abbott, A., 2007, 'Biological robotics: working out the bugs', *Nature* **455**, 250–253.

Alami, R., Albu-Schaeffer, A., Bicchi, A., Bischoff, R., Chatila, R., De Luca, A., De Santis, A., Giralt, G., Guiochet, J., Hirzinger, G., Ingrand, F., Lippiello, V., Mattone, R., Powell, D., Sen, S., Siciliano, B., Tonietti, G., Villani, L., 2006, 'Safe and dependable physical human–robot interaction in anthropic domains: state of the art and challenges', in *Proceedings of the IROS'06 Workshop on pHRI – Physical Human–Robot Interaction – in Anthropic Domains.*

Battyke, C.K., Nightingale, A., Whilles Jr, J., 1956, 'The use of myoelectric currents in the operation of prostheses', *Journal of Bone and Joint Surgery,* **37B**.

Dario, P., Carrozza, M., Guglielmelli, E., Laschi, C., Menciassi, A., Micera, S., Vecchi, F., 2005, 'Robotics as a future and emerging technology: biomimetics, cybernetics, and neuro-robotics in European projects', *IEEE Robotics and Automation Magazine* **12**(2): 29–45.

Kazerooni, H., 1990, 'Human–robot interaction via the transfer of power and information signals', *IEEE Transactions on Systems, Man, and Cybernetics* **20**(2): 450–463.

Manto, M., Rocon, E., Pons, J.L., Belda, J.M., Camut, S., 2007, 'Evaluation of a wearable orthosis and an associated algorithm for tremor suppression', *Physiological Measurement* **28**: 415–425.

Pons, J.L., 2005, *Emerging Actuator Technologies. A Micromechatronic Approach*, John Wiley & Sons, Ltd.

Rabischong, P., 1982, 'Robotics for the handicapped', in *Proceedings of the IFAC on Control Aspects of Prosthetics and Orthotics*, pp. 163–167.

Rahman, T., Sample, W., Seliktar, R., Alexander, M., Scavina, M., 2000, 'A body-powered functional upper limb orthosis', *Journal of Rehabilitation Research and Development* **37**(6): 675–680.

2

Basis for bioinspiration and biomimetism in wearable robots

A. Forner-Cordero[1], J. L. Pons[1] and M. Wisse[2]

[1]*Bioengineering Group, Instituto de Automática Industrial, CSIC, Madrid, Spain*
[2]*Mechanical Engineering Faculty, Biomechanical Engineering Department, Delft University of Technology (DUT), Delft, The Netherlands*

2.1 INTRODUCTION

In this chapter the framework for *bioinspiration* in the design of wearable robots is presented. As they are intended to be worn by humans, wearable robots must cooperate with the person. Therefore, some basic knowledge of the biological system that interacts with the robot is needed in order to understand their interactions. Cognitive and physical interactions, however, are further discussed in Chapters 4 and 5 respectively.

This chapter outlines some 'design' principles that can be found in biological systems. The focus is on the principles relating to biological motion, biomechanics and motor control since they are related to the design of wearable robots.

Biomimetics is a relatively recent term coined in the 1960s by Schmitt (1969) and Vincent *et al.* (2006). According to Webster's Dictionary, *biomimetics* is 'the study of the formation, structure, or function of biologically produced substances and materials (as enzymes or silk) and biological mechanisms and processes (as protein synthesis or photosynthesis) especially for the purpose of synthesizing similar products by artificial mechanisms which mimic natural ones'. However, the concept formalizes an idea already well established in Western philosophy, for instance by ancient Greek philosophers like Aristotle, who stated: 'If one way be better than another, that you may be sure is nature's way', or Democritus, who wrote: 'We are pupils of the animals in the most important things: the spider for spinning and mending, the swallow for building, and the songsters, swan and nightingale, for singing, by way of imitation'.

The understanding and modelling of biological systems has served as a source of inspiration in the design of different robotic systems. There are several reasons for a robot designer to study and model

Wearable Robots: Biomechatronic Exoskeletons Edited by José L. Pons

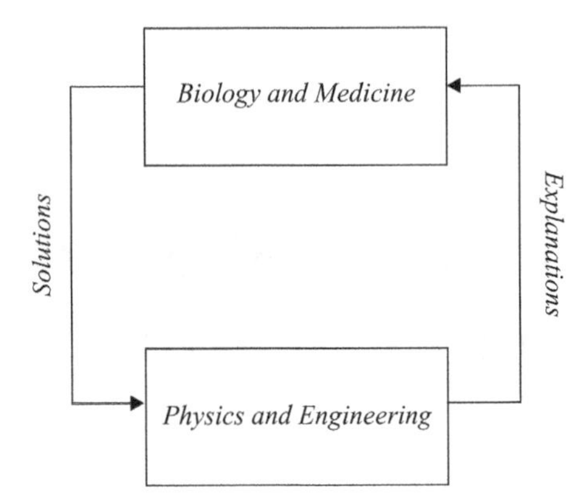

Figure 2.1 Interaction of biology and medicine with physics and engineering. Biology can provide solutions for complex engineering problems, while mathematical models provide the most precise explanation for biological systems

biological systems. One of the most important of these is the impressive performance of biological systems. Biological systems are able to deal with unpredictable situations; they can adapt, they can learn and they are robust to failure. It is therefore desirable to build artificial systems, e.g. wearable robots, with the same level of performance.

Moreover, physical or engineering models of biological systems (formalized mathematically) have proven very useful for understanding biological behaviour. Models allow some manipulation of conditions that are very difficult to establish experimentally. Therefore, there is a mutual synergy between engineering and biology (see Figure 2.1).

It will be seen at the end of the chapter that there are different levels of bioinspiration in the design of engineering systems.

2.2 GENERAL PRINCIPLES IN BIOLOGICAL DESIGN

It is a generally accepted principle in biology that certain objective functions crucial for the survival of the species must be optimized. These objective functions can be as varied as the number of existing living beings and the different behaviour patterns in each ecosystem.

In the case of the design of wearable exoskeletons, there are some characteristics that need to be optimized. For instance, it is important to minimize weight and energy consumption, or to retain adaptability to different functions, because the exoskeleton must allow for multifunctionality of the human limbs.

Biological creatures vary tremendously in appearance, for instance in size (from single-cell organisms to the largest whales) or in the environmental media where they live – land, air or water (Biewener, 2003). However, despite this diversity, virtually all biological creatures share some common principles. Biological systems are the result of millions of years of evolution in a hostile environment that has forced the natural selection of the fittest.

On the other hand, evolution is not a straightforward goal-directed process. A structure developed for a certain environment may be useless if conditions change (Shadmehr and Wise, 2005). Evolution thus builds on existing structures which may have to change their shape to perform a completely different function. The outcome of this process is complex, optimized structures in which it is difficult to determine the purpose of each part. Such optimal solutions are highly flexible and adaptable.

These optimization procedures have several objective functions stemming from the most general goal of living beings, namely survival. With respect to biological motion, there are several possible objective functions: to minimize energy consumption, to minimize damage and loads to tissues and to achieve a compromise between power and precision of movement.

2.2.1 Optimization of objective functions: energy consumption

All objective functions are ultimately oriented towards the survival of the individual or the species. Here the focus is only on one possible objective function relating to biological motion, in particular human locomotion. The reason is that it reveals some interesting aspects bearing on the design of wearable exoskeletons designed to support human motion.

Energy consumption is optimal in most of the activities of living beings, and a large portion of this energy is spent moving towards food or away from predators. Living organisms require energy to maintain their state versus the environment and to move. Most animals spend a significant percentage of their lifetimes searching for food. Therefore, the ability to perform movement in the most energy-efficient manner is an important objective in living beings. This has been demonstrated in different ways. For instance, there are several models of arm movement that use energy optimization as an objective function when simulating arm reaching tasks. The movements predicted by the model agree with the experimental results (Hogan, 1985).

When modelling the arm, there are too many degrees of freedom in the joints, and there are several muscles that can be activated to perform the same task (see the redundancy problem in Section 3.4). In this respect, the myoskeletal inverse dynamics problem, consisting of the determination of the muscle forces required to control the movement, is ill-posed in that there is no unique solution to it. However, the problem can be solved by applying a regularization procedure. In this respect, minimal energy consumption and minimizing of muscle fatigue is a good objective function that produces realistic motion (Erdemir *et al.* 2007; Koopman 1989; Pandy 2001).

One example of energy-efficient locomotion is human gait. Human gait has been defined as the translation of the body *centre of mass* (CoM) with minimal expenditure of energy (see Koopman, 1989). This idea has been illustrated with several models of gait, starting with the simplest walking model (a pelvis and two stiff legs), which generates an unrealistic compass gait with excessive oscillations of the body's CoM. It is therefore necessary to add other characteristics to make the model more realistic, with smaller oscillations of the CoM and more energy efficiency (Saunders, Inman and Eberhart, 1953). It has been shown that a natural energy-efficient CoM trajectory can be achieved by adding six characteristics (gait determinants) to the model, consisting of three pelvic movements – rotation, tilt and lateral displacement – in addition to knee flexion during stance and double ankle–knee interaction during stance. An additional feature of human gait is that the muscles are not active during the whole gait cycle but work only in certain phases of the cycle (Saunders, Inman and Eberhart, 1953). This is a highly effective energy-saving mechanism.

It has also been demonstrated that individuals select a certain speed (combination of step length and cadence) in order to minimize the metabolic energy per unit distance travelled, as shown in Figure 2.2 (Ralston, 1958).

With variations in the gait speed, some relationships can be identified between the three parameters that determine gait, namely speed, step length and cadence. It has been shown that these can be predicted through optimization of an objective function by minimizing the metabolic cost per distance travelled. Gait models based on energy optimization are thus the result of an optimal energy-efficient combination of frequency and step length for each speed (Bertram and Ruina, 2001).

There are other energy-saving mechanisms for walking in addition to kinematic adaptation to maintain the CoM following a small-amplitude sinusoidal path (Alexander, 1991). This is achieved by optimal muscle properties, adequate positioning of the resulting ground reaction forces with respect to the ankle, knee and hip joints, and by using storage release of elastic energy in the passive elements of the leg (ligaments and tendons).

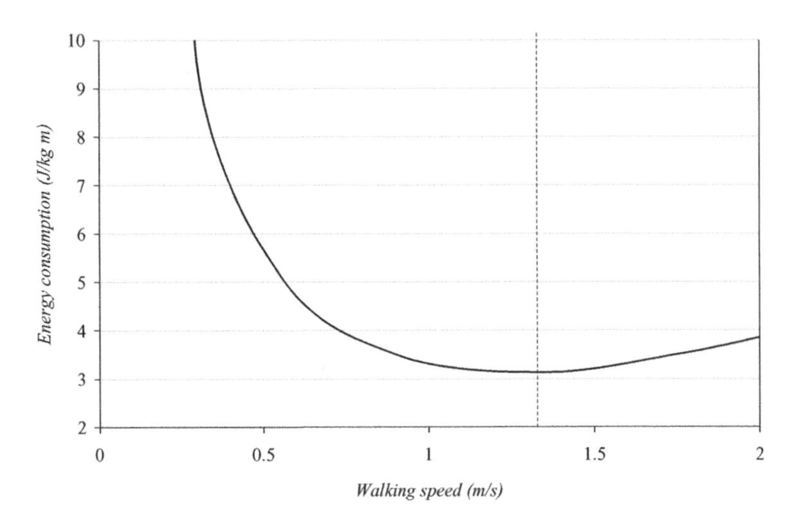

Figure 2.2 The metabolic energy consumption of gait per unit of distance travelled is a function of speed. The optimal speed is about 1.3 m/s

The design of a lower limb exoskeleton that supports gait function must take all these mechanisms into account. One example that readers will find in this book is the design of a lower limb exoskeleton – the GAIT–ESBiRRo exoskeleton – to compensate for a number of pathologies affecting gait, in particular leg weakness of the kind experienced by patients suffering post-polio syndrome (see Case Study 9.1). In this chapter, the semi-active lower limb exoskeleton is used to illustrate many aspects in the design of wearable robots. In particular, a complete explanation of energy storage and release in the GAIT exoskeleton is presented in Section 6.7, which details the biomimetic design of the exoskeleton knee actuator.

Optimal energy consumption in human gait has also inspired the development of passive robots (McGeer, 1990, 1993), capable of walking down a gentle slope without any control or actuation. The remarkable similarity of the gait of these robots to human walking patterns has inspired the design of efficient biped walkers with which it is sought to understand and imitate human gait (Collins *et al.*, 2005).

Limit-cycle controlled robots in particular are an emerging design paradigm in biped walking robots derived from the minimal energy consumption paradigm of human gait (Collins *et al.*, 2005; van der Linde, 2001; Wisse, 2005a). An example of the design of this type of bipeds in given in Case Study 2.5. In contrast to limit-cycle robots, the design of biped walkers has traditionally been based on constant torque control of the joints with a local stability criterion, usually based on the zero moment point concept (Hirai *et al.*, 1998; Vukobratovic, Frank and Juricic, 1970). These robots consume more energy than humans, as shown in Figure 2.3.

It is illustrative of the order of magnitude of the energy consumed by human gait and by a biped walking robot to refer both to the distance travelled per energy unit consumed. In this case, the distance considered was that travelled with the energy contained in a slice of bread, as shown in Figure 2.3(a). The result of the approximate calculations is quite surprising: while a human can walk 1.4 km on only one slice of bread, a torque-controlled robot from the previous generation (P2 from Honda) can only walk 31 m. The energy consumption of limit-cycle walkers, on the other hand, resembles human walking more, as can be seen in the bar graph inset in Figure 2.3(b) (Collins *et al.*, 2005). Energy consumption, which is an extremely critical issue in the design of wearable robots, is further illustrated in Case Study 2.5.

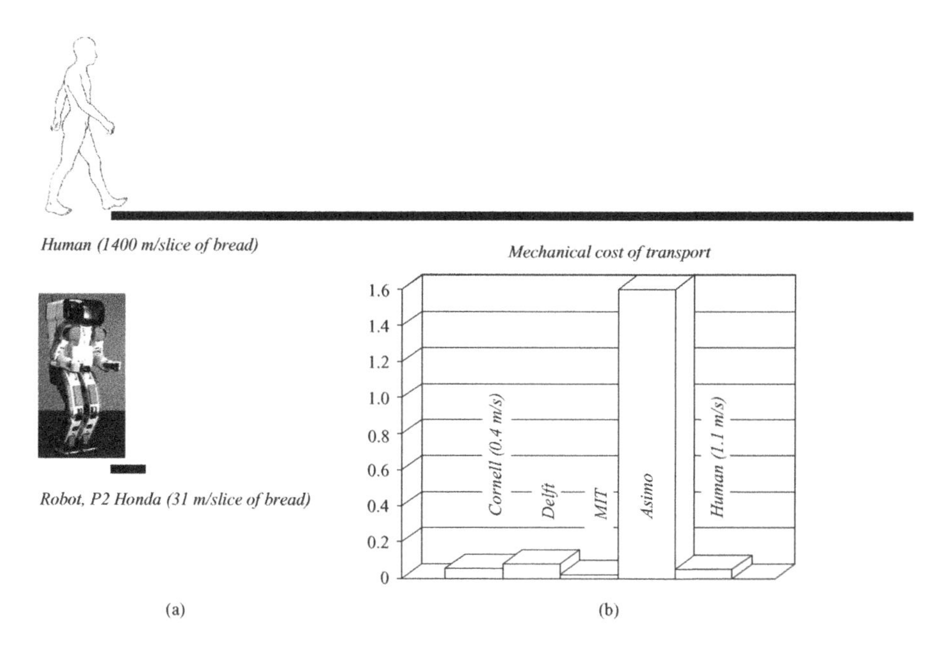

Figure 2.3 Energy consumption during bipedal walking: (a) the energy consumption of gait per unit of distance travelled by a human (for the optimal walking speed) and by an old robot (Honda P2), adapted from van der Linde (2001), and (b) comparison of the mechanical energy consumption of different types of biped robots and humans. The bipeds from Cornell, Delft and MIT are based on the limit-cycle control concept, while the ASIMO from Honda has a torque-controlled structure (Collins *et al.*, 2005). Reproduced from Collins *et al.* 2005

Finally, it should be noted that minimizing damage is another principle in biological design. This idea is illustrated in Chapter 5 when dealing with physical interaction between human and exoskeleton. For instance, the discomfort caused by excessive pressure can lead to changes in movement patterns, thus overruling energy criteria. Another example is the presence of a threat to the individual, for instance running from a predator or executing the recovery reaction from a stumble to avoid a fall, both of which involve high energy consumption (Forner-Cordero, Koopman and van der Helm, 2005). This may be seen as an optimal adaptive control mechanism that switches to the most suitable objective function depending on the environmental conditions.

2.2.2 Multifunctionality and adaptability

Biological systems are able to carry out different tasks with an optimal performance level within their capabilities (stability, fatigue, etc.). This ability is especially apparent when they are forced to perform completely new tasks. For instance, human upper limbs can perform tasks for which evolution has not prepared them, such as typing on a computer.

Recent work has shown that biological motor control can be approached in terms of optimal control problems. An adequate choice of objective functions (such as energy consumption) has made it possible to address some of the questions raised by the behavioural study of biological movement, such as the redundancy problem.

This functionality is related to the number of degrees of freedom (DoFs) of the limbs, usually exceeding the requirements of a three-dimensional working space. For instance, leaving aside the wrist and fingers, the joints of the upper limb have five DoFs (see Section 3.3), i.e. two at the elbow (flexo extension, prono-supination) and three at the shoulder joint (flexo-extension, ab-adduction, endo-exorotation). However, only four degrees of freedom are needed in a three-dimensional space. This overcompleteness or *redundancy* has advantages, such as robustness and variability in performing the same task, and also some disadvantages. There is the motor control problem of redundancy, called the Bernstein problem (Bernstein, 1967), in the selection of movement patterns. Then it is necessary to choose a combination of joints and degrees of freedom (DoFs) to achieve the desired task goal while leaving some uncontrolled in such a way that they do not affect the performance of the task. This has been enunciated as the uncontrolled manifold hypothesis (Scholz and Schoner, 1999).

From an exoskeleton design perspective, there are two different sources of variability in biological motor control:

1. Variability due to the number of DoFs. This problem is addressed in Chapter 3.
2. Variability in the execution of the movement using the same DoF. This variability is related to adaptation and learning and its source is not very clear. It is partially addressed in Chapter 4 when dealing with the cognitive interactions between the robotic exoskeleton and user.

Input noise in the sensory system can enhance sensory or motor function. This surprising mechanism is known as stochastic resonance and means that sensory noise may be used to enhance perception. One application of this concept is the use of randomly vibrating insoles to improve balance control in elderly people (Priplata *et al.*, 2003). These insoles may be seen as part of a wearable robot designed to enhance sensory perception. The idea could further suggest a mechanism to enhance the transfer of information from the wearable robot to the user. This is an aspect of the *cognitive interaction* triggered by a *physical interaction* between robot and human, two issues discussed in Chapters 4 and 5 respectively.

2.2.3 Evolution

Biological evolution, combined with other mechanisms, e.g. genetic mutation, is related to survival of the species and hence to optimization of function in its broadest sense. Biological evolution is the process that describes the cumulative changes in a population of organisms. These changes have led to the development of differentiated species. The process that regulates evolution in Nature is natural selection, as proposed by Darwin in 1859. This theory states that in a certain population of a species those individuals having certain positive inherited characteristics will produce more offspring than individuals with negative ones (Darwin, 1859). In such ways, the beneficial characteristics will prevail within several generations in the population ensemble, thus altering it. In short, the concept of natural selection is related to the survival of the fittest.

All the information required to form a new individual is contained in the genes. Individuals pass on their own genetic material to their offspring. For instance, healthy human beings have 23 chromosome pairs, each contained in a long double-helix DNA chain. Each gene is encoded in small fragments of the DNA. Some living beings reproduce by exchanging genes with another individual of the species. The mechanism by which the genes are crossed is known as crossover. An individual has a 50 % chance of inheriting the good characteristics from his or her parents. A good characteristic is one that makes the individual more suited to the environment and hence improves the chances of reproduction. In this way, the species can improve from generation to generation. During this process there is a chance that some genes will be located in the wrong place, causing a mutation. This gene mutation will be expressed in the phenotype, leading to differences from the parents. In principle, these differences may be good, making the new subject better fitted, or bad in the opposite case. This trial-and-error process over millions of years has led to optimal solutions from an engineering perspective. It must be remembered that biological sensory and motor systems are products of an evolutionary process lasting millions of years and resulting in systems adapted to their environment. Evolution and related mechanisms are therefore the source of inspiration for design methods/procedures that can be applied successfully to the design of exoskeletons. One major advantage of *genetic algorithms* is that they can be used to solve certain design problems without a complete understanding of the underlying mechanisms, and this means that the designer does not need to deal with the complexity of the system in order to reach an optimal solution. This process can be seen as a structured process of learning by trial and error. The information encoded in the successful genes that are transmitted to the next generation may be seen as a learning procedure.

2.2.3.1 Evolutionary design and genetic algorithms

A genetic algorithm is a method for seeking an optimal solution inspired by evolution combined with the mechanisms of natural selection and genetics. Thus, the survival of the fittest makes it possible to exchange information in pursuit of an optimal solution that is at once structured and randomized (Goldberg, 1989). These algorithms are very robust, much like their natural counterparts. It is important to define properly the objective function to be optimized in the evolutionary process (which, as in Nature, is clearly survival). It is necessary to encode (genes) the parameters that define performance (expression of these genes). Then, these genes are combined in a random or blind process in such a way as to improve performance. There are three basic steps in this process:

1. *Reproduction*, in which individual strings of genes are copied according to their objective function values or fitness to function. In this way, the most suitable strings, which are the ones that perform best in the objective function, will have the greatest probability of reproduction.
2. *Crossover* is the random combination of several strings to produce a new one. The combination of reproduced strings will result in a new population.
3. *Mutation* is simply the random change of one gene. In the perspective of the algorithm it is a factor of instability that prevents the algorithm from remaining stuck in a local minima.

At the end of this three-step process, the new population generated must be evaluated with reference to the objective function. If the required criteria are not met, this process will be repeated recursively. There is another mechanism to favour rapid convergence to the best solution, named *elitism*. This consists in passing the best performing strings directly on to the next generation without crossover or mutation (Goldberg, 1989; Holland, 1975).

This process has been successfully applied to the design of wearable robots, for instance a wearable robot for determination of the DoF in the upper arm, (Morizono, Mimatsu and Higashi, 2005). Case Study 3.5 presents a design for a four-bar linkage joint based on genetic algorithms. This mechanism is kinematically compliant with the knee joint and has been successfully implemented in a lower limb exoskeleton (GAIT-ESBiRRo), more fully described in Case Study 9.1.

2.3 DEVELOPMENT OF BIOLOGICALLY INSPIRED DESIGNS

A model is a representation or description designed to demonstrate the behaviour, structure and working mechanism of a certain system or object. In this section the process of biological modelling development will be outlined along with a description of biological models relating to movement and motor control.

2.3.1 Biological models

There are several reasons to study biological models in connection with the design of wearable robots. One derives from the importance of the interaction between robot and human. It is necessary to know the properties of the human motor system in order to define the design requirements in wearable robotics, and with the help of a model it is possible to predict the system's behaviour. Another is that, as in reverse engineering, the sophisticated solutions found in Nature may serve to inspire engineering; moreover, as will be seen in Section 2.4, it is necessary to understand in order to imitate.

A model in biology is a theoretical representation of biological phenomena or functions. This representation illustrates the relationship linking a set of variables that describe the system. Usually, these relationships are expressed in terms of mathematical relationships between variables. With the help of the model it is possible to predict behaviour and analyse the relative effect of each variable. In this way a model provides a conceptual framework within which to understand and investigate real systems. A model in biology is developed in a series of recursive steps starting with the observation of natural phenomena. On the basis of a pure description of the observations, it is necessary to arrive at a description of the mechanisms that govern a certain phenomenon. This is done by formalizing the relationships found between the variables and compiling a descriptive model. From this description, as in the case of physiological models, it is possible to derive a mathematical model formulating assumptions and equations that describe the relationships between variables. The mathematical model has to be validated with real data and can be used to simulate the behaviour of the system under conditions that may not be easy to establish experimentally, and to produce predictions of behaviour. The model must be interpreted and reformulated in the light of new unexplained observations.

Models of biological systems are useful as inspiration for control, actuation and measurement applications. Different levels of inspiration are related to different levels of understanding of the system and the accuracy of the model. In this section some biological systems relevant to the design of wearable robots are introduced. Each system is described from a physiological point of view in order to introduce a mathematical model that has been proposed to describe it. Some of these systems work in parallel or in series with the wearable robot. Specific aspects of the interaction between human and robot are discussed in later chapters.

2.3.2 Neuromotor control structures and mechanisms as models

A good wearable robot design must take into account the system that it is to interact with. Biological systems have their own actuators and control structures, usually much more complex than those of a robot, and they may exhibit intersubject differences and changes of behaviour with time, as in the case of learning. Finally, it should be noted that, in many applications, wearable robots are worn by people suffering from a disability, as in the case of rehabilitation robotics. In such cases, the neural or the motor system may be impaired and behaviour may be very different from normal. These particularities must be considered in the design.

2.3.2.1 The nervous system

The biological nervous system is the system that processes information from the sensory system and commands the different body functions, such as the activity of muscles and of certain organs. From an engineering perspective, the nervous system is equivalent to a controller and includes sensors as well as a state estimator. To describe the anatomy and physiology of the human nervous system is no easy task. For an in-depth treatment of anatomy and physiology the reader is referred to specific books (Kandel, Schwartz and Jessell, 1995; Moore, 1992).

The nervous system of vertebrate animals is usually considered to be a highly hierarchical system. It is often divided into the *central nervous system* (CNS) and the PNS. The CNS consists of the brain and spinal cord while the PNS consists of all other nerves and neurons that do not lie within the CNS, such as the neurons and sensory receptors below the spinal cord. Despite this hierarchical organization, each structure in the nervous system possesses a high degree of autonomy. The reason is that the vertebrate nervous system has developed through a series of evolutionary steps. Each lower structure had its utility in former evolutionary stages and presents certain features that may or may not be useful in further evolutionary situations (Shadmehr and Wise, 2005).

The nervous system has a somatic part, mainly voluntary, that controls the skeletal muscle, and an autonomic nervous system, mainly involuntary. The chief focus of interest for the purposes of this book is the *somatic nervous system* as it is the one that controls voluntary movement.

The nervous system has two major classes of cells: *neurons*, excitable cells that transmit and process action potentials (see Section 4.3), and *glial cells*, which support the neuronal functions. One type of glial cell, the *astrocyte*, may also play a role in information processing by enabling or inhibiting certain synaptic junctions (Ransom, Behar and Nedergaard, 2003). The information is encoded by the neurons in the firing rate of action potentials or nerve impulses. There are many types of neurons, but in general neurons receive impulses through dendrites via synaptic connections with preceding neurons. Synapses can be inhibitory or excitatory. The cell body integrates the impulses from different inputs to produce an action potential when a certain threshold is exceeded. This impulse is transmitted through the neuronal axon. The axon may be very short or it may be more than 1 m long in order to establish synapses (there may be more than 1000 synapses) in the dendrites of other neurons. The axons of the motor neurons contact a muscle, forming a connection called a motor end plate.

Neurons can be *afferent*, i.e. relaying information from body receptors or sensors, or *efferent*, i.e. carrying the impulses to effectors such as muscles. In addition, there are many interneurons (or association neurons), which transmit and process information between neurons (Wise and Shadmehr, 2002).

From a functional point of view several parts may be distinguished in the CNS, as shown in Figure 2.4. These are, going from the bottom up (Kandel, Schwartz and Jessell, 1995):

- *Spinal cord*. The spinal cord receives and processes sensory information from body receptors and muscles. It also sends commands to the muscles, with or without input from higher levels of the CNS, as in the reflexes. There are many pathways transporting information up and down. Knowledge of all these pathways and its functional role is still incomplete. There are also interneurons that process some information and generate spinal responses, such as reflexes, which can involve several levels of the spinal cord. The spinal cord is capable of generating rhythmic motion patterns based on neural structures. These groups of neurons behave as nonlinear oscillators and are called *central pattern generators* (CPG) (Biewener, 2003). The spinal cord is not, then, a mere transmission cable that transports information between the brain and the rest of the body.

- *Brainstem*. The brainstem lies between the spinal cord and the brain. It has many upward and downward connections. It plays an important role in the control of motion and in the control of autonomic functions. It is composed of three structures, the *medulla oblongata*, the *pons* and the *midbrain*. The pons seems to have an important role in the control of posture and balance and conveys motor information from the forebrain to the cerebellum. The midbrain controls sensorimotor functions, such as the coordination of visual and auditory reflexes.

- *Cerebellum*. The cerebellum is a structure located in the back part of the brainstem and is responsible for the coordination of movement, for instance to compensate for interaction torques during movements. These are the torques that arise in distal joints as a consequence of accelerations in a proximal joint. The robotic counterparts of these interaction torques are described in detail in Section 3.2.

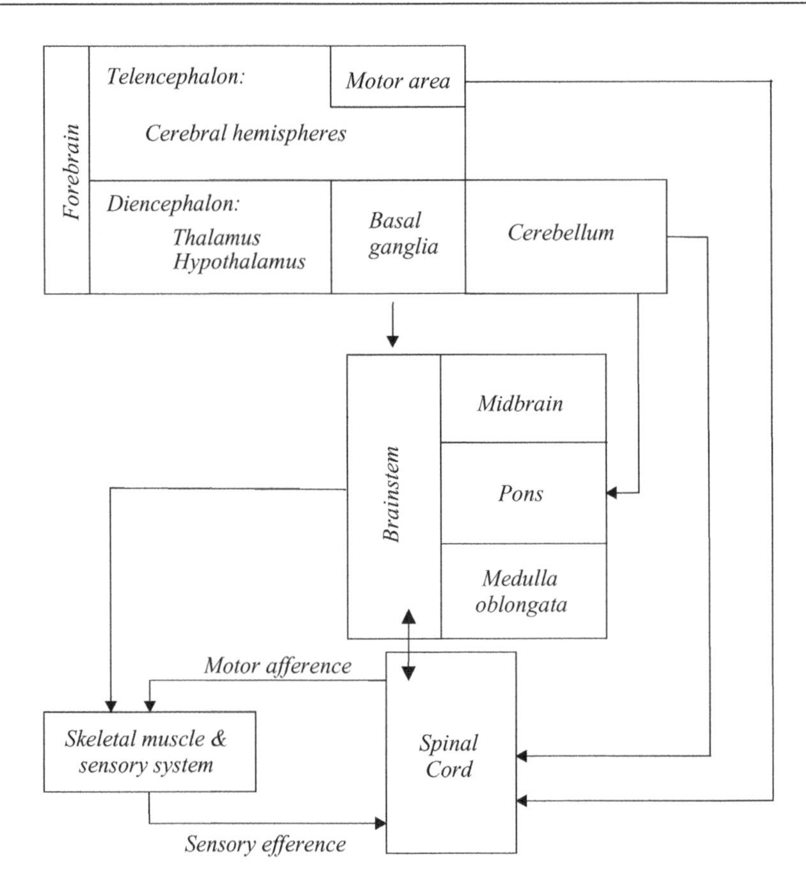

Figure 2.4 Scheme of the nervous system showing the different parts involved in motor control. Relationships are marked by arrows

- *Forebrain*. The forebrain includes the *diencephalon* and the *telencephalon*, where the cerebral cortex is located. There are two main structures in the diencephalon: the *thalamus*, which pre-processes nearly all sensory information (except the sense of smell) before it reaches the cortex, and the *hypothalamus*, which is involved in the regulation of autonomic and endocrine functions. The *telencephalon* consists of the two cerebral hemispheres, covered by a wrinkled layer that forms the cerebral cortex. The cerebral cortex is divided into lobes, which possess some degree of specialization. The frontal lobe contains the area involved in the planning of action and control of movement, while the occipital lobe processes the information from the vision system. The reader is referred to Section 4.2, which provides additional details on the anatomy and physiology of the brain in the context of the brain activity-based cognitive interaction between robot and human.

  In the parietal lobe there is activity relating to the sense of position and some vision processing for action. The temporal lobe is involved in the processing of hearing. In its deep structures lies the *hippocampus*, which plays an important role in memory processes and learning, the *amygdaloid nucleus* and the *basal ganglia*. The basal ganglia are involved in the control and selection of movement.

The brain possesses an orderly representation of the body in the primary motor and sensory areas. This representation is plastic, meaning that it can be modified by learning. For instance, the amputation of a limb would cause a reorganization of the motor and sensory areas devoted to the lost limb. Note, however, that the boundaries of this representation are not strictly defined. Finally, each submodality of sensation is represented in different areas of the cortex (Kandel, Schwartz and Jessell, 1995), which suggests that the brain includes some kind of spatial encoding of the information.

The nervous system also presents a hierarchical organization in which lower level reactions such as the reflexes can be modulated by descending inputs from higher levels. This hierarchical organization can serve as a model for control of wearable robots, as in the case of the powered MANUS prosthesis. In this example of a wearable robot there are several hierarchical levels for controlling the different grasping patterns of the hand. This is discussed in detail in Case Study 2.6.

2.3.2.2 Internal models

The action potential is the depolarization of the membrane that occurs in the neurons. When at rest, the neurons actively maintain a negative electric potential with respect to the extracellular fluid, due to differences in the ion concentration (mainly Na^+ and K^+). When the neuron is excited, i.e. there is an excitatory input from another neuron in the synaptic junction, an action potential occurs that is propagated along the membrane of the neuron and serves to transmit the information through the nervous system. This transmission is relatively slow, although the speed depends on the diameter of the axon ($\approx 0.5 - 1$ m/s). To increase speed in long axons, some neurons are covered by a sheath of myelin (Schwann cells) which isolates the membrane and inhibits the action potential. At certain distances the sheath is discontinued (Ranvier nodes) in order to allow saltatory conduction of the action potential, thus raising the speed up to 100 m/s. Nevertheless, the neurons are subject to considerable transmission delays: a feedforward anticipatory mechanism would appear to be necessary in the design (Shadmehr and Wise, 2005).

It has therefore been proposed that there are certain neural processes that replicate the dynamics of the body and its interaction with the environment (Kawato, Furukawa and Suzuki, 1987). These neural processes, called *internal models*, predict the consequences of the action; they can forecast the sensory input from the motor output that is released (Shadmehr and Wise, 2005). They can compare the sensory prediction with the actual sensory input and determine corrective actions. These internal models also provide some robustness against long transmission delays and noisy sensory information, and they are very important in the learning of a new task. These processes can be seen as a model-based feedforward control structure. In fact, it seems that the internal models generate a prediction of the sensory consequences of an action and compare the prediction with the sensory information. Some control structures used in wearable robots are presented in Chapter 5.

2.3.3 Muscular physiology as a model

The muscles are the actuators of the human motor system; they are controlled by the nervous system. Muscle contraction is triggered by depolarization of the motor neuron which innervates the muscle in the motor end-plate (Kandel, Schwartz and Jessell, 1995). This depolarization process is briefly documented in Section 4.3 in the context of establishing a cognitive human–robot interaction based on EMG. There are three types of muscles, striated muscles, which are the skeletal muscles responsible for movement, smooth muscles and heart muscles, controlled by the autonomic nervous system.

The muscles are excitable tissues that change shape, particularly length, when excited. The muscle fibres are single cells of variable length (from millimetres to centimetres) containing the contractile myofibrils. These contractile elements are composed of sarcomeres or groups of myofilaments of actine and myosin. Contraction is produced by the active (energy consumption) sliding of these myofilaments (Kandel, Schwartz and Jessell, 1995; McMahon, 1984). Each fibre of vertebrate muscle can produce a force of about 0.3 N.

A motor unit is defined as all the muscle fibres innervated by the same motor neuron. The muscle fibres are distributed throughout the volume of the muscle. The innervation ratio is the number of fibres in each motor neuron, and this determines the precision in the generation of force. A muscle whose motor units have less fibres may in principle be more precise in the generation of force. There are three types of motor unit: slow twitch fibres (Type I SO), fatigue-resistant fast twitch fibres (Type IIB FG) that can contract very quickly but cannot sustain prolonged contraction and fatigue-resistant fast twitch fibres (Type IIA FOG) that achieve a compromise between speed of contraction and fatigue.

When little force is required, small motor units containing slow fibres are excited first. When more force is required, larger motor units containing faster fibres are excited. This is explained in the *size contraction principle* described by Henneman (1957). This is one of the mechanisms for avoiding fatigue, as the slow fibres are more resistant than the fast ones. The other mechanism is motor unit rotation, in which a motor unit that has been active for a period of time is allowed to rest and another is activated. It is important to note that the muscle force output can be increased by either of two means: by increasing activation frequency and by increasing the number of active motor units. These mechanisms resemble the switched control of certain actuators used in wearable robotics, such as ultrasound motors (see Pons, 2005). Switched control is also a means of saving energy.

As regards its mechanical properties, the tension on the muscle is a nonlinear function of the length when the muscle is active. However, the passive length–tension relationship is linear and can be simulated as a spring. During contraction, the muscle also exhibits a velocity–tension relationship. Beyond a certain increase in velocity, the tension decreases, so that there is optimal muscle velocity contraction in terms of power generation. A model of the mechanical behaviour of the muscle was proposed by Hill (1970). This model consists of three passive elements and one active force generator (see Figure 2.5).

The series elastic element models the behaviour of the tendon and the connective tissues. The parallel elastic element reflects the resistance of the muscle to passive stretching, while the damper models the dynamic resistance to movement, which is speed-dependent. The elastic elements store energy. There is only one active element that models the contraction of the fibres; this is a force generator, as shown in Figure 2.5.

There are actuators whose behaviour resembles that of a muscle. They are known collectively as artificial muscles, although they can obey different actuation principles (Bar-Cohen, 2001). Artificial muscles can be implemented by means of pneumatic actuators. They have been used to make robots walk stably by means of switching control with low energy consumption (van der Linde, 2001).

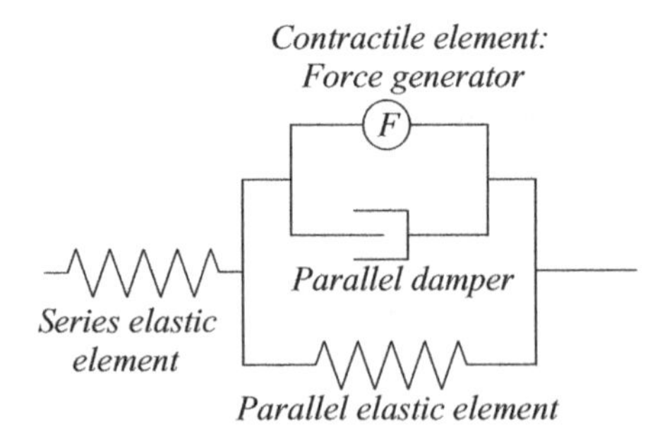

Figure 2.5 Muscle model proposed by Hill (1970). It consists of three passive elements: a series elastic element corresponding to tendon and connective tissue, a parallel elastic element and a damper. The contractile element is modelled as a parallel force generator

Another type of artificial muscles are electroactive polymers, which alter their shape as a function of the voltage applied to them. The range of forces they can provide is similar to that of biological muscles and they have an application in the field of wearable robotics (Bar-Cohen, 2001). Finally, a comparison has been done between artificial muscles and biological mammalian muscles in terms of their mechanical properties (Madden *et al.*, 2004). For additional details on actuators in the context of wearable robots, the reader is referred to Section 6.3.

The morphology of the more than 500 muscles in the body differs in terms of fibre orientation:

- *Fusiform*, all the fibres run parallel from the origin to the insertion of the muscle. This arrangement provides the maximum velocity and range of motion (change of length). One example of this type of muscle is the biceps brachii.
- *Pennate* muscles are those in which the fibres run obliquely. If all run in the same direction they are called *unipennate*, and *bipennate* if they run in two directions. There are also *multipennate* muscles in which the fibres present multiple orientations, as in the deltoid muscle. The advantage of this arrangement is that they can provide more power and stability in multiple directions, albeit at the expense of velocity and range of motion.

As regards muscle attachments, there are spurt muscles, originating far from the joint but inserted close to the joint, so that the moment arm is short. These afford fast movement but low torque. Shunt muscles originate close to the joint and are inserted far from it, so as to produce a larger moment (Freivalds, 2004).

One important issue in the production of movement is muscle configuration. Usually there are more muscles around a joint than are required to produce movement in all of its DoFs. Muscle motion occurs only in contraction, and so movement in one direction must be compensated by movement in another direction, by means of another contraction. This is achieved by agonist–antagonist pairs. In this way, while one muscle contraction causes flexion, e.g. the biceps brachii flexes the elbow, the antagonist will cause extension; in the elbow this would be the triceps brachii (see Section 3.3 for more details on muscle anatomy). In addition, there are several muscles that serve more than one joint, so that their contraction causes a combined sequence of multiarticular movements. For instance, the rectus femoris in the leg contributes to hip flexion and to knee extension.

Muscular control around a joint can be seen as a form of *impedance control* (see Chapter 5). A practical example of the implementation of knee joint behaviour during gait is given in Case Study 9.1. It describes how the behaviour of the knee during gait can be simulated by means of two springs, with high and low stiffness during stance and swing.

2.3.4 Sensorimotor mechanisms as a model

The muscles are actuators with built-in sensors that can measure length, rate of change of length and force. These sensors are part of the proprioceptive sensory system. The proprioceptive sensors are those that provide information about the position of the body. The joint capsule sensors are the ones that provide information about joint position; they are active mostly at the extremes of range of motion of the joints.

However, movement at the joints is generated by the muscles, and these are equipped with two groups of sensors:

1. *Golgi tendon organ* (GTO). The GTOs are interwoven with the fibres of the tendon that attaches the muscle to the bone. When the force on the tendon stretches the fibres they compress the

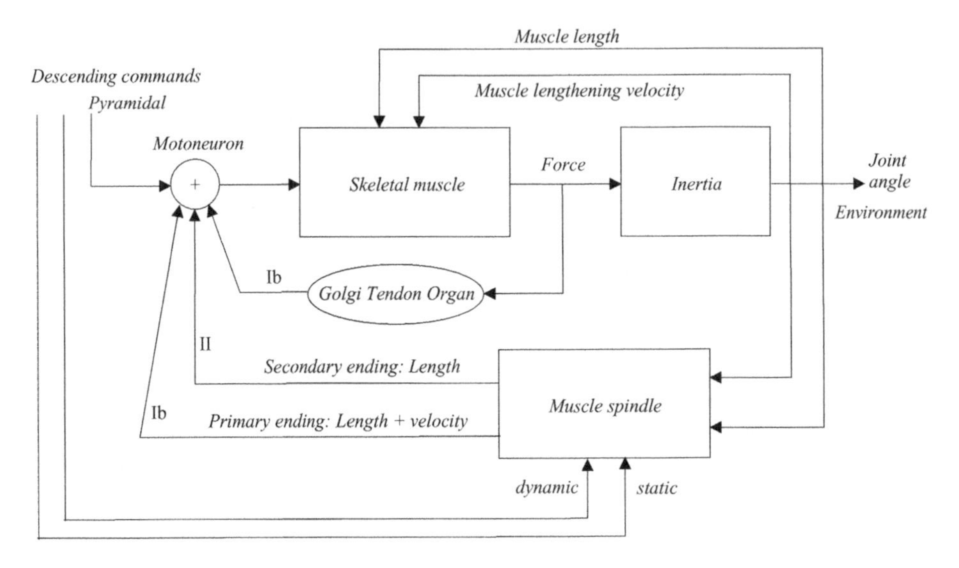

Figure 2.6 Scheme of the functional behaviour of the muscle and its sensors. The upper arrows indicate the passive muscle properties

GTO, which becomes excited and starts firing action potentials through the Ib afferents. The GTO may be described as a force sensor embedded in the muscle. Because of its ability to inhibit the α-motor neuron of the muscle, it was once believed that the role of this sensor was to limit the force developed by the muscle in order to avoid damage. However, the GTO sensor participates in an internal force-feedback loop and establishes a linear relationship between muscle force and sensor output in terms of the firing rate of the Ib afferent.

2. *Muscle spindles.* The muscle spindles are a special type of muscle fibres which are inserted in the muscle in a parallel arrangement and are covered by fusiform capsules. For that reason they are called intrafusal, while the other muscle fibres are called extrafusal. There are two types of intrafusal fibres (see Figure 2.6), namely the nuclear bag that senses the increases in length and the rate of lengthening, and the nuclear chain that senses the length of the muscle.

The muscle sensors and the neural circuitry of the spinal cord are the main components involved in execution of the lower-level reflexes, such as the *stretch reflex* and the *inhibitory reflex.* Because the muscle spindles run parallel to the extrafusal muscle fibres, they can sense a stretch in the muscle. This causes firing of the afferent neuron (Type Ia), which has a direct excitatory connection with the α-efferent motor neuron. An action potential in this motor neuron would cause a 'rapid' (around 30 ms) muscle contraction, the monosynaptic stretch reflex. It is called monosynaptic because there is only one synapse (neuron connection) involved. Nevertheless, the picture is a little more complex; the efferent motor neuron has more inputs from descending motor commands and from other neurons at the spinal cord, which may be inhibitory or excitatory, as illustrated in Figure 2.6. In addition, the GTO produces an inhibitory input to the motor neuron that is carried at a lower speed. If the sum of excitatory and inhibitory inputs goes beyond a certain threshold, the motor neuron will release an action potential. If there are more inhibitory inputs, no action potential will be generated (Kandel, Schwartz and Jessell, 1995).

2.3.5 Biomechanics of human limbs as a model

The limbs are the outcome of a process of evolution in vertebrates that developed fins and limbs in their transition from the aquatic to the terrestrial environment (Shadmehr and Wise, 2005). From the original tetrapod configuration, different types of limbs have evolved, like legs, arms and wings.

The forelimbs of most vertebrates have more DoFs than they need to move in three-dimensional space (Shadmehr and Wise, 2005). This *overcompleteness* poses a problem for motor control. In the case of human limbs this fact has repercussions for the design of wearable robots. On the one hand, the robot or exoskeleton may alter the kinematics of the limb, an issue discussed in Chapter 3. On the other hand, the robot will cause alterations to the motion patterns resulting from cognitive and physical interactions between human and robot. These interactions are described in detail in Chapters 4 and 5. For a bioinspired design of motor reflexes in the context of antislippage strategies for stable grasping with a wearable robot upper limb prosthesis, see Case Study 2.6.

2.3.6 Recursive interaction: engineering models explain biological systems

This section has shown that it is not easy to model the complexity of biological systems. Therefore, in order to develop engineering models of these systems it is necessary to simplify a number of aspects while retaining the main features of the system. In general, this limits the validity of the model to a certain set of conditions, for instance isometric (without movement) muscle contraction as in a muscle model with proprioceptive feedback (van der Helm and Rozendaal, 2000). The main advantage is that with these models it is possible to manipulate variables, even internal ones, and to perform a sensitivity analysis to identify the role of each variable in the behaviour of the biological system and characterize the biological mechanisms.

2.4 LEVELS OF BIOLOGICAL INSPIRATION IN ENGINEERING DESIGN

As noted earlier, knowledge of biological systems has inspired engineering design; e.g. the design of ultrasound motors may have been inspired by inchworm motion (Pons, 2005). The identification of different levels in this engineering imitation of Nature stems from biology and hence is also bioinspired. In ethology in particular, Tommasello distinguished three levels of imitation (Gardenfors, 2006):

1. *Strengthening of the stimulus.* An animal is attracted to use a certain tool because it has seen another animal using it. For instance, a young monkey has seen its mother using a stone to break a nut and starts playing with it. This leads to what Tommasello called 'learning by emulation'; the tools used by other animals are copied with a view to achieving the same goals, but there is no imitation of the actions or understanding of their intent or their mechanics.

2. *Mimetism.* In mimetism, the animal uses the same actions, body parts or tools with the same goals in mind, but there is no understanding of the mechanics or the consequences of the actions, e.g. birds that learn to speak.

3. *Imitation.* Finally, in imitation there is real understanding of the intent of the behaviour, and there is adaptation to the particular circumstances, so this is not mere replication of the behaviour observed previously. There are two steps in imitation; the first is to imagine the goal of the actor

and the second is to combine the sequence of actions in a manner appropriate to the new actor and to specific contexts.

It is these ideas that underlie the distinction between the different levels of bioinspiration that are found in engineering design and that are discussed in the following sections.

2.4.1 Biomimetism: replication of observable behaviour and structures

In emulation and mimetism there is no understanding of the intentions or the mechanics of the actions; there is simply copying of the tools and replication of the movements in order to achieve a certain goal. According to the definitions proposed in this book, biomimetism refers to the search for inspiration in biological systems, but taking only their external behaviour into consideration.

There are examples of biomimetism in the field of wearable robots, e.g. the HAL exoskeleton for gait support, described in detail in Case Study 9.4. This exoskeleton monitors gait patterns and then replicates the kinematics directly with a position-control system. The system measures the EMG patterns in order to detect the subject's intention and then applies torques based on the EMG. However, even when the muscles are not active, the motors must act to follow the movement of the limb, with considerable expenditure of energy. Another example is the WOTAS monocentric elbow joint, see Case Study 8.1, which does not entirely follow the kinematics of the elbow because it is implemented with a hinge joint.

2.4.2 Bioimitation: replication of dynamics and control structures

In imitation there is understanding of the intention of the behaviour of the copied system. Moreover, imitation adapts to particular circumstances, such as technological capabilities. It is not therefore a simple matter of replicating or mimetizing the behaviour.

Bioimitation consists of two steps: properly understanding the actor's goal and modelling its dynamics. In the present case it consists in developing models of biological systems and then developing an engineering model of such a system adapted to the current state of the technology.

An imitation will involve understanding the intentions behind the actions, which in turn implies understanding and replication of the internal models. The internal models can then be used to adapt the actions to the specific movements.

One example of this type of approach is limit-cycle bipeds, walking robots that are intended to copy the dynamics of human gait (Collins *et al.*, 2005). This approach is discussed in Case Study 2.5.

The GAIT-ESBiRRo exoskeleton, which is discussed in Case Study 9.1, contains examples of both approaches: biomimetism and bioimitation. On the one hand, there is an element of functional compensation that seeks to copy the function of the knee during support and swing. It emulates the behaviour of the knee musculature under certain conditions, in its goals but not in the dynamics of the muscles or in its motor control structure. This makes for a mechanism that is simple and efficient, but limited with respect to the capabilities of a healthy leg; for instance, only limited adaptation to certain locomotion situations or recovery from disturbances is possible. On the other hand, the mechanics and the kinematics of the knee joint are precisely in step; the four-bar linkage precisely follows the kinematics and the dynamics. It can therefore be said that the knee motion implemented with the state of the art in orthotics and prosthetics technology is bioimitation. Finally, the GAIT orthosis in its semi-active version has no insight into the mechanics of gait control. It simply follows the ankle joint angle and triggers a change when the angle exceeds a certain value. That is a clear example of bioemulation or biomimetism.

2.5 CASE STUDY: LIMIT-CYCLE BIPED WALKING ROBOTS TO IMITATE HUMAN GAIT AND TO INSPIRE THE DESIGN OF WEARABLE EXOSKELETONS

M. Wisse

Mechanical Engineering Faculty, Biomechanical Engineering Department, Delft University of Technology, Delft, The Netherlands

2.5.1 Introduction

Within the field of robotics, the present approach to developing human-like walking robots (see Figure 2.7) is somewhat unusual. Whereas the conventional roboticist will strive for ever greater position accuracy and more (advanced) control, the present approach is the opposite. Walking robots are built according to the concept of limit-cycle walking. With this concept the natural dynamics of the walking system is utilized while guaranteeing stability of the periodic walking motion (the 'limit cycle'). Thanks to the utilization of natural dynamics, these robots are more energy efficient than conventional walking robots by a factor of 15 (Collins *et al.*, 2005).

Because of their naturalness and efficiency, these robots achieve a close similarity to human walking; as a result they can be used to help understand human walking and to inspire the design of wearable exoskeletons. This case study provides a brief description of the approach of limit-cycle walking and the engineering solutions that give these robots their efficiency and stability.

2.5.2 Why is human walking efficient and stable?

2.5.2.1 Walking is simply a repeated forward fall

Walking may appear to be a complex motion. The upright human resembles an inverted pendulum (with the stance foot as pivot) which is difficult to stabilize. The sensation of complexity is strengthened by many present-day walking robots (Kaneko *et al.*, 2002; Sakagami *et al.*, 2002). These robots have powerful computers and motors which they use to try and maintain upright stability. This might appear to suggest that more complex and expensive wearable exoskeletons are required to serve

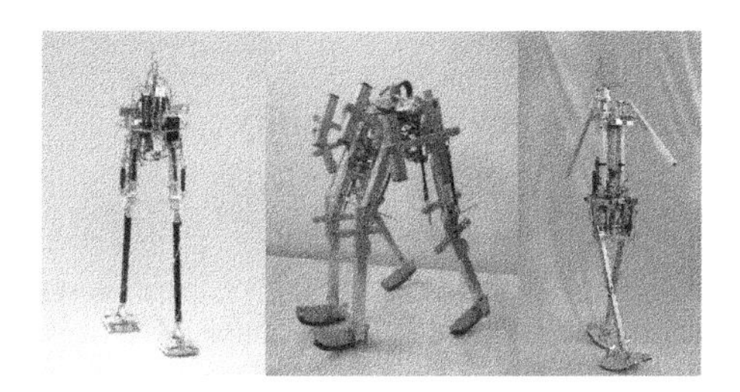

Figure 2.7 Some of the robots developed at Delft University of Technology: 'Baps' (left), 'Mike' (centre) and 'Denise' (right)

rehabilitation clients better. Fortunately, this train of thought is incorrect. The walking motion is fundamentally different from the inverted pendulum exercise. There is no need for upright stability, as each step is simply a forward fall. The instability of the fall is allowable because it alternates with stabilizing support transitions. This has been proved by the construction of 'passive dynamic walkers' (McGeer, 1990), simple walking robots that stably walk down a shallow ramp with no motors and no controls. These simple mechanical devices are not only simpler and cheaper than all other robots, but their motions are also an order of magnitude more efficient (Collins *et al.*, 2005) and appear much more natural. These results show convincingly that simpler designs may often be preferable to heavier and more complex 'high-tech' solutions.

2.5.2.2 Definition of limit-cycle walking

The notion that 'the walking motion is simply a repeated forward fall' can be formalized with the following definition of the paradigm 'limit-cycle walking' (Hobbelen and Wisse, 2007).

> Limit-cycle walking is a nominally periodic sequence of steps that is stable as a whole but not locally stable at every instant in time.

A 'nominally periodic sequence of steps' means that the intended walking motion (ideally without disturbances) is a series of exact repetitions of a closed trajectory in state space (a limit cycle) in which each of the walker's two feet is moved forward in turn. This trajectory is not locally stable at every instant in time, so that there is no need to make all points on the trajectory attractive to their local neighbourhood in state space (as is done in conventional trajectory control). The nominal motion is still stable as a whole because over the course of multiple steps, neighbouring trajectories eventually approach the nominal trajectory. This type of stability is called 'cyclic stability' or 'orbital stability' (Strogatz, 2000).

2.5.2.3 The energy benefit of limit-cycle walking

The energy efficiency of a walking motion is expressed by means of the 'cost of transport', i.e. the energy used per metre travelled per unit of weight of the walking system. For a given walking velocity, the cost of transport is a dimensionless number that can be used to compare robots and humans (see Section 2.2). Calculated from muscle work, the cost of transport of a human being at average walking speed is 0.05 (Donelan, Kram and Kuo, 2002). The cost of transport for conventional biped robots is estimated at 1.6, higher than humans by a factor of more than 30. In contrast, limit-cycle walkers have approximately the same low cost of transport as humans (Collins *et al.*, 2005).

2.5.3 Robot solutions for efficiency and stability

2.5.3.1 Arc feet for better stability

With a passive prototype (see Figure 2.8) the effect of the foot shape on stability was researched. The prototype was manually 'launched' on a slight downward ramp. The downward tilt provided sufficient energy to continue walking forever if the walking surface could have been infinitely prolonged. In reality, a sequence of about seven steps is enough to determine whether the gait is successful. By then the robot has either fallen or settled into a neat regular walking cadence. Therefore, after seven steps a disturbance was introduced. The walking surface was lowered with an adjustable offset. A search was made for the maximal allowable offset that would still permit successful walking and this was plotted as a function of the foot arc radius.

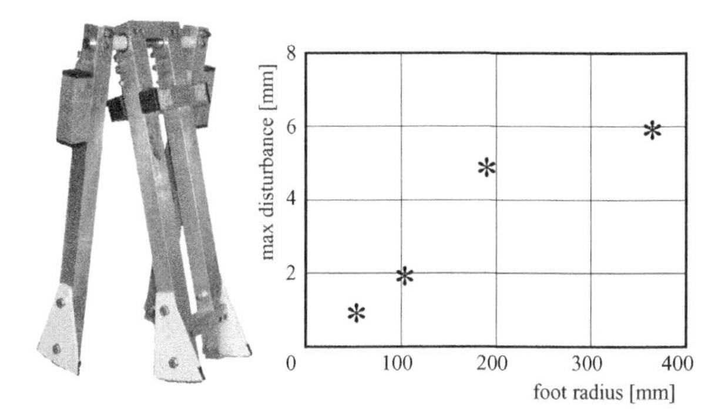

Figure 2.8 A larger foot radius provides better disturbance rejection

Figure 2.8 clearly shows the positive effect of the radius. According to this experiment (Wisse and van Frankenhuyzen, 2006) and the related simulation study (Wisse *et al.*, 2006), the maximum allowable stepdown offset is largest with a foot radius almost equal to the leg length. However, in practice the foot length (and hence the useful part of the radius) is limited by foot-scuffing and tripping problems. These robots usually end up with a foot radius of approximately one-third to one-half of the leg length. This matches the 'foot roll-over shape' (Hansen, Childress and Knox, 2000), as measured on human and prosthetic feet, surprisingly well. This research confirms what prosthesis designers and fitters already knew: the human foot size and roll-off shape are optimal for walking stability.

2.5.3.2 Ankle springs instead of arc feet

Next, it was discovered that the beneficial effect of the foot radius could equally well be achieved with ankle springs (Wisse *et al.*, 2006). The arc feet did not have an ankle joint, but in subsequent experiments a flat foot mounted on an ankle joint was used. The joint was equipped with a torsional spring and an upright stance was the equilibrium position. Springs are easily as effective as the arc foot effect. Their stabilizing capacity is equal when $k = mgr$, where k is the torsional ankle spring stiffness, m is the total mass of the person or robot, g is gravity and r is the foot radius normalized to the leg length. Note that the ideal stiffness (equivalent to a normalized radius of $\frac{1}{3}$ or $\frac{1}{2}$) is not enough to keep the system in upright balance when in quiet stance. This underlines the principal assertion; namely that the walking motion is a repeated forward fall that is stable thanks to the foot support transitions.

2.5.3.3 Lightweight legs for fast swing motion

The walking motion is a repetition of unstable forward falls. It is absolutely essential that the swing leg quickly moves forward in order to catch the fall (Wisse *et al.*, 2004). Therefore, all of the present robots use a hip motor that accelerates the swing leg. This has resulted in a significant increase in the maximum allowable disturbance (Schwab and Wisse, 2001; Wisse *et al.*, 2004), which is up to 2 % of the leg length for the 'Mike' test prototype (see Figure 2.7). A more refined study of the swing leg motion has revealed the additional beneficial effect of 'swing leg retraction'. The swing foot should have slight rearward velocity at the instant of heel strike. In human running this effect is even more pronounced (Seyfarth, Geyer and Herr, 2003). This swing leg retraction has been calculated and tested and it was found that it helps to provide stability (Wisse, Atkeson and Kloimwieder, 2006). For robot controller design, the stabilization details are important. For an exoskeleton design, all that matters is that the leg must swing forward very quickly to allow sufficient time for a retraction phase. Basically, this means that the weight of the exoskeleton, and especially the foot, must be as small as possible.

2.5.3.4 Skateboard ankles for lateral stability

In addition to the two-dimensional analysis presented thus far, the author has also studied a few three-dimensional prototypes. From one of the first prototypes ('Baps', see Figure 2.7) the important lesson was learnt that the full three-dimensional motion cannot be simplified by using two projections. A sagittal model and a frontal model were made, but it was found that the two together did not accurately describe the robot's behaviour. The reason is that there is a third projection: a top view. Although the motions in this projection are relatively small, the forces and torques are significant and influential. Several of the robots suffered from undesired yaw rotations, i.e. rotations around the vertical axis. This was solved with better foot friction, counterswinging arms (Collins, Wisse and Ruina, 2001) or a counter-rotating body (Wisse and Schwab, 2001). The implication for exoskeleton design is simply that the foot should show as little rotational slip as possible.

The most recent three-dimensional prototype ('Denise', see Figure 2.7) has one additional design feature: a skateboard-like ankle joint (Wisse and Schwab, 2004). The joint axis is directed from above the heel forward and downward through the middle of the foot sole. The effect of this joint is similar to that of a skateboard. If the robot leans to one side (after a disturbance), it must also steer to that side. The same effect can be found in skateboards and bicycles. When there is sufficient forward velocity, this lean-to-steer coupling has a stabilizing effect. Imagine the robot being pushed to the left; it will then simply steer to the left and continue its path in the new direction. Simulation studies and prototype experiments (Wisse, 2005b) suggest that the skateboard–ankle idea works well. It may well be a new design feature for future exoskeleton ankles.

2.5.4 Conclusion

So far, robot research has led to the following conclusions: simple exoskeleton designs are potentially better than high-tech looking ones; an arc radius in the foot sole helps stability; the same stability can be achieved with flat feet and ankle springs; the exoskeleton, and especially the feet, should be as lightweight as possible; and a skateboard-like ankle joint may help lateral stability. The author intends to continue this line of research to find suggestions for things such as knee design, and ultimately to help understand and diagnose human gait.

Acknowledgements

The author wishes to thank Jan van Frankenhuyzen, Arend Schwab, Daan Hobbelen, Richard van der Linde, Frans van der Helm and John Dukker. The research project was funded by the Dutch Technology Foundation STW and by NWO, the Netherlands Organization for Scientific Research.

2.6 CASE STUDY: MANUS-HAND, MIMICKING NEUROMOTOR CONTROL OF GRASPING

J. L. Pons, R. Ceres and L. Calderón

Bioengineering Group, Instituto de Automática Industrial, CSIC, Madrid, Spain

This case study is presented here to illustrate the bioinspired design of a hierarchical control architecture for efficient control of a wearable upper limb robotic prosthesis. This section also illustrates sensorimotor reflex mechanisms implemented in the robot controller to control grasp.

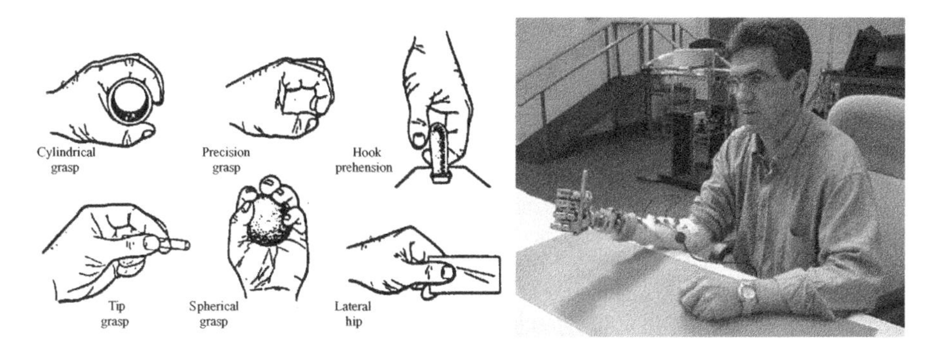

Figure 2.9 A classification of basic grasps (left) and the MANUS-HAND wearable upper limb robotic prosthesis during experimental trials (right)

2.6.1 Introduction

Wearable robotic upper limb prostheses, also known as mechatronic dextrous artificial hands, have evolved into an established field of research since the second decade of the twentieth century (Borchardt *et al.*, 1919). Ever since, research groups have addressed different aspects of the problem, from hand design to control issues (see Pons, Ceres and Pfeiffer, 1999, for a thorough review of the state of the art).

The MANUS project was set up on June 1998. The overall objective of MANUS-HAND was to develop modular hand prostheses with enhanced mobility so as to achieve the basic grasping modes, i.e. hook, precision, cylindrical or power, tip and lateral grips, of the human hand (see Figure 2.9 (left)). It was further proposed to incorporate force feedback in order to increase the patient's chances of social and professional reintegration. Since the MANUS approach is a global one, the project also proposed and developed a suitable training device. The training device serves the purpose of starting functional rehabilitation immediately after upper limb amputation. In this case study the authors address the main design aspects of the MANUS-HAND system, in particular the bioinspired control structure.

The project focused on a threefold modular approach to development of the wearable upper limb robotic prosthesis (see Figure 2.9 (right)):

- The control system is based on myoelectric signals whose number and characteristics are adaptable to users' residual capabilities.
- The number of grasping modes is also selected according to user needs, from a set of five grasping modes (cylindrical, precision, lateral, hook and tip) that cover most of the basic natural hand grips.
- The modular approach to the mechanical design allows for stepwise extension of the functionality up to the shoulder.

2.6.2 Design of the prosthesis

The following paragraphs summarize the design of the robotic prosthesis in terms of its mechanical, actuation and sensing characteristics.

2.6.2.1 Mechanical concept

The mechanical concept contains several innovations with respect to commercial and state of the art prostheses: the number of active axes is increased to 10 in the MANUS-HAND prototype whereas commercial prostheses usually have a maximum of three active axes: a crossed-tendon mechanism is used in the fingers instead of traditional bar mechanisms; the thumb movements are coupled by means of a Geneva-wheel based mechanism, which makes it possible to use one actuator for thumb movements on two planes corresponding to abduction–adduction and flexion–extension; and prototypes of ultrasonic motors were built to provide actuation technology.

Fingers having a fixed shape rotating about one joint, as usually seen in commercial prosthesis, do not move in a natural way and limit the grasping functionality. If, for example, the fingers are shaped for a cylindrical grip, the shape will not be suited to a hook grip and vice versa. Noncommercial prosthesis have been proposed with three active fingers having coupled interphalangeal joints (Kyberd, Evans and Winkel, 1998). MANUS-HAND includes fingers with three joints. A biomimicked closing pattern is imposed on the fingers by fixing the ratio between the joint angles. In the design this is achieved by means of crossed-tendon mechanisms linking the rotation of every phalanx to the previous one, as depicted in Figure 2.10 (left). In this way, both the functionality and the aesthetics of the prosthesis are improved. The proximal joints of the index and middle fingers are coupled, which means that a single actuator is driving six joints.

The thumb movement is implemented as a composition of an opposition movement (abduction–adduction) and a flexion–extension movement. MANUS-HAND introduces the concept of a neutral position, in which thumb movement around two intersecting axes is achieved by means of an intermittent mechanism (Geneva mechanism). The thumb moves from nonopposition and flexion (in which a lateral grip is available), through nonopposition and extension (in which the hook grip is achieved) until the neutral position is reached (i.e. natural-looking free hand). From the neutral position, further rotation of the motor engages the cylindrical grip zone. At the end of this grip zone, the precision grip area is reached. In this way, the closing pattern for the cylindrical and tip grip is attained by moving around the first axis. Moving around the second axis gives the closing pattern for the hook and lateral grip (see Figure 2.10 (right)). Rotation of the wrist is independent of the other movements and is therefore actuated by a separate actuator. To the authors' knowledge, no

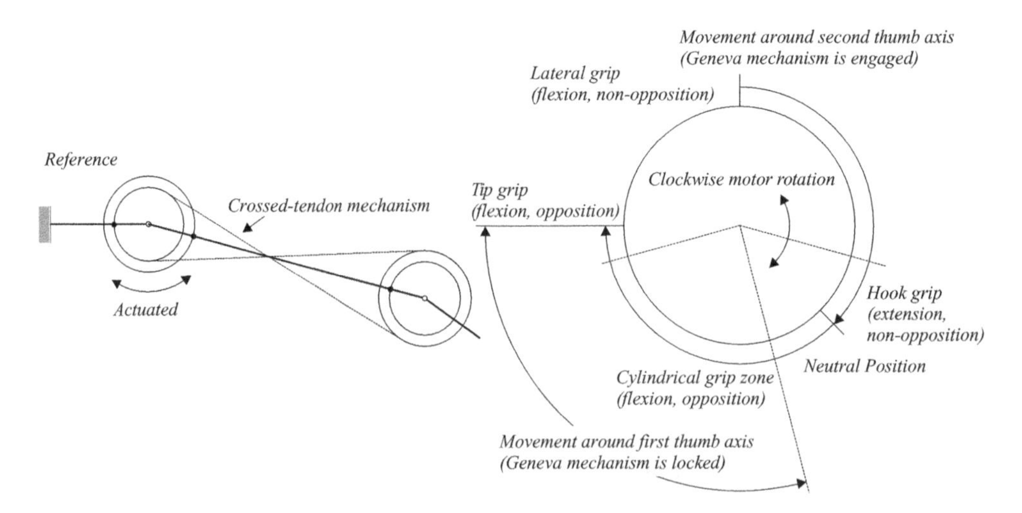

Figure 2.10 Schematic representation of the crossed tendon mechanism (left) and representation of the Geneva mechanism cycle motion (right)

commercial prosthesis uses ten joints and is capable of carrying out such a variety of grasping modes as the proposed MANUS-HAND concept.

2.6.2.2 Actuation and sensing concept

In the context of robotics as it was in April 1982, at a workshop held at MIT (Hollerbach, 1982), it was pointed out that 'the current actuation technology provides perhaps the most serious, long term impediment to artificial hand design'. The same statement still holds true today. In fact, current actuation technologies fail to provide efficient, high power density actuators suitable for artificial hand design. The lack of adequate actuation technologies directly affects the design of dextrous hands. If dexterity is to be increased, a larger number of active joints is needed and that means bulky solutions.

Two actuator technologies were selected, brushless DC motors and ultrasonic motors, although there are other novel actuator technologies, e.g. shape memory alloy actuators (see Section 6.3). Due to the limited performance of ultrasonic motors, extensive research was conducted to improve their properties. Eventually, a combination of these two technologies was implemented: DC motors drive fingers and thumb while wrist pronation–supination is based on high-torque ultrasonic motors.

Force controlled grasps require exact computation of equilibrium conditions. Such equilibrium computation relies on feedback information about points of contact between grasped object and fingers, and also on a precise estimation of grasping forces. Measuring the contact point force for a whole hand grasping operation requires a large number of force and contact sensors, making the whole system complex.

In order to keep system complexity at a manageable level, the possibility was considered of implementing stable grasps based on actively programmable finger compliance. In this approach the object is fixed to the palm via compliant linkages in such a way that each individual finger behaves as a spring with one end attached to the palm and one to the object. In such a configuration, the system presents stable equilibrium poses as long as there are enough contact points.

This approach has been shown to provide stable grasps with a minimum number of sensors. A sensor allocation scheme to minimize system complexity was adopted:

- Two force sensors are placed at the index and medium finger tips.
- One force sensor is placed at the thumb tip.
- Since finger motion is coupled and one shaft rotation determines index and middle finger placement, only one position sensor was adopted for index and middle fingers.
- One position sensor was used for the thumb since positioning in the ab–adduction and flexo–extension planes is achieved with a single shaft rotation, i.e. the thumb position is uniquely determined using a single sensor.

2.6.3 MANUS-HAND control architecture

The prosthesis prototype was integrated around a flexible electronic control architecture. The architecture has to be flexible enough to support real-time control of three active axes, including the implementation of antislippage reflexes, real-time identification of EMG commands and computation of control loops and force biofeedback. A modular architecture was selected, based on a host and three local microcontrollers. The three controllers communicate with the host controller through a fast I2B bus.

The functions supported by the host controller are: calibration of EMG signals, computation of control commands from EMG signals, computation of set points for local controllers, implementation of force biofeedback and general status and error management. The functions supported by local controllers include: low-level control of actuators, implementation of antislippage control reflexes,

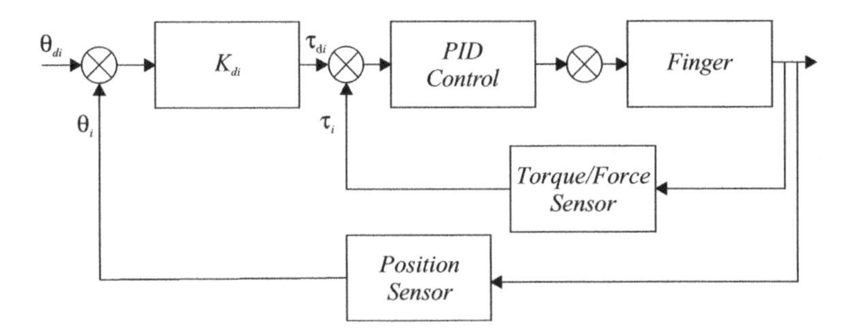

Figure 2.11 Compliance control scheme for finger control in the local controllers

monitoring of finger force and position, and providing information for force biofeedback. Every local controller is associated with an active joint or actuator while the host controller is in charge of the general functionality of the prosthesis.

Following command computation from EMG signals, a grasping mode is identified. As already described in the previous sections, a compliance control approach was proposed to obviate the need for a detailed geometrical description of the grasped object. The idea behind this approach is for the fingers and thumb to behave as springs with a neutral position in full flexion. The compliance of the springs allows the grasp to be accommodated to object size and weight. In this approach the different force levels are achieved by modifying the equivalent spring stiffness of the fingers.

From a control point of view, only information local to the finger, i.e. the current finger position, the force being exerted and the neutral position, is required for the finger to behave as a spring. Therefore, overall control of the grasp is organized around the local controllers. Each of these controllers performs the control functions depicted in Figure 2.11. The local Hall effect force and position sensors are used to close the control loop.

As Figure 2.11 shows, for each active joint i, two parameters define the grasp being performed, i.e. the neutral position, θ_{di}, and the equivalent spring stiffness, K_{di}. For instance, if θ_{di} is defined for the thumb and index to be in full flexion, this produces a cylindrical grasping pattern. The more extended the thumb and finger, the more force is applied to the object. If grasp strength needs to be increased, this is easily done by increasing K_{di}. This is the mechanism used to implement antislippage reflexes in response to vibration detected at finger tip force sensors.

The resulting control approach, taking the user into account, is an open loop in which the amputee simply commands a desired grasping mode and host and local controllers ensure this by means of a compliance scheme.

2.7 CASE STUDY: INTERNAL MODELS, CPGs AND REFLEXES TO CONTROL BIPEDAL WALKING ROBOTS AND EXOSKELETONS: THE ESBIRRO PROJECT

A. Forner-Cordero

Bioengineering Group, Instituto de Automática Industrial, CSIC, Madrid, Spain

2.7.1 Introduction

In this case study the concept design is presented of the control structure currently under development in the context of the EU-funded ESBiRRo project (Enhanced Sensory Bipedal Rehabilitation Robot:

Biomimetic actuation, sensing and control technology for limit-cycle bipedal walkers, EU FP6-2005-IST-61-045301-STP).

The development of a robotic humanoid capable of walking is one of the recurrent human dreams of all ages. Gait is the most important form of locomotion in humans. Its complexity in terms of fine coordination and stability is revealed when a disabling pathology arises or when trying to design walking robots. One of the major objectives in rehabilitation research is to restore gait function.

The goal of this project is to develop a hierarchical biomimetic limit-cycle structure to control biped walking that includes a repertoire of recovery reactions from perturbations during gait. This control structure will be applied to the design of both an autonomous walking biped and a robotic exoskeleton for gait support.

These ideas have been inspired by recent developments in different fields. Firstly, there is the emerging design of limit-cycle bipedal walking robots (Collins *et al.*, 2005), described in Case Study 2.5. Secondly, there is the development of semi-active lightweight exoskeletons such as the GAIT orthosis fully described in Case Study 9.1. Thirdly, there is growing understanding of the biological motor control system (Shadmehr and Wise, 2005) and of the biomechanics of recovery from perturbations during gait (Forner–Cordero, Koopman and van der Helm, 2003).

The development of these concepts is expected to give rise to a new generation of biologically inspired walking robots, both autonomous robots and exoskeletons for use as technical aids for the disabled.

2.7.2 Motivation for the design of LC bipeds and current limitations

There are currently several biped walking robots. In most cases control of these bipeds is based on trajectory control (TC) provided by standard motors, and their stability relies on the zero moment point. These robots have two major drawbacks: high energy consumption and small stability margins, as they are unable to walk in entirely unstructured environments. In contrast, limit-cycle (LC)-controlled robots exploit the dynamics of mechanical systems, for instance the pendulum behaviour of the swinging leg, and consume less energy (Collins *et al.*, 2005).

Nevertheless, their stability margin is comparable to that of TC robots. An important aspect of LC-controlled robots is that the global stability of the LC can be improved with the addition of a control system that includes sensors and actuators. To that end, a hierarchical control structure following the biomimetic design principles that guided the development of LC robots (McGeer, 1993) is proposed (see Figure 2.12 (left)).

2.7.3 Biomimetic control for an LC biped walking robot

Bipedal walking robots and gait support robotic exoskeletons share common problems, such as stability, energy consumption and maintenance of adequate (usually low) weight on the legs to take advantage of swing dynamics.

The initial inspiration for LC robots came from human gait. This project proposes to go one step further in the evolution of LC robotics by implementing the recovery reactions from perturbations that can be found in biological systems, like the human stumble reaction. These new-generation robots will consume less energy than their TC counterparts and be more stable. The modelling and control of a biped robot will improve understanding of human gait, paving the way for a novel wearable robot that can support gait in a reliable and efficient manner.

The global control structure proposed for the ESBiRRo designs has a bioinspired hierarchy (see Figure 2.12 (left)). The high-level controller has several functions. Firstly, it contains an internal model of the whole body. An internal model, as explained in Section 2.3, is a neural process able to predict the consequences of an action based on a model of the body and its environment (Kawato,

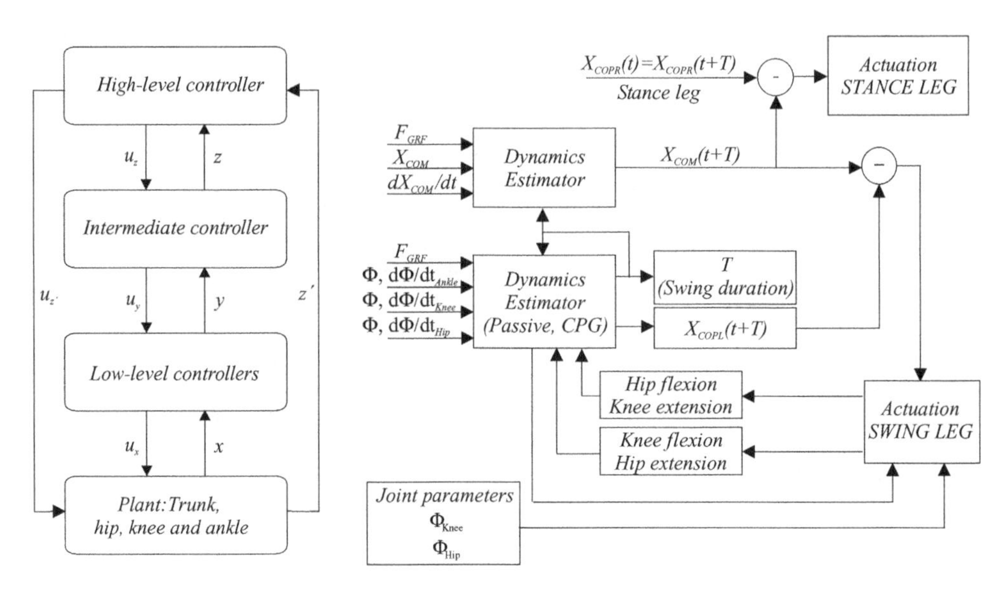

Figure 2.12 Global control structure for the ESBiRRo project (left): the high-level controller contains an internal model of the body, sets the general goals – e.g. maintenance of global stability – and triggers emergency reactions to perturbations directly in the actuators. The intermediate controller assures maintenance of the limit cycle with minimal intervention, possibly implemented with central pattern generators (CPGs). Finally, the low-level controllers are related to specific joints, performing the movements without damage and lending stiffness to counteract limit-cycle deviations. Model of recovery from human stumble based on experimental data and a biomechanical model of the recovery (right)

Furukawa and Suzuki, 1987). It is proposed that the higher-level internal model should work with a lower-dimension model, as not all the information is relevant to task performance and this can be filtered by the intermediate- or low-level controllers (Li, Todorov and Pan, 2004).

The higher-level controller sets the more abstract goal of the task. In the case of walking, this could be to reach a certain point while maintaining global stability. In addition, this controller is involved in complex reactions that implicate the whole body, such as emergency recovery reactions to perturbations during gait. The intermediate controller should guarantee maintenance of the limit cycle with minimal intervention, in the same way as optimal controllers. The objective function of this controller will be to keep the LC stable with minimum energy consumption. It is possible to implement this controller by means of CPGs, groups of neurons capable of autonomously generating a rhythmic output. The CPG can be entrained with the sensory input to control a rhythmic movement like gait. In addition, it can function as a hybrid feedback/feedforward controller (Kuo, 2002) and transmit more abstract information about the state of the system to the higher level. Finally, the low-level controllers for each joint will incorporate some of the characteristics of the low-level reflexes. They are meant to perform the movements without causing damage and increase stiffness to counteract deviations of the LC. They may be said to some extent to simulate muscle reflexes and the passive dynamics of the muscle.

Figure 2.12 (right) presents a control architecture inspired by the results of measuring recovery from a trip and a biomechanical model of the recovery (Forner-Cordero, 2003; Forner-Cordero, Koopman and van der Helm, 2004). It is assumed that the main objective function of the controller is to maintain trunk stability. This depends on the external forces and their point of application, and also on the motion of the trunk and the torques at the hip (see Figure 2.12 (right)).

2.7.4 Conclusions and future developments

Future lines of research in the design and construction of biped walking robots will depend on technological as well as scientific developments. On the one hand, improvements in sensor and actuator technologies will pave the way for the design of more efficient and robust biped walkers. The availability of smaller sensors, such as micromachined inertial sensors capable of measuring angular accelerations and velocities in the three dimensions, opens up new possibilities for the design of sensory systems for biped robots (Luinge and Veltink, 2005; Moreno *et al.*, 2005). The development of more powerful, more efficient, smaller and lighter DC motors will be crucial for the construction of lightweight biped walkers with low energy consumption. In this respect, new actuator technologies like the ones based on piezoelectric motors may well prove to have potential in applications where rotation speed is not a crucial factor (Pons, 2005).

On the other hand, advances in understanding of the biomechanics and the control mechanisms of human gait can be expected to provide a theoretical basis for biomimetic strategies to control bipedal walkers. In addition to low consumption, humans have a wide repertoire of recovery strategies following a perturbation during gait (Eng, Winter and Patla, 1994; Forner-Cordero, 2003). In order to be able to imitate these strategies, there is a need not only to describe them but also to model them and identify the objective functions of the motor control system (Forner-Cordero, Koopman and van der Helm, 2004). In general, the goal of these control algorithms will be to make it possible to absorb or generate the joint power required to bring the biped back to its stable limit cycle.

REFERENCES

Alexander, R.M., 1991, 'Energy-saving mechanisms in walking and running', *The Journal of Experimental Biology* **160**: 55–69.

Bar-Cohen, Y. (ed.), 2001, *Electroactive Polymer (EAP) Actuators as Artificial Muscles – Reality, Potential and Challenges*, SPIE Press, Bellingham, Washington.

Bernstein, N., 1967, *The Coordination and Regulation of Movement*, Pergamon Press, London.

Bertram, J.E., Ruina, A., 2001, 'Multiple walking speed–frequency relations are predicted by constrained optimization', *Journal of Theoretical Biology* **209**(4): 445–453.

Biewener, A.A., 2003, *Animal Locomotion*, Oxford Animal Biology Series, Oxford University Press.

Borchardt, M., Hartmann, K., Schlesinger, G., Schwiening, J. (eds), 1919, *Ersatzglieder und Arbeitshilfen für Kriegsbeschädigte und Unfallverletzte*, Verlag von Julius Springer.

Collins, S.H., Wisse, M., Ruina, A., 2001, 'A three-dimensional passive–dynamic walking robot with two legs and knees', *International Journal of Robotics Research* **20**(7): 607–615.

Collins, S., Ruina, A., Tedrake, R., Wisse, M., 2005, 'Efficient bipedal robots based on passive-dynamic walkers', *Science* **307**: 1082–1085.

Darwin, C., 1859, *On the Origin of Species by Means of Natural Selection, or the Preservation of Favoured Races in the Struggle for Life*, John Murray, London.

Donelan, J.M., Kram, R., Kuo, A.D., 2002, 'Mechanical work for step-to-step transitions is a major determinant of the metabolic cost of human walking', *Journal of Experimental Biology* **205**(23): 3717–3727.

Eng, J.J., Winter, D.A., Patla, A.E., 1994, 'Strategies for recovery from a trip in early and late swing during human walking', *Experimental Brain Research* **102**(2): 339–349.

Erdemir, A., McLean, S., Herzog, W., van den Bogert, A.J., 2007, 'Model-based estimation of muscle forces exerted during movements, *Clinical Biomechanics* **22**(2): 131–154.

Forner-Cordero, A., 2003, 'Human gait, stumble and... fall? Mechanical limitations in the recovery from a stumble', PhD Thesis, University of Twente, Enschede, The Netherlands.

Forner–Cordero, A., Koopman, H.F.J.M., van der Helm, F.C.T., 2003, 'Multiple-step strategies to recover from stumbling perturbations'. *Gait & Posture* **18**(1): 47–59.

Forner-Cordero, A., Koopman, H.F.J.M., van der Helm, F.C.T., 2004, 'Mechanical model of the recovery from stumbling', *Biological Cybernetics* **91**(4): 212–222.

Forner-Cordero, A., Koopman, H.F.J.M., van der Helm, F.C.T., 2005, 'Energy analysis of human stumbling: the limitations of recovery', *Gait and Posture* **21**(3): 243–254.

Freivalds, A., 2004, *Biomechanics of the Upper Limbs. Mechanics, Modelling and Musculoskeletal Injuries*, CRC Press, Boca Raton, Flouide, pp. 55–99.

Gardenfors, P., 2006, *How Homo Became Sapiens (On the Evolution of Thinking)*, Oxford University Press.

Goldberg, D.E., 1989, *Genetic Algorithms*, Addison-Wesley, Reading Massachusetts.

Hansen, A.H., Childress, D.S., Knox, E.H., 2000, 'Prosthetic foot roll-over shapes with implications for alignment of trans-tibial prostheses', *Prosthetics and Orthotics International* **24**: 205–215.

Henneman, E., 1957, Relation between size of neurons and their susceptibility to discharge, *Science* **126**: 1345–1347.

Hill, A.V., 1970, *First and Last Experiments on Muscle Mechanics*, Cambridge University Press.

Hirai, K., Hirose, M., Haikawa, Y., Takenaka, T., 1998, 'The development of Honda humanoid robot', in *Proceedings of the IEEE International Conference on Robotics and Automation*, pp. 1321–1326.

Hobbelen, D.G.E., Wisse, M., 2007, 'Limit cycle walking', in *Humanoid Robots; human-like machines* (ed M. Hackel), Advanced Robotic Systems International, Vienna. Ch. 14, pp. 277–294, ISBN: 978-3-902613-07-3.

Hogan, N., 1985, 'The mechanics of multi-joint posture and movement control', *Biological Cybernetics* **52**: 315–331.

Holland, J.H., 1975, *Adaptation in Natural and Artificial Systems*, The University of Michigan Press.

Hollerbach, J.M., 1982, 'Workshop on the design and control of dexterous hands', MIT-AI Memo No. 661, Massachusetts Institute of Technology.

Kandel, E.R., Schwartz, J.H., Jessell, T.M., 1995, *Essentials of Neural Science and Behavior*, McGraw-Hill/Appleton & Lange.

Kaneko, K., Kanehiro, F., Kajita, S., Yokoyama, K., Akachi, K., Kawasaki, T., Ota, S., Isozumi, T., 2002, 'Design of prototype humanoid robotics platform for HRP', in *Proceedings of the IEEE/RSJ International Conference on Intelligent Robots and Systems*, vol. 3, pp. 2431–2436.

Kawato, M., Furukawa, K., Suzuki, R., 1987, 'A hierarchical neural-network model for control and learning of voluntary movement', *Biological Cybernetics* **57**(3): 169–185.

Koopman, B., 1989, 'The three-dimensional analysis and prediction of human walking', PhD Thesis, University of Twente, Enschede, The Netherlands.

Kuo, A.D., 2002, 'The relative roles of feedforward and feedback in the control of rhythmic movements', *Motor Control* **6**(2): 129–45.

Kyberd, P.J., Evans, M., Winkel, S., 1998, 'An intelligent anthropomorphic hand, with automatic grasp', *Robotica* **16**: 531–536.

Li, W., Todorov, E., Pan, X., 2004, 'Hierarchical optimal control of redundant biomechanical systems', in *Proceedings of the 26th Annual International Conference of the IEEE Engineering in Medicine and Biology Society*, pp. 4618–4621.

Luinge, H.J., Veltink, P.H., 2005, 'Measuring orientation of human body segments using miniature gyroscopes and accelerometers', *Medical and Biological Engineering and Computing* **43**(2): 273–282.

McGeer, T., 1990, 'Passive dynamic walking', *International Journal of Robotics Research* **9**(2): 62–82.

McGeer, T., 1993, 'Dynamics and control of bipedal locomotion', *Journal of Theoretical Biology* **163**(3): 277–314.

McMahon, T.A., 1984, *Muscles, Reflexes and Locomotion*, Princeton University Press.

Madden, J.D.W., Vandesteeg, N.A., Anquetil, P.A., Madden, P.G.A., Takshi, A., Pytel, R.Z., Lafontaine, S.R., Wieringa, P.A., Hunter, I.W., 2004, 'Artificial muscle technology: physical principles and naval prospects', *IEEE Journal of Oceanic Engineering* **29**(3): 706–728.

Moore, K.L., 1992, *Clinically Oriented. Anatomy*, Wiliams & Wilkins, Baltimore, Maryland.

Moreno, J.C., Brunetti, F.J., Pons, J.L., Baydal, J.M., Barberà, R., 2005, 'Rationale for multiple compensation of muscle weakness walking with a wearable robotic orthosis', in *Proceedings of the 2005 IEEE International Conference on Robotics and Automation*, pp. 1914–1919.

Morizono, T., Mimatsu, N., Higashi, M., 2005, 'Downsizing of wearable HEXA for a shoulder joint of wearable robots', in *IEEE International Workshop on Robots and Human Interactive Communication*, pp. 260–266.

Pandy, M.G., 2001, 'Computer modeling and simulation of human movement', *Annual Review of Biomedical Engineering* **5**: 245–273.

Pons, J.L., 2005, *Emerging Actuator Technologies. A Micromechatronic Approach*, John Wiley & Sons, Ltd.

Pons, J.L., Ceres, R., Pfeiffer, F., 1999, Multifingered dextrous robotics hand design and control: a review. *Robotica* **17**, 661–674.

Priplata, A.A., Niemi, J.B., Harry, J.D., Lipsitz, L.A., Collins, J.J., 2003, 'Vibrating insoles and balance control in elderly people', *The Lancet* **362**: 1123–1124.

Ralston, H.J., 1958, 'Energy-speed relation and optimal speed during level walking', *Internationale Zeitschrift für angewandte Physiologie, einschliesslich Arbeitsphysiologie* **17(4)**: 277–283.

Ransom, B., Behar, T., Nedergaard, M., 2003, 'New roles for astrocytes (stars at last)', *Trends in Neurosciences* **26**(10): 520–522.

Sakagami, Y., Watanabe, R., Aoyama, C., Matsunaga, S., Higaki, N., Fujita, M., 2002, 'The intelligent ASIMO: system overview and integration', in *Proceedings of the International Conference on Intelligent Robots and Systems*, pp. 2478–2483.

Saunders, B., Inman, V.T., Eberhart, H.D., 1953, 'The major determinants in normal and pathological gait', *The Journal of Bone and Joint Surgery, American Volume* **35-A(3)**: 543–558.

Schmitt, O., 1969, 'Some interesting and useful biomimetic transforms', *Third International Biophysics Congress*, p. 297.

Scholz, J.P., Schoner, G., 1999, 'The uncontrolled manifold concept: identifying control variables for a functional task', *Experimental Brain Research* **126**(3): 289–306.

Schwab, A.L., Wisse, M., 2001, 'Basin of attraction of the simplest walking model', in *Proceedings of ASME Design Engineering Technical Conferences*, Paper DETC2001/VIB-21363.

Seyfarth, A., Geyer, H., Herr, H., 2003, 'Swing-leg retraction: a simple control model for stable running', *Journal of Experimental Biology* **206**: 2547–2555.

Shadmehr, R., Wise, S.P., 2005, *The Computational Neurobiology of Reaching and Pointing: A Foundation for Motor Learning*, The MIT Press.

Strogatz, S.H., 2000, *Nonlinear Dynamics and Chaos*, Westview Press, Cambridge, Massachusetts.

Van der Helm, F.C.T., Rozendaal, L.A., 2000, 'Musculoskeletal systems with intrinsic and proprioceptive feedback', in *Neural Control of Posture and Movement* (eds J.M. Winters and P. Crago), Springer Verlag.

Van der Linde, R.Q., 2001, 'Bi-pedal walking with active springs: gait synthesis and prototype design', PhD Thesis, Delft University of Technology, Delft, The Netherlands.

Vincent, J.F., Bogatyreva, O.A., Bogatyrev, N.R., Bowyer, A., Pahl, A.K., 2006, 'Biomimetics: its practice and theory', *Journal of the Royal Society, Interface/the Royal Society* **3**(9): 471–482.

Vukobratovic, M., Frank, A.A., Juricic, D., 1970, 'On the stability of biped locomotion', *IEEE Transactions on Bio-medical Engineering* **17**(12): 25–36.

Wise, S.P., Shadmehr, R., 2002, 'Motor control', in *Encyclopedia of the Human Brain* (ed. V.S. Ramachandran, Academic Press, San Diego, California, vol. 3, pp. 137–157.

Wisse, M., 2005a, Essentials of dynamic walking; analysis and design of two-legged robots', PhD Thesis, Delft University of Technology, Delft, The Netherlands.

Wisse, M., 2005b, 'Three additions to passive dynamic walking: actuation, an upper body, and 3D stability', *International Journal of Humanoid Robotics* **2**(4): 459–478.

Wisse, M., Atkeson, C.G., Kloimwieder, D.K., 2006, *Dynamic Stability of a Simple Biped Walking System with Swing Leg Retraction*, Springer.

Wisse, M., Schwab, A.L., 2001, 'A 3D passive dynamic biped with roll and yaw compensation', *Robotica* **19**: 275–284.

Wisse, M., Schwab, A.L., 2004, 'Skateboards, bicycles, and 3D biped walkers; velocity dependent stability by means of lean-to-yaw coupling', *International Journal of Robotics Research* **24**(6): 417–429.

Wisse, M., van Frankenhuyzen, J., 2006, 'Design and construction of Mike; a 2D autonomous biped based on passive dynamic walking', in *Adaptive Motion of Animals and Machines* (eds H. Kimura and K. Tsuchiya) Springer-Verlag, pp. 143–154.

Wisse, M., Schwab, A.L., van der Linde, R.Q., van der Helm, F.C.T., 2004, 'How to keep from falling forward; elementary swing leg action for passive dynamic walkers', *IEEE Transactions on Robotics* **21**(3): 393–401.

Wisse, M., Hobbelen, D.G.E., Rotteveel, R.J.J., Anderson, S.O., Zeglin, G.J., 2006a, 'Ankle springs instead of arc-shaped feet for passive dynamic walkers', in *Proceedings of the IEEE/RAS International Conference on Humanoid Robots*.

3

Kinematics and dynamics of wearable robots

A. Forner-Cordero[1], J. L. Pons[1], E. A. Turowska[1] and A. Schiele[2,3]

[1]*Bioengineering Group, Instituto de Automática Industrial, CSIC, Madrid, Spain*
[2]*Mechanical Engineering Department, Automation & Robotics Section, European Space Agency (ESA), Noordwijk, The Netherlands*
[3]*Mechanical Engineering Faculty, Biomechanical Engineering Department, Delft University of Technology (DUT), Delft, The Netherlands*

3.1 INTRODUCTION

This chapter presents the foundations for analysis of the kinematics and the dynamics of the human–exoskeleton ensemble, beginning with an introduction to analysis of the kinematics and dynamics of robots. The methods presented will be helpful in analysing human biomechanics later on.

It is interesting to note that, given certain reasonable assumptions, the methods used to study robots can also be used to analyse human kinematics and dynamics. For instance, one of the most common assumptions consists in modelling the human body as a chain of rigid links, where each segment has certain properties, like length or inertia, that approximate those of humans. These segments are linked by joints that imitate human ones in terms of degrees of freedom (DoFs) and ranges of motion (RoMs).

This chapter therefore addresses the mechanics of the robot first. The expression of the position and orientation of the robot is discussed in order to introduce the Denavit–Hartenberg convention algorithm, which makes a systematic method possible to describe robot position and orientation. The section ends with a discussion on manipulability of the robot from a kinematic perspective. The next section deals with the dynamics of the robot and introduces the equations for motion, the dynamics of manipulability and a description of the dynamics of the robot in state space. Then, following a review of the methodology used to characterize the motions and forces in robotics, the same problem is addressed in humans. The biomechanics of the upper and lower limbs are described

Wearable Robots: Biomechatronic Exoskeletons Edited by José L. Pons

briefly but in enough detail to show the derivation of the Denavit–Hartenberg parameters for the upper and lower limbs. That section also contains an introduction to dynamic analysis of human motion, with an overview of the different levels of modelling, and the kinematic redundancy in exoskeleton systems that occurs when there are more DoFs available than are required to perform a given task. The design aspects of wearable robotic exoskeletons are discussed in terms of the upper limb.

Finally, three case studies illustrate the topics presented in this chapter. The first introduces a biomimetic knee joint that closely follows the movement of the centre of rotation of the human knee. The second case study explains a pronation–supination joint for the forearm of an upper limb exoskeleton and the last presents an estimation of tremor power with an exoskeleton as an example of how the dynamics are calculated.

3.2 ROBOT MECHANICS: MOTION EQUATIONS

3.2.1 Kinematic analysis

Kinematics can be defined as the branch of mechanics dealing with the description of the motion of bodies or fluids without reference to the forces producing the motion. When referring to multibody, jointed mechanisms as in the case of robots, and more specifically exoskeletons, kinematics deals with analysis of the motion of each robot link with respect to a reference frame and involves:

- an analytical description of motion as a function of time;
- the nonlinear relationship between the robot end-effector position and orientation and robot configuration.

In this context, the *mobility, M,* of a robot composed of a number of serial links is defined as the number of independent parameters, $\boldsymbol{q}$, required to specify fully the position of every link (Pons, Ceres and Pfeiffer, 1999). A particular robot configuration is a vector of realizable values, $q_i, i = 1, \ldots, n$, for the independent parameters at time t. The *redundancy* of a robot is an indicator of the number of available robot configurations for a particular position of the end-effector position. High redundancy makes control complex but improves dexterity.

Focusing on the second item, there may be a forward and an inverse relationship between a robot position and orientation and its configuration. The forward kinematics problem deals with the specification of robot position and orientation, $r_j,\ j = 1, \ldots, m$, as a function of robot configuration and can be stated as follows:

$$\boldsymbol{r} = \boldsymbol{f}(\boldsymbol{q}) \tag{3.1}$$

Inverse kinematics involves the determination of robot configuration as a function of robot position and orientation. The inverse kinematics problem is not dealt with in this section, but the reader is referred to Goldstein, Poole and Safko (2002) for a detailed analysis. In general, this relationship takes the form

$$\boldsymbol{q} = \boldsymbol{g}(\boldsymbol{r}) \tag{3.2}$$

Robot position can be represented by the coordinates of the end-effector. Any type of coordinate system – Cartesian, $\{x, y, z\}$, cylindrical, $\{r, \theta, z\}$, or spherical, $\{\rho, \phi, \theta\}$ – can be used to fully determine robot position; see Figure 3.1 for a schematic representation of coordinate systems. The selection of a particular coordinate system depends on the kinematic structure of the robot.

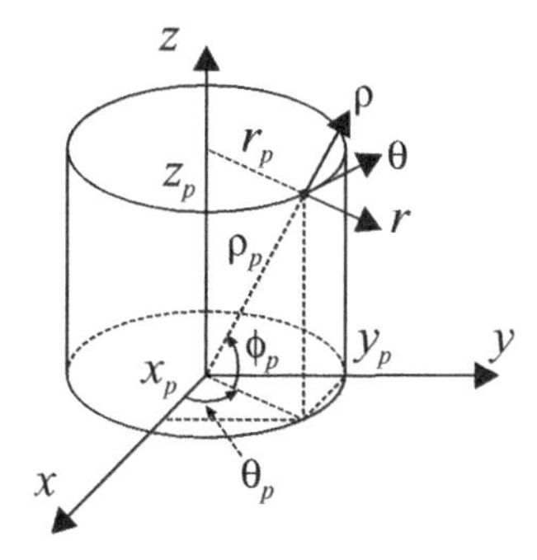

Figure 3.1 Cartesian, cylindrical and spherical coordinate systems

3.2.1.1 Euler angles

Robot orientation can be fully determined by means of either *Euler angles*, *roll–pitch–yaw (RPY)* angles, *homogeneous transformation matrices* or *quaternions*. The two most common descriptions of robot orientation based on three independent variables are Euler angles and RPY angles. If a Cartesian reference frame is considered, the so-called x convention for Euler angles, $\{\phi, \theta, \psi\}$, indicates that the first is a rotation of an angle ϕ about the z axis, the second is a rotation of an angle $\theta \in [0, \pi]$ about the x- axis and the third is a rotation of an angle ψ about the new z axis. Note that there are several conventions in common use. In the roll–pitch–yaw angle system, an orientation change is split into three rotations $\{\Psi, \nu, \Phi\}$. The first, Ψ, is a rotation about the x axis, the second is a rotation of ν about the y axis and the third is a rotation of Φ about the z axis.

In robotics, the preference is to describe the position and orientation in a more compact form based on the translation and rotation of the coordinate frame. A rotation matrix, **R**, is a transformation matrix that when multiplied by a vector has the effect of changing the direction of the vector but not its magnitude. A rotation is an orthonormal transformation for which the opposite rotation is represented by the transpose of the original matrix, $\mathbf{R}^{-1} = \mathbf{R}^{\mathrm{T}}$. Rotation matrices can be multiplied to represent the effect of a series of combined rotations. The general expression for a rotation matrix is as follows:

$$\mathbf{R} = \begin{bmatrix} n_x & o_x & a_x \\ n_y & o_y & a_y \\ n_z & o_z & a_z \end{bmatrix} = \begin{bmatrix} \boldsymbol{n} & \boldsymbol{o} & \boldsymbol{a} \end{bmatrix} \tag{3.3}$$

where, $\boldsymbol{n}$, $\boldsymbol{o}$ and $\boldsymbol{a}$ are the unit vectors of the rotated system in the original reference frame. The combination of a translation defined by a vector $\boldsymbol{p}$ and a rotation **R** can be expressed mathematically as

$$\boldsymbol{r} = \boldsymbol{p} + \mathbf{R}\boldsymbol{s} \tag{3.4}$$

Homogeneous coordinates (HCs) are a means of expressing combinations of rotations and translations in a compact form. HCs are $(n + 1)$-dimensional vectors, $\boldsymbol{r}\ (\omega x, \omega y, \omega z, \omega)$, used to represent points in an n-dimensional space, $\boldsymbol{r}\ (x, y, z)$. In homogeneous coordinates, ω is an arbitrarily valued parameter representing a scaling factor. By using the concept of homogeneous coordinates, the homogeneous transformation matrix , **T**, describing a rotation **R** followed by a translation $\boldsymbol{r}$, is defined as follows:

$$\mathbf{T} = \begin{bmatrix} n_x & o_x & a_x & r_x \\ n_y & o_y & a_y & r_y \\ n_z & o_z & a_z & r_z \\ 0 & 0 & 0 & 1 \end{bmatrix} = \begin{bmatrix} \mathbf{R} & \boldsymbol{r} \\ 0^{\mathrm{T}} & 1 \end{bmatrix} \tag{3.5}$$

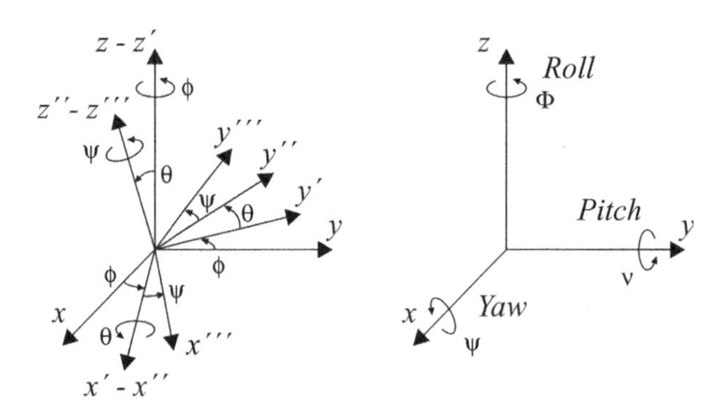

Figure 3.2 Euler angle and RPY angle representation of orientation

Complex transformations can readily be broken down into a combination of simple transformations (rotations and translations). This can be achieved by multiplying the homogeneous transformation matrices corresponding to simple transformations. When combining transformations, care must be taken to select the proper order of multiplication:

- When the transformed coordinate system, *OUVW*, is obtained through rotations and translations defined with respect to the original fixed coordinate system, *OXYZ*, the order of multiplication is the reverse of the order of transformation.
- When the transformed coordinate system, *OUVW*, is obtained through rotations and translations defined with respect to the transformed system, the order of multiplication is the same as the order of transformation.

To illustrate the role of multiplication order in deriving homogeneous transformation matrices for complex transformations, Equations (3.6) and (3.7) show the homogeneous transformation matrix corresponding to the Euler angle (x convention), $\mathbf{T}_{\text{Euler}}(\phi, \theta, \psi)$, and RPY angle, $\mathbf{T}_{\text{RPY}}(\Psi, \nu, \Phi)$, descriptions of orientation as shown in Figure 3.2.

$$\mathbf{T}_{\text{Euler}}(\phi, \theta, \psi) = \mathbf{T}(z, \phi)\mathbf{T}(x, \theta)\mathbf{T}(z, \psi)$$
$$= \begin{bmatrix} \text{c}\phi & -\text{s}\phi & 0 & 0 \\ \text{s}\phi & \text{c}\phi & 0 & 0 \\ 0 & 0 & 1 & 0 \\ 0 & 0 & 0 & 1 \end{bmatrix} \begin{bmatrix} 1 & 0 & 0 & 0 \\ 0 & \text{c}\theta & -\text{s}\theta & 0 \\ 0 & \text{s}\theta & \text{c}\theta & 0 \\ 0 & 0 & 0 & 1 \end{bmatrix} \begin{bmatrix} \text{c}\psi & -\text{s}\psi & 0 & 0 \\ \text{s}\psi & \text{c}\psi & 0 & 0 \\ 0 & 0 & 1 & 0 \\ 0 & 0 & 0 & 1 \end{bmatrix} \tag{3.6}$$

$$\mathbf{T}_{\text{RPY}}(\Psi, \nu, \Phi) = \mathbf{T}(z, \Psi)\mathbf{T}(y, \nu)\mathbf{T}(x, \Phi)$$
$$= \begin{bmatrix} \text{c}\Phi & -\text{s}\Phi & 0 & 0 \\ \text{s}\Phi & \text{c}\Phi & 0 & 0 \\ 0 & 0 & 1 & 0 \\ 0 & 0 & 0 & 1 \end{bmatrix} \begin{bmatrix} \text{c}\nu & 0 & \text{s}\nu & 0 \\ 0 & 1 & 0 & 0 \\ -\text{s}\nu & 0 & \text{c}\nu & 0 \\ 0 & 0 & 0 & 1 \end{bmatrix} \begin{bmatrix} 1 & 0 & 0 & 0 \\ 0 & \text{c}\Psi & -\text{s}\Psi & 0 \\ 0 & \text{s}\Psi & \text{c}\Psi & 0 \\ 0 & 0 & 0 & 1 \end{bmatrix} \tag{3.7}$$

The inverse transformation is represented by the inverse of the homogeneous transformation matrix. Since a transformation involving translation is not an orthonormal one, the inverse transformation matrix is not equal to its transpose. In general, it can be written as

$$\mathbf{T}^{-1} = \begin{bmatrix} n_x & n_y & n_z & -\boldsymbol{n}^{\mathrm{T}}\boldsymbol{r} \\ o_x & o_y & o_z & -\boldsymbol{o}^{\mathrm{T}}\boldsymbol{r} \\ a_x & a_y & a_z & -\boldsymbol{a}^{\mathrm{T}}\boldsymbol{r} \\ 0 & 0 & 0 & 1 \end{bmatrix} \tag{3.8}$$

3.2.1.2 Denavit–Hartenberg (D–H) convention

The analysis of robot kinematics is usually based on homogeneous transformation matrices. In jointed, multilink mechanisms like robots, the relative motion of links around a joint can be simply described by homogeneous transformation matrices. Finding the form of the forward kinematic problem of Equation (3.1) for a robot can be approached by following the Denavit–Hartenberg convention. The D–H convention establishes an algorithm for assigning a set of coordinate systems that are related through translation and rotation transformations. The transformation between successive coordinate systems takes into account the particular kinematics of robot joints. For a particular robot, joints are numbered from 1 to n starting with the base joint and finishing with the end-effector. The following steps are then required:

1. *Establish the base coordinate system* $OX_0Y_0Z_0$ with Z_0 lying on the axis of motion of joint 1.
2. *Establish the joint axis* by aligning Z_i with the axis of motion of joint $i+1$.
3. *Establish the origin of the ith coordinate system* at the intersection of Z_i and Z_{i-1} or, if they are parallel, at the intersection of the common normal to Z_i and Z_{i-1} and Z_i.
4. *Establish the X_i axis* at $X_i = \pm(Z_{i-1} \times Z_i)/\|Z_{i-1} \times Z_i\|$ or along the common normal between Z_i and Z_{i-1} if they are parallel.
5. *Establish the Y_i axis* at $Y_i = +(Z_i \times X_i)/\|Z_i \times X_i\|$ to complete the right-handed coordinate system.

The general form of the transformation matrix between two consecutive coordinate systems is given by

$$\mathbf{T}_{i-1}^{i} = \begin{bmatrix} \cos\theta_i & -\cos\alpha_i \sin\theta_i & \sin\alpha_i \sin\theta_i & a_i \cos\theta_i \\ \sin\theta_i & \cos\alpha_i \cos\theta_i & -\sin\alpha_i \cos\theta_i & a_i \sin\theta_i \\ 0 & \sin\alpha_i & \cos\alpha_i & d_i \\ 0 & 0 & 0 & 1 \end{bmatrix} \tag{3.9}$$

where θ_i is the angle of rotation from X_{i-1} to X_i about Z_{i-1} and is the joint variable if joint i is rotary; d_i is the distance from the origin of the $(i-1)$th coordinate system to the intersection of the Z_{i-1} axis and the X_i along the Z_{i-1} axis and is the i joint variable for prismatic joints; a_i is the distance from the intersection of the Z_{i-1} axis and the X_i axis to the origin of the ith coordinate system along the X_i axis; and α_i is the angle of rotation from the Z_{i-1} axis to the Z_i axis about the X_i axis as shown in Figure 3.3. The parameters θ_i, d_i, a_i and α_i are commonly referred to as *Denavit–Hartenberg parameters*.

By following the D–H approach, both the position and the orientation of the end-effector can be worked out as a function of the robot configuration $\boldsymbol{q}$. Since the transformation matrix $\mathbf{T}_{i-1}^{i}$ represents a transformation with respect to the $(i-1)$th coordinate system, the homogeneous transformation matrix corresponding to the combination of transformations imposed by successive joints can be simply derived by means of

$$\mathbf{T}_0^n = \mathbf{T}_0^1 \mathbf{T}_1^2 \ldots \mathbf{T}_{n-1}^n \tag{3.10}$$

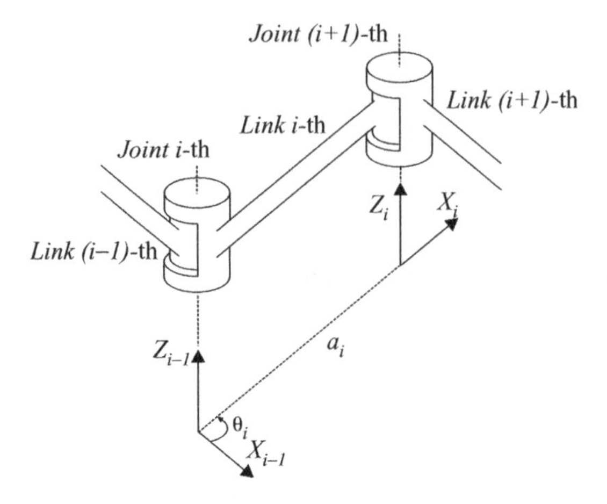

Figure 3.3 Assignment of coordinate systems in the Denavit–Hartenberg convention and D–H parameters

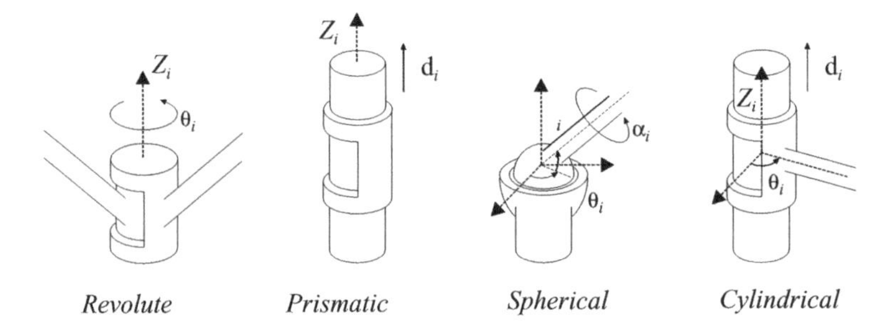

Figure 3.4 Basic robot joints

Once the homogeneous transformation matrix corresponding to a robot, $\mathbf{T}_0^n$, is obtained, the forward kinematics problem in terms of position is given by the last column of $\mathbf{T}_0^n$ and the forward kinematics problem in terms of orientation is given by the rotation submatrix of $\mathbf{T}_0^n$.

3.2.1.3 Robot configuration and end-effector position: manipulability

The four most common types of joints in classical robotics are *revolute* or *rotary* joints, *prismatic* joints, *cylindrical* joints and *spherical* joints; see Figure 3.4 for a schematic representation of these joints. On the one hand, the revolute and prismatic joints each allow only a single degree of freedom, a rotation and a translation, respectively; the transformation corresponding to revolute and prismatic joints can be directly represented by the transformation matrix from Equation (3.9). On the other hand, both cylindrical and spherical joints allow multiple degrees of freedom; the transformation imposed by one of these joints is thus obtained by combining single transformation matrices as in Equation (3.10).

Wearable robots, and in particular exoskeletons, are designed to be kinematically compliant with the human limb to which they are attached. As shown in Section 3.3, the kinematics of human anatomical joints is much more complex than that of simple one-or multi-DoF robot joints. In particular, every anatomical joint usually involves combinations of translational and rotational movements. In addition, if they are modelled as simple rotational joints, in most cases the *instantaneous centre of rotation* (ICR) migrates and generally follows a nonlinear function of joint angle. This kinematical mismatch is analysed in detail in Section 5.2, as it gives rise to undesired forces at the human–robot interface.

An important aspect of the forward kinematics problem in robots is the relationship between the velocity and acceleration of the end-effector and those of the robot joints. Starting from Equation (3.1), the relationship between end-effector $\dot{\boldsymbol{r}}$ and joint $\dot{\boldsymbol{q}}$ velocities can be expressed as follows:

$$\dot{\boldsymbol{r}} = \frac{\boldsymbol{f}(\boldsymbol{q})}{\mathrm{d}t} = \sum_{j=1}^{n} \frac{\partial \boldsymbol{f}(\boldsymbol{q})}{\partial q_j} \dot{q}_j \tag{3.11}$$

In matrix notation, this equation can be written as

$$\dot{r}_i = \frac{\mathrm{d} f_i(\boldsymbol{q})}{\mathrm{d}t} = \sum_{j=1}^{n} \frac{\partial f_i(\boldsymbol{q})}{\partial q_j} \dot{q}_j = \sum_{j=1}^{n} \mathbf{J}_{ij} \dot{q}_j \tag{3.12}$$

$\mathbf{J}_{ij}$ is the element at the ith row, jth column of matrix $\mathbf{J}$, which is known as the Jacobian matrix of the robot. There are robot configurations $\boldsymbol{q}^*$ for which rank $\mathbf{J} < m$. For these configurations, the robot is said to be in a *singular configuration*. In a singular configuration there is a subset of directions for the robot's end-effector position for which no further movement is possible.

Robot kinetic energy is proportional to $\dot{\boldsymbol{r}}^{\mathrm{T}} \dot{\boldsymbol{r}}$. It follows that since $\dot{\boldsymbol{r}}^{\mathrm{T}} \dot{\boldsymbol{r}} = \dot{\boldsymbol{q}}^{\mathrm{T}} \mathbf{J}^{\mathrm{T}} \mathbf{J} \dot{\boldsymbol{q}} > 0$, both $\mathbf{J}^{\mathrm{T}} \mathbf{J}$ and its transpose are positive definite square matrices. If a sphere is considered in the robot configuration space, $\|\dot{\boldsymbol{q}}\|^2 = \dot{q}_1^2 + \cdots + \dot{q}_m^2 \leq 1$, then it follows that $\dot{\boldsymbol{r}}^{\mathrm{T}} \dot{\boldsymbol{r}}$ is an ellipsoid. Therefore, $\mathbf{J}^{\mathrm{T}} \mathbf{J}$ represents an ellipsoid, known as the *manipulability ellipsoid*. This issue is revisited in the next section where the dynamics of robot manipulation is considered.

Workspace is a volume of space that the end-effector of the robot can reach. Workspace is also called *work volume* or *work envelope*. The workspace plays a crucial role in the particular case of wearable robots, since the motion of the human limb to which the robot is attached will be restricted by the robot in such a way that only the intersection of the robot and human limb workspaces will be reachable. If the robot's workspace is smaller than the human limb's workspace, there will be particular configurations for the human limb that will not be reachable when the robot is worn.

Even if the robot workspace contains the human limb workspace, if the redundancy of the robot is less than that of the human limb, the dexterity of the human operator is likewise restricted. If the redundancy of a wearable robot is greater than the redundancy of the human limb, then manipulability imposes dominant directions of motion. This is felt by the human operator as if there were particular directions of preferred motion, to which the robot offers less resistance than to other directions. This discussion will be continued in Section 3.3.

3.2.2 Dynamic analysis

Dynamics is the part of classical mechanics that studies objects in motion and the causes of this motion, e.g. forces. When considering multibody, jointed mechanisms like wearable robots, dynamics deals with the analysis of movement in configuration and working space as a function of internal forces (e.g. torque at each joint actuator) and external forces (e.g. interaction force with the environment).

As in the case of kinematics, two instances of the relationship between force and movement can be identified: the forward dynamics problem and the inverse dynamics problem. The *forward dynamic model* of a robot expresses the evolution of joint and working coordinates as a function of the force and torque involved. The forward dynamics of a robot is expressed as follows:

$$\begin{aligned} \ddot{\boldsymbol{r}} &= f(F, T) \\ \dot{\boldsymbol{r}} &= \int \ddot{\boldsymbol{r}} \mathrm{d}t \\ \boldsymbol{r} &= \int \dot{\boldsymbol{r}} \mathrm{d}t \end{aligned} \tag{3.13}$$

The *inverse dynamic model* describes forces and torques as a function of the evolution of joint coordinates in time and can be expressed as follows:

$$F = g\left(\boldsymbol{r}, \dot{\boldsymbol{r}}, \ddot{\boldsymbol{r}}\right) \tag{3.14}$$

In robotics, Newtonian and Lagrangian mechanics are used to derive the dynamic model of a robot. The *Newton–Euler formulation* is based on a description of mechanics in vector functions, while the *Lagrange–Euler formulation* is based on scalar functions. *Equations of motion* are equations that describe the behaviour of a system as a function of time. Sometimes the term refers to the differential equations that the system satisfies (e.g. Euler–Lagrange equations) and sometimes to the solutions to those equations. The following paragraphs look at the Lagrangian formulation for deriving the motion equations of a robot.

The robot is a jointed multibody mechanism. Each joint in the robot structure restricts the relative motion of the two consecutive jointed links. The restriction imposed by a joint is said to be *holonomic* when the constraint depends only on the system coordinates and time. It does not depend on the velocity or momentum of the system, and consequently it can be mathematically expressed by

$$f\left(r_1, r_2, \ldots, r_n, t\right) = 0 \tag{3.15}$$

where r_i is the set of coordinates describing the position of each link and t is time. In addition, a constraint is said to be *scleronomic* when in its expression the time t does not appear explicitly.

The Lagrange equations for a holonomic, scleronomic system are given by Equation (3.16). In this equation, $\boldsymbol{q}$ is a vector of *generalized coordinates*, with a minimum set of coordinates to determine fully the position of each robot's link. Likewise, τ_i are the *generalized forces*. L is the *robot Lagrangian function* and can be derived from the robot's kinetic, T, and potential energy, V: $L = T - V$. Then

$$\frac{\mathrm{d}}{\mathrm{d}t}\frac{\partial L}{\partial \dot{q}_i} - \frac{\partial L}{\partial q_i} = \tau_i, \quad \forall\, i = 1, \ldots, n \tag{3.16}$$

In matrix form, Equation (3.16) can be written as follows:

$$\mathbf{M}\left(\boldsymbol{q}\right)\ddot{\boldsymbol{q}} + \mathbf{C}\left(\boldsymbol{q}, \dot{\boldsymbol{q}}\right) + \mathbf{K}\left(\boldsymbol{q}\right) = \boldsymbol{\tau} \tag{3.17}$$

where $\boldsymbol{q}$ is the vector of joint coordinates, $\dot{\boldsymbol{q}}$ is the vector of joint velocities and $\ddot{\boldsymbol{q}}$ is the vector of joint accelerations; all three are functions of time. $\mathbf{M}\left(\boldsymbol{q}\right)$ is a square inertial matrix and represents the effect of joint acceleration on the generalized torque, $\boldsymbol{\tau}$. $\mathbf{C}\left(\boldsymbol{q}, \dot{\boldsymbol{q}}\right)$ is the vector of centrifugal and Coriolis forces and $\mathbf{K}\left(\boldsymbol{q}\right)$ is a vector of gravity-related forces.

Equation (3.17) represents joint-space dynamics, i.e. the dynamics of the chain of links described in joint space coordinates. In common practice, the robot chain is affected by both disturbances and friction. This can be taken into account by rewriting Equation (3.17) as follows:

$$\mathbf{M}(\boldsymbol{q})\ddot{\boldsymbol{q}} + \mathbf{C}(\boldsymbol{q}, \dot{\boldsymbol{q}}) + \mathbf{F}(\dot{\boldsymbol{q}}) + \mathbf{K}(\boldsymbol{q}) + \boldsymbol{\tau}_{\mathrm{d}} = \boldsymbol{\tau} \tag{3.18}$$

The friction term, $\mathbf{F}(\dot{\boldsymbol{q}})$, is given by the following equation:

$$\mathbf{F}(\dot{\boldsymbol{q}}) = \mathbf{F}_{\mathrm{u}}\dot{\boldsymbol{q}} + F_{\mathrm{d}} \tag{3.19}$$

where F_{u} is a coefficient matrix of viscous friction and F_{d} is the so-called dynamic friction. It is difficult to determine both friction coefficients for a given robot, as these only represent an approximate mathematical model for their influence. However, disturbance $\boldsymbol{\tau}_{\mathrm{d}}$ can be used to represent inaccurately modelled dynamics.

The motion equations for the chain of links can be written as follows:

$$\mathbf{M}(\boldsymbol{q})\ddot{\boldsymbol{q}} + \mathbf{N}(\boldsymbol{q}, \dot{\boldsymbol{q}}) + \boldsymbol{\tau}_{\mathrm{d}} = \boldsymbol{\tau} \tag{3.20}$$

where the term representing nonlinear effects, $\mathbf{N}(\boldsymbol{q}, \dot{\boldsymbol{q}})$, can be written as

$$\mathbf{N}(\boldsymbol{q}, \dot{\boldsymbol{q}}) = \mathbf{C}(\boldsymbol{q}, \dot{\boldsymbol{q}}) + \mathbf{F}(\dot{\boldsymbol{q}}) + \mathbf{K}(\boldsymbol{q}) \tag{3.21}$$

3.2.2.1 Dynamic manipulability

In Section 3.2.1 the concept of manipulability and the manipulability ellipsoid were introduced. These two concepts are merely kinematic and can be derived from geometric considerations. Here the concept of *dynamic manipulability* is introduced, which came into being three decades ago (Asada, 1983), when a number of figures of merit of the dynamic behaviour of robots were introduced. In this work, the author extended the concept of the inertia tensor of a solid to the case of jointed multibody structures. The kinetic energy of a robot in configuration space can be written as follows:

$$T = \frac{1}{2}\dot{q}^{\mathrm{T}}\mathbf{G}\dot{\boldsymbol{q}} \tag{3.22}$$

where $\mathbf{G}$ is an $n \times n$ symmetric and positive-definite matrix, usually referred to as the *generalized inertia tensor* of the robot.

Due to the properties of $\mathbf{G}$, the quadratic form $\dot{\boldsymbol{u}}^{\mathrm{T}}\mathbf{G}\dot{\boldsymbol{u}} = 1$ is an ellipsoid, the *robot inertia ellipsoid*. For a particular point in the manipulator space, $\boldsymbol{r}_1$, the reference frame $OX_1Y_1Z_1$ can always be considered, with the unit vectors in the direction of the principal axes of $\mathbf{G}$, as depicted in Figure 3.5. In this new reference frame, the generalized robot inertia matrix, $\mathbf{G}$, is the diagonal matrix, $\mathbf{D}_1$, whose diagonal elements are the eigenvalues, $\lambda_1, \ldots, \lambda_n$, of $\mathbf{G}$. The kinetic energy of the robot (Equation (3.22)), can therefore now be written as

$$T = \frac{1}{2}\dot{\boldsymbol{p}}^{\mathrm{T}}\mathbf{D}_1\dot{\boldsymbol{p}} \tag{3.23}$$

The motion equations for the robot in Lagrangian formulation are now written as

$$\hat{\boldsymbol{\tau}} = \mathbf{D}_1\ddot{\boldsymbol{p}} - F_N\left(\dot{\boldsymbol{p}}, \frac{\partial \mathbf{D}_1}{\partial p_i}\right) \tag{3.24}$$

The term $F_N\left(\dot{\boldsymbol{p}}, \frac{\partial D_1}{\partial \dot{p}_i}\right)$ includes all nonlinear contributions to the joint torque, $\hat{\boldsymbol{\tau}}$. It can be seen that, this term depends on the variation of the robot's inertia ellipsoid, which in turn depends on the volume variation of the ellipsoid and on the change in orientation of the ellipsoid's principal axes (see Figure 3.5). The ellipsoid represents how actions at each joint are mapped to actions at the end-effector. As long as the ellipsoid approximates a sphere, there will be no preferential directions of motion.

The concept of dynamic manipulability may be derived from the concept of kinematic manipulability linked to the manipulability ellipsoid, $\mathbf{J}^{\mathrm{T}}\mathbf{J}$. In fact, the *ellipsoid of dynamic manipulability*

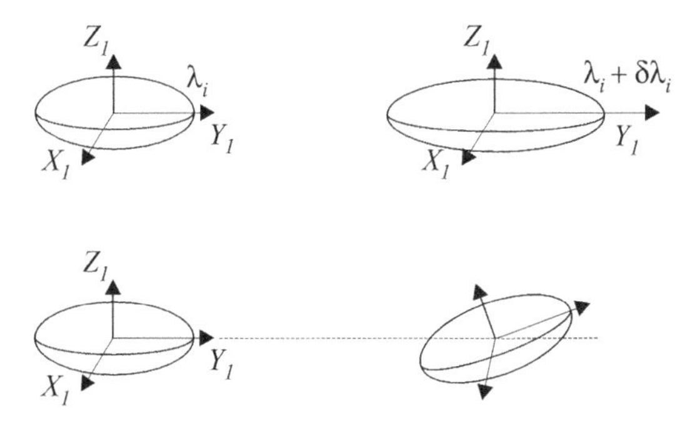

Figure 3.5 Effect of volume change and orientation change in the generalized inertia ellipsoid

represents the relationship between joint torque and end-effector acceleration. Similarly, other authors (Chiu, 1985) have introduced the *force ellipsoid* as the relationship between the joint torque and end-effector force.

The concept of dynamic manipulability plays an important role in wearable robot applications. The kinematics and dynamics of the human limb and the wearable robot are not necessarily identical. Therefore, the manipulability ellipsoids of both systems for a particular configuration will not be the same and the preferential directions of motion will not necessarily be aligned. This may lead in turn to dynamic interaction forces that will be felt by the operator as resistance to the motion.

The goal of robot control is to choose a set of time series of control torques, $\boldsymbol{\tau}(t)$, so that the robot can follow a predefined trajectory or can exert a given force on the environment. Equation (3.18) satisfies some properties that can help simplify the robot control problem:

- $\mathbf{M}(q)$ is a symmetric, positive-definite matrix which is bounded so that $\mu_1 I \leq \mathbf{M}(q) \leq \mu_2 I \ \forall q(t)$. For prismatic joints, the upper and lower bounds, μ_1, μ_2, are constant.
- The gravity vector is bounded, so that $\|K(q)\| \leq g_{\mathrm{B}}$, where g_{B} is a scalar function that may be determined for any given arm. For a revolute arm g_{B} is constant, independently of q, and for a prismatic link g_{B} may depend on q.
- The friction term is presented in Equation (3.19). Since friction is a local effect, it can be assumed that $F(\dot{q})$ is *uncoupled among joints*. A bound on the friction terms may be presented as $\|F_{\mathrm{v}}\dot{q} + F_{\mathrm{d}}(\dot{q})\| \leq \upsilon\|\dot{q}\| + k$, where υ and k are known for a specific arm.

3.2.2.2 State-variable system

Robot dynamics can be represented by linear state-space variable equations. On the other hand, the interaction between different links is described by nonlinear differential equations. It is possible to use the state-space formulation for control design, in either the nonlinear or the linear form.

In general, a nonlinear system is represented by

$$\dot{x} = f(x, u, t) \tag{3.25}$$

where $\boldsymbol{x}(t)$ is the state vector and $\boldsymbol{u}(t)$ is the control input vector. The linear system in state space formulation follows the law

$$\dot{x} = \mathbf{A}x + \mathbf{B}u \tag{3.26}$$

Depending on the state-space vector chosen to describe the robot dynamics, two main state-space formulations for the robot dynamics are commonly used:

- *Position–velocity state-space form.* The state vector is defined in this formulation as the $2n$-vector $\boldsymbol{x} \equiv [q^{\mathrm{T}}\dot{q}^{\mathrm{T}}]^{\mathrm{T}}$; the state-space equation can hence be defined as follows:

$$\dot{x} = \begin{bmatrix} \dot{q} \\ -M^{-1}(q)N(q,\dot{q}) \end{bmatrix} + \begin{bmatrix} 0 \\ M^{-1}(q) \end{bmatrix} \tag{3.27}$$

- *Hamiltonian form.* The state vector is defined as the $2n$-vector $x = (q^{\mathrm{T}}p^{\mathrm{T}})^{\mathrm{T}}$, where $p = M(q)\dot{q}$ is a generalized momentum. The state-space equation is presented as follows:

$$\dot{x} = \begin{bmatrix} M^{-1}(q)p \\ -\frac{1}{2}(I_n \otimes p^{\mathrm{T}})\frac{\partial M^{-1}(q)}{\partial q}p \end{bmatrix} + \begin{bmatrix} 0 \\ I_n \end{bmatrix} u \tag{3.28}$$

where $u = \tau - G(q)$ and $\otimes$ is the Kronecker product.

3.3 HUMAN BIOMECHANICS

Biomechanics is the application of methods and techniques from mechanical science (physics and engineering) to the analysis and understanding of biological systems. This is the area of science that analyses the mechanical properties of tissues, systems and movement of living organisms (e.g. the human body), their causes and their results. As such, it is an interdisciplinary area in the borderland between the physical and biological sciences.

From a macroscopic point of view, the human body can be modelled as a set of rigid segments. In this case, it is possible to analyse human motion according to the postulates of continuum mechanics (conservation of linear and angular momentum and conservation of mass and energy). With these rigid segments linked by joints and using these postulates with some assumptions about the particular problem, a set of equilibrium equations can be established. Considering the system as a set of rigid links, it is usually assumed that there is no linear translation in the joints (only rotational motion). In addition, it is necessary to define the external forces and moments and the point of application. If the external forces and moments and the motion of the links are known, the internal forces and moments can be derived. However, it is not possible to determine the forces produced in each muscle unambiguously because of the *redundancy problem*.

Biomechanics, as a science bordering between physics and medicine, has to deal with the technical description of motion, as explained in Section 3.2, and the anatomical description of body movements used in medicine.

From the point of view of robotics the biological motor system has too many degrees of freedom (DoFs) and too much redundancy in the number of sensors and actuators, and this makes the system's control task very complex. As mentioned in Chapter 2, this is the result of a process in which structures developed for a certain function evolve with environmental changes to serve different objectives. However, despite this apparent lack of optimality in the design, biological solutions are far more robust and efficient than the most advanced robot.

The aim of this section is to give a basic biomechanical description of the motion of the upper and lower limbs and to introduce the procedures whereby human motion can be explained with the same tools as are used in robotics. In this way, it is possible to model the human limbs as a system parallel to the wearable robot. For a more detailed anatomical description, interested readers are referred to specific anatomy books (Moore, 1992).

3.3.1 Medical description of human movements

The motion of a chain of rigid segments can be described in several ways, as explained in Section 3.2.2. The body is composed of bones linked by joints forming the skeleton and covered by soft tissues like organs and muscles. If the bones are considered to behave as rigid segments, it is possible to assume that the body is divided into segments, and so the motion between bones can be described by the same methods used for robot kinematics. However, the description of anatomical human motion commonly used in medicine explains the movement between the bones and the range of motion of each joint in three planes of the body, called anatomical planes (see Figure 3.6). These planes are defined on the basis of the body in the anatomical position, i.e. standing with the feet together, the arms rotated outwards to the side, with the head, eyes and palms of the hands facing forward. The anatomical planes that define perpendicular axes around which rotation occurs are:

- *a frontal or coronal plane* that divides the body into anterior and posterior parts;
- *a transversal plane* that divides the body into upper and lower parts;
- *a sagittal or lateral plane* that divides the body into right and left parts.

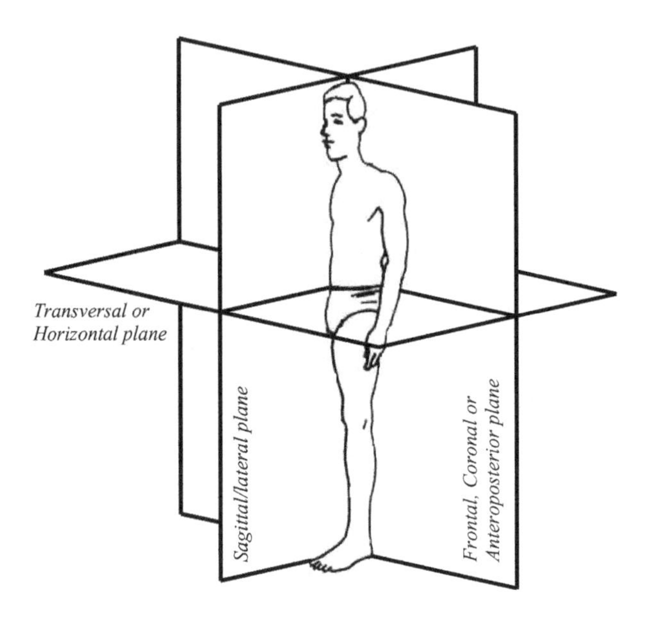

Figure 3.6 Body planes

The movement of each joint can be defined as occurring in these planes, always starting from the anatomical position. Movement in the *sagittal plane* is called *flexion–extension*:

- *Flexion* is a movement that reduces the angle between bones or parts of the body. Specific flexion activities occur only along the *sagittal plane*, i.e. from front to back and not from side to side. Flexion in the foot is usually called *plantar flexion*.
- *Extension* is a movement that increases the angle between the bones of the limb at a joint. Extension at the ankle joint is usually called *dorsal flexion* or *dorsiflexion*.

Movement in the coronal plane is called *abduction–adduction*:

- *Abduction* is an outward movement of the limb away from the median plane of the body.
- *Adduction* is a movement that brings a limb – arm or leg – closer to the body in the sagittal plane and is opposed to abduction.

Another movement is *supination–pronation*:

- *Supination* is the rotation of the forearm so that the palm position is anterior, i.e. the palm facing up.
- *Pronation* is a rotation of the forearm that moves the palm from an anterior-facing position to a posterior-facing position, i.e. the palm facing down.

Other descriptions of movement include *rotation*, either internal or external, which is the movement of a joint around the long axis of the limb in a circular motion, and *circumduction*, which is a circular movement in which flexion, abduction, extension and adduction are combined in sequence. The most commonly used example is the shoulder joint (ball-and-socket joint).

This description can be ambiguous if the movement occurs around several axes, resulting in a combination of rotations. There is no clear rule governing the order in which the rotations should be described and whether the rotation axes move with the bone or not. Therefore, this description of movement can be confusing despite its widespread use. The following sections provide the basis for a description of human movement based on the Denavit–Hartenberg convention.

The point of contact between two bones is the joint or articulation. There are several types of joints that can be classified according to the material that joins the bones: fibrous, cartilaginous or synovial. The first two have little mobility, while the most common and important from the point of view of movement between bones are the synovial joints. In these, the joint cavity is a movable space between two bones that is enclosed by the articular capsule and filled with the articular cartilage and the lubricating synovial fluid. There are ligaments around the joint that strengthen the articular capsule and limit motion in undesirable directions. There are other mechanisms that protect the joint against injury from undesired movements. For instance, protection may be provided by the bone configuration as in the elbow joint, the ligaments as in the knee, or even the muscles as in the shoulder joint.

3.3.2 Arm kinematics

The upper limb, the arm, is the region from the shoulder to the fingertips. It includes three segments: arm (upper arm – in anatomy, the region between the shoulder and the elbow), the forearm and the hand, linked by three joints: the shoulder, the elbow and the wrist. The bones that form the upper limb are the clavicle (which is joined to the trunk), the scapula, the humerus in the arm, the radius and ulna in the forearm, and the bones of the wrist and the hand: carpal bones, metacarpals and phalanges. The human upper limb is not usually engaged in heavy weightbearing, such as body weight support, and stability is compromised for the sake of increased mobility.

In studying the upper limb the hand is considered as a single segment (see Figure 3.7). On this assumption, the arm comprises three segments, upper arm, forearm and hand, linked to the torso by the shoulder joint. The elbow joins the upper and lower arm while the wrist joins the lower arm and

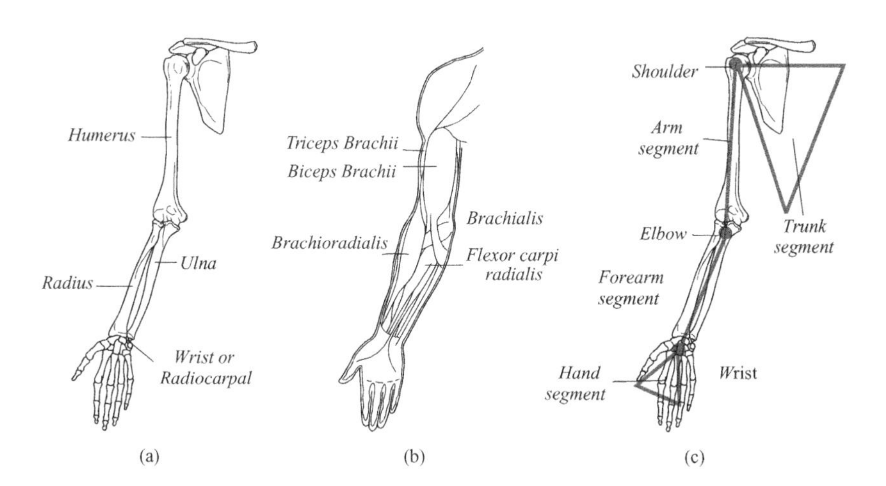

Figure 3.7 Anatomy of the upper limb. Anterior view: (a) bones, (b) muscles and (c) rigid-segment model

hand. Of course this description involves several simplifications, which are discussed in the course of the section.

3.3.2.1 The shoulder joint

The shoulder and the shoulder girdle make up one of the most complex joint groups of the human body. The hemispherical head of the humerus (upper arm) forms a ball-and-socket-type synovial joint with the glenoid cavity of the scapula. This arrangement allows three degrees of freedom and comprises three bones: clavicle, scapula and humerus.

There are up to 19 muscles with 25 joint rotation pairs around the shoulder joint and the shoulder girdle. There seems to be a compromise between the mobility afforded by the ball-and-socket configuration and the stability provided by the muscles and ligaments.

In humans the range of motion of the upper extremity provided by the combination of the motion of the shoulder joint and the shoulder girdle covers 65 % of a sphere. Circumduction is a circular movement that combines flexion, extension, abduction and adduction. In this movement, the head of the humerus is the apex of a cone and the distal end describes the base of a cone. Although shoulder mobility is the result of a combination of the motion at the glenohumeral joint and the scapulothoracic-gliding plane, it is often assumed that all rotations occur around an ideal spherical joint between the thorax and the humerus. This suggests a possible approximation for a biomechanical model consisting in three DoFs:

- *Flexion–extension.* The range of motion of the shoulder in flexion from the anatomical position is between 130 and 180°, while in extension (also called hyperextension) the range is between 30 and 80°. There are several muscles involved in stabilizing the joint during movement. Shoulder flexion is produced by combined action of the anterior part of the deltoid, the pectoralis major and the coracobrachialis. The posterior part of the deltoid also provides arm extension.
- *Abduction–adduction.* The shoulder can attain up to 180° of abduction but only 50° of adduction. The middle part of the deltoid acting with the supraspinatus abducts the arm, while the pectoralis major adducts it.
- *Rotation.* This is movement around the long axis of the upper arm segment, in this case the humerus. Rotation can be medial (internal) or lateral (external). Internal rotation can attain between 60 and 90°, while external rotation can attain 90°.

The range of motion of the shoulder for all DoFs is large considering that its structure can also bear heavy loads, as much as the body weight.

3.3.2.2 The elbow joint

The elbow joint is the one that links the upper arm and lower arm. For functional purposes it may be regarded as composed of three different joints (humero-ulnar, humero-radial and radio-ulnar). However, for simplicity's sake one joint with two degrees of freedom will be assumed:

- *Flexion–extension.* Elbow flexion is the movement whereby the forearm approaches the upper arm. The opposite movement is extension. In this case, the elbow functions as a hinge joint between the distal end of the humerus and the proximal ends of the ulna and radius. The range of flexion–extension motion varies between full extension, 0° and active maximal flexion, 140–146°. The elbow can be passively flexed up to 160°. However, the angle range in day-to-day activities varies between 30 and 130°. An important consideration in the design of exoskeletons is the orientation of the axis of rotation of the elbow. It is slightly oblique, with the medial side below the lateral side. It is inclined 5 or 6° with reference to the perpendicular to the longitudinal axis of the humerus crossing at the lateral epicondyle. The brachialis, brachioradialis and biceps brachii provide elbow flexion, while elbow extension is produced by the triceps brachii with the assistance of the anconeus.

- *Pronation–supination*. This is a rotation around the long axis of the forearm. Pronation–supination movements are defined from a starting position with the elbow flexed at 90° and the hand parallel to the sagittal plane, with the palm of the hand inwards and the thumb upwards. Pronation is the rotation that brings the palm of the hand downwards and the thumb to a more medial position. Maximum rotation is 80°. Supination is the rotation of the forearm so that the palm is upwards and the maximum rotation is 85°. The axis of rotation crosses the distal and proximal radio-ulnar joints. The biceps brachii supinates the forearm and after supination flexes it.

3.3.2.3 The wrist joint

The wrist is one of the most complex joints in the body. It has to combine high mobility with heavy loads. It is composed of two rows of carpal bones. The wrist is a multijoint complex, which may be considered as having two degrees of freedom:

- *Flexion–extension*. Flexion of the wrist is the movement around a transversal axis that allows the palm of the hand to approach the forearm; extension is the opposite. There are large intersubject differences, but generally speaking the wrist can achieve up to 90° active flexion and less extension, about 80°. The main flexor tendons are the flexor carpi ulnaris and radialis. They are inserted in the internal side of the wrist, so their contraction produces not only flexion but also adduction (ulnar inclination) of the wrist. Analogously, the main extensors, which are the abductor pollicis brevis and the extensors carpi radialis longus and brevis, are located on the outer side of the wrist. Wrist extension also involves some abduction (radial inclination). Wrist rotations always start in the distal row; the axis of rotation is not fixed but, as in the knee joint, changes with the wrist motion.
- *Abduction–adduction*. This is the movement around an antero-posterior axis that moves the hand either towards the ulna (adduction) or towards the radius (abduction). Adduction reaches up to 30 or 40°, while abduction does not exceed 15°.

The wrist has to deal with heavy loads while requiring high mobility and precision. It should be remembered that an activity requiring hand muscle contraction will generate axial compression forces in the wrist. It has been shown that for each N of force applied with the hand the carpus bears compression loads of up to 14 N (Viladot Voegli, 2001). The wrist, then, is a complex mechanism that distributes forces between the different carpal bones.

3.3.3 Leg kinematics

The lower limb (see Figure 3.8) is the part of the human body that extends from the gluteal region to the foot and is connected to the lower part of the trunk. It consists of the following parts (bones): pelvis (ilium, pubis, ischium), thigh (femur), shank (tibia and fibula) and foot (calcaneus, talus, cuboid, navicular, cuneiforms, metatarsals and phalanges). The three main joints are the hip, knee and ankle, which respectively join the pelvis to the upper leg, the latter to the lower leg and the lower leg to the foot (Moore, 1992).

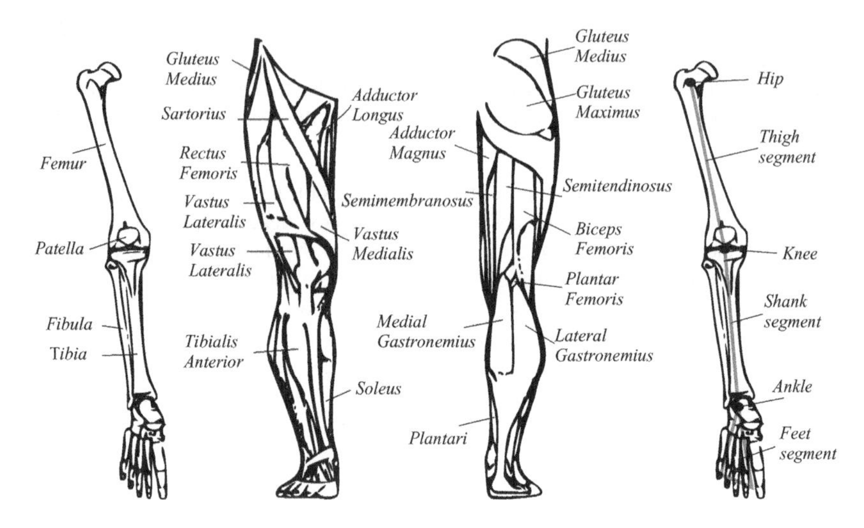

Figure 3.8 Anatomy of the lower limb

The lower limb enables humans to stand and locomote, so its functions have to combine propulsion and weight bearing.

3.3.3.1 The hip joint

The *hip joint* is composed of the cup-shaped acetabulum of the hip bone and the head of the femur, which forms two-thirds of a sphere. It is supported by several ligaments that restrain movement. The hip is a very strong and stable joint. In addition to the strong muscles that surround it and the matching surfaces of the acetabulum and femoral head, the bones are united by a dense fibrous capsule i.e. strengthened by strong ligaments, such as the iliofemoral.

The hip joint is a multiaxial ball-and-socket synovial joint, which moves in different planes that pass through the joint centre. The hip joint behaves as a spherical joint, allowing three DoFs. All these DoFs are important in order to allow stable locomotion, even in a straight line. The hip movements are (Moore, 1992):

- *Flexion–extension.* The rotating motion that brings the thigh forward and upward is flexion; the opposite is extension. The muscles contributing to flexion have insertions in the anterior part of the body. They are the iliopsoas (iliacus and psoas major), tensor fasciae latae, rectus remoris (which is biarticular and contributes to knee extension), pectineus, sartorius and adductor muscles, especially the adductor longus (which also provides adduction motion). Extension is accomplished by muscles attached to the posterior side of the leg and hip bone: gluteus maximus, semitendinosus, semimembranosus, biceps femoris and adductor magnus. The range of hip flexion is up to 120°.
- *Abduction–adduction.* In abduction, the lower limb is moved away from the mid-line of the body; in adduction, the opposite occurs. Abduction is accomplished by lateral muscles like the gluteus medius, gluteus minimus, tensor fasciae latae, sartorius, piriformis and obturator externus. Adduction is produced by muscles with medial attachments like the adductors (magnus, longus and brevis), pectineus and gracilis. The range of abduction is up to 40° and that of adduction between 30 and 35°.

- *Medial–lateral rotation.* This is rotation around the long axis of the femur. Medial rotation is accomplished by the muscles called tensor fasciae latae, gluteus medius and gluteus minimus, while lateral rotation is accomplished by the obturator internus and gemelli, obturator externus, quadratus femoris, piriformis, gluteus maximus and sartorius. The range of medial rotation is only from 15 to 30°. The range of lateral rotation is larger and can be up to 60°.

3.3.3.2 The knee joint

The *knee joint* is a synovial hinge joint, although it is considered condyloid (ellipsoidal) by some authors. The bones comprising the knee joint are the femur, the tibia and fibula, and the patella.

The knee joint has two parts: the femoro-patellar joint and the femoro-tibial joint (Moore, 1992). It is one of the joints with the most restraints in the entire body. Several movements that can damage the knee joint are blocked by powerful muscles and ligaments. Four main ligaments stabilize the knee: the medial and lateral collateral ligaments and two cruciate ligaments, anterior and posterior. These ligaments are attached to the femur and cross the knee joint to the tibia. The cruciate ligaments prevent forward or backward shifting of the tibia with respect to the femur and are crucial for knee stability. When movement occurs in a correct direction, muscles actively contract and loosen and ligaments passively extend and shrink. The most important muscle involved in stabilizing the knee joint is the quadriceps femoris. In addition, there are menisci (medial and lateral), which are plates of cartilage that act as shock absorbers.

Functionally, the knee plays a crucial role in walking and standing. This can be simplified by considering that there are two major functions. The first is *knee joint locking* in which the line of action of the ground reaction force causes an extensor moment around the knee and permits standing upright with little muscle activity. The second, which allows knee flexion, is *knee unlocking*, in which the foot is off the ground and the weight of the body is not supported.

The knee joint can be approximated by a hinge joint with the main movements occurring in the sagittal plane. This simplification is discussed in Case Study 3.5. The knee motion in the sagittal plane is:

- *Flexion–extension.* In flexion, the shank approaches the thigh while the femur and tibia remain in the same plane; the opposite movement is extension. The range of flexion is up to 120° when the hip is extended, 140° when the hip is flexed and 160° when the knee is flexed passively. The muscles involved in flexion are the semimembranosus, semitendinosus, biceps femoris, sartorius, gracilis and gastrocnemius. When the knee is extended, the patella (the sesamoid bone inside a tendon) lies loosely on the front of the lower end of the femur. At full extension, the limb is straightened and capable of weight bearing. The range of motion in that movement is 0–10°. The muscles involved in extension are the rectus femoris and the vastii (medialis, lateralis and intermedius).

Knee motion in the locked position is highly constrained. However, some rotations are possible around the long axes of the lower leg (tibia and fibula) when the knee is flexed:

- *Medial rotation.* This is an internal rotation that occurs during the final stage of extension. It brings the knee to the locked position for maximum stability. It is limited to 10° with 30° of flexion and is limited to 15° when the knee is fully flexed. The muscles involved in this rotation are the sartorius, gracilis and semitendinosus.

- *Lateral rotation.* External rotation occurs during the early stage of flexion. It is needed to unlock the joint and is limited to 30° with 30° of flexion and to 50° with 120° of flexion. The main muscle involved in this rotation is the biceps femoris.

If both medial and lateral rotations are counted, there are two DoFs available.

3.3.3.3 The ankle joint and the foot

The *ankle* is a hinge-type synovial joint and is involved in lower limb stability. The ankle and foot contain 26 bones connected by 33 joints, and more than 100 muscles, tendons and ligaments. The ankle joint is located between the lower ends of the tibia and the fibula and the upper part of the talus. This joint is supported by ligaments such as the medial or deltoid and the lateral ligaments (anterior talofibular, posterior talofibular and calcaneofibular).

The foot plays an important role in supporting the weight of the body and in locomotion. It can be divided into three parts – hindfoot, midfoot and forefoot – formed by the following bones: talus, calcaneus, navicular, cuboid and cuneiforms, metatarsal and phalanges.

The ankle comprises basically two joints: the *talocrural joint* and the *talocalcaneal joint* (Moore, 1992). However, in biomechanical modelling it is usually treated as a single joint. There are two movements that occur around a transverse axis between malleoli:

- *Dorsal flexion* (or dorsiflexion). This movement brings the foot dorsally to the anterior surface of the leg. The range of motion in this movement is up to 20°. The muscles involved in dorsiflexion are the tibialis anterior, extensor hallucis longus, extensor digitorum longus and fibularis tertius.
- *Plantar flexion*. This is the opposite movement to dorsiflexion. It occurs, for instance, when the toes are in contact with the ground and the heel is raised off the ground. The range of motion in plantar flexion is from 40 to 50°. The muscles involved in plantar flexion are the gastrocnenius, soleus and plantaris.

The foot is a complex structure with several joints that play an important role in gait and in standing stability (Moore, 1992). The foot joints with a role in stability are the *tarsal joints*: the *subtalar joint* and the *transverse tarsal joint*.

In the subtalar joint there is relative motion between the talus and the calcaneus at three different sites. The axis of the subtalar joint lies at approximately 42° in the sagittal plane and at approximately 16° in the horizontal plane. This joint enables the following movements:

- *Inversion* (heel and forefoot). This involves moving the heel and forefoot towards the mid-line, bringing the footsole towards the median plane. The range of motion for inversion is between 30 and 35°. The muscles involved in inversion are the tibialis posterior and tibialis anterior.
- *Eversion* (heel and forefoot). This consists of moving the heel and forefoot laterally placing the sole of the foot away from the median plane with a range of motion between 15 and 20°. The muscles involved in eversion are the fibularis longus and fibularis brevis.

The transverse tarsal joint is a combination of three separate joint spaces, between the calcaneus, cuboid and talus bones. The movements that this allows are:

- *Pronation*. This is a combination of forefoot inversion and abduction; the muscle involved is the fibularis longus.
- *Supination*. This is a combination of forefoot inversion and adduction, with action by the tibialis muscles (anterior and posterior).

3.3.4 Kinematic models of the limbs

The human body can be modelled as a chain of rigid segments, where each segment or link corresponds to a body segment and the attachments between links correspond to the joints in the human body. In order to control the end-effector, it is necessary to determine the relationship between joints and the position and orientation of the end-effector.

There are two possible problems to be considered. The first is *forward kinematics*, which consists in determining the body configuration and the position and orientation of the end-effector on the basis of the known angles and displacements of the links. The second is the reverse problem, *inverse kinematics*, in which angles and displacements have to be determined on the basis of the position and orientation of the end-effector, which are known.

In order to describe the rigid-body kinematics of the human body, some simplifications are often necessary, such as assuming a lower number of DoFs in the joints (e.g. by ignoring linear displacement between segments) or assuming that the bones are completely rigid while they may have some flexibility. i.e. the rationale of a method to describe the kinematics of the human body based on the Denavit–Hartenberg convention, as explained in Section 3.2.

3.3.4.1 Denavit–Hartenberg model of the arm

Recently, exoskeleton developers have proposed a 9-DoF model of the human arm, to cope with scapula motion. The model includes the sternoclavicular joint in the kinematic chain. Any ergonomic wearable robot design must address human kinematics exactly so that robot and human are kinematically compliant, as discussed in detail in Section 3.4. With this in mind, for simplicity's sake, a 7-DoF model is assumed for the upper limb comprising:

- three DoFs at the shoulder;
- two DoFs at the elbow;
- two DoFs at the wrist.

The first step was to determine the coordinate frames according to *D–H convention* for the upper limb. In order to allow for a common base frame for both arms, the base coordinate system ($X_0Y_0Z_0$) was located in the body, midway between the shoulders. Then, in one of the arms, three frames, corresponding to three DoFs, were located at the centre of the shoulder joint (see Figure 3.9):

- *circumduction*, $X_1Y_1Z_1$;
- *adduction–abduction*, $X_2Y_2Z_2$;
- *flexion–extension*, $X_3Y_3Z_3$.

In the elbow joint there are two frames corresponding to the following rotations:

- *flexion–extension*, $X_4Y_4Z_4$;
- *supination-pronation*, $X_5Y_5Z_5$.

The movement at the wrist is described by another two frames:

- *adduction–abduction*, $X_6Y_6Z_6$;
- *flexion–extension*, $X_7Y_7Z_7$.

Finally, the end-effector frame is located at the tip of the extended fingers, $X_8Y_8Z_8$, as indicated in Figure 3.9.

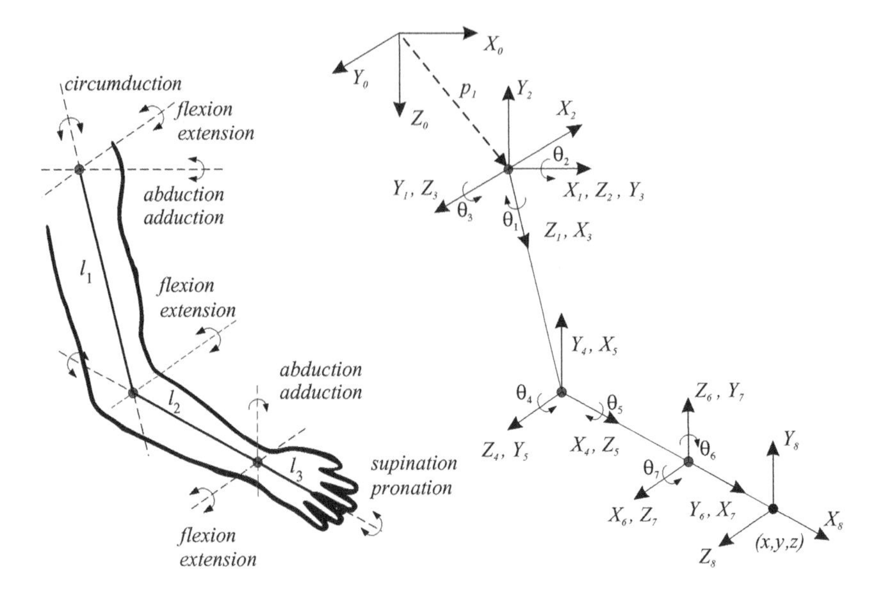

Figure 3.9 D–H notation for a human arm

Table 3.1 D–H parameters for arm segments

Joint	β_i	Number	α_i	a_i	d_i	θ_i
Base	0	$1_{(0\rightarrow 1)}$	0	a_0	d_0	0
Shoulder	(−90) medial rotation/lateral rotation (+90)	$2_{(1\rightarrow 2)}$	-90°	0	0	$\beta_1 + 90^{\circ}$
Shoulder	(−180) abduction/adduction (+50)	$3_{(2\rightarrow 3)}$	$+90^{\circ}$	0	0	$\beta_2 + 90^{\circ}$
Shoulder	(−180) flexion/extension(+80)	$4_{(3\rightarrow 4)}$	0	l_1	0	$\beta_3 + 90^{\circ}$
Elbow	(−10) extension/flexion (+145)	$5_{(4\rightarrow 5)}$	$+90^{\circ}$	0	0	$\beta_4 + 90^{\circ}$
Elbow	(−90) pronation/supination (+90)	$6_{(5\rightarrow 6)}$	$+90^{\circ}$	0	l_2	$\beta_5 + 90^{\circ}$
Wrist	(−90) flexion/extension (+70)	$7_{(6\rightarrow 7)}$	$+90^{\circ}$	0	0	$\beta_6 + 90^{\circ}$
Wrist	(−15) abduction/adduction (+40)	$8_{(7\rightarrow 8)}$	0	l_3	0	β_7

The rotation of a frame with respect to the one before it in the kinematic chain corresponds to a physiological DoF. In multi-DoF joints, e.g. the shoulder, there are several frames with the same origin. This is equivalent to considering a kinematic chain in which some of the links have a length equal to zero. These are called *virtual links*; for instance the links between $X_1Y_1Z_1$ and $X_2Y_2Z_2$ and between $X_2Y_2Z_2$ and $X_3Y_3Z_3$ are virtual links at the shoulder.

In order to complete the description it is necessary to define the D–H parameters. These are presented in the Table 3.1 and depend on the anthropometric parameters of the human body. There are always as many variables as DoFs, and so there are seven variables for the arm model presented here. The angle θ_i about axis Z_i is the variable associated with the ith DoF in the model. The range of motion for each variable in the model, θ_i, depends on the physiological range of motion (RoM) for the corresponding anatomical joint, β_i. For a summary of the kinematic D–H model of the upper limb with the physiological RoM corresponding to each anatomical joint, see Table 3.1.

The parameters a_i and d_i are the body segment lengths, which are constant for each individual. Usually these lengths scale with the total height of the person and can be approximated as in Table 3.2.

Table 3.2 Anthropometric data adapted from Winter (1990). Note that H represents the subject's body height

Body segment	Length, L	Centre of mass (% of L)	
		Proximal	Distal
Upper arm	0.186 H	0.436	0.564
Forearm	0.146 H	0.43	0.57
Hand	0.108 H	0.506	0.494
Thigh	0.245 H	0.433	0.567
Leg	0.53 H	0.433	0.567
Foot	0.152 H	0.5	0.5
HAT	0.475520 H	0.626	0.374

Once the D–H parameters are established, the transformation matrix $\mathbf{T}_{i-1}^{i}$ between two frames can be found with

$$\mathbf{T}_{i-1}^{i} = \mathbf{R}_Z(\theta_i)\mathbf{D}_Z(d_i)\mathbf{D}_X(a_i)\mathbf{R}_X(\alpha_i) \tag{3.29}$$

where $D_Z(d_i)$ and $D_X(a_i)$ are vector positions, along axes Z_i and X_i respectively.

This matrix is used to establish the homogeneous transformation matrix, $\mathbf{T}_0^n$; (see Equation 3.10). The procedure is described in Section 3.2.

3.3.4.2 Denavit–Hartenberg model of the leg

In the lower limb model, six DoFs were considered, three at the hip joint, one at the knee and two for the ankle. As in the model of the arm, first the coordinate frames needed to be defined according to the D–H convention. The base frame is positioned at the centre of the pelvis between the hips, at base $X_0Y_0Z_0$. The hip frames correspond to the following rotations, as indicated in Figure 3.10:

- *circumduction*, $X_1Y_1Z_1$;
- *adduction–abduction*, $X_2Y_2Z_2$;
- *flexion–extension*, $X_3Y_3Z_3$.

The frame located at the centre of the knee joint describes *flexion–extension*, $X_4Y_4Z_4$. There are two frames located in the ankle:

- *dorsiflexion–plantarflexion*, $X_5Y_5Z_5$;
- *inversion–eversion*, $X_6Y_6Z_6$.

The end-effector is placed at the tip of the longest toe, $X_7Y_7Z_7$ (see Figure 3.10).

The D–H parameters are presented in Table 3.3. The leg model considered has six variables, all corresponding to revolute joints. Thus, as in the model of the arm, the angles θ_i about axes Z_i are variable and dependent on the RoM β_i of each joint (see Table 3.3). The other two parameters, a_i and d_i, are the body segment lengths, which are constant and dependent on the individual human being (see Table 3.2). Finally, the transformation matrix $\mathbf{T}_{i-1}^{i}$ between frames can be derived (see Equation 3.29), to calculate the homogeneous matrix $\mathbf{T}_0^n$ (see Equation 3.10).

The angles and displacements can be deduced from the end-effector trajectory using the *Jacobian matrix*, in a process known as *inverse kinematics*. This problem is usually made more complex by redundancies, given that there may be zero, one, multiple or an infinite number of solutions, but that goes beyond the scope of this book.

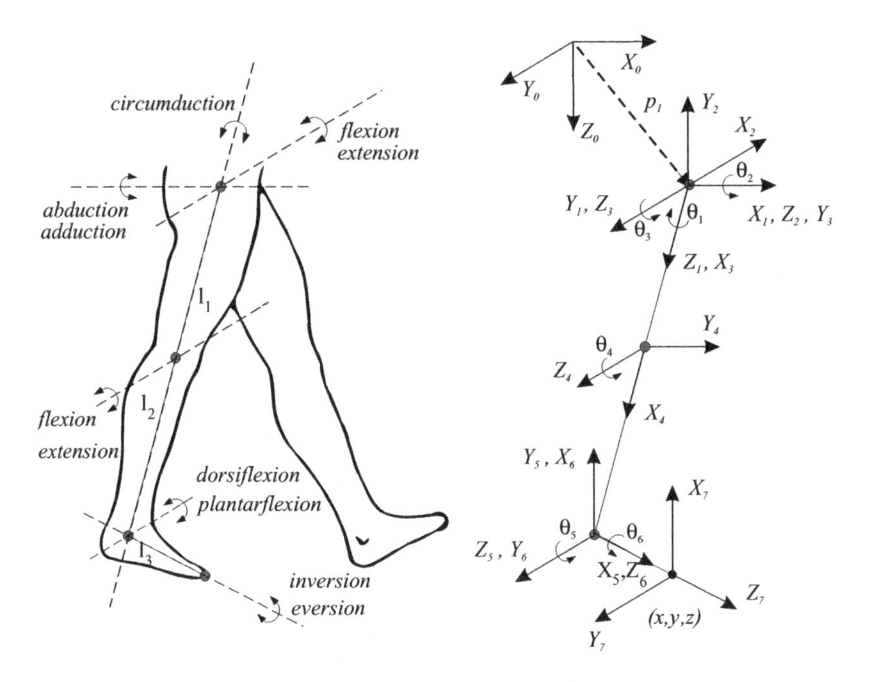

Figure 3.10 D–H notation for a human leg

Table 3.3 D-H parameters for leg segments

Joint	β_i	Number	α_i	a_i	d_i	θ_i
Base	0	$1_{(0\to1)}$	0	a_0	d_0	0
Hip	(−50) medial rotation/lateral rotation (+40)	$2_{(1\to2)}$	-90°	0	0	β_1+90°
Hip	(−20) abduction/adduction (+45)	$3_{(2\to3)}$	$+90^\circ$	0	0	β_2+90°
Hip	(−30) extension/flexion (+120)	$4_{(3\to4)}$	0	l_1	0	β_3
Knee	0 extension/flexion (+150)	$5_{(4\to5)}$	0	l_2	0	β_4+90°
Ankle	(−40) plantarflexion/dorsiflexion (+20)	$6_{(5\to6)}$	$+90^\circ$	0	0	β_5+90°
ankle	(−35) inversion/eversion (+20)	$7_{(6\to7)}$	0	0	l_3	β_6

3.3.5 Dynamic modelling of the human limbs

Following a procedure like the one presented in the previous section, it is possible to model the complexity of human dynamics in order to be able to use the same procedures as in robotics. There are different levels of detail in dynamics modelling, and each produces different results in the analysis. The first level is to calculate the forces and moments in the joints. This can be done by assuming that the human body is modelled as a chain of rigid links or segments held together by ideal joints. With this model it is possible to distinguish between the different contributions to the joint torques, e.g. from accelerations in adjacent segments.

A second level of detail in the modelling includes a functional model of joint behaviour, for instance as a second-order system; using classical system identification techniques it is possible to determine the stiffness and viscosity of the joint.

A third level of detail includes the individual muscles around each joint. This requires a definition of the geometry of each muscle involved, with the origin and insertion points, and also includes a model of the muscle dynamics. However, due to muscle redundancy (there are more muscles crossing a joint than are required to produce the prescribed motion) there will be multiple muscle force combinations that can produce the same output (motion and external forces). Using optimization techniques, like energy efficiency and maximal muscle torques (see Section 2.2), it is possible to arrive at plausible solutions for the force distribution in each muscle. A further step in the modelling could include the neuromuscular control problem; however, it is not clear how to go about finding a solution to this problem (Hatze, 2002).

Various different methods have been used to determine human body dynamics, such as forward and inverse dynamics, hybrid forward–inverse dynamics, optimization-based and neuromusculoskeletal tracking methods (Erdemir *et al.*, 2007; Li, Pierce and Herndon, 2006; Seth and Pandy, 2007; Zajac, Neptuno and Kautz, 2003). Every method has its own advantages and disadvantages depending on the specific application, experiment or simulation. This section will focus on *inverse dynamics*. This method is designed to derive the internal torques (τ) and forces in each joint from the motion data, and in the case of a closed-chain problem with additional input of external contact forces (see Figure 3.11). In addition, it is possible to distinguish between the joint torques produced by muscle activity and the ones produced by gravity and the motion of adjacent segments.

In general, three sources of information are required to apply this method (see Figure 3.11):

- Kinematic, requiring measurement of the segment motion.
- Kinetic, requiring measurement of the external force. Note that this information is not needed in the case of an open-chain problem, when there is only one point of application of the external forces to the chain. A typical instance of the closed-chain problem occurs during the double-stance phase of human walking. When both feet are on the ground it is not possible to compute the inverse

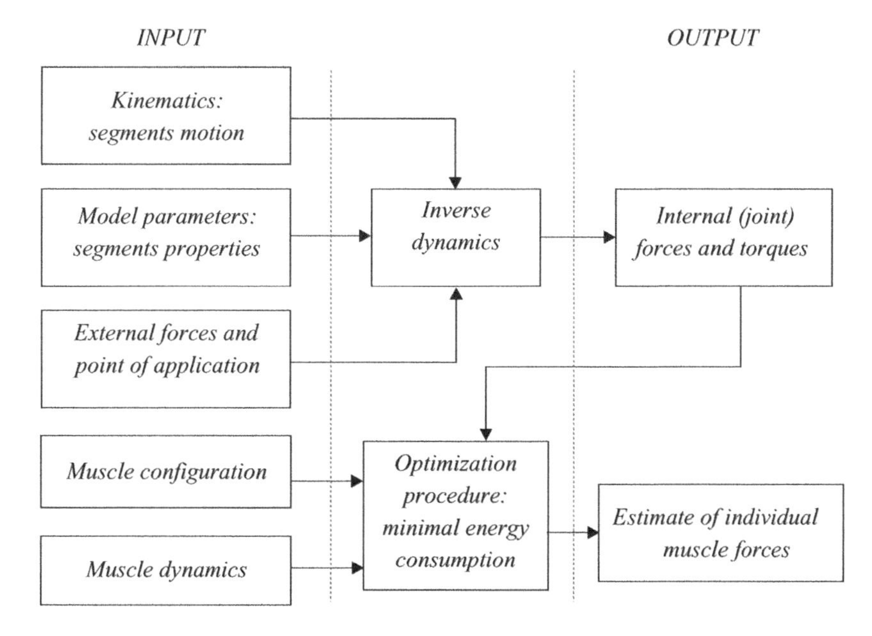

Figure 3.11 Diagram of the inverse dynamics procedure

dynamics without knowing the external forces under each foot (Forner-Cordero, Koopman and Van der Helm, 2004, 2006; Koopman, Grootenboer and de Jongh, 1995).

- Model parameters, requiring determination of the geometrical and inertial properties of the segments, the joint constraints and the ranges of motion (Koopman 1989; Winter 1990; Zajac, Neptuno and Kautz 2002).

Mathematically, body motion can be expressed as

$$\mathbf{M}(q)\ddot{q} + \mathbf{C}(q, \dot{q}) + \mathbf{K}(q) = \boldsymbol{\tau} \tag{3.30}$$

where q represents the generalized coordinates, e.g. $q = \begin{bmatrix} x & y & z \end{bmatrix}^{\mathrm{T}}$; $\mathbf{M}(q)$ is the mass moment of the inertia matrix; $\mathbf{C}(\boldsymbol{q}, \dot{\boldsymbol{q}})$, $\mathbf{K}(\boldsymbol{q})$ represent the Coriolis and centrifugal, and the gravity accelerations respectively; and $\boldsymbol{\tau}$ is the generalized torque applied on each segment. This equation is based on the Lagrangian–Euler formulation.

The Newton–Euler equations ($\boldsymbol{F} = m\boldsymbol{a}$; $\mathrm{M} = I\boldsymbol{\alpha}$) can also be used to derive the forces and torques. There are force terms due to gravity, net dynamic forces ($\boldsymbol{F} = m\boldsymbol{a}$) and net dynamic torques ($\boldsymbol{\tau} = I\boldsymbol{\alpha} + \boldsymbol{\omega} \times (I\boldsymbol{\omega})$) acting on each segment (Koopman, Grootenboer and de Jongh, 1995; Kurfess, 2005).

In order to determine the mass moment inertia matrix in Equation (3.30) (or the values of I), standard tables of anthropometric parameters are commonly used, such as Table 3.2 or references in the literature (Leva, 1996; Todd, Woods and Shanckelford, 1994; Winter, 1990). These parameters depend on the shape of the body segment and its mass distribution, and also on the axis about which the rotation takes place (Todd, Woods and Shanckelford, 1994; Winter, 1990). For a practical example of the calculation of the dynamics of the arm, see Case Study 3.7.

3.3.5.1 Muscle dynamics

Another important issue is the contribution of the muscles to movement. Muscle and other structures crossing the joint generate forces on the segments and net joint moments (Zajac, Neptuno and Kautz, 2002). The forces and torques calculated by that procedure represent the resultant action of all muscles crossing a joint. However, computation of the forces in each muscle is not immediate because of muscle redundancy; i.e. there are many muscle forces crossing a joint that can produce the same joint forces and moments. This redundancy makes it difficult to estimate individual muscle forces (Hatze, 2002; Zajac, Neptuno and Kautz, 2002).

Therefore, optimization procedures are used to derive muscle forces and torques (see Figure 3.11). These procedures consider joint forces and torques, and musculoskeletal models (including muscle configuration and its dynamics). Static optimization takes into account muscle fibre length, muscle force and muscle moment arms (defined by the path and the origin and insertion sites of the muscles) derived from musculoskeletal models. The most common target functions in this procedure are to minimize energy consumption and keep the maximal individual muscle forces bounded (Hatze, 2002; Koopman, Grootenboer and de Jongh, 1995; Zajac, Neptuno and Kautz, 2002).

3.4 KINEMATIC REDUNDANCY IN EXOSKELETON SYSTEMS

3.4.1 Introduction to kinematic redundancies

The redundancy of manipulators is task-dependent. A manipulator, for instance a robot or a human limb, is called *kinematically redundant* if it possesses more degrees of mobility than are necessary to perform a specific task. Redundancy of a manipulator always relates to the number of n joint space variables that it has, the number m of operational space variables and the number r of operational space variables required to perform a specific task. *Functional redundancy* describes a situation for which a given manipulator has more joint space variables n than task space variables, r ($n > r$). This can occur in manipulators with less than six degrees of freedom ($n < 6$) if the task space dimension r is restricted with respect to the Cartesian space ($m = 6$). A manipulator is called *intrinsically redundant* if the number of joint space variables that it has is larger than the dimension of the operational space ($n > m$). All manipulators having seven or more degrees of freedom are intrinsically redundant. In the previous sections it was seen that the human arm is an intrinsically redundant system and a model with seven degrees of freedom was proposed. If further degrees of freedom of the human arm are considered, it may even be modelled with a higher degree of redundancy. Most exoskeletons that interface with the human upper limb are also intrinsically redundant, which facilitates smooth interaction with the human arm. Whenever redundancy is mentioned in this chapter, it refers to intrinsic redundancy.

The redundancy of a manipulator can be analysed using the following differential Equation, which linearly maps the joint velocity space of a manipulator to its end-effector velocity space:

$$\boldsymbol{v} = \mathbf{J}(q)\dot{\boldsymbol{q}} \tag{3.31}$$

where, $\boldsymbol{v}$ denotes the $(r \times 1)$ vector of end-effector velocity of the manipulator and $\mathbf{J}(q)$ represents its $(r \times n)$ Jacobian matrix. The Jacobian is derived from the geometric Jacobian. Vector $\dot{\boldsymbol{q}}$ is the $(n \times 1)$ vector of joint velocities. The redundant degrees of mobility of a redundant manipulator are determined by the number $\delta = (n - r)$. It is important to note here that the Jacobian is a function of the instantaneous manipulator configuration. For some configurations, the Jacobian can degenerate, and in such cases the manipulator is in a singular configuration. The range space and null space of the Jacobian further help to understand the concept of redundancy. The range space of $\mathbf{J}(q)$ denotes a subspace $R(\mathbf{J})$ in $\Re^r$ of end-effector velocities that can be generated from joint velocities. The null-space of $\mathbf{J}(q)$ is a subspace $N(\mathbf{J})$ in $\Re^n$ of joint velocities that do not produce any end-effector velocity in the given posture. In a singularity, the dimension of the range space is reduced. A non-redundant manipulator then loses mobility; it cannot create joint velocities that produce end-effector velocities. In a singularity the dimension of the null space increases according to

$$\dim(R(\mathbf{J})) + \dim(N(\mathbf{J})) = n \tag{3.32}$$

A redundant system possesses considerable advantages thanks to the permanent existence of a null space $N(\mathbf{J})$ i.e. nonzero. This means that there are infinite solutions to the differential kinematics problem formulated in Equation (3.31) for nonsingular configurations. The same applies to the inverse problem. With a nonzero $N(\mathbf{J})$, the self-motion of the manipulator can be used to influence its geometric pose while keeping the end-effector position and orientation constant. This null-space motion of a redundant manipulator can then be used to improve its mobility. In particular, when inverting the forward kinematics, optimization techniques can be used to choose the joint space variables in such a way as to satisfy additional constraints. This is then formulated as an optimization problem; for instance, obstacles can additionally be avoided during motion. Other popular techniques are to choose a specific joint-space trajectory to avoid kinematic singularities, or to reduce the joint speeds during motion in order to minimize energy consumption. A good account of inverse kinematic procedures for redundant manipulators can be found in Sciavicco and Siciliano (2003).

In the case of a coupled human–exoskeleton system, the redundancy of the exoskeleton can be exploited to adjust the robot's geometric posture to the posture of the human arm. This is an important advantage of redundant exoskeletons and is further discussed in the following subsections.

3.4.2 Redundancies in human–exoskeleton systems

A nonredundant exoskeleton is likely to interfere with human motion in specific configurations, especially singular configurations. Despite the advantages in dexterity provided by redundant exoskeletons, they also present a number of challenges. These depend mostly on the particular case, i.e. the specific kinematic structure (the D–H parameters, so to speak) of the exoskeleton and the control architecture that has been implemented by the developers. As there is no room in this book for an exhaustive account of redundancy resolution, the reader is referred to the references on typical exoskeletons quoted in this subsection.

Sections 3.2 and 3.3 outlined aspects of modelling for both robotic manipulators and human limbs, whose kinematics and dynamics are described in the same form. If a human wearing an exoskeleton is considered, the total kinematics of the system may be described as a parallel kinematic loop consisting of two serial chains, one for the exoskeleton and one for the human limb. Both chains are attached at the base and the tip, and sometimes also at intermediate points, for instance on the upper arm and the forearm. It is important that the workspace of the exoskeleton and the human arm overlap. Figure 3.12 illustrates some typical kinematic structures of coupled human–exoskeleton systems. Drawing (a) illustrates end-point-based exoskeleton systems, which are body- or wall-grounded and are only attached to the hand of the human operator. Drawing (b) illustrates exoskeletons that are body- or wall-grounded, are kinematically equivalent to the human limb and are attached at several locations along the limb. Drawing (c) illustrates exoskeletons that are not kinematically equivalent to human limbs but are attached at several points along the user limbs.

In general, most exoskeletons feature at least seven degrees of freedom in order to achieve the dexterity and agility of the human arm movements. Note that some authors (Carignan, Liszka and Roderick 2005; Schiele and van der Helm 2006) assume high redundancy in the human arm, so it is natural that there should also be exoskeleton structures with multiple degrees of redundancy. When discussing redundancy in a human–exoskeleton system, there are two main effects worth noting besides the available working range.

Firstly, the redundancy of the human arm and the exoskeleton should be matched so that actuation torques can be delivered to the human joints unambiguously. The goal of delivering torques to the human joints is crucial for realistic force feedback in virtual reality or bilateral telemanipulation scenarios. Furthermore, inducing accurate joint trajectories, e.g. $\dot{\boldsymbol{q}}(t)$, $\boldsymbol{\tau}(t)$, to human joints or human joint groups is important for functional motor rehabilitation in human–exoskeleton systems. Joint torques to the exoskeleton are often derived from Cartesian forces by the Jacobian transpose $\mathbf{J}^{\mathrm{T}}$, according to

$$\boldsymbol{\tau} = \mathbf{J}^{\mathrm{T}}(q)\boldsymbol{\gamma} \tag{3.33}$$

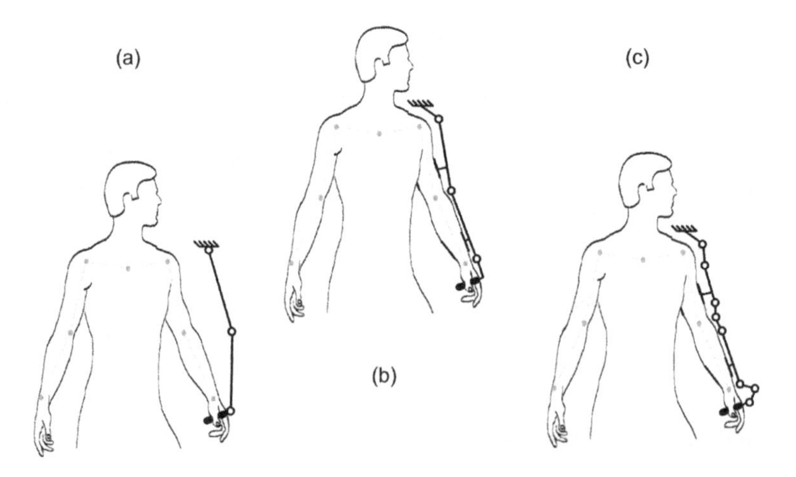

Figure 3.12 Illustration of three different types of exoskeletons: (a) end-point based, (b) kinematically equivalent to the human limb and, (c) kinematically different with respect to the human limb. Reproduced from ESA

where, $\boldsymbol{\tau}$ represents the $(n \times 1)$ vector of joint torques and $\boldsymbol{\gamma}$ represents the $(r \times 1)$ vector of Cartesian forces. The chosen dimension of $\boldsymbol{\gamma}$ is mostly 6 in order to include all Cartesian forces and torques. It is important to note that the Jacobian transpose has a range and a null space here too. For a more detailed description of the range and null space of $\mathbf{J}^{\mathrm{T}}$, see Sciavicco and Siciliano (2003). In order to match the joint space variables to the human joint space variables unambiguously, the redundancy of the exoskeleton must be constrained. This is discussed along with the second issue hereafter.

The second important role of redundancy to be considered is bilateral telemanipulation, where a second slave robot is remotely controlled by the master exoskeleton. In this case it must be possible to match the geometric pose of the exoskeleton to the geometric pose of a remotely controlled slave robot; only then can remote control of the slave be optimal. The human can then use his/her own arm redundancy to control the redundancy of the slave robot. This is important for achieving realistic situation awareness in situations where the slave robot is in contact with its environment and for avoiding obstacles with the slave during Cartesian motion. The geometric pose of the exoskeleton and remote robot can be matched using direct joint-to-joint mapping, but only if the slave robot and the exoskeleton are kinematically equivalent. If they are not, inverse kinematic algorithms are needed to pick a joint space solution so that the pose of both devices is matched to the human arm pose. One common solution is to constrain the slave robot pose to the elbow orbit angle of the exoskeleton. The elbow orbit angle is used as a constraint to choose a solution from the solution set of the redundant manipulator's inverse kinematics equation; such an algorithm has been described by Kreutz-Delgado, Long and Seraji (1992). This constraint also makes it possible to choose a meaningful set of joint torques for display to the human.

In the case of the exoskeleton depicted in Figure 3.12(a), feedback forces can be applied only to the human hand. Only operational space forces and velocities can be imposed. Such an exoskeleton has been proposed for instance in Williams *et al.* (1998) for the upper extremity and in Hesse and Uhlenbrock (2000) for the lower extremity. Those devices are unable to influence human arm redundancy and are mostly not redundant themselves. This means in practice that for all locations in operational space, the human limb has an infinite number of possible joint space configurations, while the exoskeleton does not. In functional rehabilitation, for instance, such exoskeleton systems are unable to induce joint trajectories exactly matching the human joints. It could even be dangerous to load then on to human joints if this forces motion in the human joints beyond their natural working range. To illustrate this case, consider an exoskeleton applying force to the hand of a fully stretched human arm. There is a danger of hyperextending the elbow due to the joint space ambiguity (in fact, the elbow may move in two directions). While this is a considerable drawback as regards safety, such designs have the advantage of not requiring alignment of the exoskeleton kinematic structure to the human limb structure. As long as the operational space motion is equivalent to the human arm, the physical interaction will feel good and an ample common workspace is possible. However, safety is left to the discretion of the operator, who will need to consciously avoid the workspace-end singularities of his limbs. Such designs are not optimal for bilateral telemanipulation tasks. Humans cannot control the configuration of the exoskeleton and hence cannot control the configuration of a slave robot that may be remotely linked into the control loop.

The exoskeleton class illustrated schematically in Figure 3.12(b) has exactly the same degrees of redundant mobility as the human arm, excluding the shoulder girdle, i.e. seven degrees of freedom. It is attached at the end-point, i.e. the hand, as well as to the other movable links, i.e. the upper arm and the forearm. An exoskeleton with such a kinematic structure has been proposed for instance by Bergamasco *et al.* (1994). In order to induce exact joint trajectories and to match natural redundancy, their joints must be aligned to coincide with the human joints. This is difficult, as discussed further in Section 5.2. Such exoskeletons would be able to control a slave robot with geometric pose correspondence and deliver accurate feedback forces to the human joints. This can be done, for instance, by constraining the configuration of the robot with the elbow orbiting angle, as described above. Also, safety constraints can be implemented at the level of the joint space. Nonetheless, it is important to remember that the robot axes must be perfectly aligned with the human joints. Undesired reaction forces can otherwise be created in the human joints by a kinematic mismatch (see Section 5.2 for more detail).

In order to improve the fit between the human limb and the robotic device, a class of exoskeletons is being developed with a structure similar to the one depicted in Figure 3.12(c). Such exoskeletons possess multiple degrees of redundancy to cope with interaction, not only with the human arm but also with the human shoulder and shoulder girdle. Such hyper-redundant devices have been proposed by Schiele and Visentin (2003) and Yoon *et al.* (2005). They feature a greater working range than the exoskeletons in Figure 3.12(b) (see also Case Study 8.3) and they also make it possible to match the instantaneous human joint pose to a slave robot. This can be done by optimization of the inverse kinematics. Also, explicit mapping between the exoskeleton joints and the robot joints, which is computationally more efficient, is possible for bilateral telemanipulation. The difference between this class of exoskeletons and the class depicted in Figure 3.12(b) is that no exact alignment is required between the exoskeleton and the human. Their redundancy can thus be exploited to achieve additional user comfort.

3.5 CASE STUDY: A BIOMIMETIC, KINEMATICALLY COMPLIANT KNEE JOINT MODELLED BY A FOUR-BAR LINKAGE

J. M. Baydal-Bertomeu, D. Garrido and F. Mollá

Instituto de Biomecánica de Valencia, Universidad Politécnica de Valencia, Spain

This case study is included here to illustrate concepts related to bioinspiration, both in the design of compliant kinematics for a wearable lower limb robot and in the evolutionary optimization process based on genetic algorithms. This section therefore illustrates concepts from Chapters 2 and 3. In particular, it presents the bioinspired design of the knee joint of an exoskeleton. The study first introduces models for the kinematics of the anatomical joint; then it examines a genetic algorithm approach to optimize the design of a four-bar mechanism for the robot's joint; finally, it presents the results in terms of anatomical versus artificial joint kinematics.

3.5.1 Introduction

Recent advances in electronics and control engineering have penetrated the field of medicine over the last few years. In particular, research in orthotics has led to several improvements in electronic control of orthotic joint movements. Progress in electronics has prompted the appearance of various stance control orthoses (SCOs) on the market over the last 30 years. However, most have disappeared, chiefly due to lack of robustness, unreliability and mechanical inconsistency (Michael, McMillian and Kendrick, 2003). The underlying problem is the low kinematic compatibility between orthotic and physiological knee motion.

Moreover, mismatching orthotic and natural knee joint motions can cause an unwanted interaction force, resulting in pistoning of the orthotic components on the lower limb, which in turn produces restriction of the normal range of motion, distal migrations and misalignment between the system and the body segment, and skin pressure discomfort (Lewis *et al.*, 1984). Therefore, ensuring a satisfactory alignment between the centre of rotation of the orthosis and the user's knee can potentially provide comfort and enhance the satisfactory use of any wearable robot. Furthermore, the use of hinges that are kinematically incompatible with human joints could seriously damage internal tissues and promote injuries.

This case study presents a new concept of the orthotic knee joint following the kinematics of the human knee motion.

3.5.2 Kinematics of the knee

The kinematics of the knee has historically been difficult to quantify. Anatomic rotations and translations are subject to axis alignment difficulties that inhibit comparisons between subjects. However, the bones that meet at the knee joint and the ligaments that hold the bones together can be analysed as a mechanical linkage (see Section 3.3).

The following models have been reported in the literature:

- Crowningshield, Pope and Johnson (1976). A model that assumes that during passive flexion, the femur rolls backwards on the tibial plateau, with simultaneous rotation about an axis through the medial tibial plateau.
- Wismans (1980). A three-dimensional model, with mathematical functions that fit shapes of the articular surfaces and represents the ligaments and capsule as non-linear springs.
- Walker *et al.* (1985). A model based on the geometry of the condyles.
- O'Connor *et al.* (1989). A model based on the consideration of cruciate ligament motion.

The model presented by Walker *et al.* (1985) can be used to derive the displacement of the instantaneous helical axis (IHA) in three dimensions. A kinematically compatible system must follow the path of the IHA in the plane where the joint of the wearable robot is located. This is the kind of model considered for design of a biomimetic joint. On the basis of this model, it is possible to derive an expression to determine the 'average knee motio' of flexion–extension under quadriceps action. In determining the relative motion between the femur and the tibia, the medial and lateral femoral condyles are considered as spherical surfaces. On that basis, the centres of the spheres can be used as reference points.

Following the motion study carried out by Walker *et al.* (1985), it is possible to establish the anterior–posterior translation Z_{DIS} and the distal–proximal translation Y_{DIS} of the IHA as a function of flexion angle, F:

$$\begin{aligned} Y_{\mathrm{DIS}} &= -0.05125F + 0.000308F^2 \\ Z_{\mathrm{DIS}} &= -0.0602F + 0.0000178F^2 \end{aligned} \tag{3.34}$$

Angles are expressed in degrees and displacements in millimetres. These equations are then used to define the linear motion of the femoral axis relative to the tibia on the sagittal plane (YZ) at the medial distance (X) where the external linkage would be placed, e.g. 60 mm from the intracondylar centre of origin.

As can be seen in Figure 3.13, the IHA (according to the Walker *et al.*, 1985, model) forms a three-dimensional mesh in a complete flexion–extension of the knee. The sections of this mesh, performed by sagittal planes at different distances, are the curves that a robot's joint must follow when placed at this distance. Therefore the design principle of any robotic knee should hinge on the premise that the external joint ought to describe the same motion as the curve performed by a section of the mesh at the same distance and with the same orientation as the orthotic hinge in real-use conditions.

3.5.3 Kinematic analysis of a four-bar linkage mechanism

To produce the desired motion curve, a mechanism based on a crossed four-bar linkage was selected. Similar solutions can be achieved using other mechanisms, e.g. a guide or a cam, but these present

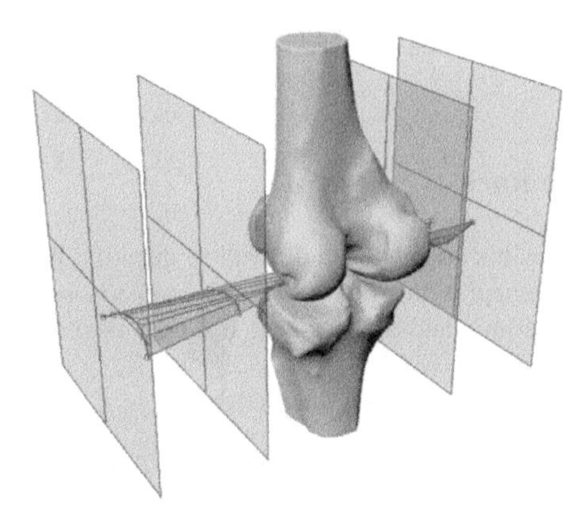

Figure 3.13 Representation of three-dimensional movement (IHA) in a complete flexion–extension of the knee joint

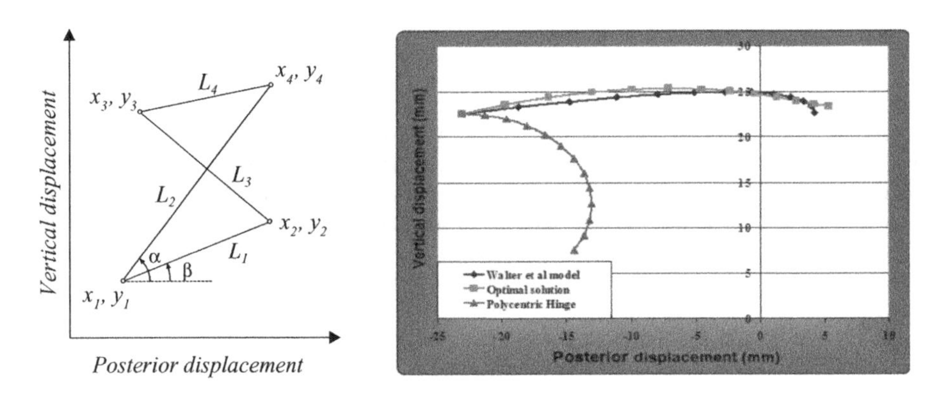

Figure 3.14 Principal parameters describing the motion of a four-bar linkage mechanism (left) and representation of the different solutions produced by the GA methodology and the human knee motion model (right)

drawbacks in terms of high manufacturing cost, resistance and pinching between the parts of the mechanism. The crossed four-bar mechanism has the advantages of simplicity, robustness and ease of design. However, its most important property is that it can be perfectly defined in an analytical way because its movement and kinematics are determined by the relative joint rotation angles.

The information from displacement analysis of the four-bar linkage can be expressed in a closed mathematical form depending on the length and location of the four bars constituting the mechanism. Another important feature of this mechanism is that the point where the links cross is the instantaneous centre of joint rotation. From there it is possible to determine the optimal mechanism that follows a particular path with a high degree of accuracy.

As can be seen in Figure 3.14 (left), a four-bar linkage mechanism is characterized by the following parameters:

- X_1, Y_1 represent the spatial position of the crossed four-bar mechanism.
- L_1, L_2, L_3, L_4 represent the length of each of the bars.

- α represents the angular rotation of the movement of the crossed four-bar linkage.
- β represents the angular position in space of the crossed four-bar linkage.

This set of parameters completely characterizes the four-bar linkage mechanism and can be used to represent it. $L4$ is called the coupler link and its movement should be totally compatible with knee flexion.

3.5.4 Genetic algorithm methodology

The design of a biomimetic joint is stated as an optimization problem, in which the IHA of the cross-bar mechanisms should be optimized to follow the path of the IHA of the knee joint in the plane where the system will be located. Since the mathematical equations for the cross-bar mechanism are already known, it is possible – at least in theory – to solve the optimization problem analytically. However, in practice the expressions obtained analytically do not serve to find the analytical solution.

Genetic algorithm (GA) methodology was therefore selected for optimization because it is useful for finding suboptimal solutions to problems where the functions are difficult or impossible to differentiate and cannot readily be solved by means of an analytical procedure (Holland, 1975). The solution chosen as optimal was the human knee motion according to the model of Walker *et al.* (1985). The optimization function was the inverse of the squared error between the trajectory of the IHA of the crossed four-bar linkage and the human model. The coordinates of the four apexes and the input angle of the crossed four-bar linkage were used as input parameters. The evaluation stage was accomplished using the equations that express the ICR position of the crossed four-bar mechanism, given the angular position of the coupled link and by comparison with the optimal solution.

Before starting the optimization process, a few constraints were implemented in order to retain the variable parameters as coordinates of a four-bar linkage:

1. The segments L_1, L_2, L_3 and L_4 are assumed to be rigid bodies. The segment L_1 is considered as the fixed bar of the four-bar linkage.
2. The length of the links must not satisfy the Grashof condition in order to avoid an indeterminate solution.
3. The motion range of the four-bar mechanism should be limited to the normal range of knee motion. The motion range was set at 120°.
4. It was assumed that the orthotic knee joint could not be larger than the human knee in order to avoid a bulky solution.

3.5.5 Final design

Genetic algorithm methodology was used to continuously explore the range of solutions, searching for the best solution through each stage of the iteration process. Eventually, a suboptimal mechanical solution was obtained.

The final design is based on the integration of two plates which act as the bars of the crossed four-bar linkage. These plates were designed in order to meet the following requirements: resistance (the orthotic knee hinge frame should be designed with adequate resistance to sustain the weight of the patient as the exoskeleton should support patients weighing up to 80–100 kg); safety (the design must prevent pinching); and range of motion (the knee joint must offer a range of motion from 0° of minimum flexion to +120° of maximum flexion and the deformation in the sagittal plane should be accommodated to a maximum of ±5°).

3.5.6 Mobility analysis of the optimal crossed four-bar linkage

Figure 3.14 (right) depicts the path performed by a polycentric hinge and the optimal crossed four-bar linkage (OCFL) compared with human knee motion. Note that the OCFL matches the human knee movement, according to the knee model, significantly more accurately than the other solutions. The motion performed by the polycentric hinge is not at all physiological and differs considerably from human knee motion.

The error, defined as the difference in position between the human knee motion and the motion performed by different hinges, was determined at different degrees of knee flexion over the whole range from 0 to 120°. This error represents kinematic incompatibility between the orthosis hinge and the human knee.

Table 3.4 represents the difference in position between each of the three different knee hinges and the human model (mm) at various different degrees over the full range of knee motion. As can be seen in this table, a common monocentric hinge has a maximum misalignment of up to approximately 27.2 mm and a polycentric hinge with a radius of 8 mm in the gears has a maximum misalignment of 23.9 mm. The maximum misalignment produced by the OCFL is approximately 2.0 mm. This is a very significant improvement, showing better adjustment of the new concept of orthotic hinge to human motion.

Monocentric and polycentric hinges enlarge the error as the knee flexes, while the OCFL hinge significantly reduces the error in the final degrees of flexion. This is particularly important as it shows that the OCFL could improve general comfort during the execution of certain activities of daily living (ADL), such as standing up or sitting down in a chair, negotiating stairs and slopes, etc. The OCFL hinge also reduces the error in the initial degrees of flexion. This will enable the user of such an optimally fitted orthosis to perform comfortably certain tasks requiring fewer degrees of flexion, one of the most important of which is walking.

As already mentioned, the comfort, safety and reliability that a leg brace provides are likely to be functions of the mechanical compatibility between the knee hinge and the knee. Therefore, since the knee hinge is of vital importance in a lower limb orthosis (Lewis *et al.*, 1984), major improvements in the kinematics of wearable robots are needed. This study broadly shows that orthoses based on monocentric or polycentric hinges, which are commonly used by most physicians, do not follow property the motion performed by the human knee. Therefore, hinges based on monocentric or polycentric mechanisms may not be the best solution, in terms of mobility, for functional orthoses or modern SCOs in which knee movement is allowed and kinematic compatibility is required.

The new concept of orthotic design will notably improve the adaptability of lower limb orthoses to the user, offering comfort (kinematic compatibility avoids relative movement between the orthosis and the lower leg, so that the knee joint provides increased comfort) and safety and ligament protection (the OCFL joint avoids the generation of unwanted pressures at the knee level, providing protection

Table 3.4 Difference in position between human motion and three different orthotic joints

Degrees of flexion	Error (mm)		
	OCFL	Monocentric hinge	Polycentric hinge
0	0	0	0
20	1.694	8.449	5.285
40	1.800	15.255	9.753
60	1.874	20.454	13.795
80	1.428	24.131	17.592
100	0.563	26.366	21.062
120	1.307	27.212	23.920

for internal soft tissues). Besides the biomechanical improvements, it should be stressed that the design is further enhanced by ease of manufacturing. Orthotic hinges based on crossed four-bar linkages can be manufactured cheaply using two lateral plates with holes at the same position as those displayed in the OCFL during the optimization procedure. These plates can also be fitted with a cover to avoid pinching caused by the free space between the parts. Design enhancement solves the problem of the excessively high accuracy i.e. required in the manufacture of polycentric and other orthotic hinges.

3.6 CASE STUDY: DESIGN OF A FOREARM PRONATION–SUPINATION JOINT IN AN UPPER LIMB EXOSKELETON

J. M. Belda-Lois, R. Poveda, R. Barberà and J. M. Baydal-Bertomeu

Instituto de Biomecánica de Valencia, Universidad Politécnica de Valencia, Spain

This case study is included to illustrate the design of a kinematically compliant joint for the upper limb exoskeleton presented in Case Study 8.1. As illustrated in Section 3.4, kinematic compatibility between the robotic and the anatomical joint is important if no kinematic constraints are to be applied to the biological motion.

3.6.1 The mechanics of pronation–supination control

The pronation–supination movement of the forearm is a rotational movement of the forearm on its longitudinal axis in which two mechanically connected joints are engaged: the upper radio-ulnar joint (which belongs to the elbow) and the lower radio-ulnar joint (which is anatomically separate from the wrist). There are two bones in the forearm that make this movement possible: ulna and radius, (see Section 3.3). Both bones are roughly pyramidal in shape and are placed in such a way that the base of the radius is at the tip of the ulna and vice versa (see Figure 3.15(left)). There is still some disagreement as to the complete description of this movement. Physiologists, e.g. Yasutomi *et al.* (2002), generally agree that this movement is achieved mainly by the rotation of the radius around the ulna; however, some authors also refer to an elusive movement of the ulna itself (Weinberg *et al.*, 2000).

Whatever description is adopted, control of the pronation–supination movement implies the fixation of a reference point (static point) and the fixation of another point in the segment of the limb i.e. to be controlled (moving point). In orthotic practice this fixation is commonly located between the humerus and the metacarpal, but that approach entails restrictions on elbow flexion–extension wrist flexion–extension and radial–ulnar deviation movements.

An alternative is to use the movement occurring between a fixation on the ulna and another fixation on the radius. The mechanism proposed in WOTAS, the exoskeleton described in Case Study 8.1, to control the pronation–supination movement is based on control of the rotation of a bar parallel to the forearm. This bar is fixed very close to the olecranon (see point B in Figure 3.15 (left)). Thus, the bar is fixed to the ulnar position at elbow level. The distal fixation of the bar is at the head of the radius, although the bar is kept on the ulnar side in order to minimize excursion of the system. The distal part of the bar has degrees of freedom in order to absorb any antero-posterior displacement of the attachments produced by the movement. The lining-up of the orthoses is done with the forearm in the neutral position of pronation–supination. At this location, the bar must not be tense. The actuator

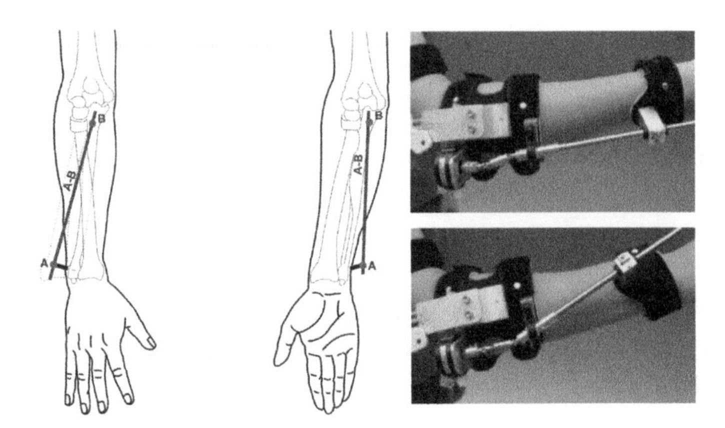

Figure 3.15 Scheme of pronation–supination control (left) and view of the resulting mechanism for controlling this movement (right)

adopted in WOTAS to activate this movement is rotational and is located on the ulnar side of the forearm, attached to the radius (see Figure 3.15 (right)).

This fixation of the mechanism is based on thermoplastic supports to fix the structure securely to the upper limb. Supports made of thermoplastics can adapt to the morphology of each user's arm, a very important characteristic for proper transfer of loads to the limb, as discussed in Section 5.3 of this book.

3.7 CASE STUDY: STUDY OF TREMOR CHARACTERISTICS BASED ON A BIOMECHANICAL MODEL OF THE UPPER LIMB

E. Rocon and J. L. Pons

Bioengineering Group, Instituto de Automática Industrial, CSIC, Madrid, Spain

This case study presents an estimation of tremor kinematic parameters, reproducing upper limb kinematics based on a biomechanical model. The method used for analysis of tremorous movements is based on a combination of solid modelling techniques with anthropometric models of the upper limb, from which a kinematic and dynamic upper limb model can be developed. This model of upper limb musculoskeletal systems can be used to estimate the force contribution of each muscle component during motion, to experiment with modifications of musculoskeletal topology and to devise complex motion coordination strategies. The input to the model is the angular position, velocity and acceleration of each joint, measured by gyroscopes placed on the upper limbs of patients suffering from tremor.

In order to investigate tremor biomechanical characteristics 31 patients suffering from different tremor pathologies were assessed. During the trials the patients were asked to perform a set of tasks selected in order to excite all different types of tremor disease.

Microeletromechanical system (MEMS) Gyroscopes were used in this application because they can measure rotation movements, and human movement can be described as rotation about joints (Moreno *et al.*, 2006). The sensor selected was the uniaxial gyroscope developed by Technaid S.L. (www.technaid.com), with dimensions of 15.5 mm × 8 mm × 4 mm. The proposed sensor is based on the combination of two independent gyroscopes placed distal and proximal to the joint of interest.

The system weighs roughly 15 g (Rocon *et al.*, 2005) and is therefore a low-mass system when compared to other sensors used in the field; for instance, Elble (2003) employed a 15 g triaxial piezoresistive accelerometer secured to a 57 g plastic splint in his experiments. It is important to use a low-mass sensor to reduce the effect of low-pass filtering on the detected signal. The main advantages of this system are that it is light, cheap and does not cause any discomfort to subjects. It is thus a powerful tool for monitoring biomechanical variables during physiological tremor movements.

Since gyroscopes provide absolute angular velocity in their active axis, a combination of two independent gyroscopes, placed distal and proximal to the joint of interest, is required. Gyroscopes placed anywhere along the same plane on the same segment give an almost identical signal. The gyroscopes can therefore be attached at positions where there is no skin and muscle movement (Moreno *et al.*, 2006). The system was configured to measure four movements of the upper limb: (1) elbow flexion-extension, (2) forearm pronation–supination, (3) wrist flexion-extension, and (4) wrist deviation. The main advantages of this system are that it is lighter, cheaper and does not cause discomfort to subjects. It could therefore be a powerful tool for monitoring biomechanical variables during tremor movements.

3.7.1 Biomechanical model of the upper arm

A biomechanical model of the upper arm in combination with WOTAS (see Case Study 8.1) was built taking into account the Leva, Zatsiorsky and Seluyanov tables (Leva, 1996) in order to define describe its kinematics and dynamics. These tables are one of the most widely accepted means of performing dynamic analysis in the field of biomechanics, in particular in sports and medical biomechanics. Leva adjustments were made in order to define accurately the anthropometric measurements required to obtain inertial parameters from Zatsiorsky tables.

A solid rigid model of the forearm was built with the information from the above-mentioned tables and parameterized following the Denavit–Hartenberg approach. In addition, a library was created to permit dynamic analysis of the system. This analysis was performed using the recursive algorithm (Fu, Gonzalez and Lee, 1987) so that the libraries could be built on a modular basis.

3.7.1.1 Denavit–Hartenberg parameterization

The model proposed considers the upper limb as a chain composed of three rigid bodies – the arm, the forearm and the hand – articulated on the rigid base formed by the trunk and linked by ideal rotational joints (Maurel, 1998). This representation relies on three assumptions: (1) the mechanical behaviour of the upper limb with respect to the trunk is independent of the rest of the human body; (2) each segment, including bones and soft tissues, has similar rigid body motions; and (3) the deformation of the soft tissues does not significantly affect the mechanical properties of a segment as a whole.

The hand was considered as a rigid extension of the forearm on the assumption that hand motion has a negligible effect on the broad motion dynamics of the upper limb. This made it necessary to determine a rigid-body equivalent to the hand and forearm assembly to be substituted in the rigid-body dynamic analysis. As a result, four rigid segments were defined in order to be able to analyse all the recorded degrees of freedom. Each segment is responsible for a degree of freedom: (1) elbow flexion–extension, (2) pronation–supination, (3) wrist flexion–extension and (4) wrist deviation.

Two of these segments are virtual (no mass and no length). Each segment has its own reference system (plus a coordinate frame for all of them) attached. Figure 3.16 shows the coordinate frames defined and the degree of freedom represented for each system.

The Denavit–Hartenberg parameters can be seen in Table 3.5. For rotary elements, the parameter θ determines the position of the joint. The table then indicates the relationship between the parameter and the physiological measured angle represented by β_i for each segment i. F_L means forearm length and H_L means hand length.

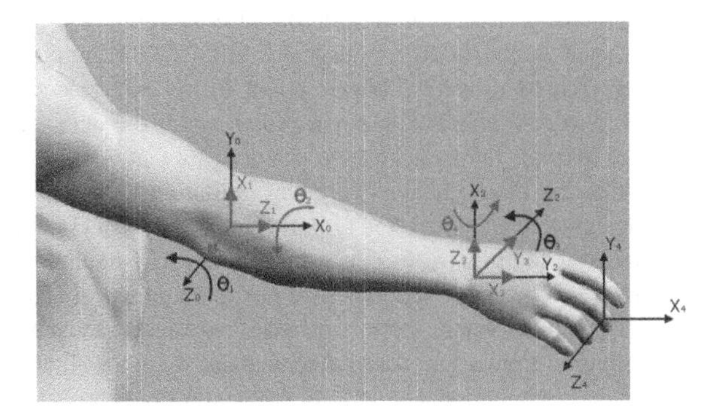

Figure 3.16 Solid model representation of the forearm

Table 3.5 DH parameters

Segment	d	a	θ	α
1. Elbow F/E	0	0	$\beta_1 + \pi/2$	$\pi/2$
2. Pronation	F_L	0	β_2	$\pi/2$
3. Wrist F/E	0	0	$\beta_3 + \pi/2$	$\pi/2$
4. Wrist Deviation	0	H_L	β_4	$\pi/2$

3.7.1.2 Biomechanical parameters per segment

Biomechanical parameters per segment were obtained from Leva (1996). All the inertial and mass parameters of a segment are defined below, using the following symbols: B_M, body mass; F_L, forearm length; F_M, forearm mass; H_L, hand length; H_M, hand mass; CoG_M, centre of gravity of each segment (obtained from Leva tables); and MI, inertia matrix. Segment 1 and segment 3 are virtual; they are only defined to cope with the degrees of freedom of elbow flexion–extension and wrist flexion–extension respectively. Therefore inertial and mass parameters for these segments are zero. However, when these segments are moved, the masses of the 'real' segments are moved. Thus

$$\begin{aligned} M_2 &= F_M = 1.6\frac{B_M}{100} \\ \mathrm{CoG}_2 &= [0, F_L(0.457 - 1), 0] \\ \mathrm{MI}_2 &= \begin{bmatrix} 0.08 & 0 & 0 \\ 0 & 0.09 & 0 \\ 0 & 0 & 0.15 \end{bmatrix} F_M F_L^2 \end{aligned} \tag{3.35}$$

$$\begin{aligned} M_4 &= M_M = 0.6\frac{B_M}{100} \\ \mathrm{CoG}_4 &= [0, H_L(0.79 - 1), 0] \\ \mathrm{MI}_4 &= \begin{bmatrix} 0.55 & 0 & 0 \\ 0 & 0.42 & 0 \\ 0 & 0 & 0.66 \end{bmatrix} H_M H_L^2 \end{aligned} \tag{3.36}$$

Table 3.6 Mean values of the torque and root mean square (RMS) value of power estimated in finger to nose and outstretched arm tasks

Movement	Finger to nose	Outstretched arm
Elbow Flexo-extension	1.9 N m, 0.2 W	1.2 N m, 0.01 W
Forearm Prono-supination	3.7 N m, 1.8 W	1.9 N m, 0.2 W
Wrist Flexo-extension	0.4 N m, 0.08 W	0.2 N m, 0.03 W
Wrist Deviation	1.1 N m, 0.4 W	0.5 N m, 0.04 W

The computational algorithm used is based on the Newton–Euler equations of motion described in Fu, Gonzalez and Lee (1987). Thanks to their recursive implementation, these equations of motion are the most efficient set of computational equations for running on a uniprocessor computer, so that implementation in real-time is possible.

3.7.2 Results

This analysis is intended to estimate the torque and power of the tremorous movement in each joint of the upper limb based on the information provided by the gyroscopes.

Active orthoses are intended to counteract tremor by applying controlled forces. Torque is an essential parameter in the choice of the actuator technology that will be used by powered orthoses. Special care should be taken with this parameter since it presents a dynamic behaviour. The actuator technology that will drive the orthosis must be able to apply the same torque characteristics. Table 3.6 summarizes the mean value of torque estimated in each joint of the upper limb for the tasks of stretching out the arm and putting finger to nose. These tasks are shown because they are the ones in which maximum values of tremor activity were registered.

The other important parameter is the power that the device can absorb. The amount of power consumed in relation to tremor is one of the key parameters that need to be taken into account in the design of these devices. The power at the joint plus the performance of the devices will also determine the battery capacity. Tremor is assumed to be a stationary movement and, leaving aside the viscous coefficient of joint braking, there is no effective work done on the joint. i.e. why the RMS value of the power estimated for the tremorous movement during the putting finger to nose and stretching out arm tasks are also presented in Table 3.6.

The results of this study show the basis of the dynamics of tremorous movement in each joint of the upper limb, information i.e. required for the design of portable active upper limb exoskeletons (see Case Study 8.1).

REFERENCES

Asada, A., 1983, 'A geometrical representation of manipulator dynamics and its application to arm design', *Journal of Dynamic Systems, Measurement and Control* **105**: 131–135.

Bergamasco, M., Allotta, B., Bosio, L., Ferretti, L., Parrini, G., Prisco, G.M., Salesdo, F., Sartini, G., 1994, 'An arm exoskeleton system for teleoperation and virtual environments applications', in *Proceedings of the IEEE International Conference on Robotics and Automation*, vol.2, pp., 1449–1454.

Carignan, C., Liszka, M., Roderick, S., 2005, 'Design of an arm exoskeleton with scapula motion for shoulder rehabilitation', in *Proceedings of the 12th IEEE International Conference on Advanced Robotics*, pp. 524–531.

Chiu, S.L., 1985, 'Task compatibility of manipulator postures', *The International Journal of Robotics Research* **7**(5): 13–21.

Crowningshield, R., Pope, M.H., Johnson, R.J., 1976, 'An analytical model of the knee', *Journal of Biomechanics* **9**: 397–405.

Elble, R.J., 2003, 'Characteristics of physiologic tremor in young and elderly adults', *Clinical Neurophysiology* **114**: 624–635.

Erdemir, A., McLean, S., Herzog, W., van den Bogert, A.J., 2007, 'Model-based estimation of muscle forces exerted during movements', *Clinical Biomechanics* **22**(2): 131–154.

Forner-Cordero, A., Koopman, H.F.J.M., van der Helm, F.C.T., 2004, 'Use of pressure insoles to calculate the complete ground reaction forces', *Journal of Biomechanics* **37**(9): 1427–1432.

Forner-Cordero, A., Koopman, H.F.J.M., van der Helm, F.C.T., 2006, 'Inverse dynamics calculations during gait with restricted ground reaction forces information from pressure insoles', *Gait and Posture* **23** (2): 189–199.

Fu, K.S., Gonzalez, R.C., Lee, C.S.G., 1987, *Robotics: Control, Sensing, Vision, and Intelligence*, McGraw-Hill Book Company.

Goldstein, H., Poole, C.P., Safko, J.L., 2002, *Classical Mechanics*, Addison Wesley, Reading, Massachusetts.

Hatze, H., 2002, 'The fundamental problem of myoskeletal inverse dynamics and its implications', *Journal of Biomechanics* **35(1)**: 109–115.

Hesse, S., Uhlenbrock, D., 2000, 'A mechanized gait trainer for restoration of gait', *Journal of Rehabilitation Research and Development* **37**(6): 701–708.

Holland, J.H., 1975, *Adaptation in Natural and Artificial Systems*, The University of Michigan Press.

Koopman, B., 1989, 'The three-dimensional analysis and prediction of human walking', PhD Thesis, University of Twente, Enschede, The Netherlands.

Koopman, B., Grootenboer, H.J., de Jongh, H.J., 1995, 'An inverse dynamics model for the analysis, reconstruction and prediction of bipedal walking', *Journal of Biomechanics* **28**(11): 1369–1376.

Kreutz-Delgado, K., Long, M., Seraji, H., 1992, 'Kinematic analysis of 7 dof manipulators', *International Journal of Robotics Research* **11**(5): 469–481.

Kurfess, T.R., (ed.), 2005, *Automatics and Automation Handbook*, CRC, Press, Boca Raton, Flocida.

Leva, P., 1996, 'Adjustments to zatsioorsky-seluyanov's segment inertia parameters', *Journal of Biomechanics* **29**(9): 1223–1230.

Lewis, J., Lew, W., Patrnchak, C., Shybut, G., 1984, 'A new concept in orthotic joint design. The Northwestern University knee orthosis system', *Orthotics and Prosthetics* **37**(4): 15–23.

Li, G., Pierce, J.E., Herndon, J.H., 2006, 'A global optimization method for prediction of muscle forces of human musculoskeletal system', *Journal of Biomechanics* **39**(3): 522–529.

Maurel, W., 1998, '3D modeling of the human upper limb including the biomechanics of joints, muscles and soft tissues', PhD Thesis, Ecole Polytechnique Federale de Lausanne, Switzerland.

Michael, J.W., McMillan, A.G., Kendrick, K., 2003, 'Stance control orthoses: history, overview and case example of improved KAFO function', *Alignment*, 60–70.

Moore, K.L., 1992, *Clinically Oriented Anatomy*, Williams & Wilkins, Baltimore, Maryland.

Moreno, J.C., Rocon, E., Ruiz, A., Brunetti, F., Pons, J.I., 2006, 'Design and implementation of an inertial measurement unit for control of artificial limbs: application on leg orthoses', *Sensors and actuators B* **118**: 333–337.

O'Connor, J.J., Shercliff, T.L., Biden, E., Goodfellow, J.W., 1989, 'The geometry of the knee in the sagittal plane', *Proceedings of the Institution of Mechanical Engineers* **203**(H4), 223–233.

Pons, J.L., Ceres, R., Pfeiffer, F., 1999, 'Multifingered dextrous robotics hand design and control: a review', *Robotica* **17**: 661–674.

Rocon, E., Ruiz, A.F., Pons, J.L., Belda-Lois, J.M., Sánchez-Lacuesta, J.J., 2005, 'Rehabilitation robotics: a wearable exo-skeleton for tremor assessment and suppression', *In Proceedings of the International Conference on Robotics and Automation*, pp., 241–246.

Schiele, A., van der Helm, F.C.T., 2006, 'Kinematic design to improve ergonomics in human machine interaction', *IEEE Transactions on Neural Systems and Rehabilitation Engineering* **14**(4): 456–469.

Schiele, A., Visentin, G., 2003, 'The ESA human arm exoskeleton for space robotics telepresence', *in 7th International Symposium on Artificial Intelligence, Robotics and Automation.*

Sciavicco, L., Siciliano, B., 2003, *Modelling and Control of Robot Manipulators*, Springer-Verlag.

Seth, A., Pandy, M.G., 2007, 'A neuromusculoskeletal tracking method for estimating individual muscle forces in human movement', *Journal of Biomechanics* **40**(2): 356–366.

Todd, B.A., Woods, H.C., Shanckelford, L.C., 1994, 'Mass centers of body segments', in *Proceedings of the 16th Annual International Conference of the IEEE*, pp. 335–336.

Viladot Voegli, A., (ed.), 2001, *Lecciones Bsicas de Biomecnica del Aparato Locomotor*, Springer.

Walker, P.S., Kurosawa, H., Rovick, J.S., Zimmerman, R.A., 1985, 'External knee joint design based on normal motion', *Journal of Rehabilitation Research and Development* **22**(1): 9–22.

Weinberg, A.M., Pietsch, I.T., Helm, M.B., Hesselbach, J., Tscherne, H., 2000, 'A new kinematic model of pro- and supination of the human forearm', *Journal of Biomechanics* **33**: 487–490.

Williams II, R.L., *et al.*, 1998, 'Kinesthetic force/moment feedback via active exoskeleton', *in Proceedings of the IMAGE Conference*, Scottsdale, Arizona.
Winter, D.A., 1990, *Biomechanics and Motor Control of Human Movement*, John Wiley & Sons, Ltd.
Wismans, J., 1980, 'A 3D mathematical model of the knee-joint', *Journal of Biomechanics* **13**(8): 677–685.
Yasutomi, T., Nakatsuchi, Y., Koike, H., Uchiyama, S., 2002, 'Mechanism of limitation of pronation/supination of the forearm in geometric models of deformities of the forearm bones', *Clinical Biomechanics* **17**: 456–463.
Yoon, S.K., Jangwook, L., Sooyong, L., Munsang, K., 2005, 'A force reflected exoskeleton-type masterarm for human–robot interaction', *IEEE Transactions on Systems, Man and Cybernetics, Part A* **35**(2): 198–212.
Zajac, F.E., Neptuno, R.R., Kautz, S.A., 2002, 'Biomechanics and muscle coordination of human walking. Part I: introduction to concepts, power transfer, dynamics and simulations', *Gait and Posture* **16**(3): 215–232.
Zajac, F.E., Neptuno, R.R., Kautz, S.A., 2003, 'Biomechanics and muscle coordination of human walking. Part II: lessons from dynamical simulations and clinical implications', *Gait and Posture* **17**(1): 1–17.

4

Human–robot cognitive interaction

L. Bueno, F. Brunetti, A. Frizera and J. L. Pons

Bioengineering Group, Instituto de Automática Industrial, CSIC, Madrid, Spain

4.1 INTRODUCTION TO HUMAN–ROBOT INTERACTION

The symbiotic relationship between humans and robots transcends the boundaries of simple physical interaction. It involves smart sensors, actuators, algorithms and control strategies capable of gathering and decoding complex human expressions or physiological phenomena. Once this process is complete, robots use the information to adapt, learn and optimize their functions, or even to transmit back a response resulting from a cognitive process occurring within the robot.

A *cognitive process* is a sequence of tasks including reasoning, planning, and finally the execution of a previously identified problem or goal. Originally this concept was restricted to living creatures, but now it can also be applied to smart robots that accomplish the above-mentioned sequence of tasks. The *human–robot cognitive interface* is the link between human and robot, in which the information regarding these processes is acquired and transmitted bidirectionally.

New human–machine interfaces have recently been developed to support all the flow of information that a smart system needs in order to perform its function. Scientists consider this cognitive interaction fundamental. Thus terms like human–machine interface (HMI), human-computer interface (HCI) and brain–computer interface (BCI) can be found, all relating to this cognitive interaction. The search for a natural communication channel is a constant object of research, especially in the area of service robots (Ratanaswasd *et al.*, 2005) and (Scholtz, 2002). In this book, the term cHRI stands for cognitive human–robot interaction and cHRi stands for cognitive human–robot interface.

New trends in cHRI are moving in the direction of sourcing the information directly from the human cognitive processes involved in the normal execution of tasks. These are called natural interfaces. The rationale for this approach comes from Sharma, Pavlovic and Huang (1998):

- *Biological reasons.* cHRI systems seek to take advantage of the natural control mechanisms fully optimized in humans. Moreover, a lot of information is lost in the translation of biologically executed tasks into discrete events, e.g. natural movements or gestures into buttons or joysticks.

Wearable Robots: Biomechatronic Exoskeletons Edited by José L. Pons

- *Practical reasons.* Delays are introduced when natural cognitive processes are encoded into an imposed sequence of tasks. In addition, a training phase is needed to teach the user to generate these nonnatural commands or to map a cognitive process into a new set of outputs. Both factors, the delays and the mapping, can induce fatigue in the user, both at a musculoskeletal level and at a mental level. These factors can be obviated if the natural outputs of a cognitive process are used for cHRI.
- *Rehabilitation.* One of the main applications of wearable robotics is rehabilitation. Interacting directly with the phenomena involved in the cognitive process is a means to excite them and assess the evolution of the rehabilitation therapy.

The design of a cHRI depends on the kind of information to be transmitted. Figure 4.1 depicts the cHRI concepts. There are unidirectional interfaces to control robots, bidirectional interfaces or closed-loop interaction, e.g. in haptic applications. Unidirectional interfaces are defined as interfaces where the subject has no information about the state of the system immediately after a command is sent. Bidirectional interfaces provide feedback immediately after the command is sent by the user. Closed-loop systems are defined as continuous symbiotic interaction between the user and robotic systems. The closed-loop interaction can enhance the user's situation awareness, increasing efficiency and achieving control in a way that is more natural to the user (Scholtz, 2002).

The role of the different components of a cHRI and the interaction environment are also depicted in Figure 4.1. The role of the sensors is to measure the human-related phenomena, while the actuators have to transmit the robotic cognitive information to the user to complement his sensory information about the task.

The interaction environment is a key issue for natural cHRIs. The aim of modern interfaces is to avoid saturating the biological cognitive channels, and therefore new ones are being explored.

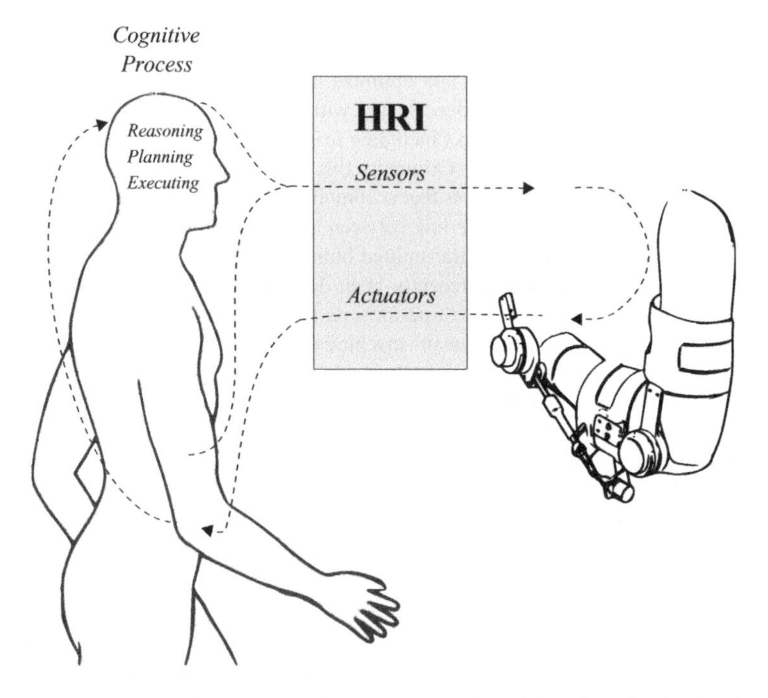

Figure 4.1 The information flow from user reasoning to final execution of the action, showing the signals acquired though the information pathway

Figure 4.1 shows three levels of interaction: one related to reasoning and planning, one related to muscle activity and one related to the wearer's motion. The planning level interaction can be accomplished by monitoring brain activity using different techniques, e.g. electroencephalography or brain-implanted electrodes (see Section 4.2 and Case Studies 4.7 and 8.3 for further information). The muscle activity level uses muscle electrical activity, i.e. EMG, to command the devices. This activity can be acquired even without limb movement (see Section 4.3 for additional details). The movement-related level of interaction uses kinematic and kinetic information from the subject as control inputs (see Section 4.4 and Case Studies 4.5, 4.6 and 4.8 for more information). Multimodal approaches propose diversified use of these channels in order to gather more realistic and robust information and to gain a better understanding of the phenomena by means of data fusion techniques.

There are many common concepts in the field of intelligent interfaces, e.g. data, information, gesture, intention and others. Data are what are obtained from an observed object or phenomenon. They comprise it as a set of descriptors that are time-correlated with the phenomenon, e.g. the sensor signal. Information is what is needed and can be extracted from the data. One example of this transformation process is the feature extraction phase in pattern recognition systems. Other terms related to this cognitive process are:

- *Feature.* Information extracted from sensor data after applying processing algorithms, which can be used to classify the dataset into different groups.
- *Event.* An event is detected by a change in a system's input or output feature, defining temporally the occurrence of a particular action (unintentionally or intentionally) taken by the user.
- *Pattern.* An ordered sequence of values or events describing the behaviour of a system or process. It can be used as a reference to describe a process and is often observed.

Biomechanical information can be either static or dynamic. A posture is defined as a static configuration of the human body segments that can be interpreted as information. In contrast, a temporal sequence of postures or a transition between positions can be interpreted as a gesture.

The complete process of implementing a cHRI is very complex. Roughly speaking, it may be said to comprise detection and processing of features, overall interpretation of the semantic meaning of the observed phenomenon, and consequent action and feedback. In some applications the robotic system tries to predict user action. The basis for detection of the user's intention can be a change of EEG patterns, an EMG signal or even the early stages of a movement.

The context is another possible input in cHRI. User conditions (often unstable), the environment and the historical evolution can all affect this interaction, so that schemes of adaptation, learning and evolution should be considered.

The aim of this chapter is to present different approaches and systems for cHRI, including different case studies. Three main levels of cognitive interaction have been identified. These levels imply three different wearable sensing technologies: EEG, EMG and biomechanical (kinetic and kinematic) sensors. The physical phenomena involved, the recommended sensing systems and the processing algorithms are presented for each interface technology. Case studies are presented at the end of the chapter to illustrate some of the concepts presented there.

4.2 cHRI USING BIOELECTRICAL MONITORING OF BRAIN ACTIVITY

EEG is a technique for recording the electrical activity of the brain using a set of electrodes positioned on the scalp. The technique was first used on humans in 1924 by the German psychiatrist Hans Berger, who called the signal record an 'Elektroenkephalogram' (Malmivuo and Plonsey, 1995).

This section discusses some aspects of the brain activity, in particular EEG as a control interface. It starts with a brief physiological description of brain activity, including its functional structure and signal generation. Following this description, the EEG signal is introduced and some of its parameters are presented. Also presented are the use of EEG as a control signal and current trends in this area.

4.2.1 Physiology of brain activity

The brain is the most complex structure in the human body. It consists of successive foldings of neural tissue resulting in a structure with specialized substructures. These substructures are interconnected with one another, forming a highly intricate circuit. They are not yet fully understood.

The functional units of the neural tissue are the neurons. The neuron structure can be seen in Figure 4.2, which presents the synapses and other structures, such as dendrites and axons.

In all cells of the human body, the potential varies across their membranes. This variation is due to differences in the concentrations of ions, which are caused by the internal activity of the cell. In excitable cells, i.e. neurons and muscles, this difference is due to the excitability of the membrane. The membrane of the neuron can receive excitatory stimulus via special structures called dendrites. The function of the neuron is to integrate the stimuli in the different dendrites and change the

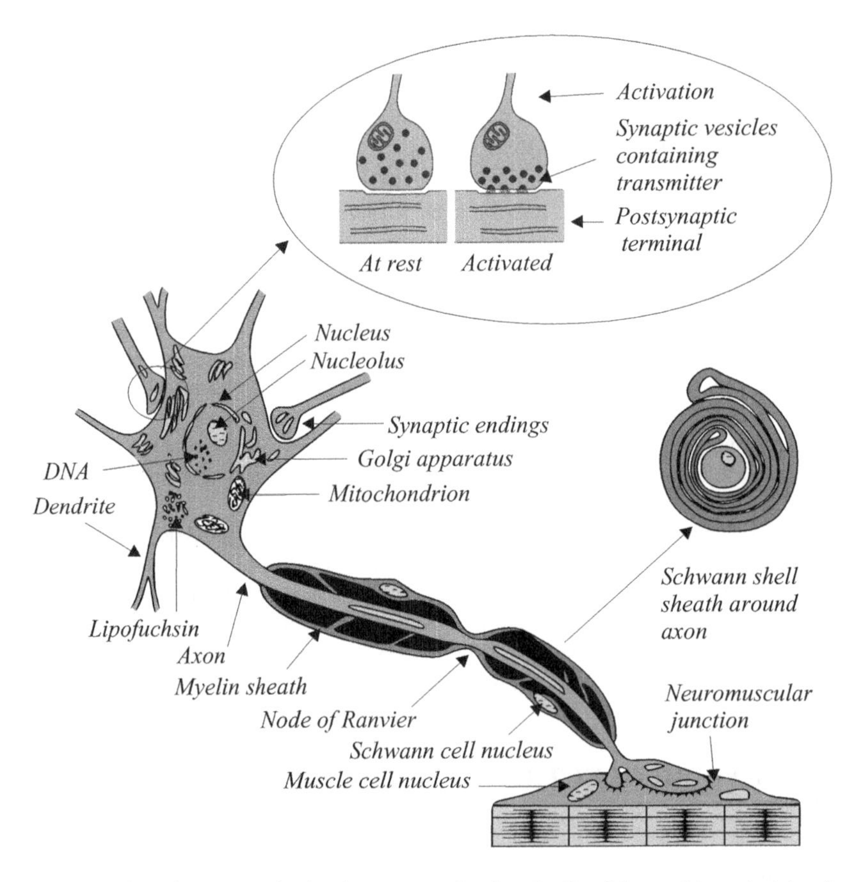

Figure 4.2 Depiction of neuron and related structures, showing details of the working principle of synaptic conduction and the Schwann cells responsible for protecting the neuron fibres

polarization of the axon according to these integrated stimuli. Stimuli between neurons are excited through a structure called a synapse. Synapses are small spaces between axons and dendrites. When a stimulus occurs, a substance called a neurotransmitter is segregated in the synapse by the axon and is perceived by the dendrite as a stimulus. The intensity of the stimulus depends on the quantity of neurotransmitter present at the synapse (Malmivuo and Plonsey, 1995).

Although the transmission of information between neurons involves neurotransmitters, it is also an exchange of ions, and this implies an electrical impulse. Propagation of the impulses in the nervous tissue generates a potential in this tissue that can be measured using electrodes.

The brain comprises two main types of nerve tissue: *white matter* and *grey matter*. The grey matter is located on the outer part of the brain and is also known as the *cortex*. The white matter is inside the brain and is responsible for interconnecting areas of the cortex and connecting the cortex to other structures of the CNS, such as the spinal cord or the cerebellum.

The cortex can be divided according to function or location. In terms of location, the *frontal*, the *temporal*, the *parietal* and the *occipital* lobes can be identified. In terms of function, the *visual cortex*, the *auditory cortex* and the *sensorimotor cortex* can be identified (see Figure 4.3). Changes in the activity of these areas are most closely related to their specific function, but due to volume conduction, the activity of one area may be spread through other areas.

The brain activity, defined by the de/repolarization of neuron groups in the mass of the brain, can either cancel out or sum up, forming electric potentials that can be observed on the scalp. Research in this area seeks to identify the structures inside the brain and their functions, as the brain is the least known organ of the human body.

EEG signals have been studied in the context of their use as control signals in technical aids for disabled people. The reason for this is that in some cases of cerebrovascular accident, spinal cord injury or degenerative diseases such as amyotrophic lateral sclerosis (ALS), the patient has no control of his/her muscular system or can only control a small muscle group, e.g. the eyes. In such cases the patient cannot communicate, even with an intact and fully functional brain. With the aid of an interface like this, a patient can communicate using a computer or even regain some independence, if the interface is used to control an intelligent environment.

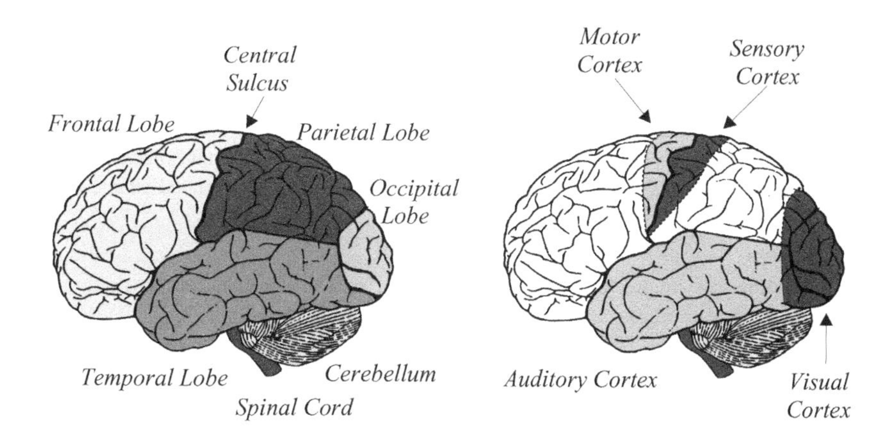

Figure 4.3 Division of the brain. The main lobes of the brain are divided according to their location. Each brain lobe has a relatively specialized function related to the main information received from the body. The central sulcus is the frontier between the frontal and parietal lobes. The occipital lobe specializes in the processing of visual information, the temporal lobe is mainly concerned with auditory stimuli and the parietal lobe is concerned with integration of the different stimuli. In the vicinity of the central sulcus are the sensory and motor cortices, the former on the parietal border of the central sulcus and the latter on the frontal border

The development of interfaces like the one described here raises operational issues, for instance the placement of the acquisition system on the user. This requires the assistance of a trained carer and long-term interface stability, long-term stability here meaning a matter of hours. The interface, which in most instances uses wet electrodes, suffers from drying of the electrolytic contact gel. There are other issues related to user safety; as the user is connected to an electronic device through electrodes that have a relatively small impedance to the skin, any fault in the power supply poses the danger of a potentially lethal electric shock. Then there are other factors that influence the signal, such as fatigue, hormone levels, the time of day, the environment itself, etc. These factors can influence the signal negatively, resulting in patterns that the interface algorithms have difficulty in recognizing.

4.2.2 Electroencephalography (EEG) models and parameters

EEG can be described using a series of parameters in both the frequency and the time domains. The models assume that the brain is a conductive mass and that neurons act as small current sources. To achieve a realistic representation of this model, the activity of millions of neurons would need to be calculated, which is not feasible since there are far more sources (neurons) than observed variables (electrode potentials). The models then use only a resultant dipole, also known as an optimal dipole, for a given distribution. In order to calculate more than one dipole, the original signals must be decomposed using a decomposition algorithm, e.g. independent component analysis (ICA). The distribution of the signals in the original space is used to calculate the dipoles for each of the independent components.

Computation of the dipoles requires a geometric and conductive model of the head and its structures. The most simplified model takes the head as a set of three concentric spheres with different conductivities. Each sphere corresponds to a different structure of the head. The innermost sphere corresponds to the brain mass, the intermediate to the skull and the outermost to the scalp. Given the symmetry of this structure, there are analytic expressions to determine the potentials in the scalp on the basis of a given dipole in a given position inside the brain mass.

Models based on clinical imaging with a more realistic depiction of the brain structure are also available. However, they are less practical in that they are typically finite element models and the model for each dipole has to be calculated in each desired position to determine the potentials. For a more detailed description of the calculation of the potentials using the three-sphere model, the reader is referred to Salu *et al.* (1990).

As noted earlier, parameters defining the brain activity are both time- and frequency-related. In terms of frequency, the EEG is divided into bands that represent a given state or are characteristic of the activity in a given area of the brain. The frequency bands and their related characteristics are as follows:

- The δ band comprises frequencies from 0.1 to 3.5 Hz. This is the predominant rhythm in deep sleep stages and is also present in lethargic and nonattentive situations.
- The θ band comprises frequencies from 3.5 to 7 Hz. This is the predominant rhythm in the rapid eye movement (REM) stage and the early stages of sleep. It also plays an important role in the memory processes, (Ward, 2003).
- The α band comprises frequencies from 8 to 13 Hz. It was first observed by Hans Berger and is the most active rhythm in relaxed states. It is predominant in the occipital and parietal areas but can also be observed throughout the brain mass as a result of volume conduction. Classically, the α band was associated with mental idleness, but recent studies claim that it is associated rather with suppression of attention in certain areas (Ward, 2003). It therefore has more to do with attention, than with idleness.

- The μ rhythm is present in the sensorimotor cortex (SMC) and shares the same frequency range as the α band, but it presents different characteristics, reflecting activity in the SMC.
- The β band comprises frequencies from 13 to about 30 Hz and is associated with attention to stimuli and with mathematics-related mental activity.
- The γ band comprises frequencies from 30 to about 70 Hz and is associated with the internal states of the brain. There is a correlation of the γ and θ bands in memory intensive processes and there is a correlation of the γ and α bands in selective attention processes (Ward, 2003).

The relationships presented above were first identified in the clinical environment and used for diagnosis. For a diagnostic approach, the frequency distributions of a patient are compared to the distributions of a nonpathological pattern. If a difference is detected, an expert clinician can identify the pathology. A relatively recent clinical use of frequency bands is so-called biofeedback, where a system presents the patient with a picture of his/her mental activity bands. These devices are intended to train the patient to reinforce certain bands such as α or the low-frequency β to compensate for some psychosomatic diseases, e.g. attention deficit disorder.

This knowledge can only be partially transferred to cHRI development, since these systems are intended to identify a mental state rather than a disease. However, new applications of the frequency bands of EEG are emerging in this scenario.

Information about frequency distribution can be used to control a device since it is roughly stable for that mental state, although this can change in response to a large number of variables, such as the ones listed in Section 4.2.1.

In addition to the frequency characteristics listed above, the EEG signal can also be characterized by its potentials. In clinical practice the EEG test includes the presentation of stimuli in order to measure a possible reaction. Potentials are more sensitive to stimuli than frequency characteristics.

There are different types of potentials, namely:

- *Event related de/synchronization* (ERD/ERS) is a change in potentials caused by synchronized polarization of a group of neurons in a certain area of the brain, triggered by an event. The event can be of any kind and the localization of the ERD/ERS depends on the stimulus.
- *Readiness potential* is a change in potentials related to mental preparation for a motor task. It occurs at the contralateral SMC about 750–500 ms before the movement onset.
- *Movement related potential* is a change in potentials directly related to the execution of motor tasks. It occurs at the contralateral SMC about 500 ms after the movement onset.
- *Error potential* is a change in potentials directly related to the conscious identification of a wrong decision by the subject. It can be visualized when the user makes a decision and realizes that it was a bad one, even when there is no instantaneous feedback. The error potential can be observed in EEG even before the EMG onset in a movement-related task (Dornhege, 2006).

The use of these characteristics as control signals and the algorithms used to identify them are addressed in detail in the following section.

4.2.3 Brain-controlled interfaces: approaches and algorithms

The main purpose of implementation of control interfaces based on EEG signals is to generate a set of features that can be used to identify the signal as a member of a class. The features can be extracted from one or more of the parameters presented in Section 4.2.2.

There are various different approaches to implementing a cHRI based on the EEG signal, but they can be roughly categorized in two main groups: those based on the subject's response to a given stimulus and those based on the subject's spontaneous activity. The former may be based either on evoked potentials following the presentation of a stimulus to the subject or on the visual cortex's ability to synchronize its main frequency to the frequency of a stimulus.

Systems based on spontaneous activity use signal characteristics that are generated spontaneously by the subject. This interface has some advantages over the first in that the moment in time when the user generates a control signal is controlled not by the system but by the subject. The response in this approach is closer to the way that a user controls a device using the normal paths of the nervous system.

In order to achieve this interaction, the characteristics of the EEG signal must be extracted from raw data. The algorithms depend on the type of characteristic to be extracted. To identify frequency characteristics, there are a number of approaches such as fast Fourier transform (FFT) coefficients, bandpass filtered signal energy or scale in a wavelet decomposition. These values provide information about the frequencies at a given electrode. To gain information about mental state, it is necessary to add the spatial information, which is obtained from the electrode position. With this information it is possible to generate a set of primitives that characterize the mental state (Bashashati, Ward and Birch 2005; García, Ebrahimi and Vesin 2003).

The use of filters is the most basic signal processing approach to separating the signal into different bands. The filters may be classified into two main filter architectures: finite impulse response (FIR) filters and infinite impulse response (IIR) filters. FIR filters use a polynomial expression based on a number of past states of the input signal. IIR filters use a polynomial based on a given number of input and output states. The polynomials can be generated using different algorithms and the magnitude and phase responses of the filter are closely related to the algorithm that is used to generate the filter coefficients.

There is another approach based on the projection of EEG signals to another space where the features are more suitable for extraction. Projection algorithms utilize a variety of techniques:

- *Principal component analysis* (PCA). The PCA algorithm is derived from the projection of the signal matrix over its own matrix of eigenvectors, suitably arranged according to their eigenvalues. The data matrix resulting from this projection is in an orthogonal space and the resulting components are ordered by variance. This transformation is sometimes used to adjust the dimensionality of the problem and so generate a smaller data matrix with almost the same variance as the original matrix. The PCA algorithm is represented in Equations (4.1) to (4.5). Equation (4.6) can be used instead of Equation (4.4) to normalize the variance of the projected signals.

$$\boldsymbol{u} = \mathbf{W}\boldsymbol{x} \tag{4.1}$$

$$\mathbf{S} = \frac{1}{(n-1)}(\boldsymbol{x}-\mu)(\boldsymbol{x}-\mu)^{\mathrm{T}} \tag{4.2}$$

$$\mathbf{S} = \mathbf{E}\boldsymbol{\lambda}\mathbf{E}^{\mathrm{T}} \tag{4.3}$$

$$\mathbf{W} = \mathbf{E}_d^{\mathrm{T}} \tag{4.4}$$

$$r = \frac{\sum_{i=1}^{d}\lambda_i}{\sum_{i=1}^{n}\lambda_i} \times 100\,\% \tag{4.5}$$

$$\mathbf{W} = \Lambda^{-1/2}\mathbf{E}_d^{\mathrm{T}} \tag{4.6}$$

where $\boldsymbol{u}$ is the new dataset, $\mathbf{W}$ is the PCA projection matrix, $\boldsymbol{x}$ is the original data vector, μ is the mean of $\boldsymbol{x}$, $\mathbf{S}$ is the covariance matrix of the zero mean matrix $(\boldsymbol{x}-\mu)$, n is the number of channels of $\boldsymbol{x}$, $\boldsymbol{\lambda}$ is the vector of eigenvalues, $\mathbf{E}$ is the eigenvector matrix, d is the number of output channels (which must be smaller than or equal to n), $\mathbf{E}_d^{\mathrm{T}}$ is the eigenvector matrix formed by d vectors, r is the amount of information retained on reduction from n to d, and $\boldsymbol{\Lambda}$ is a diagonal matrix with the values of $\boldsymbol{\lambda}$ for the d vectors.

- *Independent component analysis* (ICA). The ICA algorithm is used when separation of a matrix of mixed signals into individual component signals is required. To accomplish this, a series of statistically independent signals is generated using an iterative algorithm. The projection matrix is then used to create a series of signals that are statistically independent, although the new space is not formed by orthogonal vectors. Another aspect of the ICA transformation is that since there is no information about the transformation beforehand, the order of the independent signals resulting from the transformation is random. Also, the variance of these signal components cannot be calculated, as there is always a transformation that can be applied to the components and can adjust the variance. The algorithms assume that all the components have equal variance and the mixing matrix only gives the relationship between the signal components. Since this is an iterative algorithm, there is no analytical expression as in the case of PCA (Equations (4.1) to (4.6)). The working principle of the algorithm is to solve the following equation, where $\boldsymbol{x}$ is the acquired data and $\boldsymbol{s}$ are the independent components:

$$\boldsymbol{x} = \mathbf{A}\boldsymbol{s} \tag{4.7}$$

Since $\boldsymbol{s}$ and $\mathbf{A}$ are unknown, there are assumptions about the nature of the components:

  - The components of $\boldsymbol{s}$ are statistically independent.
  - The components have *non-Gaussian* distributions. If the components, $\boldsymbol{s}$, have Gaussian distributions, any orthogonal transformation applied to these variables will result in the same distribution and the mixing matrix will be unidentifiable. If a specific distribution is assumed for the variables, the algorithm can be simplified.
  - The unknown mixing matrix is square. This assumption is made in order to generate an independent time series from the EEG data by inverting $\mathbf{A}$, say $\mathbf{W}$, and applying the following equation:

$$s = \mathbf{W}x \tag{4.8}$$

For more information on the algorithms for determining the ICA projection matrix, the reader is referred to Hyvärinen and Oja (2000).

- *Common spatial patterns* (CSP). The CSP algorithm is a projection algorithm that generates characteristics for classification. It is used to generate variance patterns in the projected signals depending on the input classes. To create the projection matrix, the algorithm uses the covariance matrices of signals of both classes. The matrices are then summed and decomposed to generate a projection that enhances the differences between the two classes in the projected output (Dornhege, 2006). This algorithm uses the properties of two simultaneously diagonalized random variable covariance matrices (Fukunaga, 1990).

The projected signals present different variance patterns depending on the relationship of the input to the classes used to generate the projection matrix. The expressions used to generate the projection matrix are given by

$$R_a^i = \frac{(V_a^i V_a^{i\mathrm{T}})}{\mathrm{trace}(V_a^i V_a^{i\mathrm{T}})} \tag{4.9}$$

$$R_a = \langle R_a^i \rangle \mathrm{epochs} \tag{4.10}$$

$$\mathbf{R}_c = R_a + R_b \tag{4.11}$$

$$\mathbf{R}_c = \mathbf{B}_c \boldsymbol{\lambda} \mathbf{B}_c^{\mathrm{T}} \tag{4.12}$$

$$\mathbf{W} = \frac{\boldsymbol{\lambda}^{-1/2}}{\mathbf{B}_c^{\mathrm{T}}} \tag{4.13}$$

$$\mathbf{S}_a = \mathbf{W}\mathbf{R}_a\mathbf{W}^{\mathrm{T}}, \quad \mathbf{S}_b = \mathbf{W}\mathbf{R}_b\mathbf{W}^{\mathrm{T}} \tag{4.14}$$

$$\mathbf{S}_a = \mathbf{U}\psi_a\mathbf{U}^{\mathrm{T}}, \quad \mathbf{S}_b = \mathbf{U}\psi_b\mathbf{U}^{\mathrm{T}} \tag{4.15}$$

$$\psi_a + \psi_b = \mathbf{I} \tag{4.16}$$

$$\mathbf{P}^{\mathrm{T}} = \mathbf{U}^{\mathrm{T}}\mathbf{W} \tag{4.17}$$

$$Z^i = \mathbf{P}^{\mathrm{T}}V^i \tag{4.18}$$

where V_a^i is the raw data of class a on the epoch i, R_a^i is the variance-normalized data of class a on the epoch i, R_a is the mean across all epochs of class a, $\mathbf{R}_c$ is the composite matrix of the two $\mathbf{R}$ matrices for classes a and b, $\mathbf{B}_c$ and $\boldsymbol{\lambda}$ are the eigenvector and eigenvalue matrices of $\mathbf{R}_c$, $\mathbf{W}$ is a whitening transformation matrix, $\mathbf{S}_a$ and $\mathbf{S}_b$ are the whitened data matrices, $\mathbf{U}$ and ψ are the eigenvectors and eigenvalues of the whitened matrices $\mathbf{S}$, $\mathbf{I}$ is the unitary matrix, $\mathbf{P}$ is the projection matrix, and Z^i is the projected data of epoch i. The characteristics are extracted from the variance distribution of the lines of the projected data Z^i. These variances are associated with the input classes.

Projection algorithms include a static projection matrix. Since the EEG signals are not stationary, these algorithms can only be successfully applied for a limited time until the signals change and the system cannot identify the correct classes any more. In this scenario, an adaptive approach must be used, for instance a Kalman filter. Kalman filters are adaptive systems that can estimate the signal in a noisy environment on the basis of past samples of the signal and the statistical characteristics of the noise (Omidvarnia *et al.*, 2005).

Another feature that can be used is the fitting of an optimal dipole to the potentials of the scalp. Any of the models cited in Section 4.2.2 can be used for this purpose. In this approach, the task can be identified by the location of the optimal dipole in the brain mass (Qin, Ding and He, 2004).

The features extracted using frequency analysis can vary with time. The adaptation of a classifier may be more successful as the activity changes, since the algorithm used to extract is not dependent on the input classes. The use of frequency characteristics of the EEG signal to control a wheelchair is discussed in Case Study 8.3.

A series of classifiers can be used. These range from statistical classifiers like Bayesian networks (Hoya *et al.*, 2003), hidden Markov models (Obermaier, Müller and Pfurtscheller, 2003) and Gaussian classifiers (Omidvarnia *et al.*, 2005). Other approaches use linear discriminant analysis (LDA) (Obermaier, Müller and Pfurtscheller, 2003).

4.3 cHRI THROUGH BIOELECTRICAL MONITORING OF MUSCLE ACTIVITY (EMG)

The skeletal muscles are the natural actuators responsible for human motion and play a key role in this H–R interaction. That role can differ depending on the application of the WR. For instance, in the framework of empowering exoskeletons, the WR controlling function acts as a muscle function amplifier. In the context of rehabilitation robotics, the object may be to activate and train a group of skeletal muscles.

Assessment of muscle activity and understanding of neuromotor control strategies are both topics of interest in WRs. Briefly, the interest lies in:

- the potential use of PNS signals to control robots;
- a simple, optimized natural control mechanism that may be translated to robotic control;

- understanding of muscle and neuromotor activity in order to evaluate rehabilitation applications of wearable robots and HRI.

4.3.1 Physiology of muscle activity

In general, muscles are attached to the skeleton. Muscles apply force to joints and bones by contracting. Contraction may be voluntary, producing voluntary movements, or involuntary, creating movements like the one observed in pathological tremor.

Under normal conditions, muscle contraction is activated by the motor cortex. The control signals flow from the CNS to the PNS, finally reaching the muscle tissue. These signals are called action potentials (APs) and when they reach the muscle fibres, they cause the muscle to contract. A motor neuron is the structure responsible for transmitting these APs from the spinal cord to the muscle. It can innervate many muscles fibres which are activated synchronously when the motor neuron is excited. The fibres corresponding to motor neurons are located close to one another but not grouped together, as depicted in Figure 4.4. The number of fibres innervated by a single motor neuron can vary and is a factor in precision of movement. A motor unit (MU) is the combination of a single motor neuron and all the muscle fibres it innervates. Small MUs are responsible for precise movements; for instance, they can be found in eye muscles. Large motor neurons, with hundreds of innervated fibres, can be found in extremity muscles. The combination of the APs of all the muscle fibres innervated by one motor neuron is called a motor unit action potential (MUAP). A motor unit pool is the set of motor units that innervate a single muscle.

Basically, the execution of a movement consists of a neat activation sequence of several MUs innervating different muscles. The final execution of the movement depends on many MU-related parameters, e.g. twitch force, contractile speed, axonal conduction velocity, fatigue resistance, recruitment thresholds, firing rates and firing patterns, among others.

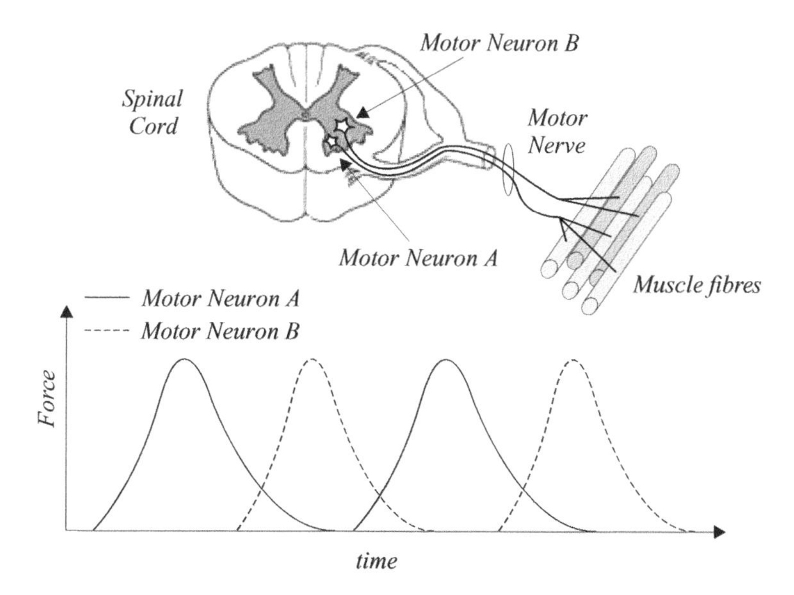

Figure 4.4 AP pathway. The sum of synchronous MUAPs can produce continuous contraction and hence force

During steady muscle contraction a group of MUs is activated. A different activation pattern is used to excite each MU. The rationale behind the activation of MUs is called recruitment strategy. There are two motor control strategies that are known to increase the force produced by the muscle contraction.

The first of these is the spatial recruitment strategy. Motor neuron excitation follows a fixed order (Henneman, 1957). When more muscle force is required, an additional MU is activated; Fuglevand presented a model describing this strategy. The model defines an activation threshold for each MU as a function of the intended force (Fuglevand, Winter and Patla, 1993). The recruitment threshold excitation (RTE) can be described by

$$\mathrm{RTE}(n) = 0.01 \exp\left[\frac{\ln(100\mathrm{MVC_{rec}})}{N_{\mathrm{MU}}} n\right] \tag{4.19}$$

where n is the n-motor neuron of the pool, N_{MU} is the total number of motor neurons in the pool and $\mathrm{MVC_{rec}}$ is the normalized percentage of the maximum contraction value (MVC) for the recruitment of the last motor neuron in the pool.

The second strategy is known as temporal recruitment. The firing rates of the active motor neurons are increased linearly with the increase in the intended force. The firing rate (FR) for each motor neuron n is described by

$$\mathrm{FR}(n) = g_e\,[E(t) - \mathrm{RTE}(n)] + \mathrm{MFR}, \qquad E(t) \geq \mathrm{RTE}_i \tag{4.20}$$

where g_e is the gain, common to all the motor neurons, and MFR is the minimum firing rate. How peak firing rates are related to $\mathrm{RTE}(n)$ is not yet known. Henneman (1965) demonstrated that MUs are recruited in order from the smallest to the largest; this is known as the size principle. However, not all muscle contractions follow this principle (Bosch and Huijing, 1996). The maximum firing rate for isometric contraction is between 20 and 45 pps as reported in Bigland-Ritchie *et al.* (1983). There are many details not considered by Equations (4.19) and (4.20), e.g. FR variability or fatigue. The neuromuscular control mechanism uses different strategies to compensate for fatigue, e.g. increasing the firing rate, additional spatial recruitment or motor unit rotation. Neuromotor control modelling is a complex task, and this section confines itself to a short introduction to the control of natural actuators to illustrate the rationale behind these natural mechanisms. For more detail see Merletti *et al.* (2001).

4.3.2 Electromyography models and parameters

Electromyography is the study of the evoked potentials generated during muscle contractions. The activity of the muscle can be assessed with this sensing technology, which is widely used to evaluate muscle and nerve conditions in the clinical field. Broadly speaking, there are two types of electromyography: surface EMG (sEMG) and intramuscular EMG (iEMG).

During the last few decades many research projects have used EMG to enable HMIs by means of robot control based on EMG signals (Arieta *et al.*, 2006; Ferris *et al.*, 2006). Most approaches have been based on sEMG, since it is a noninvasive technique, unlike intramuscular EMG. The reader will find a full description of these sensing technologies in Chapter 6.

Surface EMG signals are usually considered to be the sum of the evoked potentials of active MUs. The acquired sEMG signal is usually considered to be random. Extracting information on individual MUAPs is not trivial. Many models have been proposed for sEMG, e.g. the Anvolcon model (Blok *et al.*, 1999), the EMGsim model, (Farina, Crosetti and Merletti, 2001), the RRDsim model (Hermens *et al.*, 1992), the SiMyo model (Duchene and Hogrel, 2000) and the Fuglevand model (Fuglevand, Winter and Patla, 1993). However, these models do not include all the factors that affect sEMG signals, such as electrode configuration and location, fibre type, blood flow, subcutaneous tissue, skin, preparation, signal conditioning electronics, processing techniques and others. In the surface EMG for noninvasive assessment of muscles (SENIAM) project, most of the above-cited models were analysed and the limitations of each model discussed (Hermens *et al.*, 1999).

It may not be possible to arrive at a precise description of neuromotor mechanisms and MU electrophysiology using sEMG, but muscle activation and fatigue and force estimations can be made using different signal processing techniques. These features can be useful in WR applications, as discussed in the next sections. Despite the complexity associated with sEMG, current trends are moving towards utilization of these natural muscle activation patterns by taking advantage of simpler acquired data using intramuscular EMG, or with special electrodes at the PNS level. Nowadays, control signals can even be registered in the CNS, as illustrated in Case Study 4.7. This roadmap to biohybrid interfaces highlights the importance of neuromotor modelling.

4.3.3 Surface EMG signal feature extraction

The signal conditioning phase of sEMG signals usually comprises three stages. The first amplifies the raw sEMG signal using low-noise, high-input impedance amplifiers. Common gains are 100, 1000 and 10 000. The second stage consists of a bandpass filter. The high-cutoff frequency depends on the sampling frequency and is used to reduce aliasing phenomena. The low-cutoff frequency is generally located between 10 and 20 Hz. It should include all the harmonics of the undesired signals, e.g. movement artefacts or noise caused by skin–electrode interface variability.

Once the sEMG has been digitalized, different features can be extracted. Most of these are closely related to muscle activation. The spectral components of human movements are low-frequency, ranging from 1 or 2 Hz to 3–4 Hz for voluntary contractions and from 3–4 to 20 Hz for involuntary movements. Rapid fluctuations in sEMG signals are usually ignored as being due to random interference patterns. The signal can be analysed in the time domain or the frequency domain. In the context of WR cHRI, analysis generally yields a set of features that describe the muscle activity.

4.3.3.1 Time domain methods

Common procedures are used to detect muscle activation. These are basically described by the observable lobes appearing in the sEMG time series. There are several digital operations that can be performed to obtain the desired information.

- *Full-wave rectification.* One of the first operations performed on the sEMG signal is full-wave rectification (Hermens *et al.*, 1999). This can be done digitally by defining

$$\mathrm{sEMG}_{\mathrm{rec}}(n) = |\mathrm{sEMG}(n)| \tag{4.21}$$

  where $\mathrm{sEMG}(n)$ is the nth sample of the discrete sEMG signal.

- *Linear envelope.* After rectification, a lowpass filter can be applied to determine the envelope of the activation lobes, e.g. Butterworth or Bessel filters. The result, the envelope of the EMG (EEMG), is a measure of the contraction level of the muscle. For voluntary isometric contractions a 2 Hz cutoff frequency should be enough. For a more general detector, 6–8 Hz cutoff frequencies are used. One undesirable effect of these filters is the introduced delay in the signal. A critically damped second-order filter with this cutoff frequency would introduce a delay of approximately 80 milliseconds. However, there can also be a similar natural delay between the electrical and mechanical signals of human muscles. One advantage of the linear envelope method is that it can be used in real-time applications.

- *Integration.* Integration EMG (IEMG) was once a very common operation, but it is now obsolete. An integration may be understood as a contraction energy accumulator that can trigger an action of the controller. It is defined as

$$\mathrm{IEMG}(n) = \sum_{i=n-N+1}^{n} |\mathrm{sEMG}(i)| \tag{4.22}$$

  where an epoch of length N is considered to evaluate the energy content. In this way, the activation state of the muscle is refreshed every N samples. Optionally, a threshold can be set in such a way that the integration is reset every time it is reached. In this case, the introduced delay is reduced if the integrated signal reaches the energy threshold before the epoch length.

- *Root mean square* (RMS). This is used to calculate the amplitude of the sEMG, and therefore also as a force indicator. It is defined by

$$\mathrm{EMG}_{\mathrm{RMS}}(n) = \sqrt{\frac{1}{N} \sum_{i=n-N+1}^{n} \mathrm{sEMG}(i)^2} \tag{4.23}$$

  The method does not require prior rectification of the signal. Again, an epoch of length N has to be selected *a priori*. However, the algorithm can be implemented using sliding epochs so as to have a new output every new sample. This affects the amount of memory required for real-time applications.

- *Average rectified value* (ARV). This method is similar to the last one. It uses the rectified signal to calculate the average over an epoch, as follows:

$$\mathrm{ARV}(n) = \frac{1}{N} \sum_{i=n-N+1}^{n} |\mathrm{sEMG}(i)| \tag{4.24}$$

- *Wilson amplitude* (WAMP). This is the number of times that the difference of two consecutive samples becomes more than a preset threshold (Park and Lee, 1998). It is formulated as

$$\mathrm{WAMP}(n) = \sum_{i=n-N+1}^{n} f(\mathrm{sEMG}(i) - \mathrm{sEMG}(i+1)) \tag{4.25}$$

  where

$$f(\mathrm{sEMG}(i) - \mathrm{sEMG}(i+1)) = \begin{cases} 1 & \text{if } \ (\mathrm{sEMG}(i) - \mathrm{sEMG}(i+1)) > \ \text{threshold} \\ 0 & \text{otherwise} \end{cases} \tag{4.26}$$

  The WAMP is an indicator of MU activity and therefore can also be used as a muscle contraction indicator.

- *Autoregressive coefficients* (AR). Using an AR model, new samples are represented as a linear combination of earlier samples. The model can be represented as

$$\hat{y}(n) = \sum_{k=1}^{M} a(k)\, y(n-k) + w(n) \tag{4.27}$$

  where y is the sEMG signal, $a(k)$ are the coefficients of the model, $w(n)$ is a random white noise and M is the order of the model. The model coefficients can be used to classify the EMG activity. It has been shown that $M = 4$ is suitable for EMG signals (Graupe and Cline, 1975).

- *Cepstral coefficients.* This is a common method in speech applications. It is based on the AR model and has received increasing attention over the last 15 years (Kang *et al.*, 1993). It can be formulated as

$$c_1 = a_1, \qquad c_n = -\sum_{i=1}^{M}\left(1 - \frac{k}{n}\right) a_k c_{n-k} - a_n \tag{4.28}$$

- *Wavelet coefficients.* Wavelet transform is a very common tool in EMG analysis (Flanders, 2002). Wavelet coefficients may be represented as

$$c(n) = \sum_{k=-\infty}^{+\infty} w_i(n-k)\ \mathrm{sEMG}(k) \tag{4.29}$$

where $c_i(n)$ represents the coefficients corresponding to the w_i wavelet described as a vector of length i. The sEMG(t) can be estimated as the sum of

$$\mathrm{s\hat{E}MG}(t) = \sum_{j=1}^{P}\sum_{k=-\infty}^{+\infty} c_j(n)\, w_j(n-k) \tag{4.30}$$

where P is the number of wavelets used to decompose the signal. The discrete wavelet transform uses a limited set of scales and positions based on powers of two – a dyadic set – to optimize the analysis. This set of scales and positions acts as a bank of lowpass and highpass filters that decompose a signal into multiple signal bands. The major advantage of this method is that it is suitable for nonstationary signals. The advantage of features extracted with a nine-scale wavelet transform and with the Cepstral method for prosthetic control has been demonstrated (Boostani and Moradi, 2003). Both features exhibit the best performance in terms of computing time and cluster separability.

- *Other methods.* There are various other methods for EMG that are rarely used nowadays, e.g. zero crossings, peaks, turns and wavelength. Although obsolete, these methods may be used to complement the ones described above.

4.3.3.2 Frequency domain methods

Spectral analysis can be used to describe an EMG signal. It is commonly used in applications where oscillators or repetitive patterns are involved, for instance in the case of MU activation and pathological tremor (Brunetti *et al.*, 2005). The basis of this analysis is the discrete Fourier transform. Spectral analysis can also describe muscle fatigue during the performance of a task. This information can be very useful in WRs to evaluate HRI performance or control the WR (Arieta *et al.*, 2006). Useful frequency domain methods include:

- *Spectral analysis of random signals.* Surface EMG is usually considered as a random signal with a Gaussian distribution. The averaged periodogram is used to calculate the spectrum of a random signal. It is defined by

$$\hat{P}_{xx}(\theta) = \frac{1}{L}\sum_{l=0}^{L-1}\left[\frac{1}{N}\left|\sum_{n=0}^{N-1} \mathrm{sEMG}(n+lN)\mathrm{e}^{-\mathrm{j}\theta n}\right|^2\right] \tag{4.31}$$

where θ is the normalized discrete frequency. In practice, windows are commonly used to work on smoother functions, so that Equation (4.31) becomes

$$\hat{P}_{xx}(\theta) = \frac{1}{L}\sum_{l=0}^{L-1}\left[\frac{1}{N}\left|\sum_{n=0}^{N-1} w(n)\mathrm{sEMG}(n+lN)\mathrm{e}^{-\mathrm{j}\theta n}\right|^2\right] \tag{4.32}$$

where $w(n)$ represents the window. Windowing is a common method for smoothing the spectrum of finite length data in signal processing. The final form of the mean periodogram depends not only on the form of the window but also on N and L. Generally speaking, increasing N implies a higher frequency resolution but a sharper and more random spectrum since less segments L are used to calculate the averaging, (Brunetti *et al.*, 2005).

- *Mean frequency*. Fatigue is related to the frequency of MU activation (De Luca, 2006). The evolution of the mean frequency is used as a fatigue index. It is defined as

$$f_{\text{mean}} = \frac{\sum_{\theta=0}^{F} \theta \, Pxx(\theta)}{\sum_{\theta=0}^{F} Pxx(\theta)} \tag{4.33}$$

  where F is the Nyquist frequency.

- *Median frequency*. This is another parameter that can be used to assess muscle fatigue. The median frequency is given by the frequency that divides the power spectrum into two regions containing the same amount of power (De Luca, 2006). It can be represented as

$$\sum_{\theta=0}^{f_{\text{median}}} Pxx(\theta) = \sum_{\theta=f_{\text{median}}}^{F} Pxx(\theta) = \frac{1}{2}\sum_{\theta=0}^{F} Pxx(\theta) \tag{4.34}$$

  Median frequency is less sensitive to noise than mean frequency. It is calculated using recursive methods.

4.3.4 Classification of EMG activity

The previous section described how to identify different features to assess EMG activity. The next step includes classification of these feature patterns in order to generate the desired output (Arieta *et al.*, 2006). This task can vary from simple to complex pattern recognition algorithms. The complexity will depend basically on the number of features and the number of possible outputs or classes in the classifier.

- *Simple threshold*. This is the simplest method and is used to determine muscle activation (ON). It is a binary classifier, since the output is described using only two classes (Ferris *et al.*, 2006). The muscle activation state, MS, is calculated using a threshold, a·MVC, which expresses a percentage of the MVC. The classification rule may be expressed as follows:

$$\text{MS}(n) = \begin{cases} 1\,(\text{ON}) & \text{if ENEMG}(n) > a\cdot\text{MVC} \\ 0\,(\text{OFF}) & \text{otherwise} \end{cases} \tag{4.35}$$

  where ENEMG is the normalized and filtered (envelope) sEMG signal.

- *Double threshold*. The previous method is highly sensitive to noise. It can be improved using a second lower threshold, d, to deactivate (OFF) the MS instead of a unique ON/OFF threshold. Thus, the classifier may be written as

$$\text{MS}(n) = \begin{cases} 1\,(\text{ON}) & \text{if ENEMG}(n) > \text{a}\cdot\text{MVC} \\ \text{MS}(n-1) & \text{if a}\cdot\text{MVC} > \text{ENEMG}(n) > \text{d}\cdot\text{MVC} \\ 0\,(\text{OFF}) & \text{if ENEMG}(n) < \text{d}\cdot\text{MVC} \\ & \text{being a>d} \end{cases} \tag{4.36}$$

  where a·MVC is the activation threshold and d·MVC the deactivation threshold. Another different approach uses a double single-threshold method. The first threshold, a, is used to determine

(ON/OFF) the muscle activation, MA(n). The second threshold, b, is used to calculate the muscle state, MS, by counting the number of ON and OFF states in an M-length window (Bonato, D'Alessio and Knaflitz, 1998). Formally, it may be expressed as

$$\mathrm{MS}(n) = \begin{cases} 1\,(\mathrm{ON}) & \text{if } \sum_{i=0}^{M-1} \mathrm{MA}(n-i) > b \\ 0\,(\mathrm{OFF}) & \text{otherwise} \end{cases} \tag{4.37}$$

where

$$\mathrm{MA}(n) = \begin{cases} 1\,(\mathrm{ON}) & \text{if sEMG(n) ¿ a} \\ 0\,(\mathrm{OFF}) & \text{otherwise} \end{cases} \tag{4.38}$$

One advantage of this method is that it uses the raw sEMG signal and not the envelope, as can be seen in Equation (4.38). This reduces the delay and improves the timing precision of the detector.

- *Gaussian classifiers.* These classifiers are based on the Bayesian decision theory, which assumes that the decision problem can be expressed in statistical terms (Duda, Hart and Stork, 2001). Bayes's formula states:

$$P(\omega_j|\boldsymbol{x}) = \frac{p(\mathbf{x}|\omega_j)P(\omega_j)}{p(\mathbf{x})}, \tag{4.39}$$

where ω_j represents the class and $\boldsymbol{x}$ is the features vector. $P(\omega_j)$ represents the prior probability of occurrence of class ω_j and $p(\boldsymbol{x}|\omega_j)$ the class conditional probability function. The probability density function $p(\boldsymbol{x})$ can be seen as a scale factor which ensures that the subsequent probabilities sum to one and can be defined by

$$p(\boldsymbol{x}) = \sum_{i=1}^{N} p(\boldsymbol{x}|\omega_j)P(\omega_j) \tag{4.40}$$

where N is the total number of classes. In the case of Gaussian classifiers, it is assumed that

$$p(\boldsymbol{x}|\omega_j)P(\omega_j) \sim N(\mu, \sigma^2) \tag{4.41}$$

The parameters of the distribution may be known *a priori* or they may be estimated given a set of samples belonging to a specific class. This last type is usually known as a labelled set. By focusing on WR control based on EMG, a Gaussian probability density function can be estimated given a set of feature vectors that describe a user input or movement, e.g. muscle state or AR coefficients during a movement.

- *Neural networks.* This nonlinear method can be used as a classifier (Pattichis *et al.*, 1995). It is an interconnected group of artificial neurons that responds to mathematical models to transmit excitation values through the network. It can be organized in different forms. Generally, a final layer of artificial neurons is configured where each one represents a specific class. Again, a set of labelled samples can be used to train the artificial neural network (ANN) model. Therefore the ANN may be seen as an adaptive system whose internal structure changes in such a way as to map a set of inputs into a reduced set of outputs.

- *Other classifiers.* There are numerous methods and algorithms used for pattern recognition, e.g. linear discriminant analysis (LDA), Parzen windows, nearest-neighbour rule, hidden Markov models (HMMs) and others (Duda, Hart and Stork, 2001). The WR controller can use any of them to obtain a command from a set of features that describes the muscle activity. They may be classified as either supervised or unsupervised learning clustering or classification. Supervised learning methods use a set of labelled inputs to train the algorithm. Unsupervised methods form natural clusters while the relationship between the input and a specific class is unknown.

4.3.5 Force and torque estimation

Muscle contraction generates force. In general, this force can be translated into movement. One of the major points of interest in the use of EMG in biomechanics and WRs lies in the possibility of measuring the force and torque generated by muscles during execution of a movement. Such estimations can be used as inputs to the WR controller (Cavallaro *et al.*, 2006).

For isometric contractions, it is generally accepted that the relationship between a smoothed EMG signal and the exerted force is monotonic. However, the linearity will basically depend on the measured muscle and even more on measurement-related technical issues, which are often not awarded sufficient consideration, e.g. cross-talk, electrode location, applied filters, artefact noise and others.

The envelope of the EMG signal (EEMG), can be used as a continuous indicator of muscle activation. The force F generated by the muscle can be expressed as a function of the EEMG as follows:

$$F_n(\mathrm{EEMG}_n) = \frac{\mathrm{e}^{A_n \mathrm{EEMG}_n \mathrm{MVC}_n^{-1}} - 1}{\mathrm{e}^{A_n} - 1} F_{n,\max} \tag{4.42}$$

where index n represents a specific muscle, A_n the nonlinear factor that affects the relationship and $F_{n,\max}$ the maximum force measured during the calibration phase (Lloyd and Besier, 2003). The value of MVC has to be assessed in a similar way during the EMG calibration phase. EEMG, A_n, MVC and $F_{\max}$ are particular parameters for each n-muscle.

Now, the torque T at a k-joint can be calculated as

$$T_k = \sum_{n=1}^{N} \left(\left| \boldsymbol{I}_n \times \frac{\boldsymbol{I}_n - \boldsymbol{O}_n}{|\boldsymbol{I}_n - \boldsymbol{O}_n|} \right| \dot{F}_n \right) \tag{4.43}$$

where $\boldsymbol{O}_n$ and $\boldsymbol{I}_n$ are the points of origin and insertion of the muscle.

In an anisometric contraction, strong nonlinearities compromise the monotonic character of the relationship. Special care must be taken when dealing with these types of contractions, which occur in common activities like human gait. One way to reduce these effects is to consider short EMG signal epochs where the changes in muscle length are very small (isometric). Another option is to use nonlinear methods. An ANN can be used to estimate torque-based features such as the joint angle and the EEMG signals of the muscles involved (Cavallaro *et al.*, 2006).

4.4 cHRI THROUGH BIOMECHANICAL MONITORING

Cognitive human–robot interaction through biomechanical information is not as easily formalized as it was in Sections 4.2 and 4.3. There are different approaches to cHRI through biomechanical information that are basically application-dependent. The applications addressed in this section are empowering, telemanipulation, rehabilitation and functional compensation.

In *empowering* applications, the exoskeleton has to track the movements executed by the user. This can be achieved through a zero interaction force control strategy. In this way the robot 'follows' the human movements and at the same time empowers them, so that the power used to move a given load comes from the robot's actuators and not from the human. It can be seen that this interaction between the wearer and robot is mainly physical. However, some authors see this situation as an instance of intention detection and view it as a cognitive process.

Telemanipulation is considered less important for the purposes of this book in that it is not fully wearable. It deals only with the acquisition of biomechanical information, be it kinetic or kinematic, which is then used to control remote robots. There is therefore a cognitive process between the user and the remote robot.

In the field of *rehabilitation*, the approach is to plan beforehand certain tasks to be performed by the exoskeleton and execute them with the user, e.g. exercising certain articulations or performing

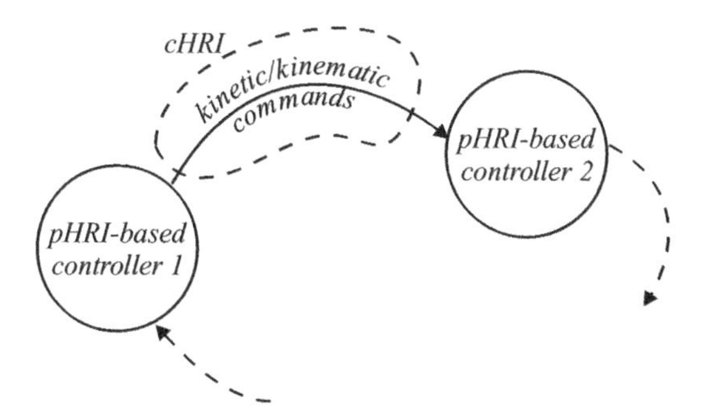

Figure 4.5 Relationship between pHRI and cHRI

physical therapy tasks. The application can be viewed as motion training. The control approach is to use a training programme in which the exoskeleton becomes a programmable resistance trainer that allows the user to exercise a certain articulation on a desired axis. The other articulations may be left free or have their motion restricted depending on the training program (Carignan, Liszka and Roderick, 2005). Again, there is no cognitive interaction, since it is only the physical interaction between the user and exoskeleton that drives the robot; the approach is very similar to the empowering application.

Cognitive human–robot interaction through biomechanical information is more apparent in applications that have to do with *functional compensation*. In such applications, the human needs robotic assistance to perform certain tasks. Although the robotic control strategies used for function compensation are again based on physical interaction, the robot needs to know when to apply them and which one the user desires at a given moment. A cognitive process is required, so that the user can generate commands and select the control strategy to be applied. Such a command generation process involves reasoning, planning and executing steps (see Figure 4.1), and is based on kinetic or kinematic information. The situation is described in Figure 4.5. In this section, the focus is on modelling such biomechanically based cHRI.

4.4.1 Biomechanical models and parameters

Chapter 3 presented human and robot kinematics and dynamics. The sensor technologies responsible for collecting kinetic and kinematic data are presented in Chapter 6. Here, the focus is on obtaining the information that is relevant or most important in order to establish the cHRI. The functional compensation devices mentioned above may be classified into applications based on cyclic and on noncyclic events. Generally, the former are related to gait.

4.4.1.1 Lower limbs

The main function of human lower limbs is to provide support, stability and mobility. For this reason, most of the wearable exoskeletons in the literature are designed to compensate gait deficiencies. Robot-assisted human gait constitutes a very close human–robot interaction paradigm. This interaction requires a critical process of sensing and actuating.

Human gait can be regarded as a cyclic process comprising a stance phase and a swing phase, as shown in Figure 4.6. This process is described mechanically by kinetic (momentum and torques)

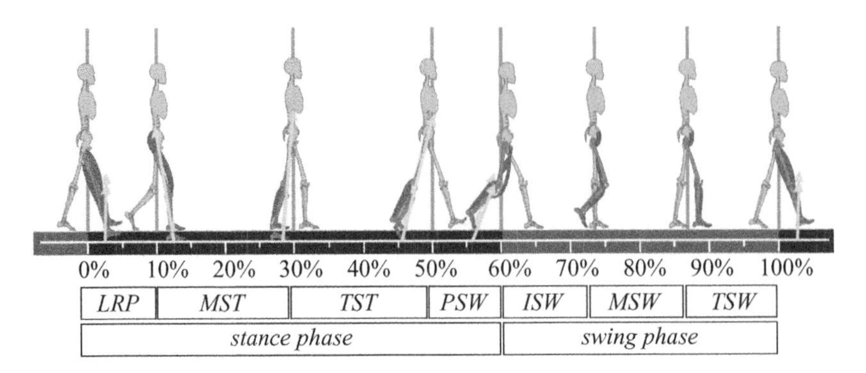

Figure 4.6 Human gait phases and periods

or kinematic (angles) variables of the human lower limb articulations. Other time-related parameters such as step length, stride length, cadence and speed are also important.

The stance phase is divided into four periods: loading response (LRP), midstance (MST), terminal stance (TST) and preswing (PSW). The swing phase in turn is divided into three periods: initial swing (ISW), midswing (MSW) and terminal swing (TSW). The beginning and end of each period are defined by specific *events*.

The stance phase begins with the *loading response* period. This period starts when the foot contacts the ground (normally heel contact). The loading response period ends with contralateral toe-off, when the opposite extremity leaves the ground. Thus, loading response corresponds to the gait cycle's first period of double limb support.

Next, the *midstance* period begins with contralateral toe-off and ends when the centre of gravity is directly over the reference foot. The next phase is the *terminal stance*. It begins when the centre of gravity is over the supporting foot and ends when the contralateral foot contacts the ground. During the terminal stance, around 35 % of the gait cycle, the heel is raised from the ground.

Preswing begins with contralateral initial contact and ends with toe-off, around 60 % of the way through gait cycle. Thus, preswing corresponds to the gait cycle's second period of double limb support. The swing phase begins with the *initial swing*. It starts at toe-off and continues up to maximum knee flexion (40–60°). Next, *midswing* is the period from maximum knee flexion until the tibia is vertical or perpendicular to the ground. Finally, *terminal* swing begins where the tibia is vertical and ends with initial contact.

To summarize, the human gait phases may be described by a series of initial events and a number of characteristics that may be reflected in the lower limb articulations, as shown in Table 4.1. A more detailed description of these parameters can be found in Lacuesta, Prat and Hoyos (1999).

Table 4.1 Gait phases and parameters

Gait phase	Initial event	Phase characteristics
Loading response	Initial foot contact	Knee flexing around 15°, plantar flexion
Midstance	Opposite toe-off	Reduction on knee flexion (around 12°), dorsal flexion
Terminal stance	Heel rise	Complete knee extension
Preswing	Opposite foot initial contact	Second period of double limb support
Initial swing	Toe-off	Ends with maximum knee flexion
Midswing	Feet adjacent	From maximum knee flexing until tibia perpendicular to the ground
Terminal swing	Tibia vertical	Change from knee flexion to complete extension

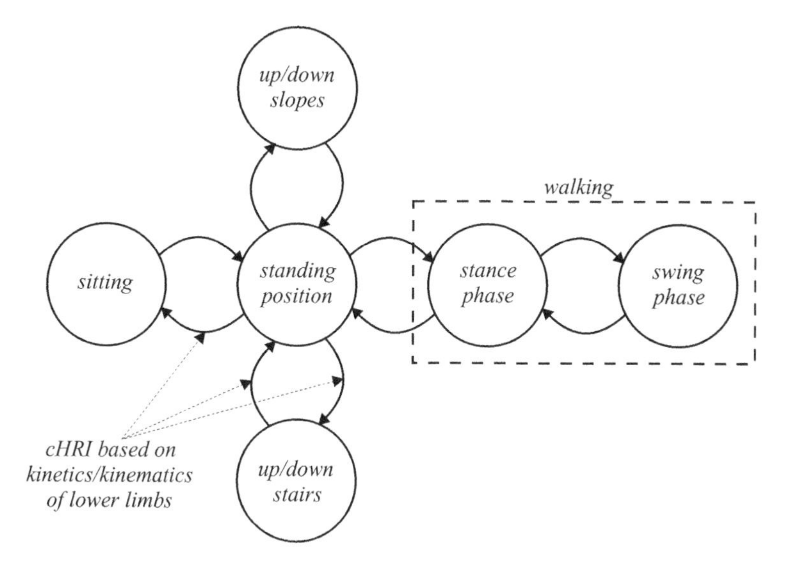

Figure 4.7 Finite state machine (FSM) implemented on lower limb cognitive control of exoskeleton

Knowledge of the different stages or phases of human gait and the corresponding events can be used to design a set of sensors that will detect events and the parameters that characterize them. Traditionally, uniaxial gyroscopes were proposed (Tong and Granat, 1999) to measure the rotation of the limb segments in combination with the insole information for ambulatory applications. A combination of gyroscopes and accelerometers has also been introduced to register biomechanical data (Mayagoitia, Nene and Veltink, 2002). Gyroscopes have been used to determine spatial–temporal information in pathological gait (Pappas *et al.*, 2001). Recently, ambulatory methods based on inertial sensors have been proposed to control and monitor intelligent orthoses and prostheses (Moreno *et al.*, 2006).

Besides the parameters relating to gait phases or states, there are noncyclical tasks such as sitting down or standing up that are important and require robotic assistance for handicapped users. The transition between these situations requires cognitive control. The user can start a transition event, in this case through the application of forces to the physical interface, in such a way that the robot understands such forces as transition commands.

Thus, the transition between states can be determined by monitoring joint angles and the forces exerted on the robot structure. Figure 4.7 shows an FSM for a lower limb cognitive controller of an exoskeleton. Using the parameters presented in this section, a rule-based algorithm is implemented to detect transitions between states.

4.4.1.2 Upper limbs

Although human *upper limbs* have an important role in human balance, especially during gait, the main function of the upper limb is connected with manipulation. The human arm has to ensure that the hand performs its functions: the hand must be able to reach any point in space, especially any point on the human body, in such a way that the person can manipulate, draw on and move objects to or from the body.

The importance of the upper limbs further derives from their capacity to execute cognitive-driven and expression-driven activities. For this reason, any alteration or pathology that affects upper limb

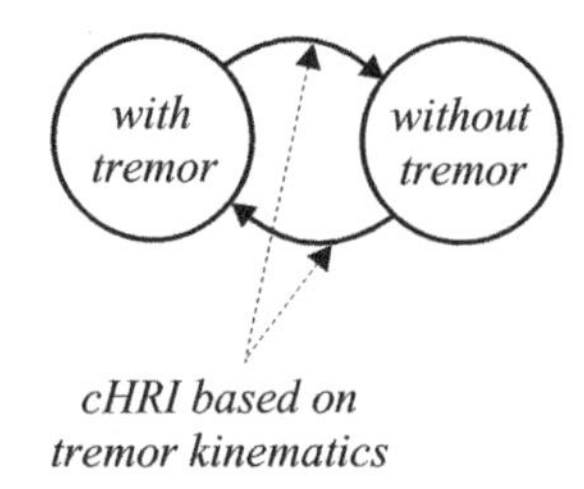

Figure 4.8 Finite state machine implemented on upper limb cognitive control of the exoskeleton

motion range, muscle power, sensitivity or skin integrity will disturb its operation. As a consequence, many of the wearable robotic solutions for upper limbs are intended for rehabilitation, be it to suppress involuntary movements or to empower voluntary/conscious movements.

Because of the complexity and the large number of different tasks that can be performed by the upper limbs, the solutions adopted for a cHRI are usually closely identified with the particular application. In this connection Rocon *et al.* (2007) presented a wearable robot, WOTAS, designed for tremor suppression. There are clearly two states: presence of tremor or absence of tremor. Taking advantage of the tremor periodicity, a pattern can be identified and used as an input to the system. The objective is to determine the tremor force, position, velocity and acceleration in order to choose a control strategy, to be applied as a function of the presence or absence of tremor. In both cases, strategies based on physical interaction and impedance control (see Chapter 5) are adopted and the exoskeleton tracks the human's voluntary movements, suppressing tremor if necessary. To identify tremor and determine its characteristics, the angular velocity of the limb segments needs to the known. The tremor information is then extracted from the overall movement and an impedance controller is used to track the voluntary or nontremor motion. Figure 4.8 presents an FSM that describes this situation. Case Study 4.6 provides additional details.

4.4.2 Biomechanically controlled interfaces: approaches and algorithms

The idea is to model the FSM and the transitions at the same time whenever the states of an FSM are not known. Here, stochastic models can be used to represent the unknown process as a series of observations. The use of *hidden Markov models* (HMM) has become very popular in these situations, thanks to the rich mathematical structure of the models and the very good results achieved in many applications (Rabiner, 1989). HMMs have gained much popularity in speech recognition; recently the technique has also been applied to computer vision recognition of human motion sequences (Yamato, Ohya and Ishii, 1992), for instance in gesture recognition using data gloves (Brashear *et al.*, 2003).

In those other situations where the states of the FSM are well defined, the algorithms are used to identify patterns or events from biomechanical data in order to define state transitions in an FSM approach. Some of the processing and classification techniques presented in the previous sections of this chapter can be used for the classification of biomechanical patterns or events, e.g. *threshold-based classifiers, Gaussian classifiers* and *neural networks*. Another common approach is to apply *rule-based algorithms*.

In this case, there is a large variety of options, but the basic idea is to define a set of rules that have to be met by a combination of values of kinetic or kinematic variables in order to trigger a transition. In this way the FSM can be driven by discrete rules, based for instance on *if* ⟨*condition*⟩, *then* ⟨*action*⟩ structures. It is important to note that in most cases the rules include redundancy for safety reasons, so the transitions depend on a sequence or the simultaneous occurrence of a group of rules based on different biomechanical data. Case Study 4.5 presents a cHRI based on kinematic and kinetic gait patterns.

4.5 CASE STUDY: LOWER LIMB EXOSKELETON CONTROL BASED ON LEARNED GAIT PATTERNS

J.C. Moreno and J.L. Pons

Bioengineering Group, Instituto de Automática Industrial, CSIC, Madrid, Spain

This case study is included here to illustrate the implementation of a cHRI system for detection of transition events in an FSM-controlled lower limb exoskeleton. Transition events are detected from biomechanical (kinetic and kinematic) data by means of a fuzzy set of rules.

4.5.1 Gait patterns with knee joint impedance modulation

The GAIT orthosis is an exoskeleton intended to provide stability and selective impedance depending on the state of the weak leg during the gait cycle. Subjects wearing such an external device need to adapt their walking strategy to force the system to successfully switch between two knee spring–damper configurations (details are presented in Section 6.7). A set of discrete rules can be determined for a given subject in order to drive the knee actuator system cyclically, based on the configuration (kinematics) of the lower leg. Adaptability to patients, flexibility and robustness of the controller are crucial design aspects for the control system. Such cyclical decision making, which is vital for patient stability, can be performed by a fuzzy inference system.

4.5.2 Architecture

The control scheme consists of a number of different modules (see Figure 4.9). A fuzzy system makes use of approximate reasoning, which refers to methodologies for describing physical systems that include complexity due to nonlinearities and uncertainties. A fuzzy inference system with two inputs and a single output node is identified and trained to map the inputs and trigger the actuator activation. Crisp output from the fuzzy inference system during each cycle is critical in achieving a transition from restrained knee flexion in stance to a free swinging leg. The activation period of

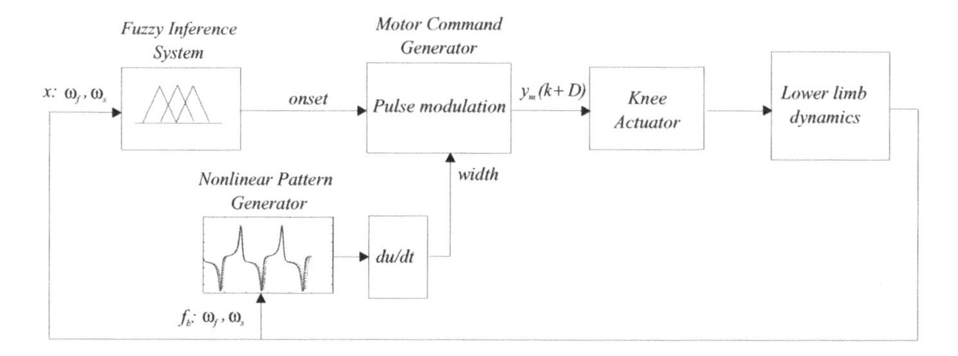

Figure 4.9 Hybrid architecture for external control of gait compensation at knee level, based on inertial sensing data. The first module contains the fuzzy inference system with a crisp output. The second module contains an oscillator as a function of gait frequency

the knee actuator (pulse width) during the swing phase is cyclically adapted by a second module consisting of an oscillator.

4.5.3 Fuzzy inference system

Conventional proportional-integral-derivative (PID) controllers have been used for the control of cyclical movements in legs of paraplegic subjects (Franken, 1995). The introduction of dynamic adaptation of the rules commanding functional electrical stimulation (FES) systems has been investigated to cope with a wider range of unsafe and uncertain situations in application of stimulation. A Sugeno system is suitable for modelling nonlinear systems. A training scheme has been combined with a fuzzy modelling network structure to develop a gait synthesis learning scheme (Horikawa *et al.*, 1990).

Suppose that the unknown system is a black box only capable of measuring a set of inputs $x_1, \ldots, x_n$ and outputs $y_1, \ldots, y_m$. The idea is to produce a fuzzy system with a crisp output and the following type of rules:

$$R_i : \text{IF } x_1 \text{ is } S_{i1} \ y \ldots y \ x_m \text{ is } S_{im}, \ \text{THEN } y \text{ is } c_i \tag{4.44}$$

The fuzzy inference system is generated by means of the grid partition method. For identification, a training dataset is generated experimentally. The identification method consists in the application of the adaptive network fuzzy inference system (ANFIS) proposed by Jang (1993) in order to build fuzzy rules and membership functions with which to generate input/output data pairs. In this case (sagittal plane), the input used is motion of the shank and foot segments, calculated from signals provided by rate gyroscopes fixed to the orthotic bars. Input parameters of the membership functions are learnt iteratively by backpropagation in an adaptive network, while the parameters of output functions are optimized by the least squares fitting method. The *adaptive network* is a feedforward multilayered network with a supervised learning scheme. The functions of given nodes in a layer are similar. For the sake of simplicity, the inference system considered here is a first-order Sugeno-type model. The kinematic inputs, the output $E(t)$ and n fuzzy rules are

$$\begin{aligned} R_n : \text{IF } \dot{\theta}_s \text{ is } A_n, \ \text{AND } \dot{\theta}_f \text{ is } B_n \\ \text{THEN } E = p_1\dot{\theta}_s + \dot{\theta}_s + t \end{aligned} \tag{4.45}$$

Gaussian membership functions were selected for smooth transition. A total of four Sugeno-type fuzzy rules were defined, with a network of 21 nodes. These rules were of the AND (minimum) antecedent type. Defuzzification to calculate the output was performed by the centroid method. The clustering radius $r = 0.2$ was adjusted for tuning. The optimization process spanned 13 epochs, with the training dataset. Figure 4.10 depicts the output surface of the system finally identified given the two inputs.

4.5.4 Simulation

Kinematics were collected for different gait speeds with manual tuning of the control system and used as the training dataset. Local minima values were detected from the output of the fuzzy system using numerical integration. The sensitivity of the local minima detector was given by δ, which corresponds to the minimum difference in amplitude from neighbouring samples. With the calibrated raw gyroscope data, $\delta = 40$ was satisfactory for all conditions. Thus, the fuzzy system provided a cyclic trigger or *onset* for activation of the knee actuator.

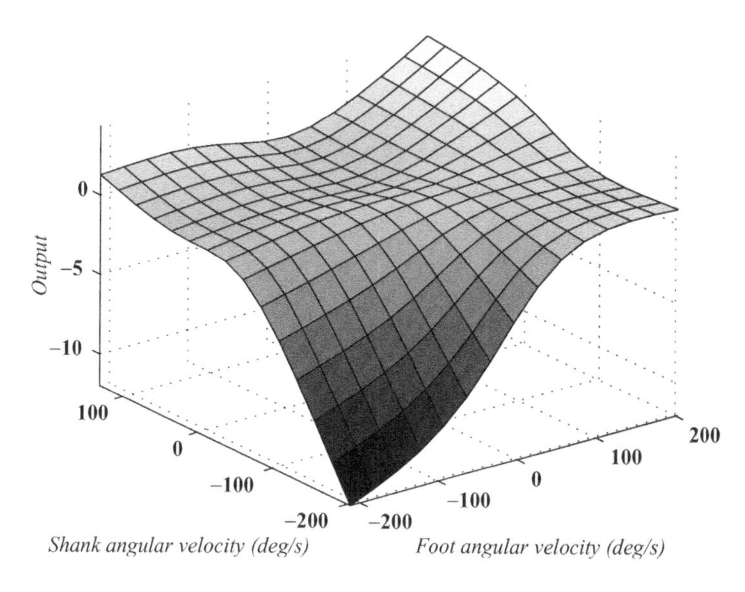

Figure 4.10 Inputs (shank and foot angular velocities) and output (knee actuator activation) surface of the fuzzy system

An example of the hybrid controller for cyclic gait at 0.94 m/s (stride length of, 1.46 m) is depicted in Figure 4.11, showing the knee angle and reference output (top), inputs extracted from gyroscope data (middle) and the fuzzy inference system outputs (bottom), which generate the triggers for activation of the actuators.

Robust performance of the fuzzy inference system was achieved with a single-step training procedure for testing data taken from variations of gait velocity and stride length. Mean errors ranging from 0.0025 to 0.15 s were found for 0.33–1.35 m/s. The fuzziness incorporated by the modelled surface provided a good response to typical variations. In the application considered in this case study, precisely timed activation of the actuator system is essential to achieve a stable stance phase and effectively free the knee for the swing of the leg.

4.6 CASE STUDY: IDENTIFICATION AND TRACKING OF INVOLUNTARY HUMAN MOTION BASED ON BIOMECHANICAL DATA

E. Rocon and J.L. Pons

Bioengineering Group, Instituto de Automática Industrial, CSIC, Madrid, Spain

This case study describes an algorithm suitable for distinguishing voluntary from tremorous movement in real time. This algorithm is the principle behind a cHRI in a tremor suppression exoskeleton. The reader is referred to Case Study 8.1 for a full description of this exoskeleton. The analysis of the tremor signal, both in terms of frequency and amplitude, is useful for assessing the stationary characteristics of tremor, i.e. frequency drift and amplitude variation. This information is important when designing control strategies to counteract tremor.

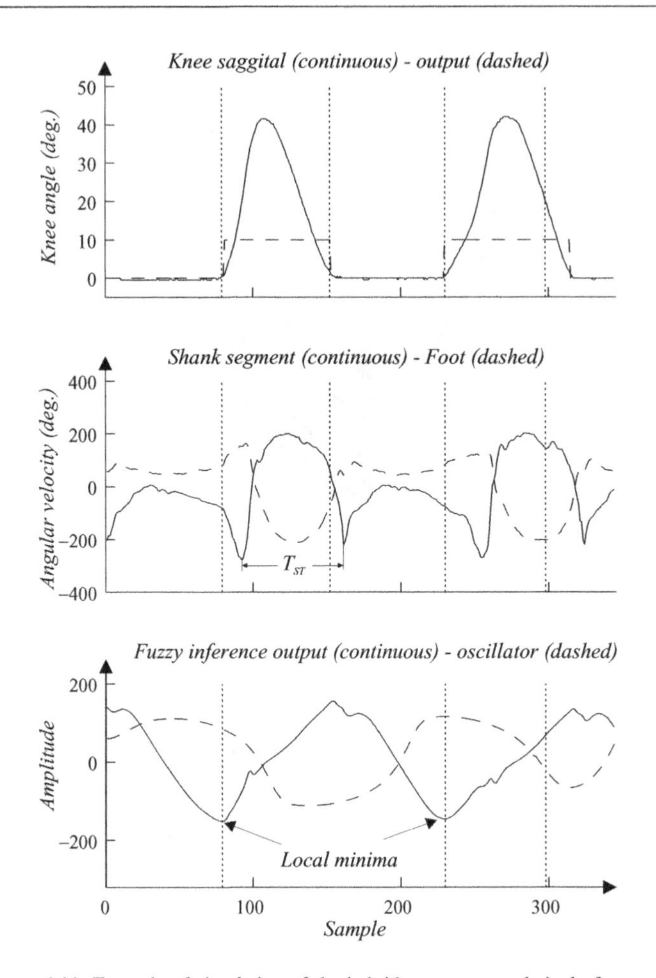

Figure 4.11 Example of simulation of the hybrid system at a relatively fast speed

A number of estimation algorithms has been developed for tremor suppression. The algorithm most commonly used to estimate tremor is the weighted-frequency Fourier linear combiner (WFLC) developed by Riviere in the context of actively counteracting physiological tremor in microsurgery (Riviere, Reich and Thakar, 1997). This algorithm is based on IEEE-STD-1057, which is a standard for fitting sine waves to noisy discrete-time observations. The WFLC is an adaptive algorithm that estimates tremor using a sinusoidal model and estimates its time-varying frequency, amplitude and phase. The WFLC can be described by

$$\varepsilon_k = s_k - \sum_{r=1}^{M} [w_{r_k} \sin(r\omega_{0_k} k) + w_{r+M_k} \cos(r\omega_{0_k} k)] \tag{4.46}$$

which assumes that the tremor can be mathematically modelled as a pure sinusoidal signal of frequency ω_0 plus M harmonics and computes the error, ε_k, between the motion, s_k, and its harmonic model.

In its recursive implementation, the WFLC can be used online to obtain estimations of both tremor frequency and amplitude (Riviere, Reich and Thakar, 1997).

$$w_{0_{k+1}} = w_{0_k} + 2\mu_0\varepsilon_k \sum_{r=1}^{M} r\left(w_{r_k} x_{M+r_k} - w_{M+r_k} x_{r_k}\right) \tag{4.47}$$

where

$$x_{rk} = \begin{cases} \sin\left(r\sum_{t=1}^{k} w_{0_t}\right), & 1 \le r \le M \\ \cos\left[(r-M)\sum_{t=1}^{k} w_{0_t}\right], & M+1 \le r \le 2M \end{cases} \tag{4.48}$$

The WFLC algorithm was evaluated in signals measured in patients suffering tremor (Rocon *et al.*, 2005). In the trials, the algorithm was able to estimate the tremor movement of all the patients with more than two degrees of accuracy in every case. The main disadvantage of the WFLC is the need for a preliminary filtering stage to eliminate the voluntary component of the movement (Riviere, Reich and Thakar, 1997). This filtering stage introduces an undesired time lag when estimating tremor movement, which may considerably affect implementation of the control strategies for tremor suppression.

The solution adopted was to develop an algorithm capable of estimating voluntary and tremorous motion with a small phase lag. The tremor literature (Elble, 1990; Riviere, Reich and Thakar, 1997), indicates that voluntary movements and tremor movements differ considerably. Voluntary movements are slower while tremor movements are jerky. This indicates that adaptive algorithms to estimate and track movement would be useful when separating the two movements with an appropriate design. The underlying idea is to design the filters so that they only estimate the less dynamic component of the input signal, which in the present case is voluntary movement, thereby filtering out the tremor movement. It was therefore proposed to develop a two-stage algorithm to estimate voluntary movement and tremor movement with a minimum time lag (see Figure 4.12).

In the first stage, a set of candidate algorithms was considered for estimation of the voluntary motion: two-point extrapolator, critically damped g–h estimator, Benedict–Bordner g–h estimator and Kalman filter. These algorithms implement both estimation and filtering equations. The combination of these actions allows the algorithm to filter the tremorous movement out of the overall motion and at the same time reduces the phase lag (Bar-Shalom and Li, 1998). The equation parameters were adjusted to track the movements with lower dynamics (voluntary movement), since tremors present a behaviour pattern characterized by quick movements (Rocon *et al.*, 2005).

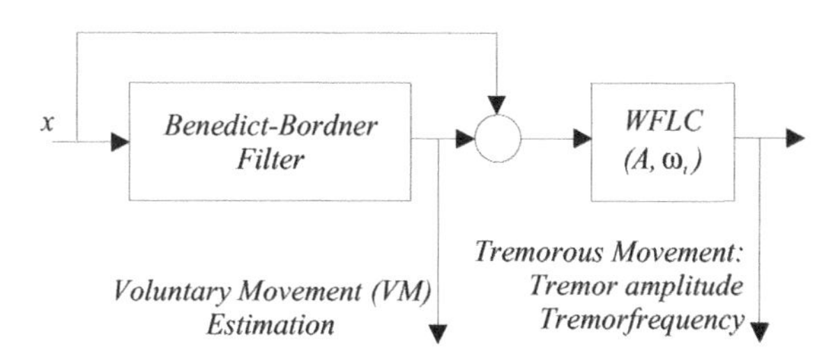

Figure 4.12 Two-stage tremor modelling: firstly, the low-frequency content voluntary motion is estimated; the voluntary motion estimation is then subtracted from the original motion; finally, tremor frequency and amplitude are determined

The algorithms evaluated were two degree-of-freedom estimators, i.e. ones that assume a constant-velocity movement model. This assumption is reasonable since the sample period is very small compared to the movement velocities (Brookner, 1998). The sampling period adopted was 1 ms and the voluntary movement estimated occurs in a bandwidth smaller than 2 Hz. The performances of these algorithms were compared on the basis of their accuracy when estimating voluntary movements in tremor time series from patients. According to the analysis, the Benedict–Bordner filter presented the best results with the lowest computational cost (Rocon *et al.*, 2005). This estimation algorithm is a g–h filter with the following tracking update equations:

$$\dot{x}^*_{k,k} = \dot{x}^*_{k,k-1} + h_k \left(\frac{y_k - x^*_{k,k-1}}{T} \right) \tag{4.49}$$

$$x^*_{k,k} = x^*_{k,k-1} + g_k \left(y_k - x^*_{k,k-1} \right) \tag{4.50}$$

and g–h prediction equations (Brookner, 1998)

$$\dot{x}^*_{k+1,k} = \dot{x}^*_{k,k} \tag{4.51}$$

$$x^*_{k+1,k} = x^*_{k,k} + T\dot{x}^*_{k+1,k} \tag{4.52}$$

The tracking update equations or estimation equations (Equations (4.49) and (4.50)) provide the joint angular velocity and position. The estimated position is based on the actual measurement as well as the past prediction. The estimated state contains all the information required from the previous measurements. The predicted position is an estimation of x_{n+1} based on past states and predictions (Equations (4.51) and (4.52)), and takes the current measurement into account by means of updated states. The Benedict–Bordner estimator is designed to minimize the transient error. As a result, it responds faster to changes in movement velocity and is slightly underdamped (Bar-Shalom and Li, 1998). The relationship between filter parameters is defined by

$$h = \frac{g^2}{2 - g} \tag{4.53}$$

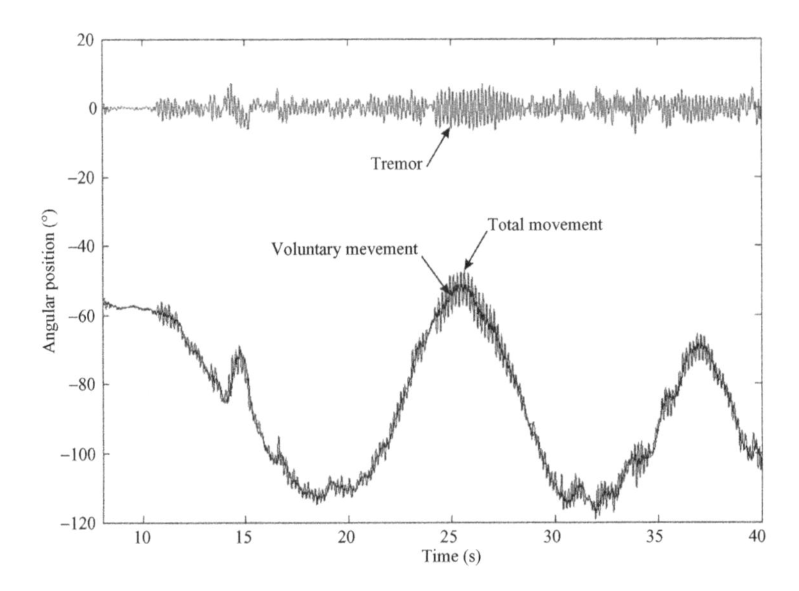

Figure 4.13 Modelling of tremor as a sinusoidal nonvoluntary motion: velocity signals (grey) obtained from gyroscopes, estimation of voluntary movement (black) and estimation of tremorous movement (top)

In the second stage, the estimated voluntary motion is subtracted from the overall motion and it is assumed that the remaining movement is tremor. The WFLC is then used to estimate tremor parameters. In this stage, the algorithm estimates both the amplitude and the time-varying frequency of the tremorous movement.

This algorithm was evaluated with data obtained from 40 patients suffering from different tremor diseases. The estimation error of the first stage was $1.4 \pm 1.3°$ (Rocon *et al.*, 2005). The second-stage algorithm had a convergence time that was always smaller than 2 s for all signals evaluated, and the mean square error (MSE) between the estimated tremor and the *real tremor* (obtained offline by means of manual decomposition based on classical filter techniques) after convergence was less than 1°. The combination of both techniques produced a very efficient algorithm, with a low processing cost, for estimating the voluntary and the tremorous components of the overall motion in real time (Rocon *et al.*, 2005). Figure 4.13 illustrates the performance of the algorithm when splitting voluntary and tremorous movements of a patient with essential tremor.

4.7 CASE STUDY: CORTICAL CONTROL OF NEUROPROSTHETIC DEVICES

J.M. Carmena

Program in Cognitive Science, Department of Electrical Engineering and Computer Sciences, Helen Wills Neuroscience Institute, University of California, Berkeley, USA

This case study illustrates possible research and development avenues for new cHRIs based on detection and analysis of cortical activity by means of implantable electrode arrays.

Research in BMIs has expanded dramatically in the last decade, with impressive demonstrations of nonhuman primates and humans controlling robots or graphical cursors in real time through signals collected from cortical areas. These demonstrations can be divided largely into two categories: either continuous control of position or velocity (Carmena *et al.*, 2003; Serruya *et al.*, 2002) or discrete control of more abstract information such as intended targets, intended actions and onset of movements (Musallam *et al.*, 2004; Santhanam *et al.*, 2006).

There are several important questions regarding the control of artificial actuators directly from brain-derived signals. These include the type of brain signals (single unit, multiple unit or field potentials) that would provide the optimal control signal for the device and the number of channels that may be necessary to operate a BMI efficiently for long periods of time. These and other questions were investigated by Carmena *et al.* (2003), where it was shown how macaque monkeys learned to use a BMI to reach and grasp virtual objects with a robot even in the absence of overt arm movement signals. Some of the findings of this study that are relevant to these two questions are summarized below.

Figure 4.14 illustrates the experimental paradigm used in Carmena *et al.* (2003). Multiple chronically implanted intracranial microelectrode arrays are used to sample the activity of large populations of single cortical neurons simultaneously in a macaque monkey. A linear filter is used to decode the intended trajectory of the arm from the combined activity of these neural signals. The decoded trajectories are used to control the movements of a robotic arm. A closed control loop is established by providing the subject with both of the visual feedback signals generated by the movement of the robotic arm (represented by a cursor on a computer screen). In this study, two macaque monkeys were chronically implanted with arrays of microwires (96 in monkey 1 and 320 in monkey 2) in several frontal and parietal cortical areas: dorsal premotor cortex (PMd), primary motor cortex (M1), supplementary motor area (SMA), primary sensory cortex (S1) and posterior parietal cortex

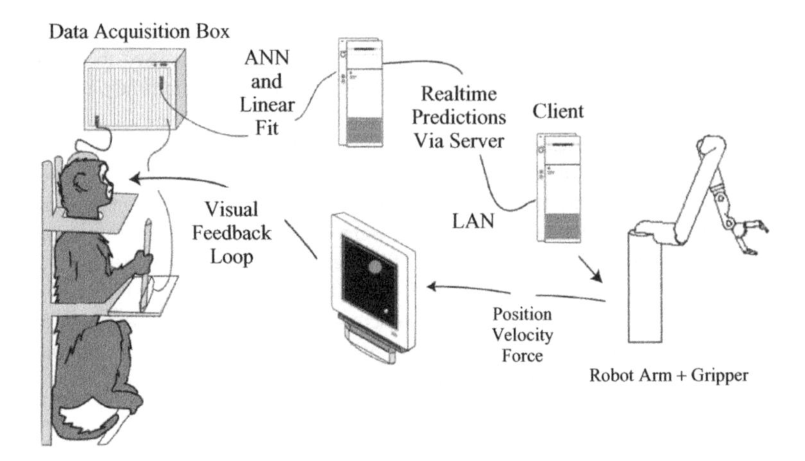

Figure 4.14 BMI experimental setup (Carmena *et al.*, 2003), including chronically implanted microelectrodes, data acquisition system, computer running linear models in real time, robot arm and visual display. Robot position was translated into the cursor position on the screen, and feedback of the gripping force was provided by changing the cursor size. Reproduced from Carmena *et al.* (2003)

(PP, MIP area). In the study, multiple linear filters were used to extract simultaneously a variety of motor parameters (hand position, hand velocity and gripping force) from the activity of cortical neural ensembles while macaque monkeys performed several motor tasks. The linear relationship between the neuronal discharges and behaviour is expressed as

$$\boldsymbol{Y}(t) = \boldsymbol{b} + \sum_{u=-m}^{n} \boldsymbol{a}(u)\boldsymbol{X}(t-u) + \varepsilon(t) \tag{4.54}$$

in which $\boldsymbol{X}(t-u)$ is an input vector of neuronal firing rates at time t and time lag u; $\boldsymbol{Y}(t)$ is a vector of behavioural variables (e.g., position, velocity, and gripping force) at time t; $\boldsymbol{a}(u)$ is a vector of weights at time lag u; $\boldsymbol{b}$ is a vector of y intercepts; and $\varepsilon(t)$ are the residual errors (for further details see Carmena *et al.* (2003) and Wessberg and Nicolelis (2004)).

Although all these parameters were extracted in real time in each session, only some of them were used to control the BMI, depending on each of the three tasks the monkeys had to solve on a given day. In each recording session, an initial 30 minute period was spent on training these models. During this period, monkeys used a hand-held pole either to move a cursor on the screen or to change the cursor size by application of gripping force to the pole. This period is referred to as the 'pole control' mode. As the models converged to an optimal performance, their coefficients were fixed and the control of the cursor position and/or size (representing gripping force) was established directly from the output of the linear models. This period is referred to as the 'brain control' mode.

During the brain control mode, animals initially produced arm movements, but they soon realized that these were not necessary and ceased to produce them for periods of time. Accurate performance was possible because large populations of neurons from multiple cortical areas were sampled, suggesting that large ensembles are preferable for efficient operation of a BMI. This conclusion is consistent with the notion that motor programming and execution is represented in a highly distributed fashion across frontal and parietal areas, and that each of these areas contains neurons that represent multiple motor parameters. It is suggested that, in principle, any of these cortical areas could be used to operate a BMI, provided that a large enough neuronal sample is obtained. This is supported by the analysis of neuron dropping curves shown in Figure 4.15(a) to (c), which indicate the number of neurons that

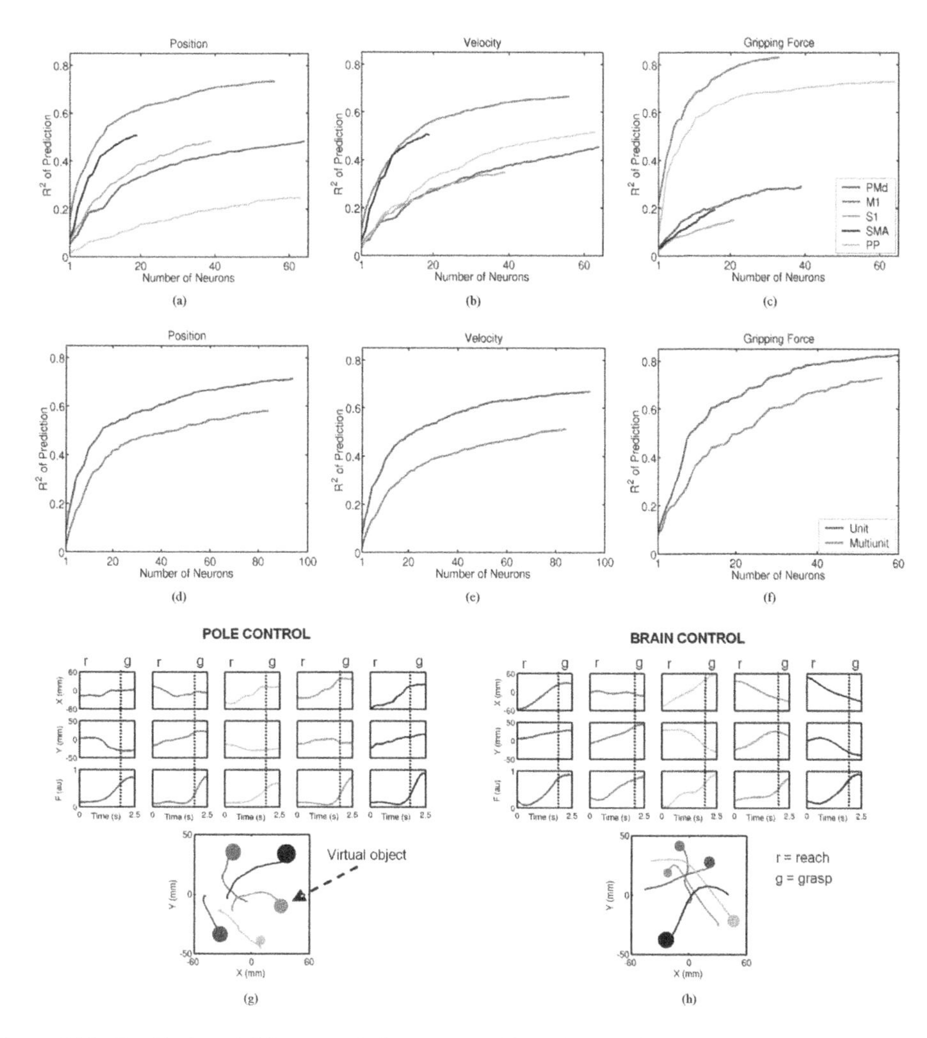

Figure 4.15 Contribution of different cortical areas to model predictions of (a) hand position, (b) velocity and (c) force. For each area, neuronal dropping curves represent average prediction accuracy (R2) as a function of the number of neurons needed to attain it. Comparison of the contribution of single units (dark) and multiunits (light) to predictions of (d) hand position, (e) velocity and (f) gripping force. Single units and multiunits were taken from all cortical areas. Representative robot trajectories and gripping force profiles during both (g) pole and (h) brain control. The bottom graphs show trajectories and the amount of the gripping force developed during grasping of each virtual object. The dotted vertical lines in the panels indicate the end of reach. Reproduced from Carmena *et al.* (2003)

are required to achieve a particular level of model prediction for each cortical area. Although all cortical areas surveyed contained information about any given motor parameter, for each area different numbers of neurons were required to achieve the same level of prediction. Although a significant sample of M1 neurons consistently provides the best predictions of all the motor parameters analysed, neurons in areas such as SMA, S1, PMd and PP also contribute to BMI performance. Another important finding of this study is that accurate real-time prediction of all motor parameters and a high level of BMI control can be achieved with multiunit signals. This observation is essential because it eliminates the need to develop elaborate real-time spike-sorting algorithms, a major technological challenge, in the design of a future cortical neuroprosthesis for clinical applications. Figure 4.15(d) to (f) shows that the linear predictions of hand position, velocity and gripping force were somewhat better when single units were used (by 17, 20, and 17 % respectively). That difference could be compensated for by increasing the number of channels. For example, in Figure 4.15(d), around 30 additional multiple units compensate for the difference in prediction of hand position by adding 20 single units. That difference was, however, not critical as the animals were still able to maintain high levels of BMI performance in all tasks using multiple-unit activity only.

These experiments demonstrated that monkeys can learn to control a BMI and produce a combination of reaching and grasping movements in order to locate and grasp virtual objects. The major challenge was to predict hand position and gripping force simultaneously using the activity recorded from the same neuronal ensemble. The monkeys' performance in brain control approximated that during pole control, with characteristic robot displacement (reach) followed by force increase (grasp). Figure 4.15(g) and (h) shows several representative examples of reaching and grasping during pole and brain control by one of the monkeys. Hand position (X, Y) and gripping force (F) records are shown. In the display of hand trajectories, the size of the disc at the end of each hand movement shows the gripping force exerted by the monkey (see Figure 4.15(g)) or by the BMI (see Figure 4.15(h)), to grasp a virtual object. The reach (r) and grasp (g) phases are clearly separate, demonstrating that the monkeys could use the same sample of neurons to produce distinct motor outputs at different moments in time. Thus, during the reaching phase, X and Y changed while F remained relatively stable. However, as the monkey got closer to the virtual object, F started to increase while X and Y stabilized to maintain the cursor over the virtual object. Note that during the brain control mode of operation, the patterns of reaching and grasping movements were preserved as in pole control.

This study also demonstrated that the initial introduction of a mechanical device such as the robot arm in the control loop of a BMI significantly impacts learning and task performance. After the robot was introduced in the control loop, the monkey had to adjust to the dynamics of this artificial actuator. As a result, there was an immediate drop in performance (data not shown). With further training, however, the animals were able to overcome the difficulties. This suggests that in order to test the limitations and challenges involved in operating a clinically useful BMI, a mechanical actuator must be incorporated to enact the subject's motor intentions, and also to train the subject to operate it.

Finally, one of the new avenues for bringing this technology into the clinical realm is the delivery of sensory feedback from the prosthetic device to the subject's brain via intracranial electrical microstimulation. While visual feedback has proven to be enough to close the loop and demonstrate the concept of brain control, it is probably insufficient for more complex motor tasks that require dexterous manipulation of objects or precise information of the limb in space. Successful encoding of proprioceptive and tactile feedback from the prosthetic device should make control of a prosthetic device more natural and therefore increase performance and reliability.

4.8 CASE STUDY: GESTURE AND POSTURE RECOGNITION USING WSNs

E. Farella and L. Benini

Computer Science and Systems, Department of Electronics (DEIS), University of Bologna, Italy

Advances in embedded systems have made it possible to design wireless sensor networks (WSNs) that are tiny, low-power, wearable and hence suitable for biomonitoring. This case study proposes a short overview of possible solutions and uses of an inertial-based wireless sensor node called WiMoCA

(Farella *et al.*, 2005), both for use alone or in a body area network to track gestures and movements for different purposes. Thanks to the flexibility of WiMoCA architecture, it was possible to implement a different node along with the ones known as 3dID glove nodes, dedicated to hand movement tracking. The general scenario is one of ambient intelligence where gestures and movements can be used as natural interfaces for human–machine interaction. Moreover, movement, posture and gait tracking may be the keys to understanding user behaviour and thus enabling seamless provision, in a smart environment, of context-aware services such as domotic applications and remote medical monitoring.

4.8.1 Platform description

Both the body area network and the glove are based on the WiMoCA wireless sensor node designed in functional layers to enhance flexibility and ease of maintenance and upgrading. The hardware layers are a power supply layer (PSL), a microcontroller and sensor layer (MSL), a wireless transmission layer (WTL) and a host interface layer (HIL) (see Figure 4.16). Each hardware layer is an independent printed circuit board with a common footprint for the cross-layer connectors. The Atmel ATMega 8 microcontroller and an MEMS three-axis digital linear accelerometer (LIS3L02DQ by ST Microelectronics) are both mounted on the MSL. On the 3dID glove five resistive bend sensors are hooked up to five of the eight available ADCs of the microcontroller to measure finger flexion. These sensors are composed of tiny patches of carbon whose resistance values change when they are bent from convex to concave shapes; average resistance values for such devices fall within the range 25–350 kΩ.

The modularity of the hardware enables different solutions to be used for wireless communication. In a first implementation, the WTL is based on the TR1001 transceiver (868 MHz) and a surface-mounted antenna. This solution is adopted when there are more than one node in use, as in body area network (BAN) applications. In this case, one node acts as a gateway by adding the HIL, which can alternatively mount a Bluetooth, RS232, USB or Ethernet interface. Thus, this node is able to communicate both with the BAN (using WTL) and with a general-purpose system (using HIL).

Whenever a single node is used, e.g. the 3dID glove as an interaction device, the WTL mounts a Bluetooth 2.0 transceiver (the WT-12 by BlueGiga, SPP interface). Since Bluetooth is now integrated in nearly all computers, a gateway can be dispensed with.

4.8.2 Implementation of concepts and algorithm

The data received from the triaxial accelerometer can be used in different ways. Integration is often used, but in this case it was avoided since it introduces measurement errors (Welch and Foxlin,

Figure 4.16 The WiMoCA layers, the WiMoCA node and the 3dID glove

2004) and has to be periodically reset, or complex filtering algorithms have to be applied (Cheok, Ganesh Kumar and Prince, 2002). Therefore the ability of accelerometers was exploited to collect the gravitational acceleration and derive the tilt of objects or body parts where the sensor is placed.

The assumption that must be made is that the dynamic component of the acceleration is negligible as compared to the gravitational component (the user moves smoothly and/or the system is in a steady state condition). Tilting of the back of the hand and finger flexion (corresponding to resistance variations in the bend sensor) can be combined to make up very simple hand gestures for use as control commands in many applications. For example, palm open and vertical can indicate that the user wants to stop something; the hand horizontal with only the second finger open can be a pointing gesture; the fist, that is all fingers closed, can be associated with a delete command or a fighting command in a game; etc. Moreover, the use of roll and pitch angles (rotation and inclination of the hand) can be mapped as horizontal and vertical two-dimensional movements of the mouse cursor, and thus the glove can substitute for the mouse on a personal computer. Flexion of a chosen finger, e.g. the thumb, can substitute for the mouse click. The 3dID glove has been used with these concepts in numerous HCI applications, e.g. in Figure 4.17 from left to right: real-time movement of a three-dimensional virtual hand; mouse substitution in typical Windows-like applications as in interaction with games such as Quake 2; use as a pointing device; static hand gesture recognition; interfacing with an MIDI application to produce a virtual instrument, together with SENIE, a Java framework implemented to support easy and fast interfacing between a personal area network and an application running on a general-purpose computer (Sama *et al.*, 2006).

Exploiting the WiMoCA's ability to collect tilt, a posture recognition application was designed for detecting user postures among a set of different possibilities. The size of this set depends on the application, which in turn affects the number of sensing nodes to be used. Each sensing module monitors the inclination of a certain part of the body. Three axis acceleration values are averaged first; then the module tilt with respect to the gravity is computed and encoded by the node itself. Thus, in this application nodes are heavily involved in the overall computation. After processing, data are sent to the gateway according to the schedule imposed by a contention-free MAC protocol implemented on each node. Each module inclination is then collected on the gateway side, frame by frame, and subsequently combined to interpret body posture. Finally, the gateway communicates the result of detection to an application running on a host machine. A sequence of postures can be used to understand user activity or to set alarms (e.g. fall detection). As only a predetermined set of postures is of interest, accuracy is not critical in this kind of application. After averaging, inclination data are only transmitted when they surpass a threshold determining the cross-point between two different postures.

In Figure 4.18 the user is equipped with three sensing units placed on the trunk, on the thigh bone and on the shin bone. This configuration can be used to determine, for example, simple postures such as sitting, standing and lying (right, left side, prone and supine). At each node data are acquired, averaged in a predefined time window and finally used to distinguish the orientation from among three different configurations for each axis: (a) in the direction of, (b) orthogonal to or (c) opposed to the gravitational acceleration.

Figure 4.17 Uses of the 3Did device based on WiMoCA technology

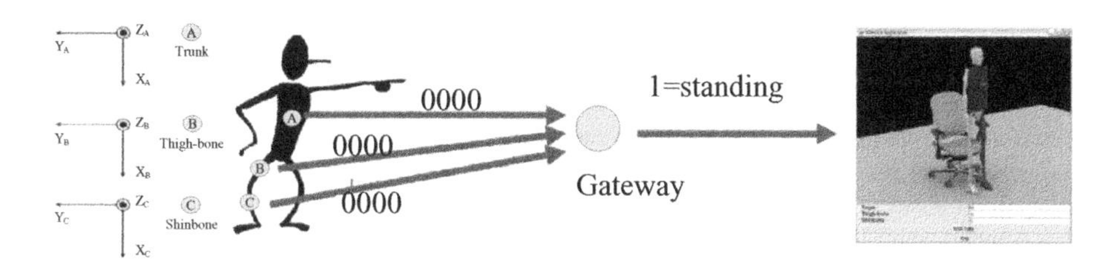

Figure 4.18 Example of posture recognition with three WiMoCA nodes

The set of three configurations is associated with a specific 4-bit code identifying the tilt of the module; for instance, in Figure 4.18 '0000' is the configuration for node in the direction of gravity. The number of possible recognized tilts can be varied according to application requirements, introducing more than three subranges for acceleration quantization. At the end of the tilt identification process, the module tilt code is sent to the gateway. On the gateway side, the firmware acquires and accumulates tilt data from each end-module for a given frame and combines all module tilts to identify a known body configuration, e.g. 'standing' as in Figure 4.18.

4.8.3 Posture detection results

In order to provide an idea of the performance attained by the WiMoCA BAN, a posture detection application based on three nodes was considered (see Figure 4.18). The total power consumption for each node at the maximum sampling rate is 46 mW in the worst case (all '1' sent in the OOK mode) and 16.85 mW in the best case (all '0'). With three nodes, a maximum sample rate of 651 positions per second (pps) is achieved (1.536 ms latency). In practical cases, the maximum frequency of human movement is 30 Hz, so that 60 pps (16.7 ms latency) is a high enough rate to detect postures without losing information. It is worth noting that this corresponds to an overall data throughput of 660 bits per second (bps) for each node.

The tracking performance of a sequence of postures was tested on 20 subjects. They were asked to repeat a sequence of five postures. One of the five was not included in the posture table in order to test system reliability against false positives. Thus, 100 recognitions were tested at 20 pps. The system successfully detected the four postures included in the pre-stored table without false positives. In the presence of an unknown position, the system returns the 'none' code.

4.8.4 Challenges: wireless sensor networks for motion tracking

With the solutions described, it is possible to embed the recognition algorithms in the nodes, and only a small amount of processing is required on a host unit receiving data from the BAN. A different choice was made in the case of a biofeedback application where the sensor network is distributed on the user's body for balance monitoring and correction (Brunelli *et al.*, 2006). This was an important feasibility study for implementation of a mobile recognition system using feedback actuation. Software architecture is characterized by nodes performing basic sensing functions with almost no on-board processing. All the computation is performed on the palmtop computer (personal digital assistant (PDA)), which interfaces with the wireless body network by means of the gateway node. The PDA is responsible for data processing and is in charge of activating the actuator that provides the user with feedback on the correct user posture. In the setup that was implemented, the system adopted was audio feedback through a headset.

This application differs from the posture recognition system in the software organization, and also because it uses accelerometers for fine position tracking. Accuracy requirements are much stricter w.r.t. posture recognition. The data that are collected and transmitted are acceleration values, which are then processed on the host side. The hardware and software architecture (communication protocols, power management policies and application-level control) have been tuned to optimize the cost, battery autonomy and real-time performance required for this application.

More complex algorithms can be implemented on a general-purpose computer; e.g. data from acceleration can be processed to extract features for training of a classification system and to implement movement or activity recognition. In this connection a gait recognition system was designed that runs on a desktop computer that processes data from a WiMoCA node placed on the ankle. Unlike posture detection systems, accelerometers are not used as inclinometers. The system recognizes movements in real time, but the processing of sensing data involves two phases: the training phase, in which the user submits a set of sample movements to train the system, and the recognition phase, in which the trained system recognizes the movements. Moreover, each phase comprises at least four processing steps: (a) windowing or segmentation, (b) pre-processing (e.g. filtering), (c) feature extraction and (d) classification.

4.8.5 Summary and outlook

In conclusion, applying sensor network technology to human motion tracking and gesture recognition imposes new design requirements. Certainly, nodes must be extremely light and wearable to avoid obstructing user movement. For this reason, and to guarantee an adequate system lifetime, there is limited room for resources on end-nodes, and so they must be efficiently managed. Yet, given the scope and variety of the motion tracking application scenario (from entertainment to healthcare and domotics, from mobile to indoor use or integration with smart environments, etc.), the tuning of design parameters is application-driven and is not trivial.

A bidimensional design space can be conceived whose axes are the bandwidth use and the level of processing distribution. Existing solutions exploit a small number of close-lying nodes, and for a limited time (less than 24 hours). Thus, data streaming is possible and processing can be centralized on the gateway side, where in many cases more resources are available. As soon as the number of nodes is increased, e.g. to monitor a large number of physiological parameters continuously for many days or to add actuation nodes, this solution ceases to be scalable and wastes energy resources. Distributed processing can be used to reduce bandwidth use and power consumption at each node.

REFERENCES

Arieta, A.H., Katoh, R., Yokoi, H., Wenwei, Y., 2006, 'Development of a multi-DOF electromyography prosthetic system using the adaptive joint mechanism', *Applied Bionics and Biomechanics* **3**(2): 101–112.

Bar-Shalom, Y., Li, X.R., 1998, *Estimation and Tracking: Principles, Techniques, and Software*, Artech House Publishers.

Bashashati, A., Ward, R.K., Birch, G.E., 2005, 'A new design of the asynchronous brain computer interface using the knowledge of the path of features', in *Proceedings of the Second International IEEE Engineering in Medicine and Biology Society Conference on Neural Engineering*, pp. 101–104.

Bigland-Ritchie, B., Johansson, R., Lippold, O.C., Smith, S., Woods, J., 1983, 'Changes in motoneurone firing rates during sustained maximal voluntary contractions', *Journal of Physiology* **340**: 335–346.

Blok, J., Stegeman, D., Freriks, B., Hermens, H., 1999, 'The SENIAM model for surface electromyogragroaphy', *SENIAM, European Recommendations for Surface Electromyography* **1**: 71–80.

Bonato, P., D'Alessio, T., Knaflitz, M., 1998, 'A statistical method for the measurement of muscle activation intervals from surface myoelectric signal during gait', *IEEE Transactions on Biomedical Engineering* **45**: 287–299.

Boostani, R., Moradi, M.H., 2003, 'Evaluation of the forearm EMG signal features for the control of a prosthetic hand', *Physiological Measurement* **24**: 309–319.

Bosch, P., Huijing, P., 1996, 'Length dependent recruitment behaviour in fast skeletal muscle', in *Engineering in Medicine and Biology Society, Bridging Disciplines for Biomedicine, Proceedings of the 18th Annual International Conference of the IEEE*, vol. 2, pp. 553–554.

Brashear, H., Starner, T., Lukowicz, P., Junker, H., 2003, 'Using multiple sensors for mobile sign language recognition', in *IEEE International Symposium on Wearable Computers*, pp. 45–52.

Brookner 1998, *Tracking and Kalman Filtering Made Easy*, John Wiley & Sons, Inc.

Brunelli, D., Farella, E., Rocchi, L., Dozza, M., Chiari, L., Benini, L., 2006, 'Bio-feedback system for rehabilitation based on wireless body area network', in *Proceedings of the Fourth IEEE International Conference on Pervasive Computing*, pp. 527–531.

Brunetti, F., Rocon, E., Manto, M., Pons, J.L., 2005, 'Instantaneous detection of neuro-oscillators using a portable tool', *Proceedings of the 2nd IEEE Engineering in Medicine and Biology Society Conference on Neural Engineering*, pp. 555–558.

Carignan, C., Liszka, M., Roderick, S., 2005, 'Design of an arm exoskeleton with scapula motion for shoulder rehabilitation' in *Proceedings of the 12th IEEE International Conference on Advanced Robotics*, pp. 524–531.

Carmena, J.M., Lebedev, M.A., Crist, R.E., O'Doherty, J.E., Santucci, D.M., Dimitrov, D., Patil, P.G., Henriquez, C.S., Nicolelis, M.A.L., 2003, 'Learning to control brain–machine interface for reaching and grasping by primates', *PloS Biology* **1**: 192–208.

Cavallaro, E., Rosen, J., Perry, J., Burns, S., 2006, 'Real-time myoprocessors for a neural controlled powered exoskeleton arm', *IEEE Transactions on Biomedical Engineering* **53**(11): 2387–2396.

Cheok, A., Ganesh Kumar, K., Prince, S., 2002, 'Micro-accelerometer based hardware interfaces for wearable computer mixed reality applications', in *Proceedings of the Sixth International Symposium on Wearable Computers*, pp. 223–230.

De Luca, C., 2006, *Encyclopedia of Medical Devices and Instrumentation*, John Wiley and Sons, Inc., pp. 98–109.

Dornhege, G., 2006, 'Increasing information transfer rates for brain–computer interfacing, PhD Thesis, University of Potsdam, Germany.

Duchene, J., Hogrel, J., 2000, 'A model of EMG generation'. *IEEE Transactions on Biomedical Engineering* **47**(2): 192–201.

Duda, R., Hart, P., Stork, D., 2001, *Pattern Classification*, Wiley Interscience.

Elble, R.J., 1990, *Tremor*. The Johns Hopkins University Press, Baltimore, Maryland.

Farella, E., Pieracci, A., Brunelli, D., Acquaviva, A., Benini, L., Riccò, B., 2005, 'Design and implementation of wimoca node for a body area wireless sensor network', in *Proceedings of the IEEE International Conference on Sensor Networks*, pp. 342–347.

Farina, D., Crosetti, A., Merletti, R., 2001, 'A model for the generation of synthetic intramuscular EMG signals to test decomposition algorithms, *IEEE Transactions on Biomedical Engineering* **48**(1): 66–77.

Ferris, D.P., Gordon, K.E., Sawicki, G.S., Peethambaran, A., 2006, 'An improved powered ankle-foot orthosis using proportional myoelectric control, *Gait and Posture* **23**: 425–428.

Flanders, M., 2002, 'Choosing a wavelet for single-trial EMG', *Journal of Neuroscience Methods* **116**: 165–177.

Franken, H.M., 1995, 'Cycle to cycle control of swing phase of paraplegic gait induced by surface electrical stimulation', *Medical and Biological Engineering and Computing* **33**: 440–451.

Fuglevand, A., Winter, D., Patla, A., 1993, 'Models of recruitment and rate coding organization in motor-unit pools', *Journal of Neurophysiology* **70**(6): 2470–2488.

Fukunaga, K., 1990, *Introduction to Statistical Pattern Recognition*, Academic Press.

García, G.N., Ebrahimi, T., Vesin, J., 2003, 'Correlative exploration of EEG signals for direct brain–computer communication', in *Proceedings of the IEEE International Conference on Acoustics, Speech, and Signal Processing*, vol. 5, pp. 816–819.

Graupe, D., Cline, W., 1975, 'Functional separation of EMG signals via ARMA identification methods for prosthesis control purposes', *IEEE Transactions on Systems, Man and Cybernetics* **5**: 252–258.

Henneman, E., 1957, 'Relation between size of neurons and their susceptibility to discharge' *Science* **26**: 1345–1347.

Henneman, E., 1965, 'Excitability and inhibitability of motorneuron of different sizes', *Journal of Neurophysiology* **28**: 599–620.

Hermens, H.J., v. Bruggen, T.A.M., Baten, C.T.M., Rutten, W.L.C., Boom, H.B.K., 1992, 'The median frequency of the surface EMG power spectrum in relation to motor unit firing and action potential properties', *Journal of Electromyography and Kinesiology* **1**: 15–25.

Hermens, H., Freriks, B., Merletti, R., Stegeman, D., Blok, J., Rau, G., Disselhorst-Klug, C., Hägg, G., 1999, 'European recommendations for surface electromyography', *Roessingh Research and Development* **8**.

Horikawa, S., Furuhashi, T., Okuma, S., Uchikawa, Y., 1990, 'A fuzzy controller using a neural network and its capability to learn experts control rules', in *Proceedings of the International Conference on Fuzzy Logic and Neural Networks*, pp. 103–106.

Hoya, T., Hori, G., Bakardjian, H., Nishimura, T., Suzuki, T., Miyawaki, Y., Funase, A., Cao, J., 2003, 'Classification of single trail EEG signals by a combined principal + independent analysis and probabilistic neural network approach', in *Fourth International Symposium on Independent Component Analysis and Blind Signal Separation*, pp. 197–202.

Hyvärinen, A., Oja, E., 2000, 'Independent component analysis: algorithms and applications', *Neural Networks* **13**(4–5): 411–430.

Jang, J., 1993, 'ANFIS: adaptive-network-based fuzzy inference system', *IEEE Transactions on Systems, Man and Cybernetics* **23**: 665–684.

Kang, W., Shiu, J., Cheng, C., Lai, J., Tsao, H., Kuo, T., 1993, 'Cepstral coefficients as the new features for electromyography (EMG) pattern recognition', in *Proceedings of the 15th Annual International Conference of the IEEE Engineering in Medicine and Biology Society*, pp. 1143–1144.

Lacuesta, J.S., Prat, J.M., Hoyos, J.V., 1999, *Biomecánica de la Marcha Humana y Patológica*, Publicaciones IBV.

Lloyd, D.G., Besier, T.F., 2003, 'An EMG-driven musculoskeletal model to estimate muscle forces and knee joint moments *invivo*', *Journal of Biomechanics* **36**: 765–776.

Malmivuo, J., Plonsey, R., 1995, *Bioelectromagnetism: Principles and Applications of Bioelectric and Biomagnetic Fields*, Oxford University Press.

Mayagoitia, R., Nene, A., Veltink, P., 2002, 'Accelerometer and rate gyroscope measurement of kinematics: an inexpensive alternative to optical motion analysis systems', *Journal of Biomechanics* **35**(6): 537–542.

Merletti, R., Farina, D., Gazzoni, M., Merlo, A., Ossola, P., Rainoldi, A., 2001, 'Surface electromyography. A window on the muscle, a glimpse on the central nervous system', *Europa Medicophysica* **37**: 57–68.

Moreno, J.C., Rocon, E., Ruiz, A., Brunetti, F., Pons, J.L., 2006, 'Design and implementation of an inertial measurement unit for control of artificial limbs: application on leg orthoses', *Sensors and Actuators B* **118**: 333–337.

Musallam, S., Corneil, B.D., Greger, B., Scherberger, H., Andersen, R.A., 2004, 'Cognitive control signals for neural prosthetics', *Science* **305**: 258–262.

Obermaier, B., Müller, G.R., Pfurtscheller, G., 2003, '"Virtual Keyboard" Controlled by Spontaneous EEG Activity', *IEEE Transactions on Neural and Rehabilitation Engineering* **11**(4): 422–426.

Omidvarnia, A.H., Atri, F., Setarehdan, S.K., Arabi, B.N., 2005, 'Kalman filter parameters as a new EEG feature vector for BCI applications', in *Proceedings of the 13th European Signal Processing Conference*.

Pappas, I., Popovic, M., Keller, T., Dietz, V., Morari, M., 2001, 'A reliable gait phase detection system', *IEEE Transactions on Neural Systems and Rehabilitation Engineering* **9**(2): 113–125.

Park, S., Lee, S., 1998, 'EMG pattern recognition based on artificial intelligence techniques', *IEEE Transactions on Rehabilitation Engineering* **6**(4): 400–405.

Pattichis, C., Schizas, C., Middleton, L., 1995, 'Neural network models in EMG diagnosis', *IEEE Transactions on Biomedical Engineering* **42**(5): 486–496.

Qin, L., Ding, L., He, B., 2004, 'Motor imagery classification by mean of source analysis for brain–computer interface applications', *Journal of Neural Engineering* **1**: 135–141.

Rabiner, L.R., 1989, 'A tutorial on hidden Markov models and selected applications in speech recognition', *Proceedings of the IEEE* **77**(2): 257–286.

Ratanaswasd, P., Dodd, W., Kawamura, K., Noelle, D., 2005, 'Modular behavior control for a cognitive robot', in *Proceedings of the 12th International Conference on Advanced Robotics*, pp. 713–718.

Riviere, C.N., Reich, S.G., Thakar, N.V., 1997, 'Adaptive Fourier modeling for quantification of tremor', *Journal of Neuroscience Methods* **74**: 7–87.

Rocon, E., Ruiz, A.F., Pons, J.L., Belda-Lois, J.M., Sánchez-Lacuesta, J.J., 2005, Rehabilitation robotics: a wearable exo-skeleton for tremor assessment and suppression', in *Proceedings of the International Conference on Robotics and Automation*, pp. 241–246.

Rocon, E., Belda-Lois, J.M., Ruiz, A.F., Manto, M., Pons, J.L., 2007, 'Design and validation of a rehabilitation robotic exoskeleton for tremor assessment and suppression', *IEEE Transactions on Neural Systems and Rehabilitation Engineering* **15**(3): 367–378.

Salu, Y., Cohen, L., Rose, D., Sxato, S., Kufta, C., Hallett, M., 1990, 'An improved method for localizing electric brain dipoles', *IEEE Transactions on Biomedical Engineering* **37**(7): 699–705.

Sama, M., Pacella, V., Farella, E., Benini, L., Riccò, B., 2006, '3did: a low-power, low-cost hand motion capture device', in *Proceedings of the IEEE Design, Automation and Test in Europe Conference and Exhibition*, pp. 136–141.

Santhanam, G., Ryu, S.I., Yu, B.M., Afshar, A., Shenoy, K.V., 2006, 'A high-performance brain–computer interface', *Nature* **442**: 195–198.

Scholtz, J.C., 2002, 'Human–robot interactions: creating synergistic cyber forces', in *Proceedings of the NRL Workshop on Multi-Robot Systems*, pp. 177–184.

Serruya, M.D., Hatsopoulos, N.G., Paninski, L., Fellows, M.R., Donoghue, J.P., 2002, 'Instant neural control of a movement signal', *Nature* **416**: 141–142.

Sharma, R., Pavlovic, V.I., Huang, T.S., 1998, 'Toward multimodal human–computer interface', *Proceedings of the IEEE* **5**(86): 853–869.

Tong, K., Granat, M., 1999, 'A practical gait analysis system using gyroscopes', *Medical Engineering and Physics* **21**(2): 87–94.

Ward, L.M., 2003, 'Synchronous neural oscillations and cognitive processes', *Trends in Cognitive Sciences* **7**(12): 553–559.

Welch, G., Foxlin, E., 2004, 'Motion tracking: no silver bullet, but a respectable arsenal', *Computer Graphics and Applications* **22**: 24–38.

Wessberg, J., Nicolelis, M.A.L., 2004, 'Optimizing a linear algorithm for real-time robotic control using chronic cortical ensemble recordings in monkeys', *Journal of Cognitive Neuroscience* **16**: 1022–1035.

Yamato, J., Ohya, J., Ishii, K., 1992, 'Recognizing human action in time–sequential images using hidden Markov model', in *Proceedings of the IEEE Conference on Computer Vision and Pattern Recognition*, pp. 379–385.

5

Human–robot physical interaction

E. Rocon[1], A. F. Ruiz[1], R. Raya[1], A. Schiele[2,3] and J. L. Pons[1]

[1]*Bioengineering Group, Instituto de Automática Industrial, CSIC, Madrid, Spain*
[2] *European Space Agency (ESA), Mechanical Engineering Department, Automation & Robotics Section, Noordwijk, The Netherlands*
[3] *Mechanical Engineering Faculty, Biomechanical Engineering Department, Delft University of Technology (DUT) Delft, The Netherlands*

5.1 INTRODUCTION

One important specific feature of wearable robotics is the intrinsic interaction between human and robot. This interaction, in its simplest manifestation, implies a physical coupling between the robot and the human, leading to the application of controlled forces between both actors. The actions of the two agents must be coordinated and adapted reciprocally since unexpected behaviour of one of them during interaction can result in severe injuries. A classic example of physical interaction is exoskeleton-based functional compensation of human gait. Here, the robotic exoskeleton applies functional compensation by supporting human gait, i.e. by stabilizing the stance phase. The fact that a human being is an integral part of the design is one of the most exciting and challenging aspects in the design of wearable robots. It imposes several constraints and requirements in the design of this kind of device. In this book, the term pHRI refers to human–robot physical interaction and pHRI refers to the human–robot physical interface. Both terms are defined in Chapter 1. In this chapter the analysis of pHRI and pHRI will focus on exoskeletons. The authors consider that, without losing sight of the general picture, the design challenges for successful physical interaction are more illustrative in robots of this kind.

The application of force and pressure to the human body is a special concern in pHRI. The next section introduces some physiological aspects related to pHRI and briefly describes the physiology of the human touch sensory system, in order to explain how humans react to a robot-generated force.

Wearable Robots: Biomechatronic Exoskeletons Edited by José L. Pons

5.1.1 Physiological factors

Humans have a force and pressure sensory system composed of a number of sensory receptors. The sensors directly implicated in touch perception are mechanoreceptors, which are stimulated by mechanical forces. Mechanoreceptors can detect pressure, touch, vibration, strain and tactile sensation and are mostly located in the human skin (Despopoulos and Silbernagl, 1991).

Pressure and tactile sensors are basically displacement sensors and are sensitive to skin deformation. Tactile receptors are located more superficially. Pressure receptors are located deeper in the skin structure.

The dermis of the skin houses sensory receptors for touch, pressure, pain and temperature. The number of receptors varies according to the region of the skin. The fingertips have numerous touch receptors. Skin has five principal types of receptors: free receptors (nerve endings), Meissner corpuscles, Merkel discs, Paciani corpuscles and Ruffini corpuscles (see Figure 5.1).

Nerve endings are located near the skin surface. These free endings respond to temperature and pain sensations. Meissner corpuscles are velocity sensors and react to light touch. Merkel discs detect pressure and vibration and are the most sensitive corpuscles to low-frequency vibrations, around 5 to 15 Hz. Pacinian corpuscles act as acceleration sensors, detecting pressure changes and high-frequency vibrations, around 200–300 Hz, deep in the skin. Finally, Ruffini corpuscles react to pressure and skin stretch (Despopoulos and Silbernagl, 1991).

There are other important mechanoreceptors, the so-called proprioceptors. These provide information about a person's state rather than about interaction with the environment. There are proprioceptors in muscles, joints, tendons and several other parts. Proprioception provides information on the orientation of limbs with respect to one another, whether still or in motion, and the relative speed. Proprioceptors generate a sense of position, a sense of movement and a sense of force. That sensitivity is essential to orientate movements and to be aware of the position of the limbs when exploring objects. In human–robot interaction, haptics refers to the use of robotic interfaces and devices for

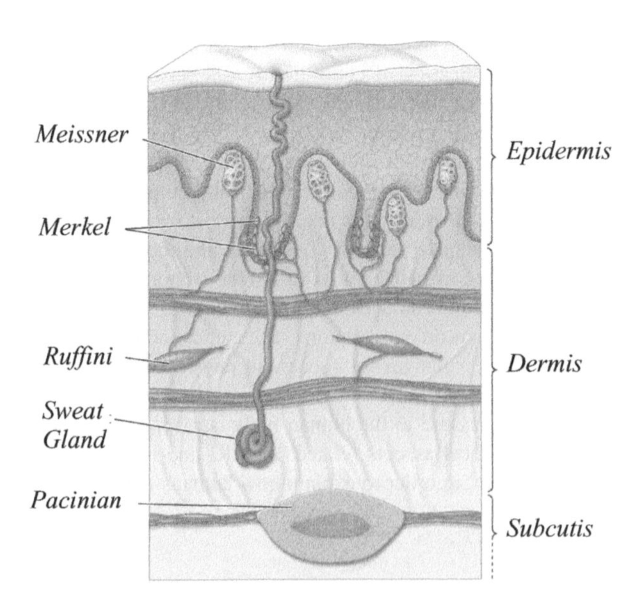

Figure 5.1 Sensory receptors in the human skin

force feedback in human–computer interfaces. Proprioceptive and touch senses have an important role in the context of haptic applications.

Force-reflecting interfaces are robotic devices used to display touch- or force-related sensory information to the user from a remote or virtual environment. This is achieved by delivering feedback to the various joints of human limbs, and tactile sensations such as texture, surface roughness, etc., to the hand. Touch and proprioceptive feedback from a haptic environment impose very different sets of device design criteria.

The physiological factors discussed above, such as human force sensing, pressure perception and proprioception, enter into the design phase of wearable robots. For instance, psychophysical experiments have been carried out to provide information for the design of force-reflecting haptic displays. The following section looks at a number of other design aspects.

5.1.2 Aspects of wearable robot design

pHRI represents the most critical form of interaction between humans and machines. Any motion of the robot that occurs in contact with a human, and any force exerted by the robot, has to be soft and compliant and must never exceed the force exerted by the human to protect himself/herself. In addressing this topic, the design of the robot structure, sensors, actuators and control architecture need to be considered all together, from the specific viewpoint of interaction with humans.

With regard to pHRI, the design goals of any wearable robot (WR) should generally consider:

- *Safety*. It is paramount for pHRI to guarantee safe operation. The WR should avoid unnatural or arbitrary movements, for instance excessive excursions that could hyperextend or hyperflex human joints.
- *Actuator performance*. WRs serve a large number of applications. The particularities of each application define the constraints in the design of the actuator system. For instance, rendering haptic sensations imposes strict requirements on the actuator design.
- *Ergonomics and comfort*. One of the challenges in the design of wearable robots is the ability to adapt to the specific needs and ergonomic particularities of humans (Schiele and van der Helm, 2006). The WR and the human's biological joints must be exactly aligned for proper operation. Misalignment of joints could generate interaction forces and may produce pressure sores on the skin of the wearer. Moreover, the robot's operational workspace should be compatible with the natural limb workspace. This problem calls for the design of truly ergonomic wearable robot structures and is analysed in Section 5.2.
- *Application of loads to humans*. pHRI causes the transmission of loads to the human musculoskeletal system through soft tissues. This raises the question of the intensity, the mode and the areas on the human body where it is possible to apply loads. It is a topic that requires special attention since it defines how the wearable robot is to be coupled to the human limb; it is addressed in detail in Section 5.3. Moreover, the application of controlled forces between the human and the robot requires the development of advanced control schemes (Lemay, Hogan and Van Dorsten, 1998), an issue that is analysed in detail in Section 5.4.
- *Control strategies*. pHRI in wearable robots involves the cooperation of two dynamic control systems, i.e. human motor control and robot control, in a closed loop system. Both systems should be able to adapt to each other in order to achieve a common goal stably. This issue is crucial in the design of WR systems and is dealt with in detail in Section 5.4.
- *Ease of use*. Final solutions should be easy to don, adjust, use and remove. This imposes constraints with regard to the size and weight of the WR.

The aim of this chapter is to evaluate the influence of pHRI in the design of wearable robots. The chapter discusses the following key aspects in detail: the kinematic compatibility between humans and robots, the application of loads to humans and control strategies for better human physical interaction. Finally, four case studies are presented to illustrate the different aspects analysed in this chapter.

5.2 KINEMATIC COMPATIBILITY BETWEEN HUMAN LIMBS AND WEARABLE ROBOTS

A good wearable exoskeleton design starts with the choice of a suitable kinematic structure for the device. That is to say, even before implementing actuation and control of a device, the purely mechanical structure must enable wearability, ease of use and operator comfort. Comfort depends partly on the ability of the exoskeleton to interact smoothly with the human; in other words, the exoskeleton must not limit the range of motion of the operator and must otherwise be as transparent as possible. For simplicity's sake, this chapter focuses on arm exoskeletons but is applicable to other devices, for instance for the human leg, torso or fingers.

In Section 3.4 different types of existing exoskeleton kinematic structures were introduced, which were (a) *end-point* or *end-effector based*, (b) *wearable* and *kinematically equivalent* to the human arm or (c) *wearable* but *kinematically not equivalent* to the human arm through inclusion of extra degrees of freedom (see Figure 3.12 for a schematic illustration).

In Section 3.3 an approach to modelling the human limb kinematics and dynamics were presented. There, the mechanical structure of the human limb was modelled in a formalism usually applied to robots, which was reduced to describe a chain of rigid links that behave like a serial manipulator with a total of 7 DoFs. The serial chain properties were then parameterized by means of the Denavit–Hartenberg convention.

For the process of exoskeleton design it is important first of all to have a good model describing the human limb. This model, which here will be called the human limb model, can be used in simulations to test proposed exoskeleton mechanisms for their ability to interact with the human limb. It therefore makes good sense to use the same modelling approach for both the human limb and the exoskeleton. Computer aided design (CAD) tools or other computer simulations can then be used to quantify the human–machine interaction before the hardware is built. However, in order to design an exoskeleton that is truly compatible with human movement and is ergonomic, safe and comfortable once it is built, there should be awareness of the fact that human biomechanics does not actually behave like a conventional serial chain robot.

5.2.1 Causes of kinematic incompatibility and their negative effects

There are a great many effects that can contribute to kinematic incompatibility between a wearable exoskeleton and a real human limb, once an exoskeleton has been implemented in hardware. The reason for that is the real-life variability of biomechanic parameters between subjects, and also the variability of some parameters within individual subjects during movement. Unpredictability of joint axis locations and body segment sizes, for instance, can disturb the interaction between an exoskeleton and the human operator, depending on the exoskeleton's kinematic design. This especially applies to exoskeletons that are wearable and kinematically equivalent to the human arm.

Typical biomechanical effects that cannot easily be captured within a human arm model used for exoskeleton development include:

- The intersubject variability of human limb link parameters (D–H parameters such as length of bones, distances between rotation axes, orientations of rotation axes).

- The variability within an individual subject of joint centres of rotation during movement. The ICRs of each anatomical joint move very little during joint motion.
- The intersubject variability of body segment dimensions: mass, size, volume and so forth.

These anatomical variations make it difficult to produce an exoskeleton design that fits a wide range of users without problems. Another challenge to the pHRI design is the fact that there are no precise anatomical data for the operators. All joints are covered with soft tissue, cartilage and muscles, so the ICRs are not visible and are hard to measure by other *in situ* means.

Causes of kinematic incompatibilities between humans and exoskeletons can then be classified into two groups, (1) macromisalignments and (2) micromisalignments, between human joint axes and exoskeleton joint axes:

1. Macromisalignments can occur if exoskeleton joints for interaction with specific human joints or joint groups are oversimplified. Oversimplification in this case essentially means that an exoskeleton joint or joint group has less degrees of freedom than the corresponding human joint or joint group. This applies to all wearable exoskeletons featuring seven degrees of freedom. Probably the best example of this arises when investigating interaction with the human shoulder. Shoulder movement typically involves not only motions in the glenohumeral joint, which is often modelled as the 3 DoF shperical joint base in the human arm. It also involves movement of the shoulder girdle, to which the clavicle and the shoulder blade belong. The movement of the shoulder girdle in turn influences the location of the glenohumeral joint. Figure 5.2 illustrates the vertical translation of the glenohumeral ICR caused by shoulder elevation in a sagittal plane. A significant shift of the glenohumeral ICR occurs in elevations exceeding 90°, i.e. already in the mid shoulder joint range. Pure vertical shift is still simplified but helps to illustrate the concept. In reality, there is also a horizontal translation of the joint (for other movements). If an exoskeleton kinematic structure for the shoulder is now designed in order to align to the glenohumeral joint only, the corresponding

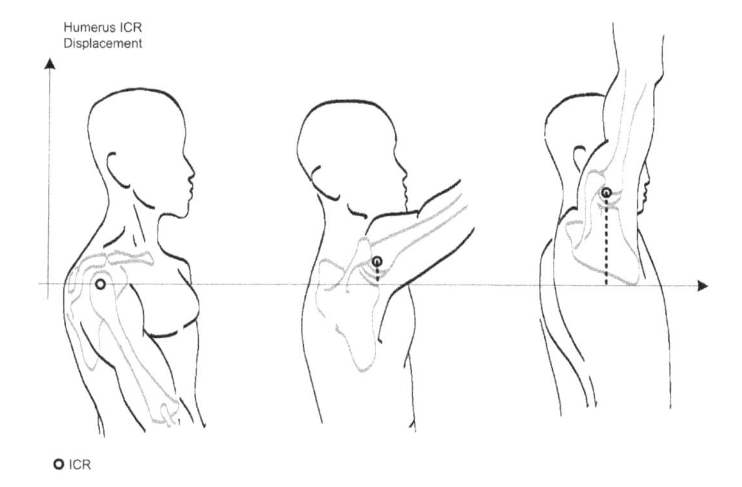

Figure 5.2 Translation of the instantaneous centre of rotation of the glenohumeral shoulder joint during shoulder elevation in a sagittal plane. This shows that human joint centres of rotation are nonstationary. An ergonomic exoskeleton will have to cope with such alterations of the human anatomical structure during movement. Reproduced from ESA

human joint will increasingly translate away from the exoskeleton joint ICR during motion. This will result in large offsets between the exoskeleton joints and the anatomical joints.

Macromisalignments are thus induced by a mismatch between the degrees of freedom of human limb motion and exoskeleton link motion. In end-point-based exoskeletons, the negative effect of such macromisalignments is not significant. In wearable exoskeleton interfaces, however, they impose severe restrictions on the common available workspace shared with the human limb. To the authors' knowledge, at present most arm exoskeletons that feature a spherical joint set for the shoulder have a restricted shoulder workspace.

2. Micromisalignments are less obvious but occur in all wearable exoskeleton designs. They occur even if the number of degrees of freedom between the exoskeleton and the human joints is correct – for instance, if an exoskeleton has joints to track shoulder girdle movement by aligning two additional axes to the sternoclavicular joint (the joint that connects the shoulder girdle to the torso). Micromisalignments are still caused by noncoincident joint rotation axes between the exoskeleton and the human limb. This is almost always the case because it is not possible to align an exoskeleton perfectly to the human joints, due to intersubject variability and coverage of the joints as explained above. Now, imagine an elbow exoskeleton as shown in Figure 5.3(a), i.e. worn by an operator on his elbow. The device's ICR will always present a small offset toward the human elbow ICR. This is, firstly, because the operator's elbow ICR is not exactly known, so perfect manual alignment of the exoskeleton to this joint is impossible, and, secondly, because the biological joint surfaces are not ideally circular. This means that the ICR shifts during motion. Micromisalignment can furthermore be caused by slippage of the exoskeleton attachments on the human skin during motion. Slippage-induced offsets have been reported in the literature (Colombo, Wirz and Dietz, 2001) for the LOKOMAT gait orthosis. There, it caused greater misalignments between the orthosis joints and the human joints, leading to stumbling during test sessions with patients. A significant negative effect of micromisalignments is the creation of interaction forces,

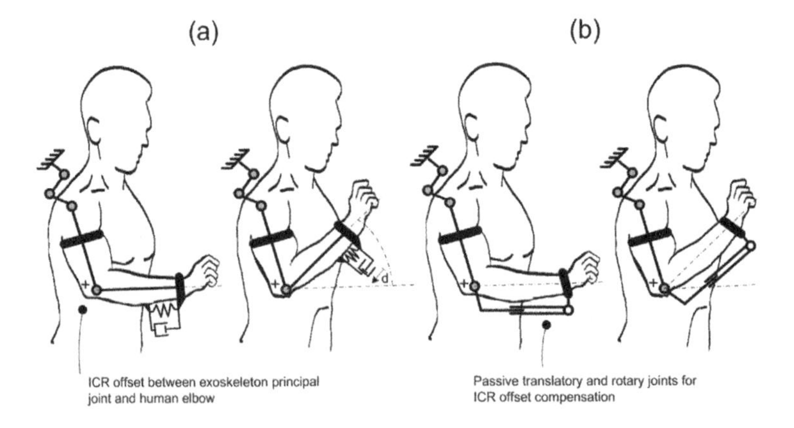

Figure 5.3 (a) Illustration of the creation of an interaction force as a consequence of joint misalignments between the exoskeleton and the human limb. During motion, the exoskeleton slides on the human limb. (b) If passive joints are added into the structure (Schiele and van der Helm, 2006), these forces are not created and the joints compensate for the slipping of the device. Reproduced from ESA

such as shear forces between the exoskeleton attachment point and the human limb. These forces are created by displacements d (Figure 5.3(a)) of the exoskeleton along the limb. In the figure, the stiffness and damping between the human and the exoskeleton robot are represented by a Voight element. Its parameters determine the magnitude of such interaction forces.

Hidler and Wall (2005), show that interaction forces alter the natural motion patterns in combined human–exoskeleton motion. Case Study 5.5 in this book quantifies those constraint displacements and shows typical magnitudes of measured interaction forces in an exoskeleton interface with non-ergonomic setting.

Macro- as well as micromisalignments contribute to operator discomfort and can limit and alter the natural movement of a human limb inside an exoskeleton. This raises the question of how a kinematic structure for an exoskeleton should be designed to provide a maximum of comfort and ease of use despite the variability of anatomic properties within and between users.

5.2.2 Overcoming kinematic incompatibility

While a purely end-effector-based exoskeleton is independent of joint alignments, no exact motion can be induced in the human joints due to the natural redundancy of the human limbs (see Section 3.4). As explained earlier, such devices cannot resolve natural human limb redundancy and can therefore potentially apply harmful loading to human joints. Recall the application of a force to an outstretched human arm in the direction of the shoulder (in line with the arm). Such a force could move the operator's joints in arbitrary directions, producing extension or hyperextension, and the operator could suffer considerable harm. A safe, ergonomic interface should therefore not be end-effector-based.

The first ergonomic design criteria for exoskeleton-based human–robot interaction were presented by Schiele and van der Helm (2006). For a truly ergonomic exoskeleton design to be feasible, the authors postulate that an exoskeleton must be able to:

- Interact with the complete functional workspace of the human limb of interest;
- Induce exact torque, position and velocities to the human joints (and hence be able to resolve the redundancy);
- Must not cause discomfort or safety hazards for the user.

This can be achieved by applying the ergonomic design paradigm presented in the referenced article:

- An exoskeleton should be wearable, not end-point based.
- An exoskeleton should never have more then 6 DoFs between two consecutive attachments.
- An exoskeleton kinematic structure must explicitly not copy the kinematic structure of the adjacent human limb. Thus, the exoskeleton should be *not kinematically equivalent* to the human limb it interacts with.

The purpose of the last point is to deal with negative effects stemming from Micromisalignments. If an exoskeleton has a kinematic structure i.e. different from the human arm but can provide the same Cartesian motion between two attachments, no misalignment can negatively affect motion or user comfort. Additional passive compensation joints can be incorporated into an exoskeleton kinematic structure to compensate constraint displacements between the device and the human limb. Figure 5.3(b) illustrates an elbow articulation with such passive joints. An exoskeleton designed with such a kinematic structure will be easy to use, and adjustments to individuals will no longer be required. The exoskeleton introduced in Case Study 8.3 follows such an ergonomic design approach.

5.3 APPLICATION OF LOAD TO HUMANS

The function of most wearable robots relies on the application of loads to the human musculoskeletal system through soft tissues. Inadequate application of forces can cause problems such as fatigue to the user. Fatigue is defined as a temporary loss of strength and energy resulting from hard physical or mental work. It has been demonstrated that a continuous application of mechanical loads to human limbs originates from a loss of endurance, and this must be taken into account. Quantifying fatigue is a subjective task. Some authors use a methodology consisting in moving a body part while subjected to load and measuring the effects of fatigue in the movement (Sasaki, Noritsugu and Takaiwa, 2005).

The application of loads to humans raises two main concerns:

- the human's tolerance of pressure;
- the mechanical characterization of the soft tissues.

5.3.1 Human tolerance of pressure

Excessive pressure is one of the main concerns related to the application of loads to the body. The application of loads by the robot to the skeleton produces contact pressures that can compromise safety and comfort. In this regard, two aspects relating to the pressure applied have been defined: pressure distribution and pressure magnitude. The former relates to comfort while the later relates more to safety. Regarding safety, the usual guideline is to avoid pressure above the ischemic level, i.e., the level at which the capillary vessels are unable to conduct blood, thus compromising the tissue. This level has been estimated at 30 mmHg (Landis, 1930).

The relationship between applied pressure and comfort is complex. Comfort is defined as a state of being relaxed and feeling no pain, but in fact there is no objective way to quantify comfort. Some authors use a combination of peak pressure, pressure gradients and contact area to quantify discomfort (Krouskop *et al.*, 1985). Touch receptors are sensitive to deformation of the layers of tissue where they are located (Dandekar, Raju and Srinivasan, 2003) and therefore the perception of pressure is indirect: pressure deforms tissues and this deformation is sensed by skin receptors. Furthermore, the type, density and distribution of skin receptors vary significantly from one part of the body to another. Finally, the skin receptors respond dynamically to excitation (receptor adaptation). This dynamic response means that pressure perception is dependent on the dynamics of the process whereby the pressure is applied.

The literature describes three parameters for measuring human tolerance of pressure:

- *Pressure pain threshold* (PPT). PPT is defined as the limit of pressure above which a person feels pain. Perceived pain caused by a high local external pressure is often a limiting factor during work and activities of daily living, and for that reason PPT is a fundamental factor.
- *Maximum pressure tolerance* (MPT). MPT is defined as the ratio between force applied and probe area. Since maximum pressure tolerance depends on the contact area, at high force levels a larger area may cause greater discomfort than a smaller area when stimulated with the same magnitude of pressure (Goonetilleke and Eng, 1994).
- *Pressure discomfort threshold* (PDT). PDT is the limit of pressure above which uncomfortable sensations are felt. Various different methodologies can be used to gauge this parameter. Generally, external pressure is applied on the body part and the threshold when the pressure sensation becomes uncomfortable is recorded. This kind of measurement is useful for calculating the best possible pressure distribution.

There are two basic strategies for managing an external load: to concentrate loads over a small area with high tolerance of pressure or to distribute the load over as large an area as possible to reduce the pressure. The latter is generally considered adequate to prevent pain or injury and is therefore commonly adopted as the right one. Nevertheless, this approach does not guarantee comfort. For instance, it has been shown that mattresses with uniform pressure distribution can cause restlessness (Krouskop *et al.*, 1985). Other authors have shown that there must be a threshold beyond which it is better to concentrate than to distribute forces (Goonetilleke and Eng, 1994). This is explained by the spatial summation theory, which establishes that as the area of contact increases, the number of excited skin receptors also increases and consequently comfort perception becomes worse.

Regardless of the magnitude of applied load, there are considerable differences between possible points of application. As a result, not all parts of the body are suitable for transmitting loads to the skeleton. There are a number of recommendations:

- Allow a free area around joints so that they have their full range of movement.
- Avoid bony prominences, bony processes and tendons since bones in these areas can act as stressors and increase the likelihood of suffering injury.
- Avoid areas with surface vessels or nerves so as to limit the likelihood of injuries.
- Avoid highly irrigated and enervated areas such as axilla and the groin so as to avoid pain or discomfort.

An analysis of pressure distribution is presented in Case Study 5.6. This case study presents practical information such as tolerance areas, in both upper and lower limbs, for load transmission. It also presents an estimation of pressure tolerance.

Along with applied pressure, another major cause of pain or trauma is the exposure of the soft tissues to shear force. This force is produced when there is movement between two surfaces that are pressed together. The shear stress is proportional to the pressure applied.

5.3.2 Transmission of forces through soft tissues

Contact stiffness between the WR and the body is a key factor in the transmission of loads from the robot to the human. This transmission is mediated by soft tissues between the human skeletal system and the robot. Consequently, soft tissues play an important role in the performance of the WR. The soft tissues most commonly implicated in the transmission of forces in WR applications are basically muscles, skin, fat, ligaments, blood vessels, nerves and tendons. These are quasi-incompressible, non-homogeneous, anisotropic, nonlinear viscoelastic tissues when subject to major deformation (Maurel, 1998).

In mechanical terms, soft tissues may be described as a combination of nonlinear elastic and viscoelastic elements (Maurel, 1998):

- *Nonlinear elasticity.* Under uniaxial tension, parallel-fibred collagenous tissues respond with a nonlinear stress–strain relationship which is characterized by a region of increasing modulus, a region of maximum constant modulus and finally a region of decreasing modulus where the rupture of the tissue occurs (see Figure 5.4 (left)) (Kwan and Woo, 1989). The low-modulus region is attributed to elimination of the undulations of collagen fibrils that normally occur in a relaxed tissue. The modulus of the tissue increases in response to the resistance presented by the fibrils. The maximum-modulus region occurs when the fibrils come under tension. After that, groups of fibrils begin to fail, causing the modulus to decline until complete tissue rupture occurs.

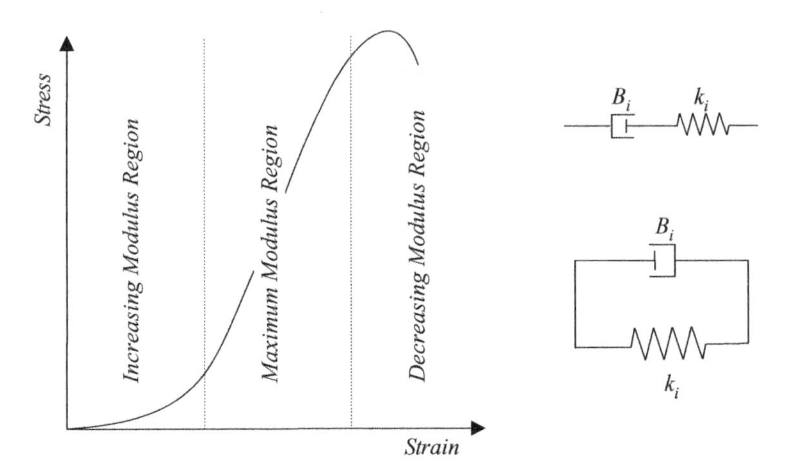

Figure 5.4 Load extension curve (left). Maxwell element (top-right) and Voight Element (bottom-right)

- *Viscoelasticity.* Viscosity is the resistance of a material to flow under stress. Elasticity is the physical property whereby materials instantaneously undergo strain when a tension load is applied and return to their initial state when that tension is removed. Soft tissues present both viscous and elastic characteristics. Viscoelasticity features several mechanical properties (Fung, 1993): *stress relaxation* (tendency for the stress in tissues to gradually decrease with time under constant strain); *creep* (tendency of the rate of lengthening of tissues to diminish with time under constant tension); and *hysteresis* (tendency for deformation of tissues to follow one path upon loading and a different path upon unloading).

There are many approaches to mechanical modelling of soft tissues. One classification splits them into uniaxial or multidimensional models; moreover, both can be divided again into elastic or viscoelastic models (Maurel, 1998). The following are some of the most representative uniaxial models:

- *Elastic Models.* These models consist in uniaxial relationships between the Lagrange stress, T, which is the force divided by the original cross-section area, $T = F/S_0$, and the infinitesimal strain, $\varepsilon = L - L_0/L_0$, or the extension ratio, $\lambda = 1 + \varepsilon$. In the case of tendons in particular, there is a polynomial relationship between stress and strain (Maurel, 1998):

$$\varepsilon^2 = aT^2 + bT \tag{5.1}$$

In elastic fibres (Carton, Dainauskas and Clark, 1962), the strain is an exponential function of the tension applied:

$$\varepsilon^2 = a(1 - e^{bT}). \tag{5.2}$$

Finally, skin may be modelled as (Kenedi, gibson and Daby, 1964)

$$T = a\varepsilon^b \tag{5.3}$$

- *Viscoelastic models.* Linear viscoelastic behaviour can be represented by an infinite number of Maxwell elements in parallel. In this connection, the continuous relaxation spectrum of soft tissues has been described by the combination of an infinite number of Voight and Maxwell elements (Buchthal and Kaiser, 1951). A Maxwell element is a series connection of a spring (k) and dashpot (B).

The equation for a single Maxwell element (see Figure 5.4 (top right)) and Ferry (1980) is

$$k(t) = k_i e^{-t/\lambda_i} \tag{5.4}$$

where $\lambda_i = B_i/k_i$. However, creep-related linear viscoelasticity cannot be described by Maxwell elements. To address this issue, Voight elements need to be introduced. The equation for a single Voight element (see Figure 5.4 (bottom-right)) is

$$J(t) = J_i e^{-t/\lambda_i} \tag{5.5}$$

where $J_i = 1/k_i$.

Multidimensional models are more complex than uniaxial models and can be divided into the same two groups: elastic and viscoelastic. One of the most interesting multidimensional theories proposed models based on microstructural and thermodynamic considerations (Lanir, 1983). Other authors proposed that tissues are composed of several networks of different types of fibres embedded in a fluid matrix and present a strain energy function including angular and geometrical nonuniformities (Maurel, 1998). For additional information on tissue models the reader is referred to Fung (1993), Maurel (1998) and Tong and Fung (1976).

The models presented in this section are widely used in medicine, surgery and other fields. In wearable robotics, the aim is to find the relationship between force applied and deformation of soft tissues. This information may be useful in the design of control strategies for pHRI.

The basic properties of soft tissues, such as nonlinear elasticity or viscoelasticity, are considered in the design of WR supports. Nevertheless, in practice, several simplifications of these models are used to address this problem, as many design details depend on the particularities of each application. For instance, in the context of evaluating the effect of upper limb soft tissues on control strategies for exoskeleton-based tremor suppression, a study was conducted to characterize the soft tissues at different points on the upper limb (Rocon *et al.*, 2005). According to this study, the equivalent stiffness of the tissue increases as the stress increases. A force–deformation curve of the soft tissues at a particular point of the forearm is shown in Figure 5.5. It can be seen that, the behaviour of the curve is highly nonlinear and is characterized by hysteresis. Moreover, the study concluded that the soft tissues of the forearm could be modelled by a third-degree polynomial that describes the deformation of the tissue as a function of the applied force.

Although this particular case illustrates the development of a model for a specific application, some principles can be generalized for wearable robots and used in the design of their supports, as described in the next section.

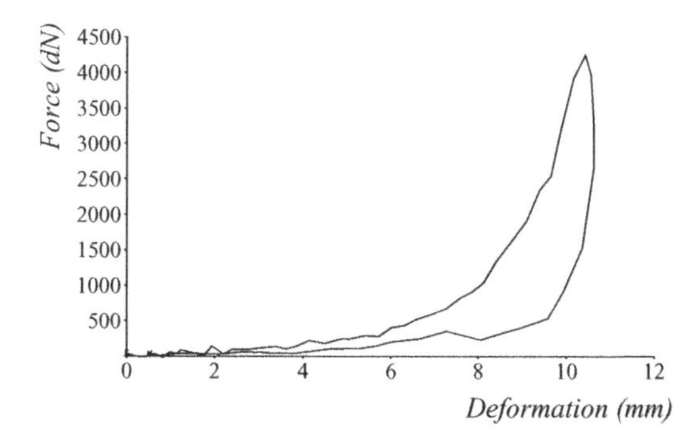

Figure 5.5 Force–deformation curve for a particular point of the forearm

5.3.3 Support design

The function of mechanical supports in wearable robots is to implement the physical interface between human and robot. In this respect, they are responsible for transmitting loads to the human and for transmitting kinetic information to the robotic device.

In order to perform its function properly, the design of these elements has to fulfil several requirements (Rocon *et al.*, 2005):

- *They must ensure the correct placement of the exoskeleton.* The fixation system should keep the robot in place. This will prevent misalignments between the robot and user. Loss of alignment occurs because the fixation points tend to rotate under interaction forces, which can induce the generation of shear forces. To deal with this problem, each fixation should have several contact points on the attachment.
- *They are designed to maximize the transmission of forces.* The supports should increase the stiffness of contact between the human and robot. Low stiffness reduces the effectiveness of force transmission. In order to avoid such loss of effectiveness, the impedance of contact must be increased. This can be achieved by making the support compress human soft tissues, since the stiffness of the limb tissues will be increased if tighter supports are used.
- *They should be comfortable and easy to use.* The supports should be designed so as to avoid the application of inadequate forces, by controlling the distribution pressure and the points of application. Both parameters were discussed in the previous section and are illustrated in Case Study 5.6. Finally, supports should be easy to wear, adjust, use and remove.

In order to illustrate how these requirements affect the support design, the following paragraphs describe the development of the support system of two particular applications.

The design of the support system of an exoskeleton for tremor suppression illustrates these considerations in a particular application (see Case Study 8.1) (Rocon *et al.*, 2005). This example is illustrative since tremor suppression by means of robots requires extremely precise application of forces in view of the intrinsically dynamic characteristics of tremor. Different types of materials for the support elements were considered and it was decided to use supports made of thermoplastic materials. The supports are thus adapted to the morphology of the wearer's arm. Each support has at least three contact points per segment, thus avoiding misalignments between orthosis and limb (see Figure 5.6 (top left)). Velcro straps are used to tighten the support to the arm (see Figure 5.6 (bottom left)). In addition, a textile substrate is used to compress the soft tissues and enhance performance of the fixation supports.

In the exoskeleton presented in Case Study 9.1, fixation to the limb is implemented by pellote carriers. These elements were designed on the basis of the results presented in Case Study 5.6. Their fixation to the orthotic bars is mobile to assure good contouring to the body shape (see Figure 5.6 (right)). The role of these fixation systems is to keep the leg exoskeleton in place.

5.4 CONTROL OF HUMAN–ROBOT INTERACTION

The mechanical interaction between the wearable robot and the human is most often solved through impedance control. This section focuses on how both systems, human and robot, work together to achieve a common goal. Firstly, how a human understands a wearable robot coupled to him/her is investigated. Then, the main issues concerning the development of advanced control strategies for better user interaction are examined. This is followed by an assessment of the interaction of the two control systems. Finally, stability issues relating to this cooperation are discussed.

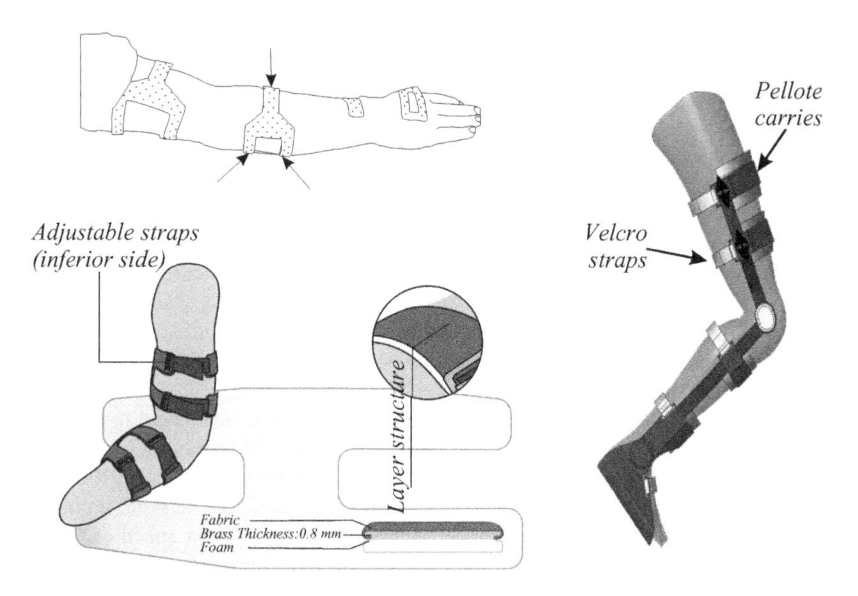

Figure 5.6 Three points of contact in attachment (top left), fixing straps (bottom left) and support elements of the GAIT exoskeleton (right)

5.4.1 Human–robot interaction: human behaviour

Whatever the application for which a wearable robot is used, the transmission of mechanical power is always a principal concern in human–robot interaction. When humans interact with an exoskeleton, they perceive it as an environment or external force. The CNS tries to model that force and compensate for the interaction.

Several works on human motor control have demonstrated that when humans are subjected to a force field that systematically disturbs arm motion, they are able to recover their original kinematic patterns (Shadmehr and Mussa-Ivaldi, 1994). This is accomplished by the adaptation of torques at their joints to compensate for the perturbing forces (Debicki and Gribble, 2004; Gribble and Ostry, 2000). When the perturbation force is abruptly removed, they exhibit error due to adaptation. Thus, there is a basic compensation and learning mechanism that exploits the viscoelastic properties of the neuromuscular system. The learning process is implicated in the development of internal models in the cerebellum (Kawato, 1999).

Control of the impedance of the neuromuscular system is a form of adaptive behaviour that the CNS uses to accommodate perturbations from the environment (Hogan, 1984). The CNS avoids the kind of complex computation involved in motor control through the viscoelastic properties of the neuromuscular system, the muscles and the reflex loops. This viscoelasticity is able to generate restitution forces against external perturbations and may be considered as a control gain in the peripheral feedback. Humans can tune viscoelasticity by regulating the level of muscle co-contraction and the reflex gain.

Each object or environment interacting with a human creates forces with different space and time characteristics, such as gravitational forces, viscous forces and accelerative forces (Shadmehr and Mussa-Ivaldi, 1994). The human is able to compensate for such dynamic environments. The dynamics of the neuromusculoskeletal system can be seen as a feedback mechanism to overcome the action of internal and external perturbations, such as gravity, the intrinsic dynamic of human masses or intersegmental torques.

Numerous studies have approached the dynamic behaviour of human body segments as a mechanical impedance (Dolan, Friedman and Nagurka, 1993; Hogan, 1985; Tsuji, Morasso and Ito, 1995). Mechanical impedance in this context may be defined as the dynamic relationship between small forces and position variations. In fact, the modulation of human impedance provides the basis for several theories in human motor control such as the α-model and $\boldsymbol{\lambda}$-model equilibrium point theories (Feldman, 1986), the virtual trajectory theory (Flash, 1987) and dynamic interaction in manipulation (Hogan, 1985).

The equilibrium point theory proposes that by exploiting viscoelasticity the brain can control the limbs by simply commanding a set of stable equilibrium positions and tuning the mechanical properties of muscles and reflexes (Feldman, 1986). In manipulation, the mechanical impedance of the neuromuscular system determines the reaction forces in response to perturbations from the grasped object. Choosing the mechanical impedance may be one of the ways whereby the CNS controls the behaviour of the complete system of human plus robot (Hogan, 1984).

Several studies have assumed a second-order mass–damping–stiffness model to describe the dynamics of the human arm. These models are based on mass, M, damping, B, and stiffness, K, parameters, which provide a direct physical interpretation (Kearney and Hunter, 1990).

The model is nonlinear and varies depending on factors such as torque bias and posture. Thus, in order to fit the data to a second-order linear model, an operating point must be specified. The operating point consists of constant posture, constant force and nonfatiguing contractions for a particular task. The parameters in the model change as the operating point changes. The combination of linear models estimated over a range of operating conditions may be thought of as defining a quasi-static model of arm dynamics and can be defined by the linear equation

$$F(t) = I\frac{\partial^2 X(t)}{\partial^2 t} + B(\delta)\frac{\partial X(t)}{\partial t} + K(\delta)X(t) \tag{5.6}$$

where $F(t)$ and $X(t)$ represent the force and the displacement respectively, and δ defines the operating point. The quasi-static model cannot be used when the operating point of the system is changing dynamically.

The human mechanical impedance can be changed voluntarily over a wide range (Mussa-Ivaldi, Hogan and Bizzi, 1985). One way to change impedance is through the level of muscle contractions. The relationship between the neural input to a muscle and its subsequent mechanical behaviour is extremely complex. For a given neural input the contractile force of a muscle depends on the length of the muscle, its velocity of shortening, the type of muscle, its state of fatigue and its history of exercise (or of electrical stimulation), among others. However, one fundamental observation is that the neural input to a muscle simultaneously determines the force and the stiffness of the muscle, i.e. its resistance to stretch (Hogan, 1984).

The physical interaction leads to a learning process in which human motor control is adapted. In the learning process there is a reduction of the average neural input signal. This reduction is to a large extent due to a decrease in the coactivation of antagonistic muscles. Therefore, the learning process implies adaptability of the mechanical impedance. There are several scenarios of HRI in which the exoskeleton controller must take into account the high variability of the impedance delivered by humans.

5.4.2 Human–robot interaction: robot behaviour

The basic control approaches to management of pHRI can be classified into two groups: feedforward control and feedback control. Feedforward systems execute control action using a model-based estimation. This requires the capacity to anticipate the action necessary to accomplish the system's goal. Errors in models or unexpected perturbations can only be handled by feedback systems. In feedback

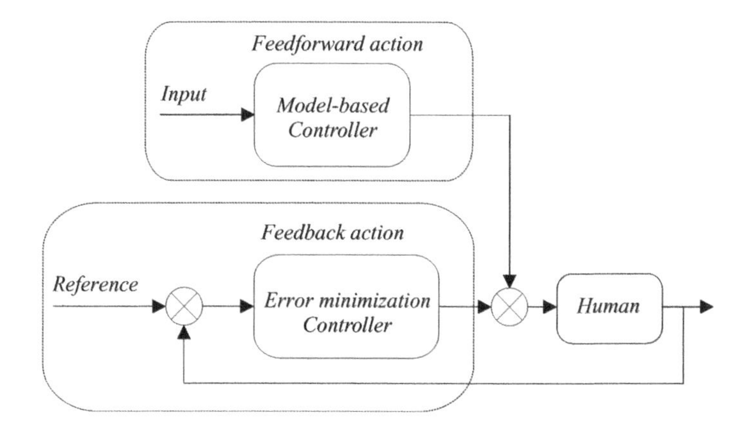

Figure 5.7 Schematic representation of feedforward and feedback control concepts

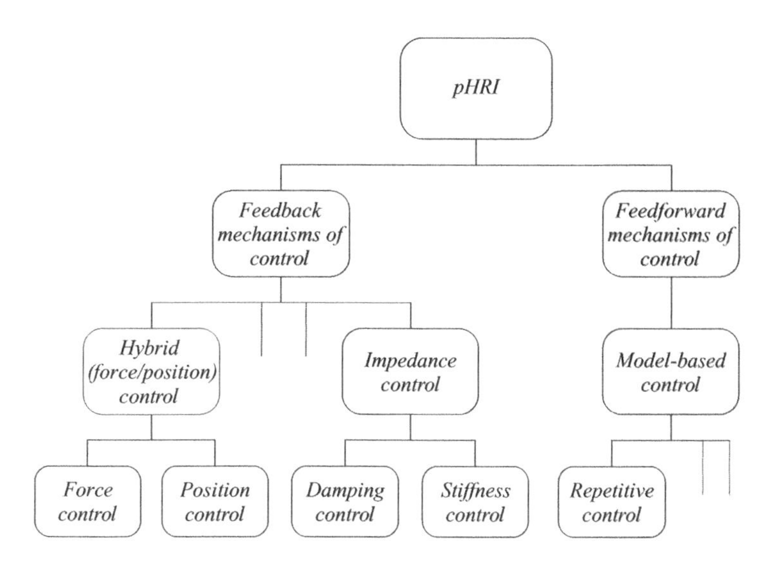

Figure 5.8 Fundamental pHRI control strategies

control, the error is used to determine the control action see Figure 5.7. Both approaches can be implemented by several types of control strategy; these are summarized in Figure 5.8.

Recent years have seen a great deal of research into strategies to control the physical interaction between human and robot (Debicki and Gribble, 2004; Hogan, 1985; Ikeura and Inooka, 1995; Tanaka *et al.*, 2007; Tsuji and Tanaka, 2005; Vukobratovic, 1997). Advances in mainstream robot technologies are now moving from pure position control towards mechanisms that either possess an inherent ability to control contacts with the environment or have sufficient auxiliary sensing integrated in their low-level control loops to control force.

Various control concepts and schemes have been established and proposed to control the relationship between robot motion and interaction force. Of these, two approaches in particular are worth mentioning (Roy, 2001): hybrid (force/position) control and impedance control. In hybrid control,

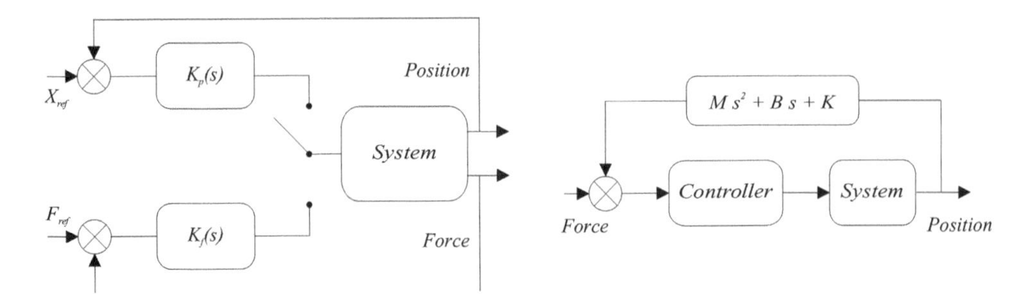

Figure 5.9 Hybrid control scheme (left) and impedance control scheme (right)

the end-effector force is explicitly controlled in selected directions and the end-effector position is controlled in the remaining (complementary) directions (see Figure 5.9 (left)) (Mason, 1981; Raibert and Craig, 1981). In impedance control, a prescribed static or dynamic relationship between the robot force and position has to be maintained (see Figure 5.9 (right)) (Hogan, 1985). This is a generalization of damping control (Whitney, 1977), and stiffness control (Salisbury, 1980).

In wearable robotics, the most common approach to control of the interaction forces between robot and human is impedance control (Vukobratovic, 1997). One of the main advantages of impedance control is that a compliant behaviour of the wearable robot can be implemented, producing a more natural physical interaction and reducing the risk of damage. The basic principles and considerations regarding impedance control were addressed by Hogan (1985). In his paper, Hogan explained the conditions for causality in the treatment of dynamic interaction between the manipulator and environment. He used the concept of mechanical impedance to address the mechanics of musculoskeletal systems, dealt with the implementation of this control approach and finally addressed the selection of appropriate impedance for a given application.

Impedance control is a strategy for constrained motion rather than a concrete control scheme. The basic idea of this approach is to have a closed-loop control system whose dynamics can be mathematically described by the following equation (Hogan, 1985):

$$F = M(\ddot{q} - \ddot{q}_0) + B(\dot{q} - \dot{q}_0) + K(q - q_0) \tag{5.7}$$

where M, B, K represent the inertia, damping and stiffness of the interactive system respectively.

Inertia, damping and stiffness can be adjusted by the control system according to the goals pursued (impedance may vary in the different task space directions, typically in a nonlinear and coupled way). The interaction between the robot and a human then produces a dynamic balance between these two 'systems'. This balance is influenced by the mutual weight of the human and the robot compliant features. In principle, it is possible to reduce the robot's compliance so that it dominates in the pHRI and vice versa. In addition, cognitive information can be used to set the parameters of robot impedance dynamically in the light of task-dependent safety issues.

The most important consequence of a dynamic physical interaction between two systems is that one must physically complement the other. If one is an impedance – accepts effort inputs, i.e. forces, and yields flow outputs, i.e. motion – the other must be an admittance – accept flow inputs, i.e. motion, and yields effort outputs, i.e. forces – and vice versa.

When an exoskeleton is mechanically coupled to its environment (which is an admittance as it imposes positions), to ensure physical compatibility the exoskeleton should assume the role of an impedance. In addition, since the human behaves as a variable impedance, the behaviour of the exoskeleton should be adaptable. Thus the controller should be capable of modulating the impedance of the robot to adjust to a particular impedance of the human.

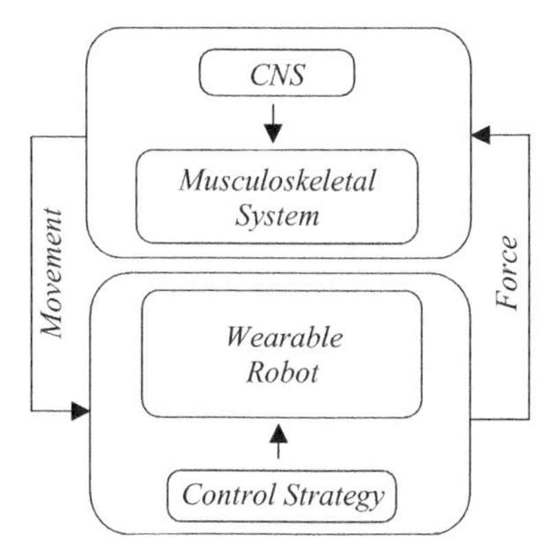

Figure 5.10 Human–robot loop interacting dynamically

5.4.3 Human–robot closed loop

There is a combination, in a closed loop, of two dynamic control systems: the human motor control system and the robot controller (see Figure 5.10). The human and the wearable robot interact through their sensory and motor channels. The human receptors record the physical state of the human body and its environment. The sensory information is perceived by the CNS and interpreted by cognitive processes to deliver motor actions. Similarly, the robot's sensors detect the state of the machine and its environment. Again, the environment is interpreted by the robot control system in order to command the actuators.

The key issue in this interaction is the capacity of the two systems for mutual adaptation. As noted in the previous section, human characteristics may be approximated by a variable impedance model. In addition, robot controllers typically run impedance control strategies to manage the pHRI. Such a system is illustrated in Figure 5.11.

On the one hand, the robot controller should be able to adjust the impedance of the robot to a particular impedance of the human, depending on the goal of the robot. On the other hand, humans adjust their control properties to the robot's impedance in order to maintain the dynamic properties of the overall system (Tsuji and Tanaka, 2005). This process is illustrated in Case Study 5.7.

The problem of designing controllers for this hybrid system has been formulated as follows (Colgate, 1988):

$$\|Z_c(jw) - Z_t(jw)\| < \epsilon \qquad \text{for } \omega < \omega_b \tag{5.8}$$

where $Z_c(jw)$ is the closed-loop impedance, $Z_t(jw)$ is the target impedance, ϵ is an arbitrarily small number and ω_b is the desired bandwidth over which $Z_c(jw)$ and $Z_t(jw)$ should match.

According to Equation (5.8), a specific impedance $Z_t(jw)$ must be defined for each application. There are several studies that address the impedance selection problem (Ikeura and Inooka, 1995; Tsumugiwa, yokogawa and Hara, 2002). Selection of the variable impedance is strongly related to the application of the wearable robot. Depending on the application, different control approaches may be adopted:

- *Empowering*. The main purpose of a WR in this application is to amplify the physical capacities of a human. To achieve this, the person delivers control signals to the exoskeleton, while the device delivers mechanical power in order to accomplish a particular task. In this application, the target impedance is calculated in order to minimize the interaction force between the human and exoskeleton.

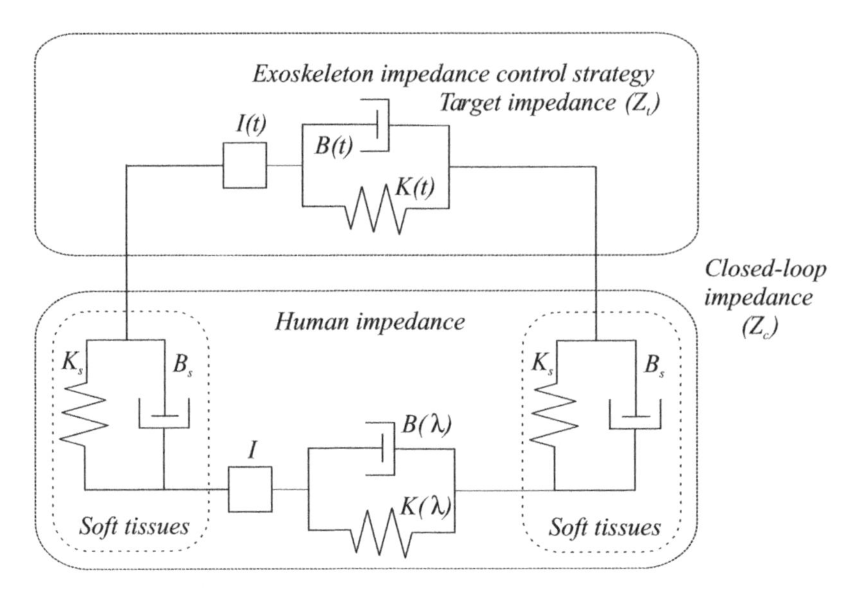

Figure 5.11 Schematic view of the human–robot control strategy

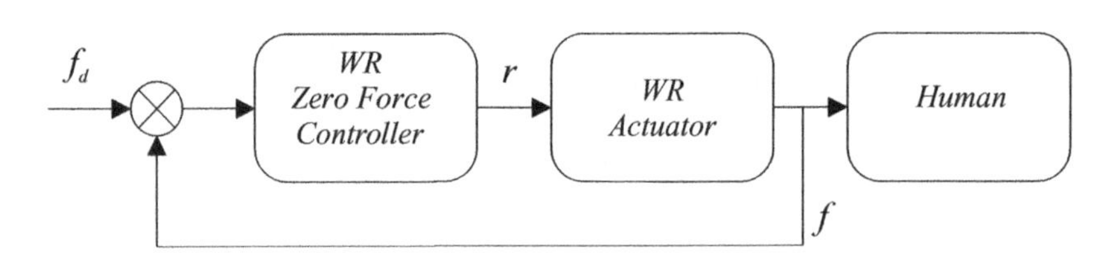

Figure 5.12 Block diagram of the 1 DoF zero interaction force controller

The simplest strategy to achieve this goal is a zero interaction force controller. Figure 5.12 illustrates a 1 DoF block diagram of this control strategy, where f represents the interaction force between the user and the exoskeleton and f_d is the reference force, which in this case is set to zero. The corresponding control law is simply

$$r = g(f_d - f) \tag{5.9}$$

where the control action r is a function of the error in the measured force.

Other studies propose more elaborate control laws to minimize the interaction force (Banala, Kulpe and Agrawal, 2007). The basic principle of a control strategy is its ability to respond quickly to the wearer's voluntary motion. The control strategy should be able to block the influence of external forces (not generated by the wearer). In order to address this issue, Kazerooni recently presented a new strategy for the control of interaction forces in empowering systems that requires no direct measurement of the human–machine interaction force; i.e. no force sensors are needed (Kazerooni *et al.*, 2005).

- *Telemanipulation.* This application comprises the set of technologies that enable tasks to be executed remotely. A robotic exoskeleton acts as a master device in a teleoperation system. In bilateral

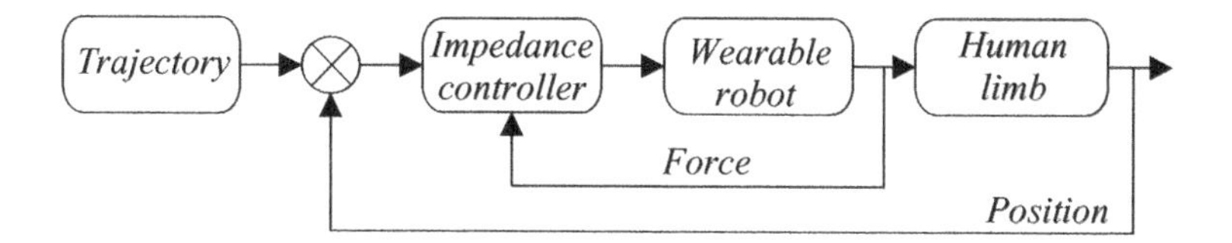

Figure 5.13 Impedance control schema for a wearable robot for rehabilitation

control mode, it allows the operator to control a remote robotic arm (slave). Interaction forces between the remote robot arm and its environment are fed back to the master and applied by the exoskeleton to the human arm (Cavusoglu, sherman and Tendrick, 2002). The classic control architectures in telemanipulation are the position–position and the position–force approaches. The control information flows bidirectionally between the master and slave.

- *Rehabilitation.* Robot-assisted physical therapy has been shown to aid in rehabilitation following neurological injuries. For rehabilitation applications, the exoskeleton provides both active and passive assistance in several therapies. The exoskeleton adopts the role of a therapist during the treatment. Sensors attached to the exoskeleton can assess forces and movements of the patient. This gives the therapist quantitative feedback on the patient's recuperation.

Several control algorithms have been developed that automatically adapt the reference trajectory and the impedance of the controller to the training motion of a particular patient (Jezernik, colombo and Morari, 2004). The involvement of human control inevitably leads to promoted patient activity during the exercise and to increased patient motivation, and therefore the outcome of training with adaptive control of pHRI may be expected to improve the outcome of the rehabilitation therapy.

Figure 5.13 presents an impedance control scheme for a wearable robot for rehabilitation. The reference to the controller is the trajectory defined by the physiotherapist. During physiotherapy, the wearable robot replicates specific trajectories.

- *Functional compensation.* In this application the WR is an enabling technology for physically disabled people. In this context, the controller must meet special requirements such as robustness, reliability and safety. The robot must identify the user's intention, analyse the information in real time and compute control actions according to the appropriate functional compensation strategy.

An illustrative example of a control strategy for functional compensation is the one developed for active tremor suppression by means of exoskeletons (see Case Study 8.1). This strategy is based on a feedforward controller. It drives the exoskeleton to generate a motion equal but opposite to the tremor, based on a real-time estimation of the involuntary component of motion. This is possible thanks to the particular rhythmic characteristics of tremor. In this strategy, repetitive control was selected to handle periodic (repetitive) signals and disturbances. Repetitive control may be regarded as a subset of learning control since the control action is determined from a model of the repetitive behaviour of the plant. The control strategy is illustrated in Figure 5.14.

This control strategy is based on a dual control loop. The upper control loop is responsible for tremor suppression while the lower control loop is meant to minimize the influence of the control strategy on voluntary movement by means of zero interaction force control (Rocon *et al.*, 2007). In the upper loop, the control action is delayed by M seconds according to the tremor model. M is defined by the following equation, where ω_s is the sample frequency of the control strategy and ω_t is the estimated tremor frequency:

$$M = \frac{\omega_s}{\omega_t} \tag{5.10}$$

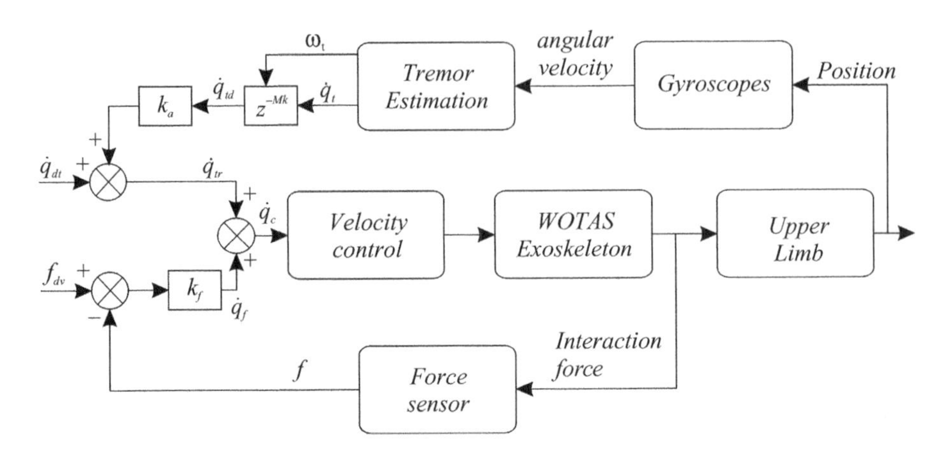

Figure 5.14 Feedforward tremor suppression control strategy

The following equation defines the upper loop control law:

$$\dot{q}_{\rm c} = \dot{q}_{\rm f} + \dot{q}_{\rm tr} = k_{\rm f}(f_{\rm dv} - f) + (\dot{q}_{dt} + z^{-M_k}\dot{q}_{\rm t}) \tag{5.11}$$

5.4.4 Physically triggered cognitive interactions

In general, physical HR interaction will trigger cognitive processes on either side of the HR interface. These cognitive processes will in turn affect motor control both in the human and in the robot, all in a context of mutual adaptation during interaction. This adaptation mechanism has not been clearly formalized in the literature, most probably due to the complexity of the CNS and related motor control mechanisms.

Nevertheless, physically triggered cognitive interactions can be illustrated through the example of a wearable robot interacting physically with a human in order to suppress dynamically pathological tremor (see Case Studies 8.1 and 5.7). When tremor is counteracted by the interaction between the human and robot, the following phenomena are observed:

- The visual feedback of a limb with a reduced tremor amplitude has a positive impact on the human control system (Rocon *et al.*, 2007). The human motor control is reinforced (a sort of positive gain feedback) and residual tremor is further reduced as a result.
- The reduction of tremor in one joint following interaction with the robot results in perturbed human motor control, possibly leading to increased tremor intensity at the adjacent proximal joints. This phenomenon is called the distal to proximal tremor shift (DPTS) (Rocon *et al.*, 2007).

Two adaptive control strategies, impedance control (see Case Study 5.7), and model-based repetitive control (see Section 5.4), were proposed. The following approach was adopted to manage both of the physically triggered human motor control phenomena.

- *The control strategies exhibit an adaptive behaviour.* In the case of impedance control, the combined impedance of each articulation of the HR system is calculated on the basis of a real-time estimation of the tremor amplitude. In the case of repetitive control, the amplitude of the force applied to counteract tremor is also based on the continuous estimation and tracking of tremor frequency and amplitude. Thus the system can respond to the changes that take place in the human motor system. The outcome of the adaptive approach to both control strategies is mutual adaptation, both physical and cognitive.

- *The implementation of the control strategies is based on individual independent control loops in each joint of the exoskeleton.* This minimizes the DPTS problem: if cancellation of the tremor in one joint increases the tremor in the other joint, the algorithm responsible for controlling the adjacent joint will automatically identify the increased tremor and try to reduce the tremor migration. The aim is to achieve equilibrium between the active behaviour of tremor reduction in either joint, thereby reducing the coupling effects of the upper limb joints.

5.4.5 Stability

The stability of a HR closed loop is a major concern in the design of controllers since it has a direct bearing on safety (Colgate, 1988). If the robot becomes instable, it might hurt the human. Moreover, stability analysis is useful to determine the impedance parameter values of the robot control system. This is particularly helpful in that there is no specific guideline for designing and adjusting the values of impedance parameters in the HR cooperative task system (Vukobratovic, 1997).

Two main sources of instability in coupled human–robot systems are firstly, instability in the control of physical interaction and, secondly, instability due to time delays in the human-in-the-loop control. These two sources are discussed in the following paragraphs.

5.4.5.1 Instability in control of the pHRI

The state-of-the-art function treats a robot's environment as a source of disturbances rather than as a dynamic system. The dynamic properties of the environment have an important role in the stability of the overall dynamic system, i.e. the robot interacting with the environment (Roy and Whitcomb, 2002; Vukobratovic, 1997). In wearable robotics, the robot's environment consists of the wearer. As presented in Section 5.4, the human behaviour is commonly considered passive from an energetic point of view, notwithstanding the neural feedback in the arm. This has a fundamental implication for the stability of the WR. If the human behaviour can be regarded as passive, it is sufficient to ensure that the robot behaves in a passive way in order to assure the passivity of the overall system. In control theory, passivity is a more restrictive condition than stability, and a passive system usually also behaves stably under the right conditions.

The term *coupled stability* has been defined as the stability of a system composed of the robot and environment coupled together mechanically (Fasse, 1987). A system is said to exhibit coupled stability if:

- The system is stable when isolated.
- The system remains stable when coupled to any passive environment that is also stable when isolated.

Several assumptions are usually made when studying the coupled stability of robotic systems (Colgate, 1988; Vukobratovic, 1997):

- The robot is composed of rigid links with torque sources at the joints.
- Additional dynamic effects (e.g. actuator bandwidth limitations or transmission dynamics) are important but are included in the robot model.
- Some modelling uncertainties, such as unmodelled dynamics or parameters errors, need to be taken into account.

Given these assumptions Hogan (1987), has shown that, so long as the term B in Equation (5.7) is chosen in such a way that the robot is capable of stably positioning an arbitrary small mass, then it can interact stably with any dissipative environment describable in Hamiltonian terms. Extensions of this analysis can be found in Fasse (1987). Kazerooni puts forward a similar argument for the case where both the robot and the environment can be described in a second-order linear system with positive definite inertia, damping and stiffness matrices (Kazerooni, 1985).

Nevertheless, controllers do not in practice succeed in driving the target dynamics robustly and stably. The reason is that real robots do not present the simple behaviour that these methods assume. Actuator and transmission dynamics, joint and link flexibilities, and computational delays are some of the nonlinearities of the closed-loop system. In this connection Kazerooni (1985) developed a method for guaranteeing the stability of closed-loop designs of arbitrary complexity coupled to environments that are characterized by the magnitude of their input/output mapping. This method was developed on the basis of multivariable analysis techniques. Kazerooni has also shown that it can be extended via the *small gain theorem* for application to a broad range of nonlinear systems.

Fasse developed a method for guaranteeing the stability of closed-loop systems coupled to passive environments (which may be nonlinear) (Fasse, 1987). The basis of this method is the selection of an appropriate Lyapunov function for the controlled system. Fasse shows several examples of linear systems for which such Lyapunov functions can be found by restricting the feedback gains in appropriate ways. Moreover, he addresses the robustness of coupled stability to various types of modelling errors. In this analysis, he deals with the standard implementations proposed by Hogan (1985), but subject to nonideal behaviour such as sensor gain errors, actuator dynamics base dynamics, friction and gravity.

5.4.5.2 Instability due to time delay in both cHRI and pHRI

In addition to the stability problems relating to the application of forces between humans and robots, stability issues also arise from the fact of having a human-in-the-loop system, with the human taking part in the control strategy. In this connection, Tsumugiwa experimented with the stability of impedance control of a specific robot system in a human–robot cooperative task system (Tsumugiwa, yokogawa and yoshida, 2004). In that study the authors considered that stability is affected by conditions such as the robot's characteristic impedance, the time delay of the human operator in control of the robot, the time delay introduced by the robot controller and the contact compliance between the robot and human.

Figure 5.15 shows a block diagram used for stability analysis of the human–wearable robot system. In this block diagram, the dynamics of the robot is modelled as a second-order system. A first-order lag element is used to represent each element of the time delay. According to this model, two time delay elements are considered in the control loop:

- the dead-time attributable to the controller and to the movement delay of the robot;
- the reaction time of the human operator in the HR system.

The stability of the proposed model was investigated by root locus analysis using the pole locations of the characteristic equation (Tsumugiwa, yokogawa and Hara, 2003). This stability analysis revealed that the parameters most significantly influencing the stability of the impedance control system are the characteristic impedance of the robot, the time delay of the human operator, the time delay of the robot controller, the compliance of interaction between the human and robot, and pHRI stiffness. In particular, pRHI stiffness influences the stability of the system more than any other parameter. This is important in that HR systems are most often coupled through soft tissues and hence present low contact stiffness (see Section 5.3).

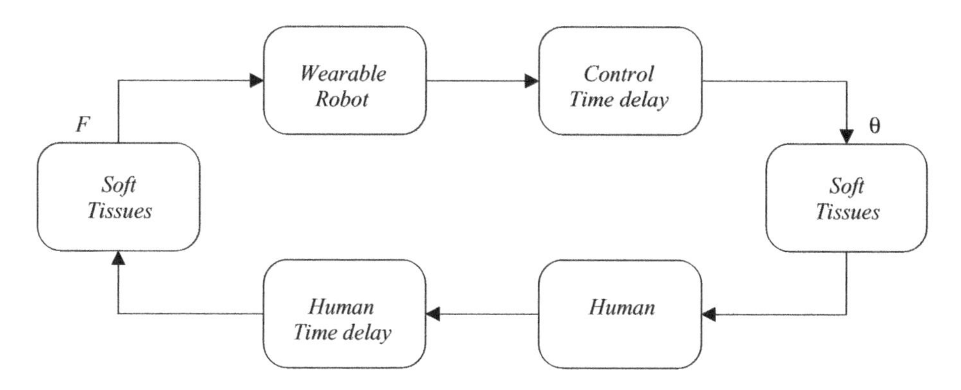

Figure 5.15 Block diagram of the stability analysis model

5.5 CASE STUDY: QUANTIFICATION OF CONSTRAINT DISPLACEMENTS AND INTERACTION FORCES IN NONERGONOMIC pHR INTERFACES

A. Schiele[1,2]

[1] *Mechanical Engineering Department, Automation & Robotics Section, European Space Agency (ESA), Noordwijk, The Netherlands*
[2] *Mechanical Engineering Faculty, Biomechanical Engineering Department, Delft University of Technology (DUT), Delft, The Netherlands*

In this case study the constraint displacements and interaction forces that are present in some wearable exoskeletons during motion are quantified. As explained in detail in Section 5.2, constraint forces are created between a wearable exoskeleton and a human limb if the exoskeleton is intended to be equivalent to the corresponding human limb in terms of kinematic structure. Exoskeletons that imitate human limb kinematics must be perfectly aligned with the human axes. Such perfect alignment is not in fact achievable given the variability of human anatomical properties between different users and during motion. There are therefore constraint displacements of the exoskeleton attachment points on the human limb.

This case study begins by presenting a theoretical analysis of the problem. An analytical model is established, which can be used to predict constraint displacements of an exoskeleton attachment fixture along a human limb. Such displacements are a function of the respective ICR offsets of the exoskeleton and human arm and the limb rotation angle. Analytical results are complemented by an extract of experimental data. Measured interaction forces are shown; these were acquired during a test campaign with 14 subjects of varying stature and mass. The exoskeleton device used for this campaign was the one presented in Case Study 8.3. The ergonomic ESA exoskeleton is a device that does not usually require alignment to the human joints. It incorporates passive compensatory joints to cope with offsets of the ICRs of the exoskeleton and human limb and therefore fits a wide range of users without leading to constraint displacements and undesired interaction forces. For the experimental campaign, the passive compensatory joints of the exoskeleton were locked to emulate an exoskeleton that requires alignment.

5.5.1 Theoretical analysis of constraint displacements, *d*

For simplicity's sake, this case study concentrates on a 1 DoF motion. A combined physical human–robot interaction system can be simplified as illustrated in Figure 5.16. A human operator wears an exoskeleton i.e. attached at two locations along the arm, i.e. the upper arm and the forearm. The focus in this situation is on elbow movement. In the drawing on the left, the estimated kinematic structure of the human limb is shaded in light grey.

The estimated, but unknown, position of the elbow rotation axis, or ICR_h, is indicated in the figure by a grey dot and black cross. The exoskeleton structure is shown schematically. Its elbow joint does not align well with the operator's elbow. For this analytical model, a rigid fixation of the exoskeleton on the upper arm and a soft, compliant fixation of the exoskeleton on the forearm are assumed. The soft fixation can be modelled, for instance, with a Voight element, which simulates viscoelastic properties. It is important to note, however, that in reality the upper arm fixation also influences the interaction forces between the robot and the human limb – an aspect that is not dealt with in this chapter. Figure 5.16(a) abstracts this situation to a static mechanical problem. Any offset x or y between the two joints will create a constraint displacement d of the distal exoskeleton fixture along the operator's forearm (during movement). A force will be created in the Voight element that balances the force induced by the constraint displacement. The constraint force F_d is velocity-dependent and determined by

$$F_d = kd + b\dot{d} \tag{5.12}$$

where k denotes the stiffness and b the damping constant between the exoskeleton attachment and the human bone. All muscle, tissue and skin parameters, as well as the soft pads of the exoskeleton fixation, are lumped into those parameters.

In order to understand the magnitude and behaviour of the interaction force, first the magnitude is determined of the constraint displacement d of the distal attachment point of the exoskeleton on the human forearm. This reduces the model to a purely kinematic one. The known parameters are typically the human limb angle α, the exoskeleton joint angle β, the offsets x and y between the human limb instantaneous centre of rotation ICR_h and the exoskeleton centre of rotation CR_e, and the length of the exoskeleton link l_{ex} (which is a D–H parameter of the exoskeleton) between the CR_e and the attachment on the forearm. The angles α and β are geometrically related, as will be seen. From Figure 5.16(a), first the starting angle β_0 is determined of the exoskeleton joint when the

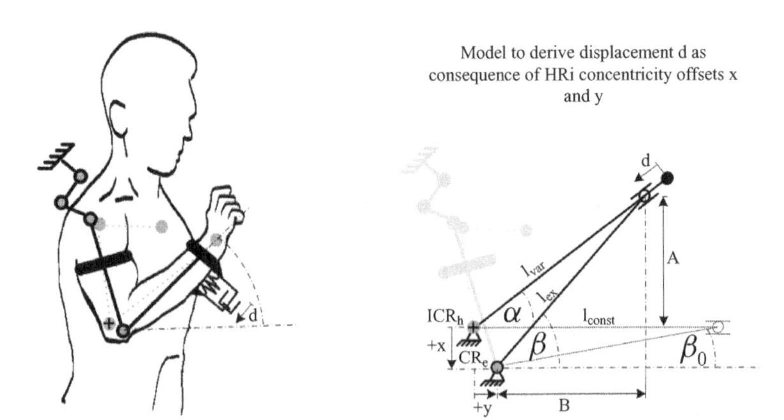

Figure 5.16 Analytical model to assess the magnitude of constraint displacements d caused by offsets of an exoskeleton joint CR_e and the corresponding human limb joint ICR_h

human limb is fully stretched, $\alpha = 0$. Consequently,

$$\beta_0 = \arcsin \frac{x}{l_{ex}} \tag{5.13}$$

Then the nominal distance l_{const} from the distal attachment of the exoskeleton on the human limb to the ICR_h is given by

$$l_{const} = y + l_{ex} \cos \beta_0 \tag{5.14}$$

In movement, the displacement d on the operator's forearm can be determined by:

$$d = l_{const} - l_{var}(\alpha, x, y, l_{ex}) \tag{5.15}$$

When d is known (see Schiele 2008 for a detailed description on the calculation of d), the interaction force can be calculated. A model of the human tissue and the muscle stiffness and damping is then required; for instance one of the models presented in Section 5.3 could be used. The detailed estimation of forces is complex and is not germane to this monograph. The following paragraph shows the theoretically determined values of constraint displacements d caused by offsets between the exoskeleton CR_e and the human elbow ICR_h, which were derived by resolving Equation 5.15, as in Schiele 2008.

Figure 5.17 (top) shows a three-dimensional plot of the constraint displacement at the forearm attachment of an exoskeleton dependent on y and α. Only offsets in direction y are shown, with x constant and equal to zero. It can be seen that with increasing elbow flexion angle α, the displacement peaks at 180°. On the right hand side of Figure 5.17, an exemplary curve is shown for a pure offset of $y = 10$ cm. The length of the exoskeleton forearm link was set at 30 cm, which is realistic. It can be seen that offsets in the direction $+y$ strongly influence d. Theoretically, displacement amplitudes of up to 20 cm are possible, but in reality elbow rotations of about 90° are more realistic. Nevertheless, displacements can reach up to 10 cm, which is definitely excessive.

Figure 5.17 (bottom) shows the same plots, with the difference that x is the variable and y is constant and equal to zero. It can be seen that offsets in the direction $+x$ cause the constraint displacement d to peak at about 90°. The vertical lines in the 2D graphs of both figures show the 95th percentile of maximum elbow flexion that US men and women can exert. Both of the above figures clearly show that offsets can cause relatively large constraint displacements within nominal working ranges.

Having shown the theoretical displacements dependent on the geometric parameters, let us compare measured interaction forces with this model.

5.5.2 Experimental quantification of interaction force, F_d

In an experiment in which 14 test subjects participated, the ESA EXARM exoskeleton was used to quantify the interaction forces. Although the EXARM exoskeleton is an ergonomic device (see Case Study 8.3 and Schiele and van der Helm 2006), in this experiment its passive joints were locked so as to emulate a nonergonomic wearable exoskeleton that imitates the human kinematic structure. By locking its passive joints, the exoskeleton becomes sensitive to misalignment. The exoskeleton was dressed to each subject of the group identically. The mean stature of the test subjects was 1.75 m ± 0.09 and the mean body mass was 68.7 kg ± 12.8.

Figure 5.18 shows a test subject wearing the exoskeleton. After donning the device, the test subjects were asked to track a multisine position signal on a computer screen by moving their elbow. The elbow movement was also displayed on the screen. All forces between the human forearm and the exoskeleton were measured with a 6 DoF force and torque sensor (ATI Nano Series) built into the exoskeleton's mechanical structure. The tracking signal contained a range of varying random frequencies from 0.05 to 0.5 Hz and amplitudes from 0 to 90° (elbow flexion–extension). The

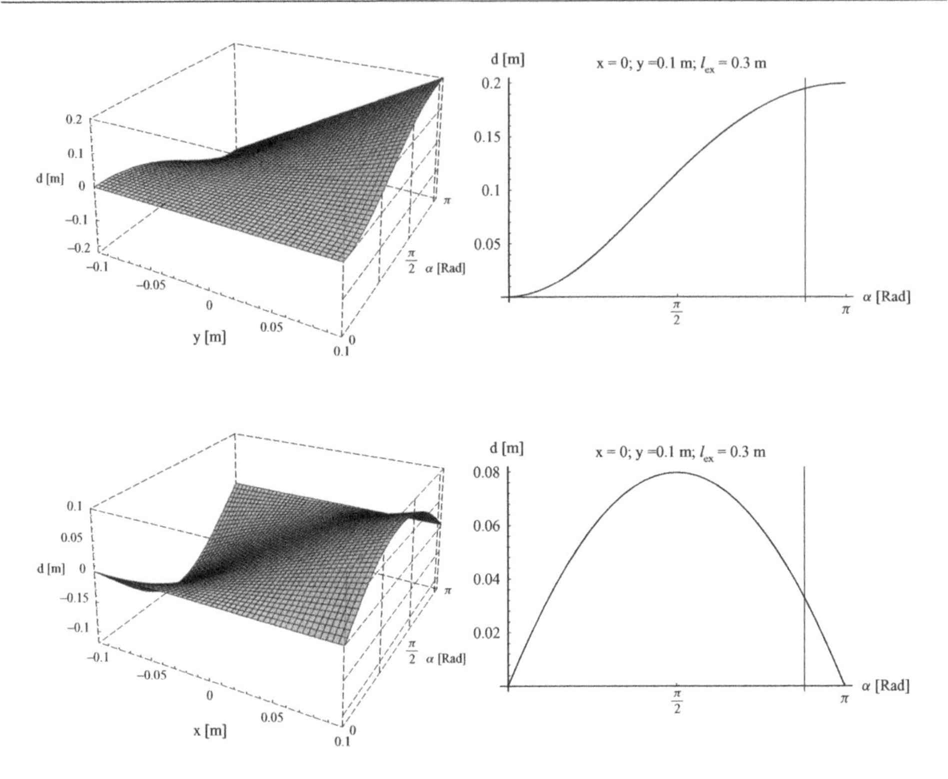

Figure 5.17 Constraint displacement d dependent on offset in direction y and elbow rotation angle α (top left) and on offset in direction x and elbow rotation angle α (bottom left). An example of a typical constraint displacement d depending on α (left). An exoskeleton of length $l_{ex} = 0.3$ m has no offsets in x but an offset of 10 cm in direction y (top right), while the exoskeleton has an offset of 10 cm in x but no offset in direction y (bottom-right). The line on the α coordinate denotes the maximum elbow extension available to the 95th percentile of the male population (159°)

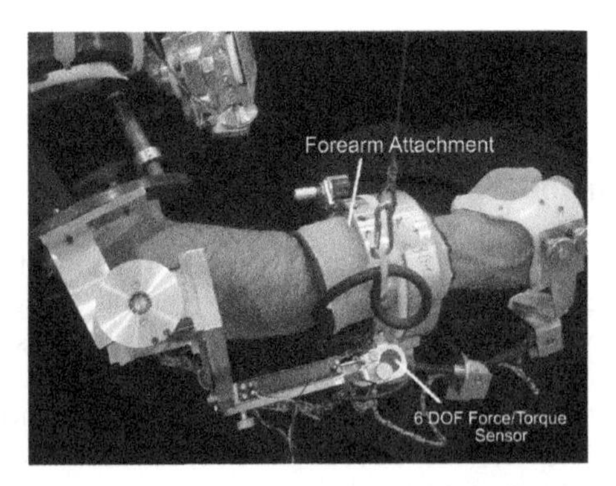

Figure 5.18 EXARM Exoskeleton used for the experiment. Passive compensatory joints were locked

pressure in the exoskeleton's forearm fixation cushion was set to 20 mmHg, which was rated the most comfortable pressure by the group.

Figure 5.19 shows an extract of the data. Only the force component in the longitudinal direction of the forearm is shown. This shear force F_d is caused by the constraint displacement d during

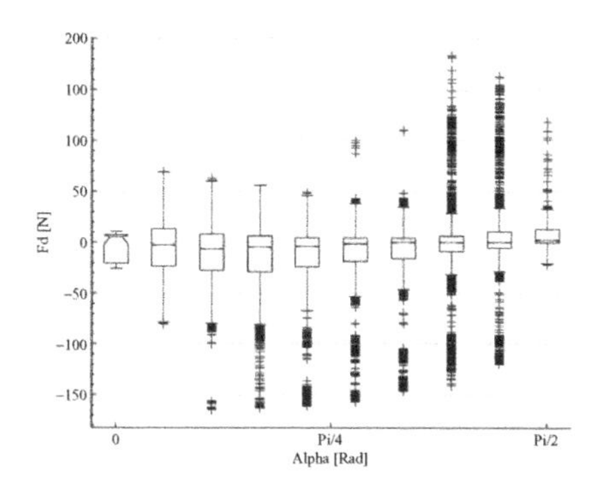

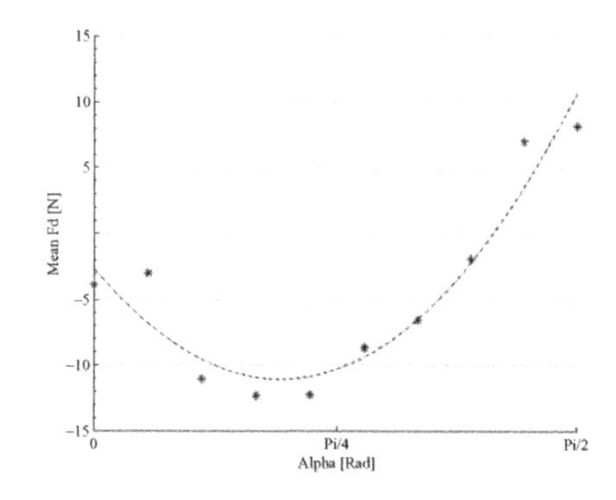

Figure 5.19 Measured constraint force F_d at the forearm attachment of an elbow exoskeleton depending on the elbow extension angle α. The boxplot (left) shows accumulated data at 10° intervals for 14 subjects, with the mean force F_d for each 10°. (Right) The interval, along with a quadratic fit over the means (right)

elbow rotation. Because the force is measured inside the exoskeleton mechanism, a negative sign of the measured force corresponds to a positive displacement of d. On the left is a boxplot of all accumulated measured forces F_d of the 14 subjects over their forearm rotation angle α. Forces are shown over rotation bins of 10° each, spanning a range from 0 to 90°. Most of the interaction force values lie within a band spanning from approximately −25 to +20 N. However, peak forces higher then 180 N were measured for some subjects. The mean absolute force over the entire duration of the experiment was 18 N. The trend of force creation over the elbow angle is difficult to appreciate from the boxplot.

The graph on the right of Figure 5.19, shows the mean force values of each angle bin over the elbow flexion angle. It also shows a quadratic fit through the mean force values. It seems surprising at first sight that the constraint forces should have a negative sign and peak at about 30°; they drop to 0 N at about 70° and continue to rise in the positive direction thereafter, up to 90°. No such trend was apparent in the previous figures.

This trend is typical of constraint displacements caused by combined offsets in the negative x and positive y directions. Figure 5.20 shows a corresponding graph generated by the analytical model, for which the offsets were $x = -3$ cm and $y = 5$ cm. It is interesting to note here that the model output matches the measured data quite well in terms of its characteristics. The amplitudes are different, of course, and are determined by the interface properties.

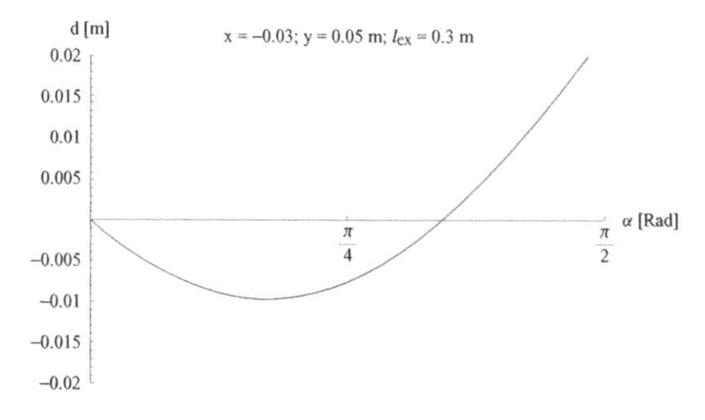

Figure 5.20 Model output for constraint displacement d if offsets are negative in direction x and positive in direction y

In the experimental campaign, the displacements were also measured when the passive joints of the exoskeleton were not locked. The rotational velocities of the elbow articulation were measured as well. These data and the force measurements can be used to determine the human–exoskeleton interface stiffness. This is useful information for a better analysis of the haptic controller i.e. being implemented; however, these results are presented elsewhere.

5.6 CASE STUDY: ANALYSIS OF PRESSURE DISTRIBUTION AND TOLERANCE AREAS FOR WEARABLE ROBOTS

J. M. Belda-Lois, R. Poveda and M. J. Vivas

Instituto de Biomecánica de Valencia, Universidad Politécnica de Valencia, Spain

The design of wearable robots implies the design of structures intended to apply loads to the human skeleton through the layers of soft tissue between the supports of the robotic system and the bones of the user. There are two main aspects that must be taken into account when designing support systems for wearable robots, which are the anatomical areas and structures able to support effective loads and the maximum levels of pressure that these structures can handle without raising issues of safety and comfort. These principles are addressed in Section 5.3. Figure 5.21 shows how these principles affect the upper limb (left) and lower limb (right). Joint movement areas, bony prominences, surface tendons and surface nerves or vessels are highlighted.

The main structures requiring protection in the upper limb are: (1) elbow movement area; (2) medial epicondyle; (3) lateral epicondyle; (4) wrist movement area; (5) radial styloid process; (6) ulnar styloid process; (7) Guyon tunnel; (8) carpal tunnel; (9) thumb movement area; (10) finger movement area; (11) metacarpal heads. On the lower limb, the main structures requiring protection are: (1) head of the fibula; (2) patella; (3) knee condyles; (4) tibial process; (5) ankle malleolus; (6) trochanter; (7) Achilles tendon; (8) quadriceps tendon; (9) ischiotibial tendons; (10) groin; (11) popliteal cavity; (12) hip movement area; (13) knee movement area; (14) ankle movement area.

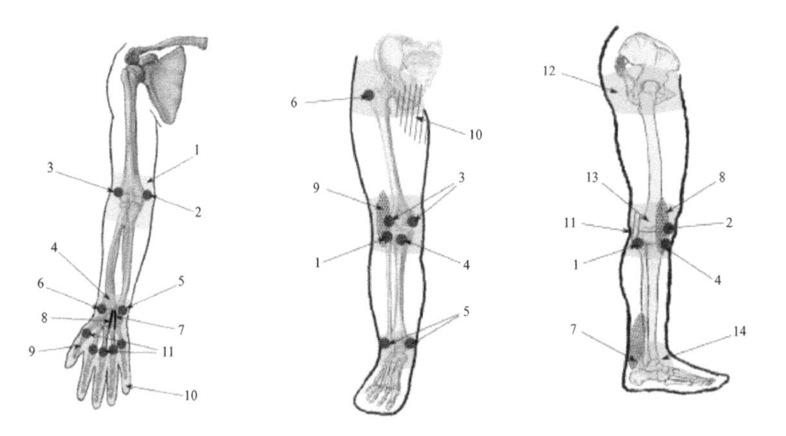

Figure 5.21 Areas to be avoided in the upper limb (left) and in the lower limb (right)

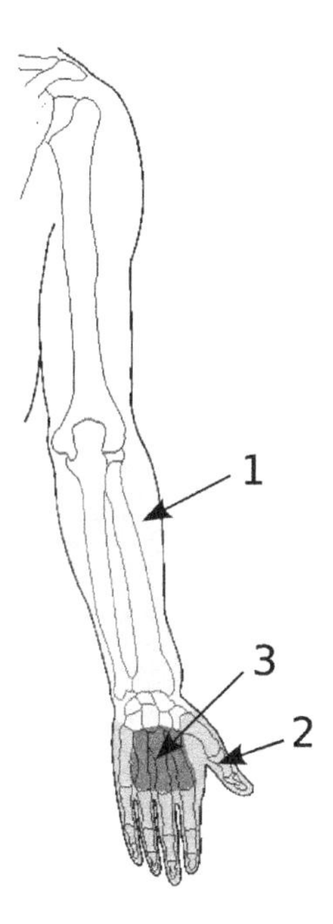

Figure 5.22 Areas of sensitivity to pressure in the hand and forearm: (1) Low-tolerance area (average ca. 450 kPa), (2) Middle-tolerance area and (3) High-tolerance area (average ca. 950 kPa)

5.6.1 Measurement of pressure tolerance

Pain is a warning sign of damage caused by overpressure, and likewise a good indicator of potential cell damage and death (Fransson-Hall and Kilbom, 1993). It is also a good indicator for comparing the relative sensibility of different parts of the body to pressure. The point at which the user begins to feel pain is known as the PPT or MPT and can readily be measured by means of an indenter. These parameters are described in Section 5.3.

Unfortunately the PPT is largely determined by the shape and size of the indenter, and therefore these indicators can only be used for intrasubject comparison and for stabilizing pressure maps of the body like the one in Figure 5.22 (Bystrom *et al.*, 1995). Nevertheless, some rough figures can be calculated. PPT in the hand has been estimated at between 6 and 11 percent of average skin strength (7640 kPa) (Yamada, 1970). This figure is larger than the maximum pressure required to prevent tissue ischemia. However, it has also been found that sustained external pressure corresponding to 50 % PPT becomes painful in a few minutes.

Another study investigating possible differences in pressure tolerance in the lower limb showed that there were some differences. Measurement points were located taking into account the common placement of load transmission elements of lower limb wearable devices (see Figure 5.23 (left)). Significant differences between these points were found (significant level: 0.0029). Points 1 and 6 proved more sensitive to external loading (see Figure 5.23 (right)).

It is essential to bear in mind that the external pressures mentioned above are punctual, instantaneous forces. Therefore the limits are not valid for a sustained external loading, but they do give an indication of the behaviour of the different points on the lower leg (see Table 5.1). Three homogeneous groups have been found, namely firstly, a high-sensitivity group, secondly, a medium-sensitivity group (can cope with pressures up to 416 kPa) and, lastly, a low-sensitivity group (can cope with pressures up to 557 kPa).

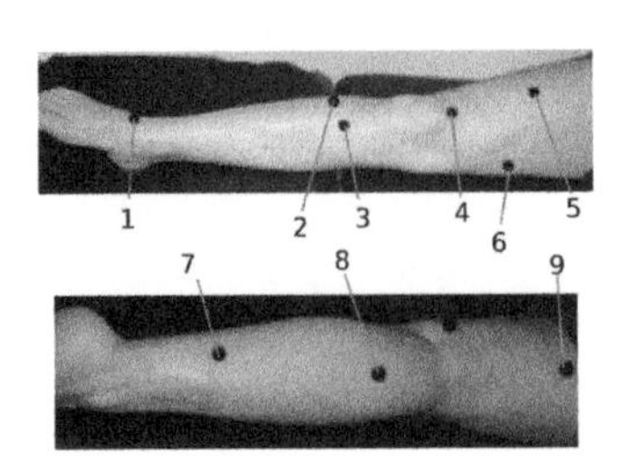

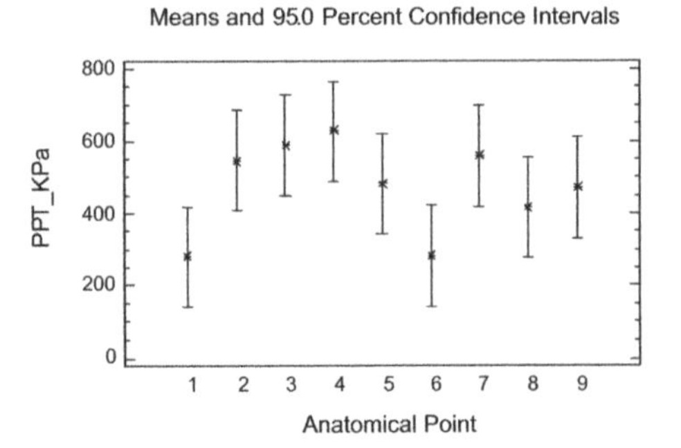

Figure 5.23 Points for the analysis of PPT in the lower limb

Table 5.1 Homogeneous groups of pressure sensitivity in the lower limb

Anatomical point	PPT (kPa)	Homogeneous group
P1	281.7	1
P2	545.5	3
P3	588.1	3
P4	628.1	3
P5	482.7	2
P6	281.9	1
P7	557.7	3
P8	416.6	2
P9	470.5	2

5.7 CASE STUDY: UPPER LIMB TREMOR SUPPRESSION THROUGH IMPEDANCE CONTROL

E. Rocon and J. L. Pons

Bioengineering Group, Instituto de Automática Industrial, CSIC, Madrid, Spain

This case study concerns the application of impedance control with the wearable orthosis for tremor assessment and suppression (WOTAS), an upper limb exoskeleton for tremor suppression described in Case Study 8.1. Tremor is characterized by involuntary oscillations of a part of the body. The most widely accepted definition is as follows: 'an involuntary, approximately rhythmic, and roughly sinusoidal movement' (Rocon *et al.*, 2004). Tremor, the most common of all involuntary movements, can affect various body parts such as the hands, head, facial structures, tongue, trunk and legs. There is evidence that an alteration of all three components of upper limb natural impedance (apparent inertia, viscosity and stiffness) modifies the biomechanical characteristics of tremor in the upper limb (Adelstein, 1981).

In this approach, the musculoskeletal system (each joint of the upper limb that contributes to the tremor) is modelled as a second-order biomechanical system (Adelstein, 1981). It is known that the frequency response of a second-order system presents the behaviour of a lowpass filter. The cutoff frequency of this filter is directly related to the biomechanical parameters of the second-order system. The proposed approach consists in selecting the right inertia and damping parameters so that the cutoff frequency, f_c, of the musculoskeletal system is just above the maximum frequency of voluntary movement Figure (5.24). Unlike other approaches in the literature, the control scheme is conceived in such a way as to minimize the effect of the suppression load on voluntary motion.

Control is based on a dual control loop, Figure (5.25). The value of torque applied by the exoskeleton to the upper limb, τ_d, is calculated on the basis of the combined effects of both loops:

$$\tau_d = f_{dt} - f_{mt} - \tau = f_{dt} - K_m \ddot{q}_t - K_d \dot{q}_t - \tau \tag{5.16}$$

This closed-loop control architecture uses information from gyroscopes, $\dot{q}$, on the *data treatment block* to distinguish between voluntary and tremor motion, $\dot{q}_t$, in the overall movement. The function of this block, which is to implement a cHRI based on biomechanical data, is explained in more detail in Case Study 4.6. The angular velocity information from the estimated tremorous component, $\dot{q}_t$, is subsequently multiplied by the coefficients K_m and K_d, which describe the reference inertia and damping characteristics of the upper limb. This process defines the actual impedance force, fm_t, of

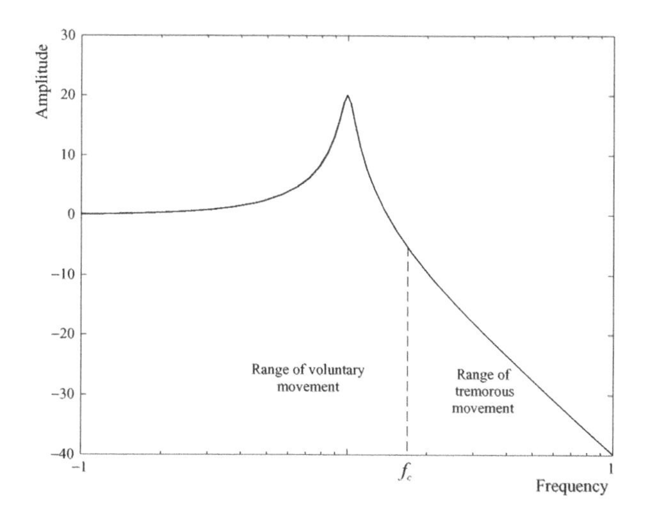

Figure 5.24 The musculoskeletal system is modelled as a second-order biomechanical system. The robotic exoskeleton is used to modify the apparent biomechanical characteristics of the upper limb so that the cutoff frequency, f_c, lies between the frequency ranges of voluntary and tremor motion

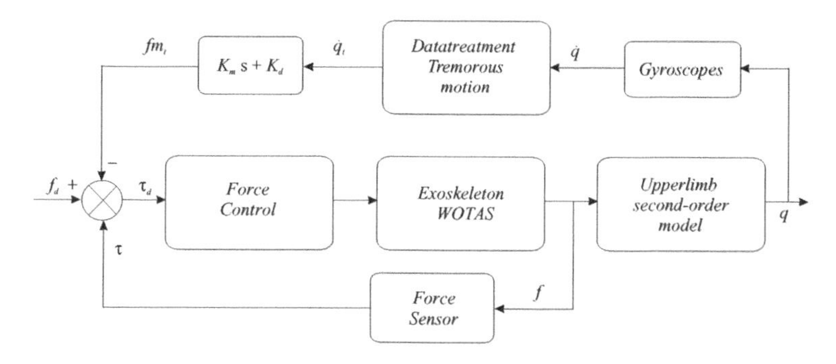

Figure 5.25 Control strategy to modify the biomechanical parameters of the upper limb

the system which is set to reduce the tremor. This impedance force should tend to vanish as tremorous motion is suppressed.

The lower loop of Figure 5.25 serves the goal of minimizing the effect of the exoskeleton on voluntary motion. In this case, force sensors measure the interaction force between orthosis and limb, f. Under ideal conditions, a user free from pathologic tremor should feel no opposing force from the exoskeleton; i.e. there should be no loading produced by voluntary motion. In order to achieve this, the interaction force, f, is filtered so that only the force opposing voluntary motion, τ, is fed back into the lower branch of the control loop.

The proposed control strategy envisages an adaptive behaviour, so it is constantly (in real time) updating the estimated tremor amplitude. In this way the system can respond to the changes that the control strategy effects in the tremor (Rocon *et al.*, 2007).

To summarize, the operation of WOTAS is based on the identification of tremorous motion from among the motion measured by kinematic and kinetic sensors. Adaptive algorithms help to identify tremorous motion and distinguish it from the voluntary kind. This information is then used to establish an actual physical interaction (modification of the combined human–exoskeleton articular impedance), which should reduce the tremor.

Another very important feature of this implementation of the proposed control strategies is that the basis of the approach is articular control, because this is simpler and also makes it possible to implement individual control loops in each joint, with a high dynamic range. Yet another interesting aspect is that each exoskeleton joint attempts to suppress the tremor generated in its corresponding anatomical joint and thus assure tremor reduction in each human joint. In this way the problem of coupling the tremor between the upper limb joints can be successfully tackled (Rocon *et al.*, 2005).

Previous studies by the authors evaluated the behaviour and contribution of each joint in upper limb tremor (Rocon *et al.*, 2007). This work has shown that in most patients the tremor movement migrates along the kinematic chain of the arm when its effects are reduced (by applying biomechanical loads) at one of the arm joints. However, tremor behaviour has yet to be properly explored when its effects are cancelled in different arm joints (Rocon *et al.*, 2004). To deal with this aspect, active and independent control strategies have been devised in each joint. This means that if cancellation of the tremor in one of the joints increases the tremor in the other joint, the algorithm responsible for controlling the adjacent joint will identify the increased tremor and try to reduce the tremor generated by coupling of the upper limb joints. The aim, then, is to achieve equilibrium in the active tremor reduction in each joint, thereby reducing the coupling effects on the upper limb joints.

During validation of the control strategies some patients spontaneously reported that they perceived a reduction in the amplitude of their tremorous movement and thus felt more confident about executing the task. The visual feedback from a smooth movement therefore has a positive impact on the user. These factors illustrate how the robot control strategy and the human internal control strategy can adapt to one another. These aspects are discussed in Section 5.4.

5.8 CASE STUDY: STANCE STABILIZATION DURING GAIT THROUGH IMPEDANCE CONTROL

J. C. Moreno and J. L. Pons

Bioengineering Group, Instituto de Automática Industrial, CSIC, Madrid, Spain

Where there is proximal leg weakness, the knee joint can be externally stabilized by means of an exoskeleton or orthotic device. A novel solution for unilateral cases has been conceived within the framework of the European GAIT project (the GAIT exoskeleton is fully described in Section 9.1). This section presents the application of intermittent control of impedance of the knee joint in order to provide a stable stance phase. The approach adopted makes use of a custom actuator system adapted to a KAFO (the actuator design presented in Section 6.7): the GAIT orthosis. Such an integrated orthotic system is a novel technique for functional compensation of muscle weakness, consisting of a wearable solution selectively applying springs of different stiffnesses in order to achieve more natural gait patterns and avoid collapse of the knee and the risk of falling.

5.8.1 Knee–ankle–foot orthosis (exoskeleton)

The exoskeleton is designed with a single-sided frame (unilateral bar), built as a four-point support. The frame is single-sided at knee level and double-sided at ankle level. A single-sided frame at knee level is preferred for lightness, size, adaptability, comfort and wearability. A double-sided frame is preferred at ankle level to add stability in the frontal plane. The knee hinge is performed by a four-bar mechanism, following the displacement of the instantaneous helicoidal axis of the knee on the lateral side (see Case Study 3.5 for more details). Ankle hinging is performed by a single hinge placed on the malleolli.

5.8.2 Lower leg–exoskeleton system

The lower leg system encompasses the thigh, shank and foot segments, as shown in Figure 5.26. The weak quadriceps group provides only partial or no torque at the knee. The kinematics of the exoskeleton hinge restricts the system to transmitting torque in the sagittal plane. The ankle actuator in the exoskeleton is designed as a passive compensator, with two springs applying different stiffnesses according to the direction of rotation of the foot. The stiffness of the spring elements is customized using a functional model of the knee and ankle joints developed with average walking data from Winter (1990). The objective of the controller in a gait cycle in terms of functional compensation is to approach normal kinematic and kinetic gait patterns at joints. The passive ankle actuator controls foot

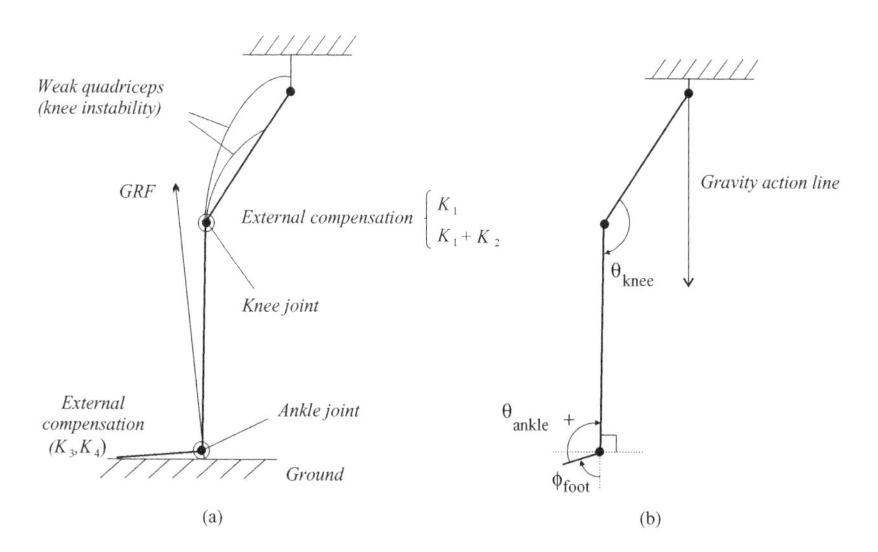

Figure 5.26 (a) Scheme of a weak (quadriceps) lower leg with the exoskeleton. The knee actuator applies K_1 in the stance phase for a given period of time to provide joint stability; during the swing phase the actuator applies K_2 ($K_1 \gg K_2$) to store and recover spring energy and assist leg extension prior to heel contact. (b) Relative joint angles in the system. Tilt angle of the foot, ϕ_{foot}, with respect to the line of action of earth gravity

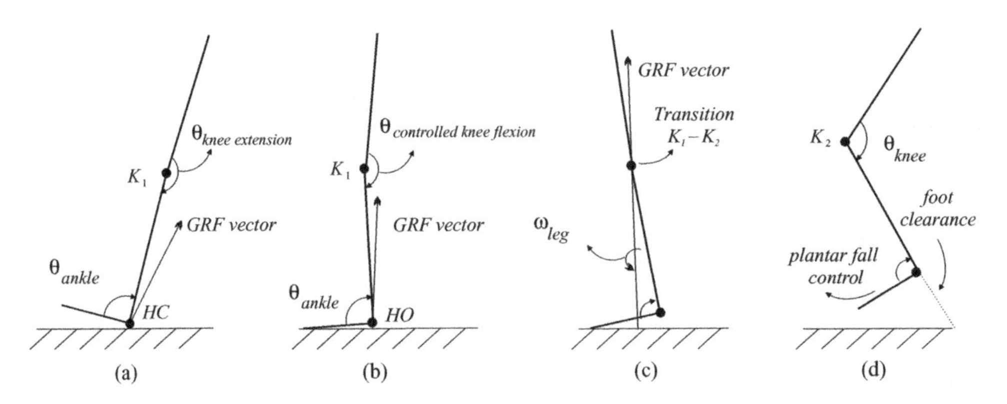

Figure 5.27 Partial control objectives of the knee in gait cycle: (a) heel contact (HC) with foot fall control followed by stabilized knee extension through K_1 at terminal swing; (b) heel off (HO) during the stance phase after controlled flexion of the knee; (c) change of sign of the rate of turn of the shank during pre-swing, where K_1 to K_2 transition releases the knee; (d) shank rotation during knee flexion, where K_2 (partially charged by inertia) assists extension at the end of the cycle

rotations by means of compensations provided by a spring device with a rate of K_3, compressed to store energy and control dorsiflexion in the stance phase. A spring with a rate of K_4 applies external stiffness to provide foot clearance during swing. The knee actuator applies K_1 for a given period of time in the stance phase to provide joint stability. During the swing phase the actuator applies K_2 ($K_1 \gg K_2$) to store and recover spring energy to assist leg extension prior to heel contact. The transition from K_1 to K_2 frees the knee joint, which acts like a flexing leg instead of a stiff, restricted hinge. Partial control objectives of the knee are depicted in Figure 5.27.

During cyclical walking, an intermittent mechanism features two discrete states in the actuator, for stance and swing respectively. Opportune transition of the system between states permits locomotion with a safe stance phase and a free-swinging leg and allows functional compensation. The transition from stance control to swing control is achieved by the activation, at a given *onset time*, of a linear pulling solenoid. The onset time for activation is estimated on the basis of inertial sensing information (Moreno *et al.*, 2006).

The transition from stance control to swing control is performed automatically by mechanical means when full extension of the leg is completed at terminal swing.

5.8.3 Stance phase stabilization: patient test

The subject is a patient (body weight, 78 kg; height, 150 cm; age, 42) affected by leg poliomyelitis. The subject daily wears a knee and ankle orthosis, which enables gait with a permanently locked knee. The orthosis with the controllable system was fitted to the patient. The stiffness of the springs in the actuator system were customized to suit the subject's body weight, assuming no remaining muscular capacity. For more details of the customization procedure see Case Study 6.7. The control algorithm was tuned to his self-preferred speed. After a number of trials, the user learned to drive the control system by adapting the profile and speed. The resulting pattern in this case was characterized by initial contact with slight knee flexion, accompanied by a self-adopted strategy of compensation through forward motion of the hip in order to extend the leg fully and lock the knee. Once the patient had adapted to the system, the ground reaction force was measured.

The calculated mean values show a reduction of mediolateral reaction forces in the case of this subject. This then reduces the need for lateral movement common in post-polio syndrome patients (see Figure 5.28). The profile of vertical ground reaction forces with the controllable system approximates

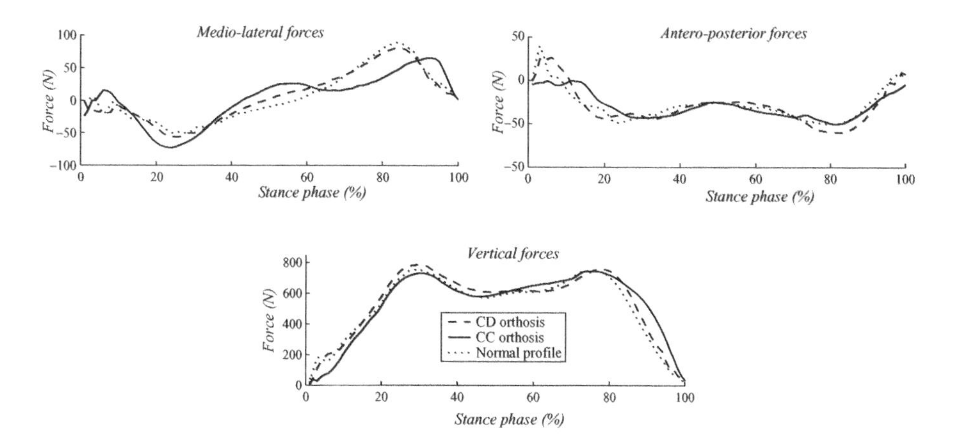

Figure 5.28 Ground reaction forces (mean values): evaluation data of S1 under the test conditions and average normal profile of walking with the orthosis calculated for the weight of the subject

to the normal profile with a correlation factor of 0.94. The calculation of average forces during braking, propulsion and push-off revealed no significant differences in terms of shock absorption. The evidence of a stable, noncollapsing knee during stance suggests that this is a promising approach that may provide a practicable means of improving lower limb exoskeletons (knee–ankle–foot orthoses, or KAFOs).

REFERENCES

Adelstein, B.D., 1981, 'Peripheral mechanical loading and the mechanism of abnormal intention tremor', PhD Thesis, MIT, Cambridge, Massachusetts.

Banala, S.K., Kulpe, A., Agrawal, S.K., 2007, 'A powered leg orthosis for gait rehabilitation of motor-impaired patients', *in Proceedings of the International Conference on Robotics and Automation 2007*, pp. 4140–4145.

Buchthal, F., Kaiser, E., 1951, 'The rheology of the cross-striated muscle fibre with particular reference to isotonic conditions', *Det Kongelige Danske Videnskabernes Selskab Biologiske Meddelser* **21**(7): 233–291.

Bystrom, S., Hall, C., Welander, T., Kilbom, A., 1995, 'Clinical disorders and pressure-pain threshold of the forearm and hand among automobile assembly line workers', *Journal of Hand Surgery* **20**(6): 782–790.

Carton, R.W., Dainauskas, J., Clark, J.W., 1962, 'Elastic properties of single elastic fibers', *Journal of Applied Physiology* **17**: 547–551.

Cavusoglu, M.C., Sherman, A., Tendick, F., 2002, 'Design of bilateral teleoperation controllers for haptic exploration and telemanipulation of soft environments', *IEEE Transactions on Robotics and Automation* **18**(4): 641–647.

Colgate, J.E., 1988, 'The control of dynamically interacting systems', PhD Thesis, MIT, Cambridge, Massachusetts.

Colombo, G., Joerg, M., Schreier, R., Dietz, V., 2000, 'Treadmill training of paraplegic patients using a robotic orthosis', *Journal of Rehabilation Reasearch and Development* **37**: 130–134.

Dandekar, K., Raju, B.I., Srinivasan, M.A., 2003, 3-D finite-element models of human and monkey fingertips to investigate the mechanics of tactile sense', *Journal of Biomechanical Engineering* **125**: 682–691.

Debicki, D.B., Gribble, P.L., 2004, 'Inter-joint coupling strategy during adaptation to novel viscous loads in human arm movement,' *Journal of Neurophysiology* **92**: 754–765.

Despopoulos, A., Silbernagl, S., 1991, *Color Atlas of Physiology*, 4th edition, Thiem Medical Publisher.

Dolan, J.M., Friedman, M.B., Nagurka, M.L., 1993, 'Dynamic and loaded impedance components in the maintenance of human arm posture', *IEEE Transactions on Systems, Man and Cybernetics* **23**: 698–709.

Fasse, E.D., 1987, 'stability robustness of impedance controlled manipulators coupled to passive environments', MSc Thesis, MIT, Cambridge, Massaechusetts.

Feldman, A.G., 1986, 'Once more on the equilibrium point hypothesis (*lamba* model) for motor control', *Journal of Motor Behaviour* **18**: 17–54.

Ferry, J.D., 1980, *Viscoelastic Properties Polymers*, John Wiley and Sons, Ltd.

Flash, T., 1987, 'The control of hand equilibrium trayectories in multi-joint arm movements', *Byological Cybernetics* **57**: 257–274.

Fransson-Hall, C., Kilbom, A., 1993, 'Sensitivity of the hand to surface pressure', *Applied Ergonomics* **24**(3): 181–189.

Fung, Y., C., 1993, *Bioviscoelastic Solids. Biomechanics: Mechanical Properties of Living Tissues*, Springer.

Goonetilleke, R.S., Eng, T., 1994, 'Contact area effects on discomfort', *in Proceedings of the 38th Human Factors and Ergonomics Society Conference*, pp. 688–690.

Gribble, P.L., Ostry, D.J., 2000, 'Compensation for loads during arm movements using equilibrium-point control', *Experimental Brain Research* **135**: 474–482.

Hidler, M., Wall, A.E., 2005, 'alterations in muscle activation patterns during robotic-assisted walking', *Clinical Biomechanics* **20**: 184–193.

Hogan, N., 1984, 'Adaptive control of mechanical impedance by coactivation of antagonist muscles', *IEEE Transactions on Automatic Control* **29**: 681–690.

Hogan, N., 1985, 'Impedance control: an approach to manipulation: Part I – Theory, Part II – Implementation, Part III – Applications', *Journal of Dynamics Systems, Measurement and Control* **107**: 1–24.

Hogan, N., 1987, 'Stable execution of contact tasks using impedance control', in *Proceedings of the IEEE International Conference on Robotics and Automation*, pp. 1047–1054.

Ikeura, R., Inooka, H., 1995, 'Variable impedance control of a robot for cooperation with a human', in *Proceedings of the International Conference on Robotics and Automation 1995*, pp. 3097–3102.

Jezernik, S., Colombo, G., Morari, M., 2004, 'Automatic gait-pattern adaptation algorithms for rehabilitation with a 4-DOF robotic orthosis', *IEEE Transactions on Robotics and Automation* **20**(3): 574–582.

Kawato, M., 1999, 'Internal models for motor control and trajectory panning', *Current Opinion in Neurobiology* **9**(1): 718–727.

Kazerooni, H., 1985, 'A robust design method for impedance control of constrained dynamics systems', PhD Thesis, MIT, Cambridge, Massaechusetts.

Kazerooni, H., Racine, J.L., Huang, L., Steger, R., 2005, 'On the control of the Berkeley lower extermity exoskeleton (BLEEX)', in *Proceedings of the IEEE International Conference on Robotics and Automation*, pp. 4353–4360.

Kearney, R.E., Hunter, I.W., 1990, 'System identification of human joint dynamics', *Critical Reviews on Biomedical Engineering* **18**: 55–87.

Kenedi, R.M., Gibson, T., Daly, C.H., 1964, 'Bioengineering studies of the human skin', in *Biomechanics and Related Bioengineering Topics* (ed. R.M. Kenedi), Pergamon Press, Oxword.

Krouskop, T.A., Williams, R., Krebs, M., Herszkowicz, M.S., Garber, S., 1985, 'Effectiveness of mattress overlays in reducing interface pressures during recumbency', *Journal of Rehabilitation Research and Development* **22**(3): 7–10.

Kwan, M.K., Woo, S.L., 1989, 'A structural model to describe the non-linear stress–strain behavior for parallel-fibered collagenous tissues', *Journal of Biomechanics Engineering* **111**: 361–363.

Landis, E., 1930, 'Micro-injection studies of capillary blood pressure in human skin', *Heart* **15**: 209–228.

Lanir, Y., 1983, 'Constitutive equations for fibrous connective tissues', *Journal of Biomechanics* **16**: 1–12.

Lemay, M., Hogan, N., van Dorsten, J.W., 1998, 'Issues in impedance selection and input devices for multipoint powered orthotics', *IEEE Transactions on Rehabilitation Engineering* **6**(1): 102–105.

Mason, M., 1981, 'Compliance and force control for computer controlled manipulators', *IEEE Transactions on Systems, Man, and Cybernetics* **11**(6): 418–432.

Maurel, W., 1998, '3D modeling of the human upper limb including the biomechanics of joints, muscles and soft tissues', PhD Thesis, Ecole Polytechnique Federale de Lausanne, Switzerland.

Moreno, J.C., Rocon, E., Ruiz, A., Brunetti, F., Pons, J.L., 2006, 'Design and implementation of an inertial measurement unit for control of artificial limbs: application on leg orthoses', *Sensors and Actuators B* **118**: 333–337.

Mussa-Ivaldi, F.A., Hogan, N., Bizzi, E., 1985, 'Neural, mechanical, and geometric factors subserving arm posture in humans', *The Journal of Neuroscience* **5**: 2732–2743.

Raibert, M.H., Craig, J.J., 1981, 'Hybrid position/force control of manipulators', *Trans. ASME, Journal of Dynamic Systems, Measurements and Control* **102**: 275–282.

Rocon, E., Belda-Lois, J.M., Sánchez-Lacuesta, J.J., Pons, J.L., 2004, 'Pathological tremor management: modelling, compensatory technology and evaluation', *Technology and Disability* **3**: 3–18.

Rocon, E., Ruiz, A.F., Pons, J.L., Belda-Lois, J.M., Sánchez-Lacuesta, J.J., 2005, 'Rehabilitation robotics: a wearable exo-skeleton for tremor assessment and suppression', in *Proceedings of the International Conference on Robotics and Automation 2005*, pp. 241–246.

Rocon, E., Manto, M., Pons, J.L., Camut, S., Belda-Lois, J.M., 2007, 'Mechanical suppression of essential tremor', *The Cerebellum* **6**: 73–78.

Roy, J., 2001, 'Force controlled robots: design, analysis, control and applications', PhD Thesis, John Hopkins University, Baltimore, Maryland.

Roy, J., Whitcomb, L.L., 2002, 'Adaptive force control of position/velocity controlled robots: theory and experiments', *IEEE Transactions on Robotics and Automation* **18**(2): 121–137.

Salisbury, J.K., 1980, 'Active stiffness control of manipulator in Cartesian coordinates', in *Proceedings of the 19th IEEE Conference on Decision and Control*, pp. 95–100.

Sasaki, D., Noritsugu, T., Takaiwa, M., 2005, 'Development of active support splint driven by pneumatic soft actuator (ASSIST)', in *Proceedings of the 2005 IEEE International Conference on Robotics and Automation*, pp. 520–525.

Schiele, A., van der Helm, F.C.T., 2006, 'Kinematic design to improve ergonomics in human machine interaction', *IEEE Transactions on Neural Systems and Rehabilitation Engineering* **14**(4): 456–469.

Schiele, A., 2008, 'An explicit model to predict and interpret constraint force creation in p-HRI with exoskeletons', in *Procc. of International Conference on Robotics and Automation (ICRA), IEEE* (in press).

Shadmehr, R., Mussa-Ivaldi, F.A., 1994, 'Adaptative representation of dynamics during learning of a motor task', *Journal of Neuroscience* **14**: 3208–3224.

Tanaka, Y., Onishi, T., Tsuji, T., Yamada, N., Takeda, Y., Masamori, I., 2007, 'Analysis and modeling of human impedance properties for designing a human–machine control system', in *Proceedings of the International Conference on Robotics and Automation 2007*, pp. 3627–3632.

Tong, P., Fung, Y.C., 1976, 'The stress–strain relationship for the skin', *Journal of Biomechanics* **9**: 649–657.

Tsuji, T., Morasso, P.G., Ito, K., 1995, 'Human hand impedance characteristics during maintained posture', *Biological Cybernetics* **74**: 475–485.

Tsuji, T., Tanaka, Y., 2005, 'Tracking control properties of human-robotic systems based on impedance control', *IEEE Transactions on Systems, Man, and Cybernetics – Part A: Systems and Humans*, **35**(4), pp. 523–535.

Tsumugiwa, T., Yokogawa, R., Hara, K., 2002, 'Variable impedance control based on estimation of human arm stiffness for human–robot cooperative calligraphic task', in *Proceedings of the 2002 IEEE International Conference on Robotics and Automation*, pp. 644–650.

Tsumugiwa, T., Yokogawa, R., Hara, K., 2003, 'Measurement method for compliance of vertical-multi-articulated robot – application to 7-DoF robot PA-I0', in *Proceedings of the 2003 IEEE International Conference on Robotics and Automation*, pp. 2741–2746.

Tsumugiwa, T., Yokogawa, R., Yoshida, K., 2004, 'Stability analysis for impedance control of robot for human–robot cooperative task system', in *Proceedings of the 2004 IEEE/RSJ International Conference on Intelligent Robots and Systems*, pp. 3883–3888.

Vukobratovic, M., 1997, 'How to control robots interacting with dynamic environment', *Journal of Intelligent and Robotic Systems* **19**: 119–152.

Whitney, D.E., 1977, 'Force feedback control of manipulator fine motions', Trans. *ASME, Journal of Dynamic System, Measurement and Control* **99**: 91–97.

Winter, D.A., 1990, *Biomechanics and Motor Control of Human Movement*, John Wiley & Sons, Ltd.

Yamada, H., 1970, *Strength of Biological Materials*, The Williams & Willkins Company, Baltimore, Maryland.

6

Wearable robot technologies

J. C. Moreno, L. Bueno and J. L. Pons

Bioengineering Group, Instituto de Automática Industrial, CSIC, Madrid, Spain

6.1 INTRODUCTION TO WEARABLE ROBOT TECHNOLOGIES

This chapter reviews the key technologies relevant to the current and future development of wearable robots. The interface between human and robot can exchange signals in order to drive an action, provide feedback for human motor control and monitor the status of the HRI and its surroundings.

The following sections review the main hardware sensor, actuator and battery technologies with reference to practical applications. *Wearability* within an application imposes a number of particular requirements on sensor, actuator and energy storage technologies. A comparative analysis of the state-of-the-art technologies is given for each category, providing an overview of the current achievements in technology development.

When defining reliable sensors for a wearable application it may be useful to analyse a wide range of candidate measurement devices. Measurement requirements for a system may consider or combine accurate tracking of movement or force, quantification of the status of the HR interface, acquisition of a physiological signal for feedback, etc.

In order to equip a wearable robot with a measurement system, the designer must unavoidably accept tradeoffs between functional versatility and simplicity of implementation. In this connection, Section 6.2 presents motion, bioelectrical activity, force and pressure sensing technologies.

Using the natural body as a sensing mechanism can be an elegant solution to enhance the usability of a WR and also to overcome particular challenges imposed by applications. This chapter describes several methods that are being investigated to measure biological muscle and brain activity signals for the purpose of controlling and providing feedback to a WR.

Actuators may be required at the level of the human joints or limbs, to respond to signals from the human body and from the environment. Whether a given actuator technology is considered for integration will depend on the type of physical interaction with the limb that is required in the given range of applications, e.g. damping, modulation of resistance, powering, etc. The review of actuator technologies in Section 6.3 focuses on principles, practical availability and limitations.

Wearable Robots: Biomechatronic Exoskeletons Edited by José L. Pons

Section 6.3 analyses and compares the most suitable portable energy storage technologies to enable WR technologies.

Finally, case studies are presented, dealing with sensing of microclimate conditions in a human–robot interface, the fusion of inertial sensor data in a leg exoskeleton and the biologically based design of a knee actuator system.

6.2 SENSOR TECHNOLOGIES

The measurement of kinematic (position, acceleration) and bioelectrical activity is critical in wearable robot applications. Force and pressure sensor systems to collect information about HRI interaction (force, torque, pressure) may be vital for the application of an exoskeleton.

6.2.1 Position and motion sensing: HR limb kinematic information

The measurement of angular position or linear displacements of a given joint or segment is a fundamental requirement. The sensing technology that is selected for a wearable robot depends heavily on the specifics of the target application. Various techniques can be considered to build sensors for joint and segment positions in wearable robots. This section discusses a wide range of sensor technologies suitable for wearable applications, including encoders, magnetic sensors, potentiometers, linear variable differential transformers (LVDTs), electrogoniometers and MEMS inertial sensing devices.

6.2.1.1 Encoders

Linear or rotary encoders are electromechanical transducers that measure absolute or relative motion. Linear motion is converted into rotary motion via toothed belts, pinion gearings or cable control. Encoders are classified as incremental or absolute. A relative encoder (also called an incremental encoder) typically uses an optical switch to generate an electrical pulse when radial lines in a disc pass through its field of view (Figure 6.1). External electronic circuits are required to count the pulses

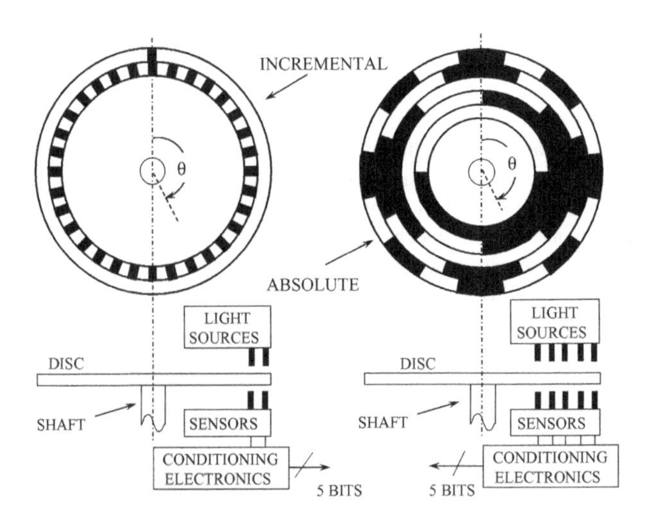

Figure 6.1 Incremental and absolute encoders. Reproduced by permission of Analog Devices

and determine the relative angle. This transducer cannot determine the direction of rotation without placing additional sensors. It is more suitable for applications where reliability and resolution are not critical.

Absolute rotary encoders produce a unique digital code for each position of a shaft. Absolute encoders can be optical, mechanical, fibre optic or magnetic. With an optical encoder, incremental angular counting is achieved by a light emitter and receiver. The state of alternating opaque and transparent sliding contacts with respect to a fixed part defines a unique code for each angle (see Figure 6.1). This solves the requirement of moving to a calibration point when reactivation of the voltage supply is required. Different absolute encoding methods can be applied. Binary coding is a technique that can produce large errors for every bit of incorrect interpretation. Grey coding is a technique that minimizes the error in a single bit (to 1 LSB (least significant bit)).

The output of relative encoders ranges from 60 to 1000 pulses per revolution, with a maximum response frequency up to 1000 Hz. Absolute encoders using grey code range from 5 to 16 bits of absolute positioning information, and single-turn transducers typically offer up to 20 Hz response frequency.

Encoders are common in wearable robot systems; examples are found in the measurement of robot kinematics for teleoperation (Sooyong *et al.*, 1999), tracking of human posture and motion with encoders in wearable support systems, and lower extremity exoskeletons to assist human walking and load carrying, equipped with encoders to capture joint information.

6.2.1.2 Magnetic (Hall effect) sensors

A Hall effect sensor is a transducer that produces a variation in output voltage in response to changes in magnetic field density. Hall sensors are noncontact sensors that can be used for proximity, positioning and current sensing. When electrons flow through a conductor, a magnetic field is produced. Electrical potential is developed between two edges of a current-carrying conductor (see Figure 6.2(a)). The Hall element is the basic magnetic field sensor. In an angular position sensor setup, a rotating magnet generates one sinusoidal wave per revolution. This type of setup can only cover a range of 90°. The range up to about ±45°, with additional sensors, can be used for accurate angle measurements (see Figure 6.2(b)). Hall effect linear position sensors can respond to a wide range of positive or negative magnetic fields and provide high sensitivity. Linear electrical outputs feature linearities close to ±2 %.

Appropriately isolated, Hall effect devices are immune to environmental factors and thus solve the problems posed by optical and electromechanical sensing. A noncontact solution is a cost-effective means of increasing the load life to tens of millions of life cycles. The main drawback is the possibility of cross-coupling when multiple sensors are required.

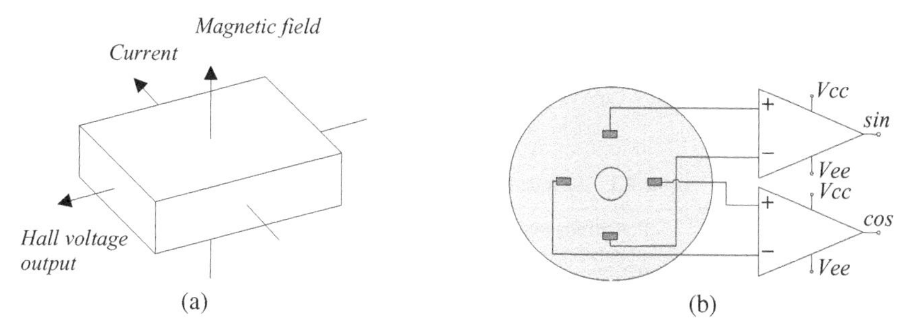

Figure 6.2 (a) Hall effect principle and (b) Integrated Hall effect rotary transducer with differential amplifiers. Reproduced from Searle and Kirkup (2000)

Integration of Hall-effect-based transducers has been proposed for wearable robots. Examples of applications include tracking hand position (Ferrazzin *et al.*, 1999), measuring position for active joints in biomechatronic hands (Carrozza *et al.*, 2001) and upper limb prostheses (Kyberd and Pons, 2003).

6.2.1.3 Potentiometers and LVDTs

Composed of a variable resistive material, potentiometers are the simplest position transducers. An electrical contact causes variation of the measured voltage potential. Rotary potentiometers are suitable for direct measurement of a joint angle with an analogue output. Potential dividers can be used for signal conditioning. The advantage of the potential divider as opposed to a variable resistor in series with the source is that dividers are able to vary the output voltage from maximum voltage to ground within the mechanical range of the potentiometer. Problems of signal quantization and sliding noise are the main drawbacks in precision rotary potentiometers. One example of integration in a wearable device is the force controllable ankle foot orthosis to assist drop-foot (Blaya and Herr, 2004).

Another candidate for position sensing within an exoskeleton is the linear variable differential transformer (LVDT), which is a relatively simple electrical transducer with high resolution and reliability. It is an absolute position sensor with a measurement range for linear displacements from micrometres to several centimetres; however, it is less cost-effective at stroke lengths greater than approximately 7 cm. Induction of current through a secondary coil caused by current driven through a primary coil generates a differential voltage (see Figure 6.3). A conditioning circuit (voltage regulator and sine wave generator) is required to drive the primary coil. LVDTs can be configured as rotary devices and are typically available for full-scale travel of up to 120° of rotation. Several conditioning solutions are commercial. The main drawback of an LVDT is the nonlinearity of the output signal versus the input measurand. Examples of LVDT applications include measurement of probe deflection for teleoperated nanomanipulation (Sitti, 2003) and spring length measurement for force estimation in a gait rehabilitation robot (Veneman *et al.*, 2006).

Table 6.1 summarizes the main comparative features of the sensor systems described for measurement of joint position.

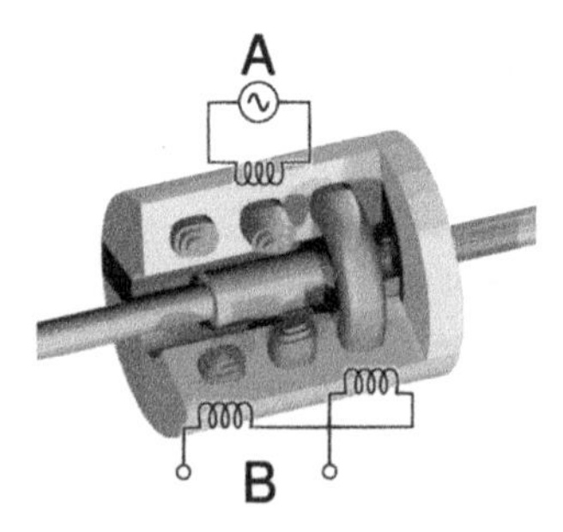

Figure 6.3 Cutaway view of an LVDT. Reproduced from Eric Pierce, GNU FLD

Table 6.1 Comparison of joint position transducers

	Potentiometer	LVDT	Hall effect transducer	Encoder
Linearity(%)	0.2–2	0.1–0.25	1–2	0.01
Resolution(μm)	5	0.25	0.1	0.25
Cost	Low	Medium	Low	High
Life	Low	Medium	High	High
Robustness	Medium\low	Medium	Medium	High

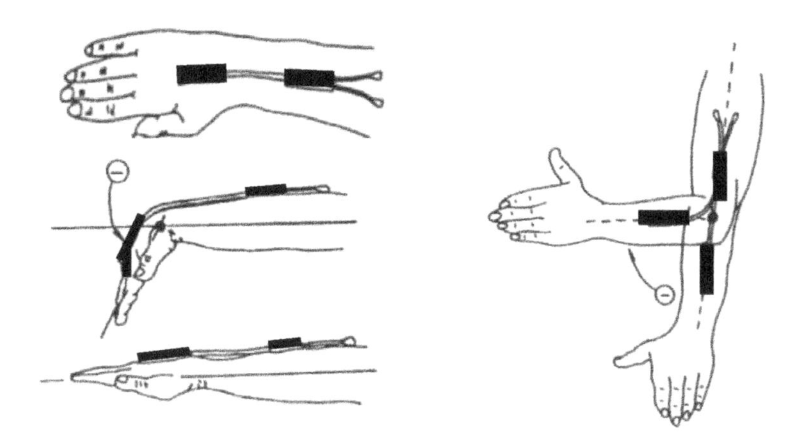

Figure 6.4 Dual-axis electrogoniometers attached to the upper limb. The example presents instrumentation to measure wrist flexion–extension/radio–ulnar deviation (left) and elbow flexion–extension (right)

6.2.1.4 Electrogoniometers

Another widely used technique in human biomechanics studies is electrogoniometry, which provides an accurate means of measuring joint movements. Electrogoniometers consist of one or two resistance strain sensing elements separated by a flexible film between two bars. An electrogoniometer is based on the variation of voltage that occurs depending on the angle between the bars crossing a given joint. Electrogoniometry systems are easy to set up for measurement of human joint angles (see Figure 6.4), but are relatively costly and cumbersome solutions. Some applications can be ruled out given the need to cross the joint concerned. An example of an application in control is an hybrid orthotic system for gait improvement (Gharooni, Heller and Tokhi, 2000). Electrogoniometers have been widely used in gait analysis under controlled conditions and more recently have been proposed for everyday applications, e.g. a wearable biomechatronic system for analysis of leg movements (Micera *et al.*, 2004).

6.2.1.5 MEMS inertial sensing technology

MEMS inertial sensors are suitable for tracking changes in velocity, position and orientation. In the past, accelerometers have been built with large mechanical masses and gyroscopes with multiple mechanical gimbals and bearings. Recent advances in microelectromechanical system (MEMS) technologies have made miniaturized inertial sensors possible. These devices are currently an exciting alternative to motion capture in wearable applications. Thanks to relatively low power consumption and portability in addition to low cost and size, developments that integrate complementary technologies as wearable means of extracting body kinematic information can now be seen. Based on micromachined accelerometers, orientation and position can be extracted by combining rate velocity estimations from rate gyroscopes. Three-dimensional static orientation can be derived in a very direct way using accelerometer or magnetometer data.

6.2.1.6 Accelerometer

An accelerometer can be used to measure the acceleration of a rigid body. The basic configuration of a single-axis accelerometer consists of a mass attached to a spring (Luinge, 2002). The displacement

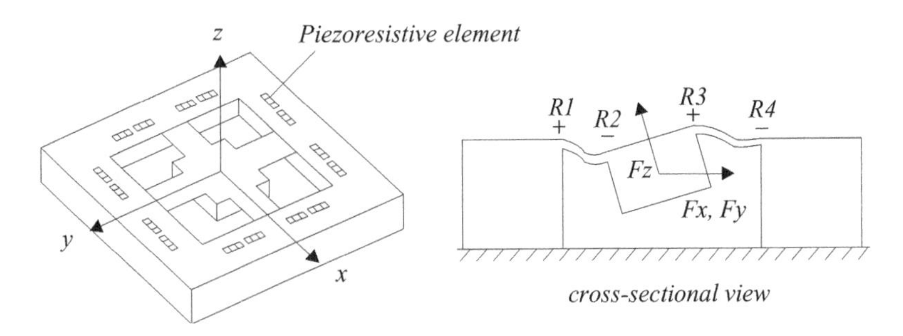

Figure 6.5 Accelerometer design for three-dimensional measurements. A full bridge circuit of piezoresistors detects unbalanced voltage. Reproduced from www.hitachimetals.com

of the mass is a measure of the difference between the acceleration $\boldsymbol{a}$ of the body and the acceleration $\boldsymbol{g}$ due to the earth's gravity acceleration, projected on to the sensitive axis of the sensor, which is represented by the unitary vector $\boldsymbol{n}$. The electrical output signal is represented by the magnitude of the vector $\boldsymbol{S}$, as

$$\boldsymbol{S}_{A,n} = k_{A,n}(\boldsymbol{a} - \boldsymbol{g}).\boldsymbol{n} + o_{A,n} \tag{6.1}$$

where $k_{A,n}$ represents a scale factor and $o_{A,n}$ the offset. A three-dimensional accelerometer can be produced by combining three accelerometers with orthogonal axes (see Figure 6.5). Three-dimensional configurations to measure body inclination are typically achieved with piezoresistive sensors. Other means of producing symmetrical capacitive triaxial accelerometers have also been proposed (Lotters, 1998).

Piezoelectric accelerometers are active devices that generate power under a mechanical excitation and are therefore not suitable for measurement of accelerations produced by static forces. Capacitive pendulous accelerometers typically respond to motion of the proof mass with a change of voltage across a capacitive sensor element. Capacitive accelerometers feature high sensitivities (in the range of 10 mV/g).

6.2.1.7 Gyroscope

Gyroscopes are rate sensors that commonly use vibrating mechanical elements (proof-mass) to sense rotation. Gyro technologies include mechanical, optical and vibrating types. The operation is typically based on the transfer of energy between two vibration modes of a structure, caused by Coriolis acceleration, which occurs in a rotating frame and is proportional to the rotational velocity. In MEMS vibrating gyroscopes, a comb structure is formed on the silicon substrate to detect Coriolis acceleration effects. When the sensor experiences a rotation about its sensitive axis (see Figure 6.6), the vibrating element experiences proportional Coriolis forces in a direction tangential to the rotation. The magnitude of this force is given by

$$f = 2mv\omega \tag{6.2}$$

where m is the mass of the vibrating element, v the linear velocity of the element and ω the angular rate.

Analog Devices has been actively developing tuning fork-based designs for MEMS gyroscopes for several years. Fully integrated circuits (ICs) combining mechanical structures are being developed, containing signal processing circuitry for calibration and tuning. Measurement ranges of available MEMS gyroscopes go from 75 to 300°, with sensitivities that can range from 5 to 25 mV/degree s.

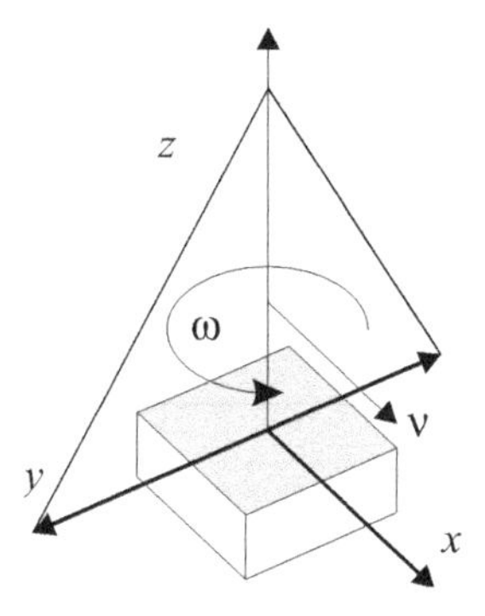

Figure 6.6 Operation of a vibrating gyroscope

Optical gyros are also available; these are based on the detection of phase differences in light waves travelling counter to the direction of rotation. Optical gyroscopes offer higher bias stabilities than vibrating gyroscopes but are more costly.

Recent advances in wireless technologies are being combined with MEMS inertial sensors to configure novel motion capture platforms (Brunetti *et al.*, 2006). In particular, accelerometers have been judged suitable for compensation of user motion, e.g. in closed-loop configurations for wearable vision platforms (Mayol, Tordoff and Murray, 2000) or tremor compensation in instruments for microsurgery (Riviere, Ang and Khosla, 2003). Examples of applications of rate gyroscopes include active compensation of upper limb tremor by means of motorized exoskeletons (Rocon *et al.*, 2005), control and monitoring of leg orthoses (Moreno *et al.*, 2006), neural vestibular prostheses (Liu *et al.*, 2003) and others. Case Study 6.5 describes a method of estimating limb orientation in exoskeleton devices by means of inertial sensors fusion.

6.2.2 Bioelectrical activity sensors

According to Malmivuo and Plonsey (1995), the human body can be modelled as a volume conductor since it is composed mainly of electrolytes and thus has some conductivity. In human cells there is a natural voltage difference between the inside and outside of the membrane. Depolarization of the motor unit causes depolarization of the muscle cell. This depolarization produces an electric impulse that travels through the volume conductor. The same operating principle can also be applied in the case of brain signals, where the impulses produced by depolarization of the neuron cells in the brain tissue travel through the volume conductor and can be measured on the scalp.

To record these electrical potentials, a set of electrode leads must be placed as close as possible to the sources. There are two main approaches: one using surface electrodes on the skin and another using internal electrodes placed near the active cells.

Amplification of these signals raises a number of issues, chiefly concerning safety and noise. In order to acquire the signals with a high signal-to-noise ratio, a better contact between the lead and the tissue is required. A better contact ensures less noise in the acquired signal, but on the other hand it entails a greater risk of electric shock in the event of failure of the power supply system, as the impedance contact between the user and the leads is relatively low. This risk is even higher in the case of implanted electrodes, where the leads are directly connected to the internal tissue.

The materials used for the electrodes are highly dependent on the type of contact, e.g. silver/silver chloride (Ag/AgCl) or stainless steel. These different materials present different electrical responses, which are determined by their electrical properties. The electrodes can be used on dry or wet contacts, or again they may be insulated. Insulated electrodes typically pose problems with the time constant

Table 6.2 Comparison of the contact impedance of different electrode materials, according to Searle and Kirkup (2000). Ag/AgCl electrodes are gelled, while others are dry electrodes

	Stainless steel	Aluminium	Titanium	Ag/AgCl
Impedance at $t = 0$ s	2 MΩ	3 MΩ	2 MΩ	0.18 MΩ
Impedance at $t = 400$ s	0.7 MΩ	0.9 MΩ	0.5 MΩ	0.16 MΩ
Impedance at $t = 800$ s	0.4 MΩ	0.4 MΩ	0.4 MΩ	0.15 MΩ

of the contact, as it resembles an RC circuit. The reader is referred to Searle and Kirkup (2000) for a detailed discussion of the differences between dry, wet and insulated surface electrodes.

Silver/silver chloride electrodes are usually used with an electrolyte gel and provide a better signal-to-noise ratio. However, there are some situations where a wet electrode is not desirable, e.g. in a cHRI for a prosthesis where a simpler interface is required and a wet electrode might compromise the signal quality. The contact quality of wet electrodes is not constant. If the conductive gel dehydrates, the contact impedance will increase, and in some applications reapplication of gel is not feasible (Searle and Kirkup, 2000). Table 6.2 shows the electrode–skin impedance of different electrodes, presented at different time instants.

A differential amplifier architecture can be used to minimize noise issues. The differential architecture is depicted in Figure 6.7 along with other possible configurations. In this electrical configuration, two or more signals are acquired and subtracted in order to minimize common noise on the electrode sites.

Another way to improve signal quality is to place the amplification circuitry as close as possible to the electrodes. This approach is currently used in most surface EMG (sEMG) systems, e.g. commercial systems by DELSYS™ (http://www.delsys.com). In these systems, the so-called active electrode consists of two or more dry electrodes with amplification circuitry in a small box. The individual electrodes are placed on the active electrode at a fixed distance, which also simplifies mounting of the sensors. The expressions in Figure 6.7 show the output generated on these active sensors.

The amplitude of the acquired signal for the sEMG is typically around 10 mV (peak to peak) (Luca, 2002). With such low amplitude, active electrodes are recommended as they are less sensitive to noise than systems with electrodes connected to long leads. The surface EMG usually presents a bandwidth between 0 and 500 Hz (Luca, 2002). This bandwidth needs to be reinforced with filters in order to eliminate noise at frequencies outside this band. This filtering stage is necessary to avoid acquisition problems caused by aliasing of the signal.

The positioning of the sensors strongly influences signal quality; a badly positioned or oriented sensor can produce significant variations in the signal quality and/or amplitude. The SENIAM project (Hermens *et al.*, 1999), made a series of recommendations regarding the placement of sEMG electrodes for a set of muscle groups.

Many of the techniques proposed for the capture and amplification of sEMG are also applicable in the case of EEG. Both signals are generated inside the human body and are acquired using a set of electrode leads attached to the skin. For EEG acquisition, the most commonly used electrode material is Ag/AgCl, and electrode contact is typically wet. The EEG signal has a weak amplitude and is subject to heavy noise contamination from both external and internal sources. Given the signal's weaker amplitudes, the gains have to be higher, and noise from nearby muscles, e.g. facial muscles, and electrooculography (EOG) may be present. In EEG, the background brain activity may also be perceived as noise; this makes it more difficult to process and classify the signals, as some of the characteristics of the noise can be very similar to those of the desired signal.

As in the case of EMG, brain activity can also be used as an interface signal to control a device. Brain activity can be acquired using both surface and implanted electrodes. The implementation of a brain–machine interface using implanted electrodes is presented in Case Study 4.7.

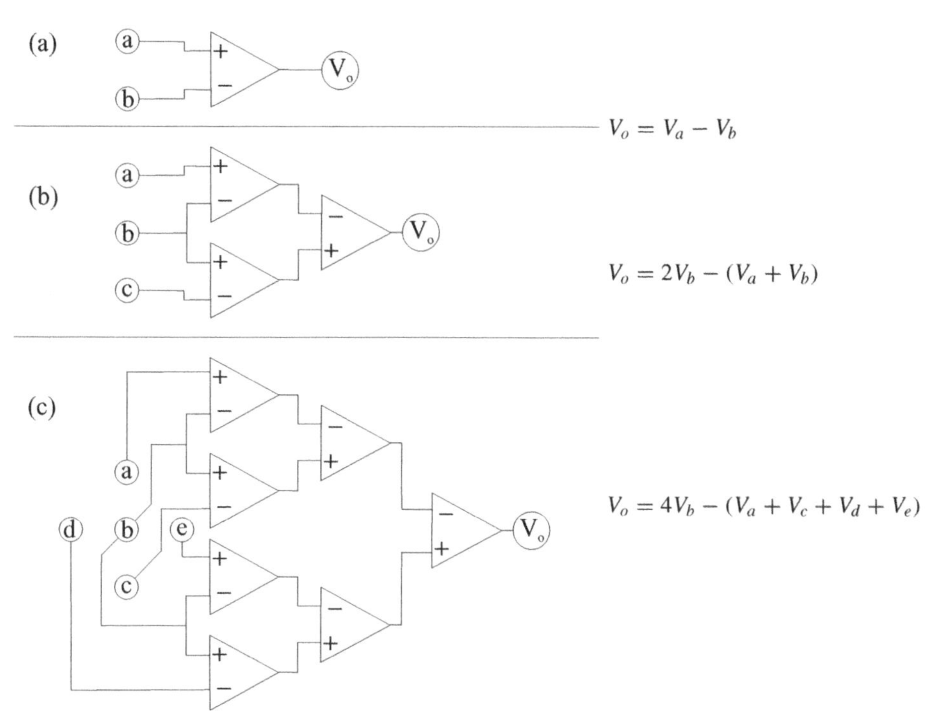

Figure 6.7 Schematic representation of three differential amplifier architectures, used to acquire bioelectric signals, and their resulting equations on (a) a single differential amplifier, on (b) a double differential amplifier and on (c) a normal double differential amplifier

EEG signals present lower amplitudes when compared to sEMG signals. While an sEMG signal has a peak-to-peak amplitude of around 10 mV, the EEG signal is around 100 μV (Malmivuo and Plonsey, 1995). EEG signals also present lower characteristic frequencies, while EMG signals have a bandwidth between 0 and 500 Hz. Classical EEG analysis considers that its characteristic frequencies are from 0 to 40 Hz, but recent studies have shown that there is significant activity from 0 to about 250 Hz (González *et al.*, 2006).

Clinical EEG systems do not normally use differential amplifiers. Instead, they use a single reference, placed close to the electrodes but with no nearby neural or muscular activity, e.g. the ears.

EEG amplifiers demand increased safety levels. Clinical amplifiers use isolation amplifiers and insulated power supplies to provide galvanic insulation of the electrode leads to the power mains.

Noise is a major issue in processing the EEG signal. Noise sources can be classified as internal and external with respect to the user. Internal noise sources are other bioelectrical sources inside the body, such as muscle activity, ocular activity or mental activities. External noise sources refer to induced noise on the body or on the electrode leads, such as power line induction.

Since the most commonly used electrode type is wet Ag/AgCl, the main problem with electrode contact, under controlled conditions, is drying of the gel. Under noncontrolled conditions, normally entailing more active use, other contact issues can compromise the signal, for instance electrode movement or sweating.

EEG electrode placement is important in that the location of the electrode over a given area determines which area of the brain will be monitored. However, as EEG potentials are spread over a fairly large area, minor changes in electrode position have smaller consequences than in the case of sEMG.

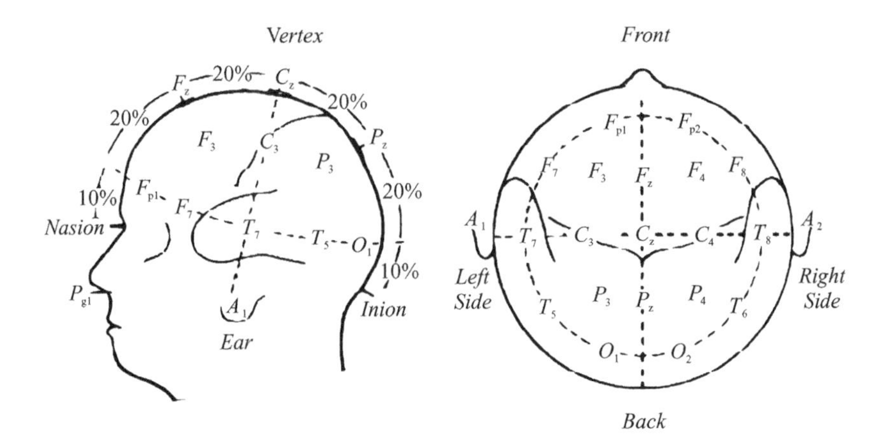

Figure 6.8 Human head with the landmarks used to identify electrode location according to the 10/20 system. The figure depicts the *central sulcus* (divide the anterior and posterior halves) and the *posterior ramus* (temporal structure of the brain). Electrodes T_3, T_4, T_5 and T_6 are also referred to as T_7, T_8, P_7 and P_8 respectively

Electrode positioning is normalized according to the international standard 10/20, which is based on relative distances on the head of the subject. The base measurements are the distance from inion to nasion and the perimeter of the head. These landmark points are depicted in Figure 6.8. The electrode positions are named after the lobes where they are placed: temporal, parietal, occipital, frontal and central. As there is no central lobe in the brain, this nomenclature is used only as a reference for electrode positions over the *central sulcus*. The initial is followed by an identifier referring to the position. The letter z is used for electrodes located over the *medial longitudinal fissure*. Electrodes located on the left side of the head receive odd numbers, in ascending order of distance from the vertex. Those on the right side receive even numbers, following the same ordering principle as the left side.

Research groups worldwide are working on the use of the EEG signal as a source of information for control of robotic and automated systems. Section 4.2 provides an overview of BCI principles. For some approaches on BCI systems based on EEG, the user is referred to Case Study 9.6.

sEMG and EEG present some similarities as regards signal digitalization. Both types of signal must be acquired with at least 12 bits of resolution. At this level of resolution it is possible to identify signal characteristics, since the acquisition noise is negligible. The main difference between acquisition of EEG and acquisition of sEMG lies in selection of the sampling frequency. The maximum signal frequency for sEMG is around 500 Hz and therefore a minimum sampling frequency of 1 kHz is recommended. The maximum frequency for EEG, according to González *et al.* (2006), is 250 Hz, which means that the sampling frequency must be at least 500 Hz. However, the classic EEG band is limited to 40 Hz, as featured in typical clinical acquisition systems, where a minimum sampling rate of 80 Hz is applied.

Electrode leads can also be used as stimulation devices. In this case, a given voltage is applied to the leads and, since human tissues are conductive, this electric stimulus depolarizes the muscle or muscle group below the leads, with attendant contraction of the muscle. Applications of functional electrical stimulation (FES) include rehabilitation devices (Ferrario, 2006) and movement restoration (Smith *et al.*, 2005). Like EMG electrodes, FES systems can be either superficial or intramuscular. Implanted FES uses a set of electrodes implanted inside the muscles close to the innervation point. Stimulation systems have been used for some time in cardiac pacemakers. More recently, other types of electric stimulators have appeared, e.g. deep brain stimulators (DBSs), which can be used to alter the rhythmic behaviour of the central nervous system, and cochlear implants, used to restore hearing in cases where the auditory neuron is not damaged. Some groups have reported the use of DBS in various areas including obesity control (Covalin, Feshali and Judy, 2005) and movement disorders (Caparros-Lefebvre, Blond and Vermesch, 1993).

Implanted FES activation patterns resemble the neuronal pattern since they use a pulsed current as a stimulus to control muscle contraction. Surface FES uses a pair of electrodes placed over the muscle or over a neuron and induces conduction on the nerve or muscle through an induced current (Ferrario, 2006).

6.2.3 HR interface force and pressure: human comfort and limb kinetic information

Chapter 4 presented human–robot physical interaction according to the functionality of the robot. It is a good idea that the capabilities of the biological force and pressure sensory systems be included in the HRI. The exoskeleton interacting with a human limb needs to be able to sense static or dynamic forces, either at a single point or in a distribution area. Such an estimation of the applied force may be required to close an exoskeleton control loop or to monitor the level of forces applied in the limb or loads on external structures. Direct measurements of forces of the WR are also required for parameter optimization or weight measurement, to name only a few instances. For example, force transducers may be required on a leg exoskeleton robot to determine the amount of load that the system can support.

Depending on the application, it may be necessary to configure force sensors to measure forces and torques along defined axes, typically referred to as *X*, *Y* and *Z*. This can mean six independent measurements: three force channels (F_x, F_y and F_z) and three torque or moment channels (M_x, M_y and M_z). A load along any measurement axis will produce a cross-talk that can vary between 1 and 5 %. There are various ways to overcome this problem, e.g. compensation using matrix methods (Schrand, 2007), simultaneous measurements of linear forces along orthogonal axes (Richards, 2003), hybrid sensor configurations, etc.

Wearable robots are attached to human body parts to apply loads to the human for a particular purpose. Such application of loads on humans is discussed in Section 5.3. During this physical interaction, the forces applied to the mechanical structure of the robot produce stress and strain. The stress is the internal distribution of force per unit area that balances and reacts to external loads applied to a body, while the strain is the deformation that takes place. Assuming uniform distribution of internal resisting forces, stress σ can be calculated by dividing the force F applied by the unit area A:

$$\sigma = \frac{F}{A} \tag{6.3}$$

The strain is the amount of deformation per unit length and can be calculated if the original length is known:

$$\epsilon = \frac{\Delta L}{L} \tag{6.4}$$

Values of less than 0.01 cm/cm are common for strain in force measurement applications and are typically expressed in microstrain ($\epsilon 10^6$) units. Transduction mechanisms using mainly currents, voltages or magnetic fields can be applied to convert mechanical forces to electrical energy. Such transduction mechanisms will determine force sensor resolution. A force or torque sensor may experience a variation of capacitance, inductance or resistance proportional to the strain undergone. The most attractive materials for force transduction in human–robot applications are piezoceramics, piezopolymers and magnetostrictive materials.

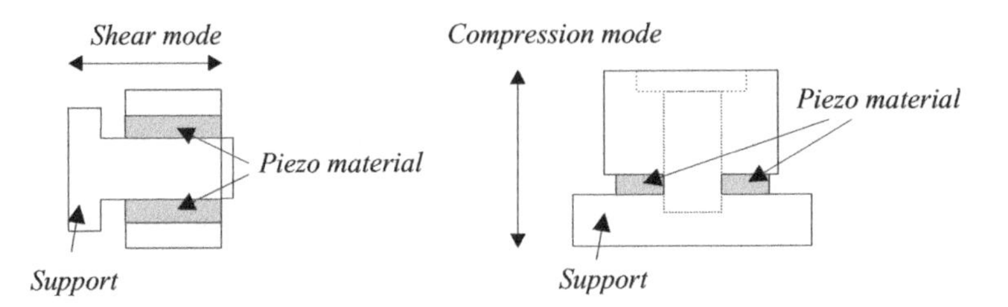

Figure 6.9 Piezoelectric sensor setups. Examples of shear mode and compression mode assemblies

6.2.3.1 Piezoelectric sensors

Adequate detection of acceleration and dynamic force can be achieved with a piezoelectric sensor, which delivers a voltage proportional to an applied 'squeezing' force. Piezoelectric sensors are only suitable for detection of dynamic forces, since the voltage generated by the material decays rapidly after the application of force. Applications can include the measurement of sliding friction or shear forces, bending or compression forces (see Figure 6.9). Because of their stiffness, piezoelectric materials are suitable for insertion in structures. For a basic force sensor configuration, the maximum force range depends on the mechanical limitations, e.g. maximum allowable stress on the piezoelectric material and other components. As a general rule, the voltage across the piezoelectric material (sensor sensitivity) will be a ratio between the electrostatic charge generated and the total capacitance across the piezoelectric element.

Examples of applications include haptic devices, e.g. tactile display units composed of piezoelectric bimorphs (Yun *et al.*, 2004), finger posture and shear force measurement at fingertips (Mascaro and Asada, 2001), and human power extenders (Kazerooni, 1998). Various synthetic polymers also exhibit piezoelectric properties, e.g. polyvinyldifluoride (PVDF). Although existing piezopolymers are not generally stiff enough for most applications, they are flexible and manufacturable enough for use as thin-film contact sensors (Fletcher, 1996), in human–technology interfaces.

6.2.3.2 Capacitive force sensors

A capacitive force sensor typically comprises a base containing a capacitance electrode, a cover with a second capacitance electrode and a spacer establishing a gap between the electrodes. When a compressing force is applied to reduce the gap, this alters the capacitance. Capacitive force sensors can be found in the form of single-cell sensors or of tactile sensor matrices in commercial devices, such as pointing sticks, vacuum gauges and high-resolution pressure sensors. The simplicity of a capacitor element allows for a great deal of flexibility in both design and construction. Repeatability and durability are also important features of these technologies. Due to the high sensitivity required by the electronics, capacitive sensor systems are very sensitive to electromagnetic interference. The advantages of capacitive force sensors can be harnessed in low-power applications, e.g. a sensor interface for human gait assessment (Bilas *et al.*, 2001).

6.2.3.3 Strain gauges

Strain can be measured by a strain gauge or a resistance gauge, both static force detection devices. These are commonly implemented to measure strain in a number of applications; examples include evaluation of contact forces in orthotic devices for the treatment of malformations (Hanafusa *et al.*,

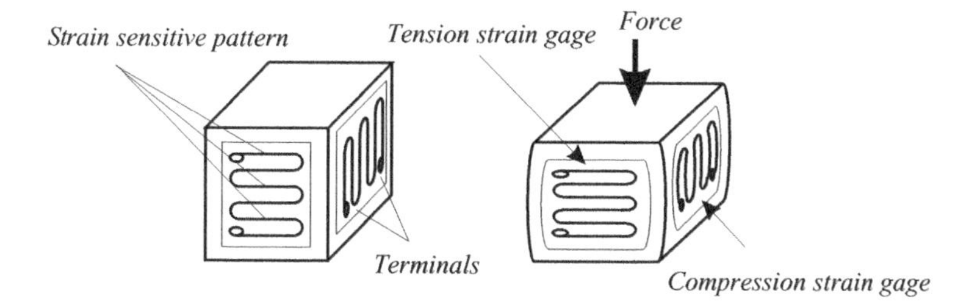

Figure 6.10 Working concept of strain gauges under bending

2002), and measurement of joint torques in upper limb exoskeletons for tremor suppression (see Case Study 3.7). The resistance gauge consists of a thin metallic or semiconductor grid that is bonded to the surface of the element. The bonded device is fundamentally designed to convert mechanical motion into an electrical signal (see Figure 6.10). Under a given torsional stress the wired grid undergoes a change in length and cross-section which produces a proportional variation of resistance, R_G. The gauge's strain sensitivity or gauge factor, GF, is given by

$$\mathrm{GF} = \frac{\Delta R/R_G}{\varepsilon} \tag{6.5}$$

where ΔR is the change in resistance caused by strain and ε is strain. Sensor stability and temperature sensitivity are also considerations in the selection of a strain gauge. For long-term applications, temperature and drift compensation is required. Typical materials are constantan, nichrome V, platinum alloys, isoelastic, or Karma-type alloy wires, foils or semiconductor materials. The most popular alloys are copper–nickel and nickel–chromium.

Strain gauge configurations are based on the concept of a Wheatstone bridge. A Wheatstone bridge is a network of four resistive elements, in which at least one can be an active sensing element. The electrical equivalent of a Wheatstone bridge are two parallel voltage divider circuits (see Figure 6.11(a)). The Wheatstone bridge configuration is useful for measuring small variations in resistance and can be configured as a quarter-bridge, half-bridge or full-bridge. The orientation and number of active elements determines the type of bridge configuration.

An example of a basic Wheatstone bridge configuration is depicted in Figure 6.11(a). Full-bridge gauges for bending-beam transducers can be found as combined patterns. The example in Figure 6.11(b) presents single-surface gauging, which simplifies the construction. Once the signal V_0 is acquired, this voltage can be amplified and converted to strain units with the appropriate conversion equations. Signal conditioning can be performed by means of operational voltage difference amplifiers, with adequate high impedance inputs.

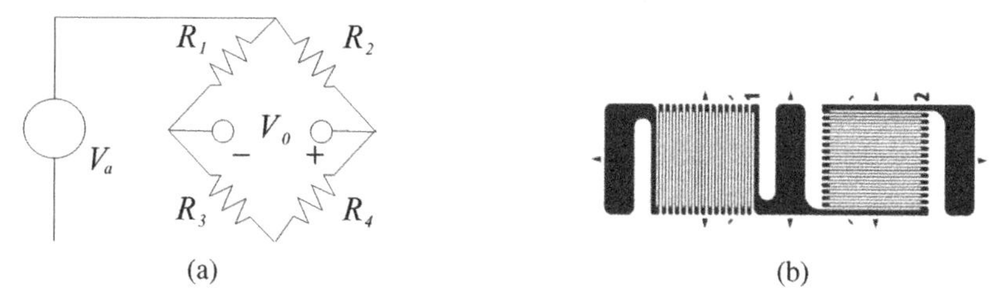

Figure 6.11 (a) Basic Wheatstone bridge configuration and (b) example of bridge tee Rosette gauge pattern. Reproduced by permission of Vishay Inter-technology, Inc

Strain gauges are frequently found in wearable exoskeletons, e.g. providing torque feedback in powered orthoses (Rocon *et al.*, 2005) and also on the characterization of interaction forces for therapy assistance (see Reed *et al.*, 2005).

6.2.3.4 Piezoresistive polymers

Piezoresistive polymers are attractive materials for force transduction. Polymers of this type can modulate an electric current via a force-dependent electrical resistance; they can be used to measure compression or extension and offer higher sensitivity than traditional metallic gauges. The piezoresistive deformation type comprises a conductive strip whose resistance varies when the cross-section is deformed by an external force. These materials ameliorate the problems of noise and hysteresis associated with conduction mechanisms that are typical of piezoresistive materials.

6.2.3.5 Pressure sensing

Pressure is calculated in units of force per unit area. The magnitude of the pressure is reduced if a given force is applied over a wider area. Pressure applied to the skin needs to be carefully processed and analysed in an HRI. Pressure information is desirable to ensure a good fit of an external device or exoskeleton to the human body, while excessive pressure on the skin can cause tissue damage. Furthermore, there are exoskeletons where control of pressure variations may be necessary, e.g. smart orthoses for the treatment of scoliosis (Lou *et al.*, 2005). Quantification of the load transfer in prosthetic sockets is one of the main applications for which pressure measurements have been developed over the last 50 years or so (Mak, Zhang and Boone, 2001).

Measuring contact pressures in wearable interfaces is a challenging task involving several factors which compromise the reliability of the measurements. Reliability of current commercial devices is an open issue that has been raised in a number of cases. For example, the F-Scan sensor system (Tekscan ®), an untethered in-shoe plantar pressure measurement system, has reported significant intrasensor variation before the warm-up period (Nicolopoulos *et al.*, 2000) but good sensor reliability afterwards.

Among the sensor working principles, similar approaches and materials as for direct force and torque measurements are used to estimate punctual and distributed pressures in robot–human interfaces. Fibre optic, mechanical deflection, piezoresistive (semiconductor), vibrating element and variable capacitance devices are the most mature technologies. Commonly implemented as low-cost solutions, diaphragm constructions are used with bonded strain gauges to act as resistive sensors under pressure-induced strain. A comparison of the main properties of pressure sensing technologies is summarized in Table 6.3.

Table 6.3 Comparison of properties of pressure sensing technologies

	Capacitive	Potentiometric	Resistive	Piezoelectric	Inductive
Maximum range	Good	Excellent	Excellent	Good	Excellent
Minimum size	Good	Good	Excellent	Poor	Poor
Sensitivity	Excellent	Good	Poor	Good	Excellent
Repeatability	Excellent	Poor	Good	Good	Excellent
Temperature stability	Excellent	Excellent	Excellent	Poor	Excellent
Hysteresis	Low	High	High	Medium	Low
Installation	Good	Good	Poor	Good	Poor

6.2.4 Microclimate sensing

This section contains a brief discussion of the technologies used for sensing of climate variables as they relate to the contact interface between wearable robot and human limbs. These climate conditions are important in order to improve user comfort and minimize the risk of injuries.

The microclimate-related variables are *temperature* and *relative humidity*. Following is a brief discussion of the physical principles of sensors for measuring these variables, and also of how each of these variables affects comfort.

6.2.4.1 Humidity sensors

Relative humidity is defined as the ratio of the actual partial water vapour pressure to the saturation vapour pressure at a given temperature. In order to measure this variable, a sensor must measure a temperature-related difference in the vapour concentration.

There are three main principles in the measurement of relativity humidity. It is known that the concentration of water vapour in the air alters some properties of the air, such as dielectric capability and thermal conductivity. It is also known from the osmosis principle that the water concentration in two media separated by a membrane tends to reach a concentration equilibrium point. These properties are used by humidity sensors, and sensor structures differ according to the principle that is followed. As to the change in dielectric properties, a planar capacitor with air or a porous dielectric material can be used to measure the amount of water vapour in the air. For an example of the structure of such a sensor see Figure 6.12. Case Study 6.6 presents a more detailed discussion on this type of sensor and addresses the identification of microclimate characteristics in a wearable device.

Following the osmosis principle, a hygroscopic medium, such as a conductive polymer, is placed in the atmosphere and the impedance of the medium is measured using electrodes. As the concentrations in the medium and the atmosphere equalize, the electrical resistance of the medium changes accordingly. This type of sensor features a relatively high capacitance. In order to eliminate capacitive charge, an alternating current is always used for excitation. In fact the resistive sensor could also be called an impedance humidity sensor. The structure of the sensor is shown in Figure 6.12(c).

Water vapour presents much higher thermal impedance than dry air. On the basis of this principle it is possible to measure the absolute humidity of the air using a pair of thermistors. The sensor consists of a bridge with two thermistors, one of which is kept in a dry atmosphere and the other in the open air. Two fixed resistors are used to complete the bridge. When the thermistors are heated, the temperature of the thermistor in the dry atmosphere will be higher since dry air has less thermal impedance than humid air. The absolute humidity can be read from the current in the internal branch of the bridge. The structure of this sensor can be seen in Figure 6.12. The performance characteristics of the different humidity sensors are summarized in Table 6.4.

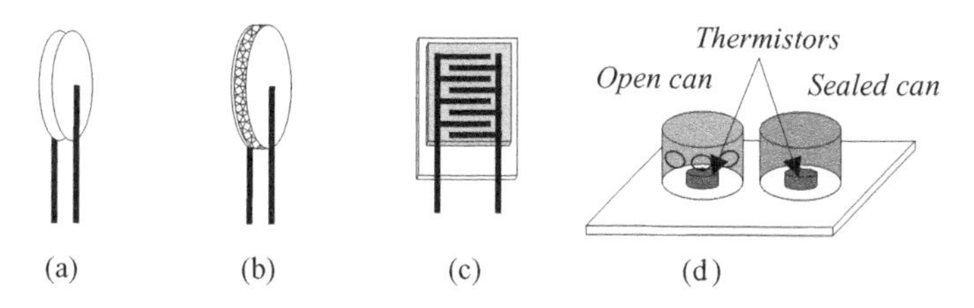

Figure 6.12 Schematic representation of humidity sensors: (a) planar capacitor sensor with air dielectric; (b) planar capacitor sensor with hygroscopic polymer dielectric; (c) resistive sensor; (d) thermal conductivity sensor, showing the thermistors inside sealed and open cans

Table 6.4 Comparison of the characteristics of three types of relative humidity sensors

	Capacitive	Resistive	Thermal conductivity
Range	0–100 %	20–95 %	20–100 %
Accuracy	1–5 %	1–10 %	3–4 %
Stability	0.2–1 %/year	3–5 %/year	0.1 %/year
Linearity	1–5 %	Poor linearity	Poor linearity
Response time	50–90 s	2–5 min	2–min

Resistive and capacitive sensors are available in small packages, similar to the ones used in capacitors or transistors, and hence can be used in wearable devices. Soft-MEMS techniques can be applied to configure wearable humidity sensors. An example of an application with a porous membrane using soft-MEMS techniques is presented in Miyoshi *et al.* (2005).

The purpose of a relative humidity sensor in the case of an interface is to detect perspiration, and so active cooling is required. A thermal conductivity sensor cannot be used for this application because of its high working temperature.

6.2.4.2 Temperature sensors

Temperature sensors measure changes in temperature produced by power dissipation. There are a variety of technologies that can be used to measure this variable. Silicon-based sensors utilize the fact that the reverse current in a PN junction of a semiconductor changes with temperature. Sensors of this type can include electronic circuits to process the signal and generate an output signal linear to the temperature within a specific voltage range.

Other temperature-sensing technologies use resistive materials as sensors. The resistance of a given material is directly related to its temperature. This principle is applied in *positive temperature coefficient* (PTC) thermistors. Some special materials present an inverse response to temperature and are used in devices known as *negative temperature coefficient* (NTC) thermistors. These thermistors are used in a variety of applications, especially when a semiconductor sensor will not do. PTC thermistors are mostly used in protection circuits as their resistance characteristics are highly nonlinear.

A third technology used for temperature sensing is the *thermocouple*. These devices utilize the property that comes into being when two different materials are put together. A small potential difference occurs in the two leads, and this potential changes according to the temperature of the metal junction. Devices with thermocouples are mostly used in industrial systems as they can withstand higher temperatures than the semiconductor or resistive approaches. Other approaches used in industrial systems to measure high temperatures are not addressed here, but the reader may refer to Omega Engineering (2006). The main characteristics of the temperature sensors presented here are summarized in Table 6.5.

Table 6.5 Comparison of the characteristics of four types of temperature sensors

	Semiconductor (analogue)	Semiconductor (digital)	Resistive (NTC)	Thermocouples
Range (°C)	−55–125	−55–125	−50–150	−250–1250[a]
Accuracy (maximum)	± 5 C	± 5 C	± 2 %	3–4 %
Linearity	0.4 C		Poor linearity	Poor linearity

[a] These limits are dependent on the thermocouple materials

6.3 ACTUATOR TECHNOLOGIES

6.3.1 State of the art

Pioneering work to develop the concept of man–amplifiers as manipulators to enhance the strength of human operators started in the 1960s at Cornell Aeronautical Laboratories. This may be regarded as the first implementation of a wearable robot. In 1962 they established the technological limitations affecting development of the concept at that time, which related to servos, sensors, mechanical structure and design. Then, in 1964, hydraulic actuator technology was identified as an additional limiting factor. Later on, in 1982, in the context of a Workshop on Robotic Dextrous Hands held at Massachusetts Institute of Technology, it was pointed out that the limiting factor at that time was the actuator technology (Hollerbach, 1982). That workshop focused on dextrous robotic hands, i.e. wearable robots of a kind, but in fact much the same can be said now in the context of wearable robotics as a whole; current actuation technologies fail to provide efficient, high-power density actuators suitable for wearable robot design.

In the context of wearable robotics, actuators drive a robot interacting with a human according to control inputs. They are used to impose controlled action on the WR in accordance with the reference trajectory. Imposing a state on a robot raises a number of issues:

1. *Univocal correspondence between control action and imposed system variable.* Ideally, there should be a unique output value corresponding to the control action.
2. *Linearity.* The above univocal correspondence will not generally be linear, but linearity is always desirable.
3. *Stability.* The correspondence between input and output should not be influenced by external perturbations or drifts.

Whenever the actuator is a component of a wearable robot, issues related to safety and dependability in the pHRI head the list of requirements of the actuator system. In selecting actuators for a particular application, a number of requirements may arise. These include power or force density, efficiency, size and weight, and cost.

In general, actuators in wearable robot applications are used under dynamic operating conditions. Dynamic operation usually produces changing conditions in the amount of power flow across the pHRI, and consequently across the actuator (power requirements), in the relative value of the actuator variables (velocity and force) and in the efficiency of transduction between input and output energy. The key figures of merit defining actuator performance are thus:

- *Power density*, P_V, which is the ratio of the maximum available mechanical output power, P_{out} to the volume of the actuator V:

$$P_V = \frac{P_{out}}{V} \tag{6.6}$$

If the ratio of output mechanical power to the weight, ρV, of the actuator is considered, this defines *specific power density*, P_ρ:

$$P_\rho = \frac{P_{\text{out}}}{\rho V} \tag{6.7}$$

- *Work density per cycle*, W_V, which is defined as the amount of mechanical work that an actuator can deliver during an actuation cycle and is defined by the ratio of output work to volume:

$$W_V = \frac{W_{\text{out}}}{V} \tag{6.8}$$

Likewise, *specific work density per cycle*, W_ρ, is defined as the ratio of maximum available output mechanical work per actuation cycle to the weight of the actuator:

$$W_\rho = \frac{W_{\text{out}}}{\rho V} \quad (6.9)$$

- *Bandwidth*. The available bandwidth of the actuator is defined by the cutoff frequency, which in turn is linked to the actuator's time constant. The time constant and the maximum available frequency of an actuator are related by the following expression:

$$f = \frac{1}{2\pi\tau} \quad (6.10)$$

- The *efficiency*, η, which in an actuator is defined as the ratio of the output mechanical energy, W_{m}, to the input electrical energy, W_{e}:

$$\eta = \frac{W_{\text{m}}}{W_{\text{e}}} \quad (6.11)$$

In many instances, wearable robot control strategies require force-controlled actuators. An ideal force-controllable actuator would be a perfect force source, delivering exactly the commanded force independent of load movement. All force-controllable actuators have limitations that result in deviations from a perfect force source. These limitations include impedance, stiction and bandwidth. Dynamic performance of actuators is important in any wearable robot, but it is critical in empowering applications and in functional compensation for the disabled. Within the latter two application domains, lower limb exoskeletons are the ones that impose the strictest requirements in terms of power and torque delivery. This is illustrated further in the next section; here, discussion will be confined to a selection of examples illustrating the state of the art.

In wearable robotics, traditional actuator technologies, e.g. pneumatic, hydraulic and electromagnetic actuators, are commonly used. Hydraulic and pneumatic actuators are known for their high force density and high force or torque characteristics, and have been used in a number of applications. Case Study 8.6 presents a soft-actuated exoskeleton for use in upper limb physiotherapy and training. The system is based on pneumatic muscle actuators and benefits from the compliant actuation characteristics of pneumatic actuators. This compliant behaviour is basically due to the high compressibility of the actuator fluid.

Direct drive actuators are a good approximation of an ideal force source. However, they are too large for robots that must support the wearer's weight in addition to the weight of the actuators. Therefore, their use is limited to applications where the actuator can be placed in a nonmoving base of the robot, thus strongly constraining wearability and portability. Transmission stages are then required.

A gear reduction in general introduces significant friction and increases the reflected inertia at the output of the gearbox. Friction can become practically too large (nonbackdriveable gear reductions) in transmissions with large reduction factors. Such a system would result in extremely poor force fidelity. Cable drive transmissions, on the other hand, have low stiction and low backlash and their dynamics are fairly linear. However, they require large pullies if a high transmission ratio is required. In order to mitigate the effects of friction and inertia introduced in conventional geared actuators, a load cell and a feedback control algorithm can be used. The load cell measures the force imparted on the load by the actuator. The feedback controller calculates the error between the measured force and the desired force and applies the appropriate current to the motor to correct any discrepancies.

Pneumatic actuators are also used to drive lower limb exoskeletons. In particular, Case Study 9.2 presents an ankle–foot orthosis powered by artificial pneumatic muscles. In this case the pneumatic actuator drives the ankle joint of the exoskeleton so that plantar and dorsal flexion torques can be applied. Thanks to their compliant behaviour, pneumatic actuators are especially suited to such high-torque applications where energy storage and delivery is important. In this application, the authors

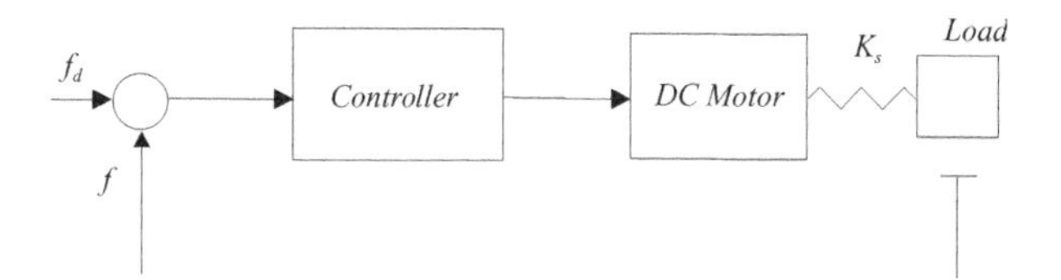

Figure 6.13 Control loop of a series elastic actuator

claim that the pneumatic actuator delivers a maximum torque of 70 N m in plantar flexion for an exoskeleton weight of only 1.6 kg, including the actuator but excluding the energy source. Also in Case Study 9.5, the author introduces a pneumatically driven full-body exoskeleton. A detailed description and characterization of this type of actuator can be found in Davis *et al.* (2003).

Other authors (Blaya and Herr, 2004) use conventional actuator technologies, e.g. electromagnetic DC motors, to build series elastic actuators. This is done in the context of a variable impedance exoskeleton for functional compensation of gait in wearers affected by *drop foot*. A *series elastic actuator* in an exoskeleton control loop is depicted schematically in Figure 6.13. The actuator is used to drive the exoskeleton in the sagittal plane and is set to control the applied force and the compression in the spring through the motor's angular rotation. The resulting actuator exhibits low impedance, low friction and an acceptable dynamic range.

As reported by Pratt, Krupp and Morse (2002), series elastic actuators exhibit numerous benefits:

1. The actuators exhibit lower output impedance and backdriveability, even in hydraulic systems. The dynamic effects of the motor inertia and geartrain friction (or fluidic inertia and seal friction) are nearly invisible at the output. In traditional systems, the actuator dynamics often dominate the mechanism dynamics, making it difficult to accomplish tasks that require high force fidelity.
2. The force transmission fidelity of the gear reduction or piston is no longer critical, allowing inexpensive gear reduction to be used. Gears typically transmit position with much higher fidelity than force. The series elasticity serves as a transducer between the gear reduction output position and load force, greatly increasing the fidelity of force control.
3. The motor's required force fidelity is drastically reduced, as it is the motor shaft's position, not its output torque, that is responsible for the generation of load force.
4. Force control stability is improved, even in intermittent contact with hard surfaces.
5. Energy can be stored and released in the elastic element, potentially improving efficiency in harmonic applications.

6.3.2 Control requirements for actuator technologies

In general, the function of an exoskeleton when coupled to the human can be threefold:

- *The actuator imparts an arbitrary motion to the wearable robot joints.* This is generally done with the aim of assisting joint motion, be it to empower a healthy wearer in performing a power-demanding task or to compensate for lost or weak limb function.
- *The actuator establishes a relationship between applied force and resulting limb motion.* As detailed in Chapter 5, human limbs behave like a controllable impedance whenever they interact with the environment. When actuators implement this function, they mimic human limb behaviour.

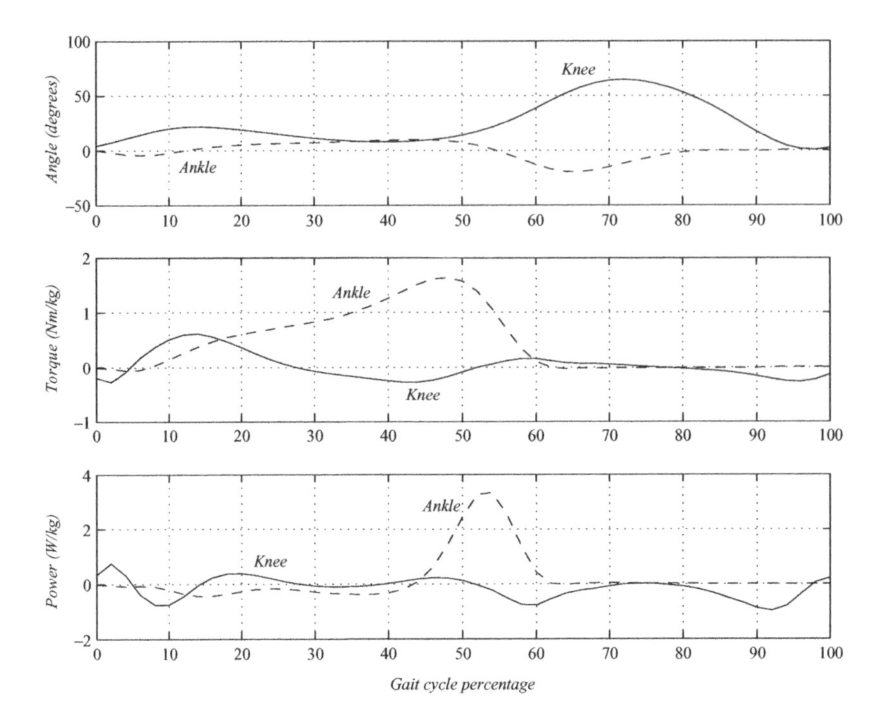

Figure 6.14 Kinematic and kinetic information at knee and ankle level during human gait

- *The actuator system is set to block or allow joint motion.* This function is typically implemented in functional compensation exoskeletons, e.g. in the control of knee motion during stance to support wearers with weak muscle activity.

Human actuators (i.e. muscles) are able, in combination with the anatomical kinematic structure of human limbs, to impart any of the three actuator functions. Details of the anatomical kinematic structure of human limbs can be found in Chapter 3. For the actuation characteristics of human muscles, it can be said that they allow an energy density delivery of around 0.07 J/cm^2, that their response time is around 100 ms and that they can contract by up to 40 % of their original length. These high capacities, combined with the anatomical kinematic structure, result in high actuation efficiency and performance.

To illustrate this, Figure 6.14 shows kinematics and kinetics of human gait for the ankle and knee joints. Data in Figure 6.14 are relative to body weight. From this it is clear that for an exoskeleton wearer with a weight of 80 kg, muscle torque may be as high as 50 N m at the knee during stance. The figures for the ankle are even higher; for instance, during plantar flexion for push-off (around 45 % of gait cycle), muscle torque is around 130 N m for the same wearer. The figures in terms of power to be delivered are also very high.

On the other hand, if reference is made to demands on the actuator system in wearable robots for the upper limb, torque and power to be delivered are, generally speaking, significantly lower. If, for instance, the upper limb exoskeleton for tremor suppression, described in Case Study 8.1, is taken, the estimation of required torque and power for the actuation system (see Case Study 3.7) gives a maximum torque of about 3.7 N m and a power of 1.8 W for the elbow joint. It is clear that while in lower limb wearable robots the basic requirement is weight support, in upper limb wearable robots dexterity and manoeuvrability are the main issues. As a result there are significant differences in the control and power requirements for actuators depending on whether they are for use in upper or lower limbs.

6.3.3 Emerging actuator technologies

This section provides a brief introduction to new actuators with plausibility in the area of wearable robots. These are electroactive polymer (EAP) actuators, Electro- and magnetorheological fluid (ERF and MRF) actuators and shape memory alloy (SMA) actuators.

6.3.3.1 Electroactive polymer actuators

The family of electroactive polymer (EAP) actuators is a broad set of dissimilar technologies combining various different transduction phenomena on different substrate materials and producing diverse actuation characteristics. Most of the phenomena underlying the actuation process are not fully understood. EAP are classified into wet (ionic) and dry (electronic) polymer technologies.

EAP actuators are the newest actuator technologies and consequently are subject to continuous research. Actuation performance is also very dissimilar among types of EAP actuators. In general, it can be said that wet EAP actuators are limited in their response time by ion diffusion mechanisms, and therefore the dynamic range is very low. On the other hand, dry EAP actuators are fast but require high driving voltages. The reader is referred to Bar-Cohen (2001) or Pons (2005) for additional information on actuator characteristics and applications.

6.3.3.2 Electro- and magnetorheological fluid actuators

Electro- and magnetorheological fluid (ERF and MRF) actuators are semi-active technologies. They essentially differ from all other actuator technologies in that they can only (actively) dissipate the energy of the wearable robot they are coupled to. ERF and MRF are field-responsive fluids whose rheological properties undergo abrupt changes when they are subjected to an external electric or magnetic field. This alters the apparent viscosity of these fluids.

ERF and MRF actuators can basically be built for three operating modes: shear mode actuators, flow mode actuators and squeeze mode actuators (see Figure 6.15 for a schematic representation). The reader is referred to Pons (2005) for details on principles, actuation modes, control aspects and applications of these technologies.

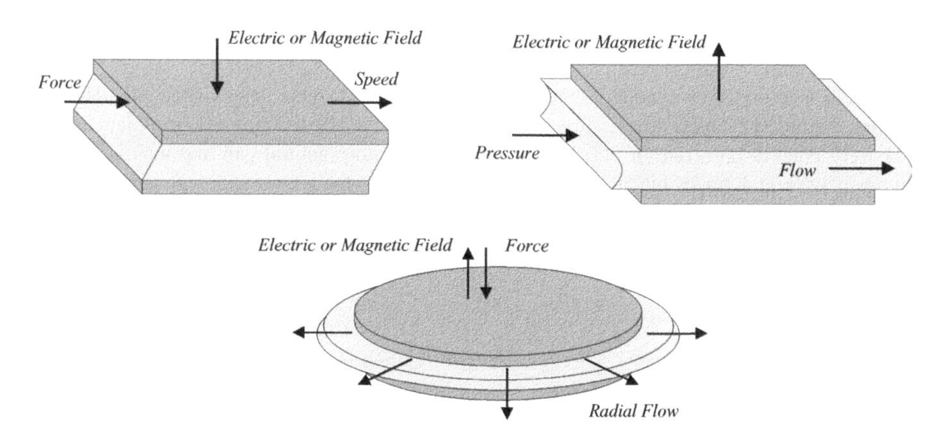

Figure 6.15 Schematic representation of ERF and MRF actuation modes

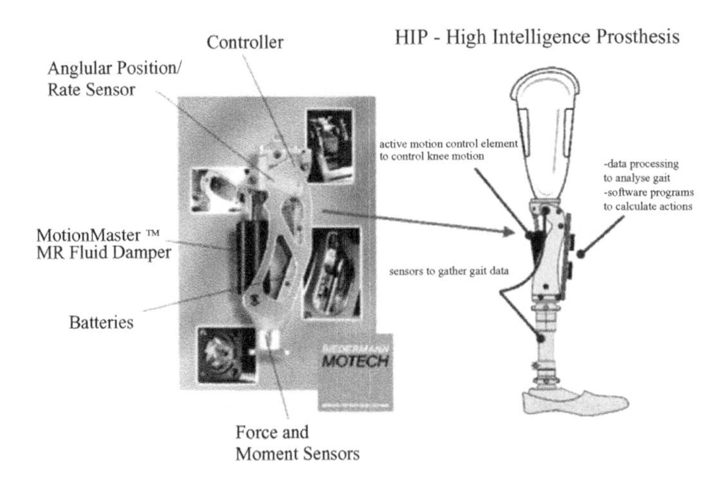

Figure 6.16 Schematic representation of the Prolite™ Smart Magnetix™ above-the-knee (AK) prosthesis. Reproduced from Pons 2005

As an example of a state-of-the-art active prosthesis, the Otto Bock C-Leg® prosthetic knee includes multiple sensors. These transmit information at a speed of 50 Hz, allowing the feedback controller to operate its mechanical and hydraulic systems. Two strain gauges measure pressures on the leg and determine how often the heel strikes (thus giving an estimation of the walking cadence); magnetic sensors report changes in knee angle.

The use of field responsive fluids in orthotic and prosthetic applications is a logical step in replacing traditional hydraulic or pure electromechanical systems. One recent introduction is the Prolite™ Smart Magnetix™ above-the-knee (AK) prosthesis. The Prolite™ Smart Magnetix™ is manufactured by Biedermann OT Vertrieb, a German maker of prosthetic components, and was developed jointly with Lord Corporation. Part of the information in this section was provided by Lord Corporation. The pictures in this section are by courtesy of G. Hummel and L. Yanyo, Lord Corporation.

Figure 6.16 shows a schematic representation of the Prolite™ Smart Magnetix™ system. The system includes both kinetic (force and torque) and kinematic (angular position and rate) sensors. The sensors are used to adapt the rheological characteristics of a modified Lord RD-1005 MR fluid damper (Pons, 2005).

The system incorporates controllers that adapt the damping characteristics of the MR damper to the walking conditions. Thanks to the fast response time of the MR technology, this adaptation can be made very quickly (at a rate of 500 Hz), allowing for a more natural gait and making climbing up and down stairs and slopes much easier. Moreover, it makes for a more efficient walking pattern, which is one of the most serious problems suffered by users of passive prostheses.

Figure 6.17 shows a prosthesis user walking down a slope with the Prolite™ Smart Magnetix™ fitted. It is claimed that after the adaptation process, users can ride a bicycle, carry heavy objects and walk or run with varying gaits.

6.3.3.3 Shape memory alloy actuators

Shape memory alloy (SMA) actuators are devices that make use of a thermally activated *martensitic transformation*. In shape memory alloys there are two stable phases: *martensite* (also referred to as α-phase), which is stable at low temperature, and the *parent phase* or *austenite* (β-phase), which is stable at low temperature.

Figure 6.17 A Prolite™ Smart Magnetix™ above-the-knee (AK) prosthesis user walking down a slope: the characteristics required for the knee prosthesis will generally be different from other walking conditions. The MR technology provides the desired adaptation. Reproduced from Pons 2005

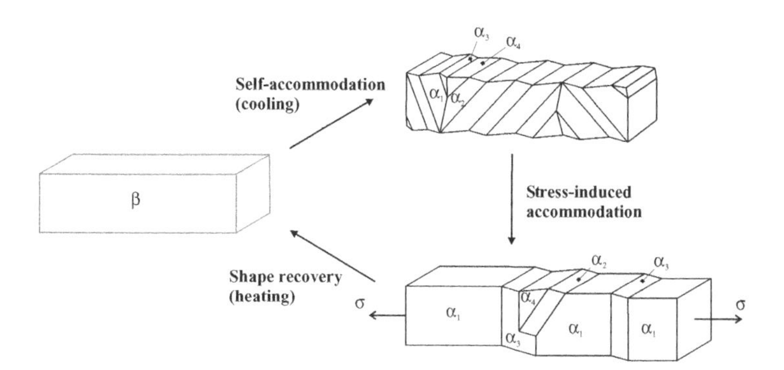

Figure 6.18 Illustration of the self-accommodation process upon cooling from the parent phase, the stress-induced accommodation of variants in the martensite phase and shape recovery after heating from the martensite phase

The shape memory effect is schematically illustrated in Figure 6.18. When a mechanical stress is applied to the alloy in the martensite phase, there is stress-induced accommodation of variants causing a reorientation of twin related variants and resulting in a macroscopic deformation (see the schematic in Figure 6.18). When the material is heated again, the martensite variants that were accommodated when stress was applied will revert to the original orientation in the parent phase and the original shape will be recovered. The result is thermally triggered macroscopic recovery of a stress-induced material deformation. When this is applied to recovery from a tensile deformation of a wire, the resulting actuation principle is similar to the muscle function.

Shape memory alloy actuators are mostly tailored to a specific application. In general, tensile, bending or torsion deformations, or a combination of the three, can be recovered in the implementation of an actuator. The main actuation characteristics of SMA actuators are summarized in Table 6.6. The chief advantage of SMAs as compared to other technologies is the high force achieved in tensile

Table 6.6 Operational characteristics for SMA actuators

Figures of merit	SMA wire	SMA bender	SMA spring
Force, F	High, ≤ 250 MPa	$\approx$ zero	Low
Displacement, S	Up to 6–8%	High	High
Work density, W_V	Theory: $10-10^2$; Practice: $10^{-1}-1$ J/cm^3		
Power density, P_V	Theory: 10^2-10^3; Practice: $10^{-1}-1$ W/cm^3		
Bandwidth, f	Up to 5 Hz		
Efficiency, η	Theory limit 10 %; Practice: ≤ 3 %		

deformation configurations. Limiting factors for the use of this technology include a long response time (as they are thermally driven) and fatigue-limited actuation cycles.

Endoscopes are used in colonoscopy and other techniques to allow diagnosis through imaging or biopsy. They are usually equipped with conduit wires so that they can be inserted in the colon and stomach. The process of guiding the endoscope is difficult both for the physician and the patient.

During the early 1980s, the team led by Prof. Hirose at Tokyo Institute of Technology developed a prototype of a cable-driven serpentine robot. The prototype allowed the operator to actuate independent bending segments remotely so that the shape of the so-called 'ELASTOR' could be adapted to the task.

The next step in the evolution towards a minimally invasive tool was achieved by implementing shape memory alloy actuators to bend the different segments in the active serpentine actively.

SMA actuators exhibit relatively low electrical resistance, which is utilized through Joule heating to induce a phase transformation. Prof. Hirose's team introduced a novel configuration of the various different SMA actuators by placing them mechanically in parallel and electrically in series. In this way, the electrical resistance of the actuators can be increased so that a higher voltage, low-current power source can be used to drive the serpentine robot.

In controlling the SMA-driven active endoscope, use is made of this technology's intrinsic sensor capability. The electrical resistance of the different SMA wires is fed back to the controller for optimum driving, in what is a very compact solution. Additionally, the SMA-driven serpentine endoscope is mounted on a linear-displacement servo drive. In this way, the endoscope can be axially positioned as well as being adaptable in shape.

The active endoscope prototype has an overall diameter of just 13 mm and a total length of 250 mm. Five active segments are implemented. The controller attains the reference bending angle with respect to the first segment and synchronously drives the following segments at the insertion speed determined by the linear displacement servo drive, thus ensuring a smooth operation. The final configuration of the system can be seen in Figure 6.19.

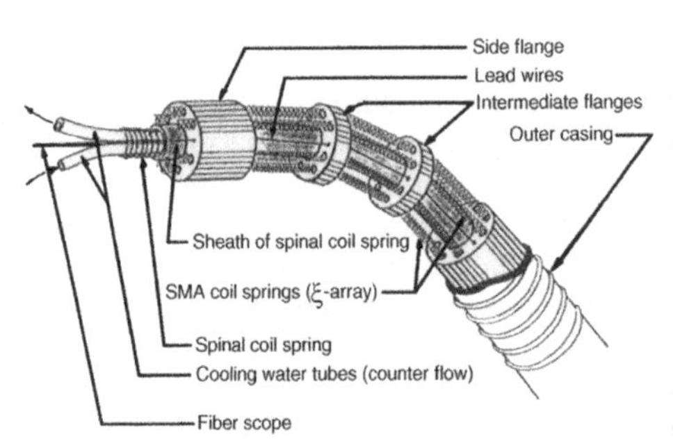

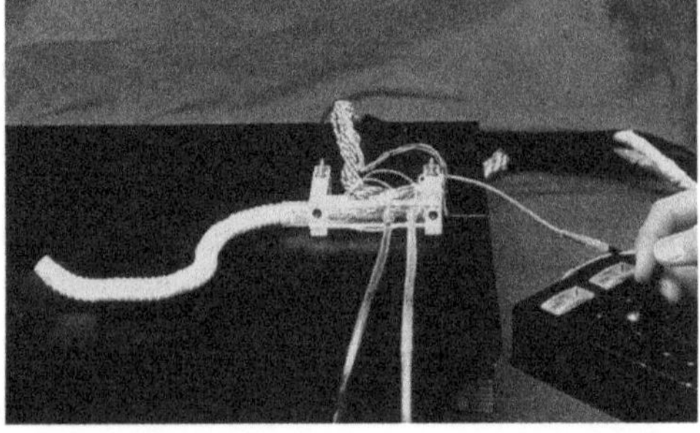

Figure 6.19 Schematic configuration of the SMA-driven active endoscope and picture of the final prototype. Reproduced from Pons 2005

6.4 PORTABLE ENERGY STORAGE TECHNOLOGIES

Energy storage systems are technologies that enable the application of wearable devices by extending and enhancing their portability. The energy storage device must satisfy the power and energy requirements of wearable robots. The key technologies for portable power sources, which are required to successfully achieve a wearable robot application, include batteries, fuel cells and hybrid sources.

Different types of batteries are commercially available as portable energy solutions. The main issue with battery technologies is the ability to meet power and energy requirements while minimizing the weight of the energy storage device. This requirement will be a major factor in the selection of a given actuation technology and in the practical application of the WR for interaction with a human being. Battery systems range from reliable technologies, such as lead–acid, that have been proven and developed over many years, to various newer designs that are currently under development. Commercial solutions include lithium–ion, sodium–sulfur and sodium–nickel chloride. Table 6.7 summarizes the main features of the principal battery storage systems.

For example, as an alternative for exoskeleton applications, NiMH batteries offer power densities of 1 kW/kg for up to 10 s with a consequent drop in storage efficiency to only 2.8 W h/kg. This is useful for meeting peak loads in a hybrid system. For a number of reasons, but chiefly because of the memory effect and primary energy density in nickel–cadmium batteries, lithium–ion batteries are regarded as superior and are the most popular in current commercial portable devices.

6.4.1 Future trends

Other technologies are needed to overcome the main drawbacks of existing solutions. Lithium polymer batteries, with a specially formed dry polymer, currently offer the advantage of unrestricted shape and can therefore be thinner than the lithium–ion-based design. New materials are also being investigated with which to build improved batteries. Zinc–silver batteries are a promising example: they provide double the energy density of lithium-ion batteries and are safer, i.e. non-flammable.

Supercapacitors can be used in a hybrid system to meet high peak power demands. For example, a supercapacitor can provide a relatively high specific power of 7 kW/kg and a relatively low specific energy of 3.5 W h/kg. However, supercapacitors are insufficient in themselves as an energy source (Jansen *et al.*, 2000).

Fuel cells have improved in the last few years, with a significant reduction in costs. Available fuel cells in the range of 500 W to 2 kW can provide specific power of the order of 50–150 W/kg. Fuel cells potentially offer up to 10 times the life of a lithium–ion battery, but they need to be refuelled. A generation of fuel cells for mobile devices is currently being developed by technology companies like Toshiba, Hitachi, NEC and Intel, and is expected to become available in the near future.

Table 6.7 Comparison of battery energy storage systems

	Specific power (W/kg)	Cycle life	Upper limit power	Specific energy (W h/kg)
Nickel–cadmium	75	2500	10 MW	Low
Lead–acid	35–50	500–1500	10 MW	Medium
Sodium–sulphur	150–240	2500	MW scale	High
Sodium–nickel	125	2500	10^2 kW	High
Lithium–ion	150–200	1–10000	10 kW	High

6.5 CASE STUDY: INERTIAL SENSOR FUSION FOR LIMB ORIENTATION

J. C. Moreno, L. Bueno and J. L. Pons

Bioengineering group, Instituto de Automática Industrial, CSIC, Madrid, Spain

In order to calculate the kinematic configuration of the segments for a given wearable exoskeleton, it is desirable to configure a noncumbersome solution that allows free movement regardless of the context of application. A wearable motion capture system can be built of small inertial measurement units mounted on the exoskeleton structures.

Inertial sensors measure mostly linear accelerations and angular rates. Rate information estimated from inertial sensors may be integrated to obtain displacements or orientations, and therefore small errors in the rate information typically result in unbounded errors in the integration results. Gyroscopes experience bias, and also a drift in the bias that may be caused by the output electronic offsets or bearing torques. Another requirement for integration of orientation from rates is an initial orientation. The gravity component of an accelerometer signal indicates the inclination of the sensor, but the device output includes both the gravitational and the acceleration component. Therefore, an accurate estimation of inclination can only be achieved during static or very low frequency conditions. To provide an accurate estimation of orientation at all times, the sensor acceleration must be calculated, which can be done by fusion of the accelerometer and rate gyroscope information. In order to compensate for drift errors, different types of complementary filters have been proposed to bound the orientation drift errors from gyroscopes outputs, aided by additional sensors.

An extended linear Kalman filter can be used to estimate the orientation of a body segment by fusing sensor information (Marins *et al.*, 2001). Gyroscope information can be integrated to track high-frequency orientation components, and quaternion measurements can be used to track low-frequency components and compensate the bias. Quaternions have gained popularity as an alternative method of orientation representation and are more suitable than Euler angles.

In some applications where there are periods of static or quasi-static conditions, it is possible to eliminate the drift during integration of gyroscopes if it is calculated in short time periods. This is applicable in the case of human walking, where it may be desirable to identify the beginning and the end of the gait cycle in order to reset and thus eliminate the signal drift. It has further been proposed to identify gait events from inertial sensor signals in free human walking (Pappas *et al.*, 2001) and with exoskeletons (Moreno *et al.*, 2006).

It is possible to have two different measurement sources to estimate one variable where the noise properties of the two measurements are such that one source gives good information only in the low-frequency region while the other is good only in the high-frequency region. In such cases, complementary filters are used. In the case of a gait exoskeleton, estimation from the leg joints will normally be used for both control and monitoring purposes. The attachment of rate gyroscopes and biaxial accelerometers is shown in Figure 6.20.

Estimation of the exoskeleton kinematics is based on a complementary filter approach: highpass filtering is used to correct the drift for measurement of the shank and thigh inclination angle signals based on the gyros; the sensitivity of the accelerometers is used to detect orientation with respect to gravity during quasi-static conditions. Gyroscopes yield absolute orientations of the leg segments. An example of estimation of absolute orientation of the shank segment during walking is presented in Figure 6.21. Angular acceleration of the involved segment can be calculated by numeric differentiation of the angular velocity.

The method for estimation of the knee angle along the sagittal plane during walking with the GAIT exoskeleton considers subtraction of the absolute orientation of each segment, α_i (see Figure 6.20(b)), crossing the orthotic hinge. The rate gyroscopes in each segment require an integration offset and

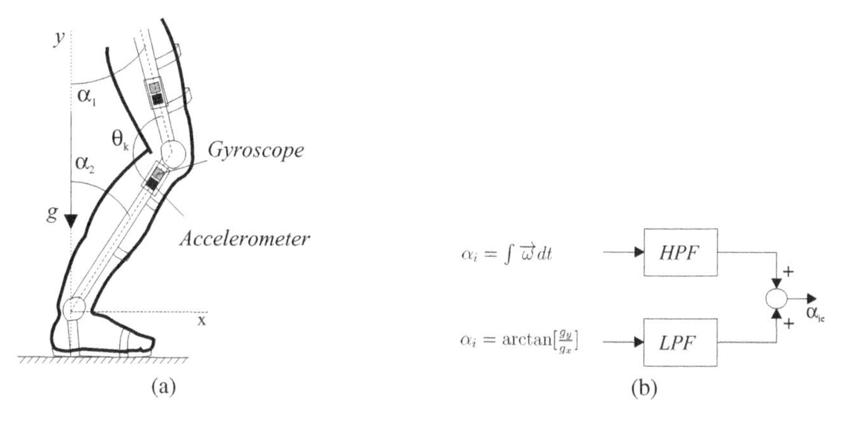

Figure 6.20 (a) Leg exoskeleton with rate gyroscopes and biaxial accelerometers and (b) complementary filter configuration for estimation of joint angle information

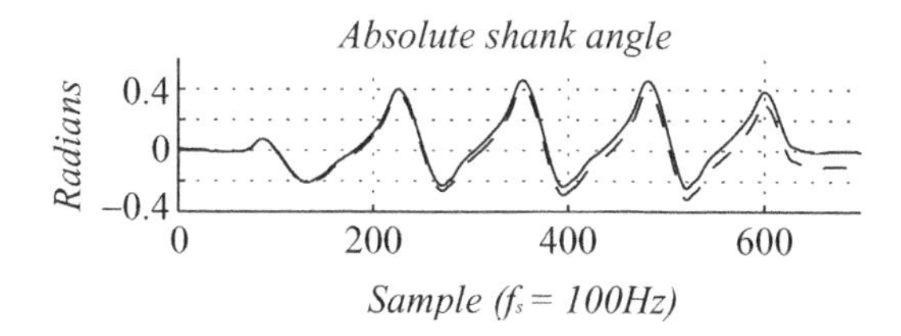

Figure 6.21 Example of absolute segment angle estimation for a walking trial. The dashed line is the absolute shank angle of the exoskeleton from the gyroscope and the continuous line is the compensated absolute angle

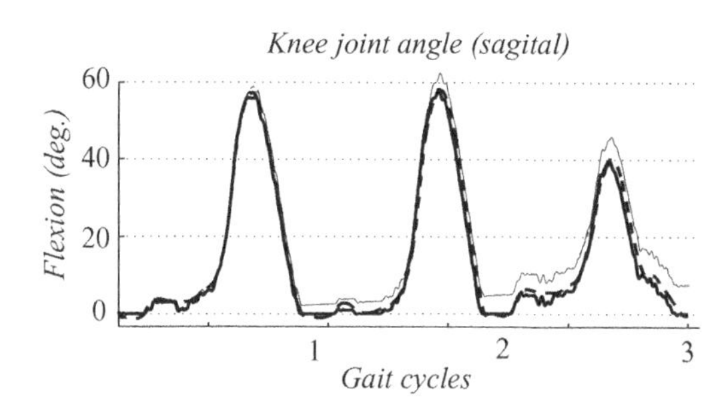

Figure 6.22 Knee joint angle during a three-cycle walking trial with the GAIT exoskeleton. The thick line is the calculated knee angle after integration. The thin line is the calculated knee angle after filtering. The dashed line is the reference from a goniometer

the reference angles (orientation during low frequency) are calculated using the arctangent function. Figure 6.22 shows the effect of drift correction performed by highpass filtering of the calculated knee angle, compared to a goniometer, during walking. The resulting correlation coefficient was 0.996 for estimation of the knee joint angle in a walking trial with three complete cycles. Triaxial configurations may be required for further analyses of estimations during turning or free movement of the exoskeleton segments (e.g. upper limb motion).

6.6 CASE STUDY: MICROCLIMATE SENSING IN WEARABLE DEVICES

J. M. Baydal-Bertomeu, J. M. Belda-Lois, J. M. Prat and R. Barberà

Instituto de Biomecánica de Valencia, Universidad Politécnica de Valencia, Spain

6.6.1 Introduction

Humans are homeotherm beings and are therefore able to control thermal interaction with the environment to keep their body temperature steady. This is one of the reasons for the success of human adaptation to very different climatic conditions from the tropics to the poles. In addition, humans have developed behavioural and social adaptations to different weather conditions. One of these adaptations is the use of clothing. Clothing isolates the body from the environment, creating a microclimate between fabric and skin.

When a device, e.g. a wearable robot, is placed on the skin, the climate conditions are altered in the same way as with clothing. These climate conditions are especially important when the device is designed to transmit loads to the human body. In conditions of high temperature and humidity, skin can macerate, increasing the likelihood to breakdown (Bader and Chase, 1993).

However, in the above-mentioned conditions, the first effect noticed by the user would be a sense of discomfort produced by adverse microclimate conditions. A feeling of discomfort is in many cases the first warning of adverse conditions that can ultimately lead to injury. Therefore, in the design of wearable devices it is important to take microclimate conditions between the device and the skin into account in order to prevent discomfort and safety issues.

6.6.2 Thermal balance of humans

Humans are able to keep their body temperature steady in a wide range of environments. On the one hand, human skin regulates heat exchange with the environment through various heat transfer mechanisms: convection, radiation, evaporation and to a lesser extent conduction. On the other hand, the human body can increase its internal temperature through the metabolism and by physical activity.

Humans use several strategies to raise the heat-loss rate when the environment is hotter than the body temperature, among others sweating evaporation or increasing the blood flow in superficial vessels. The human body can also reduce the heat-loss rate when the environment is colder than the body temperature, among others, by means of increasing physical exercise, shivering or reducing the blood flow in superficial vessels.

Several complex thermal indices were developed in recent decades to describe and quantify the thermal environment of humans and the energy flux between body and environment. One of the most widely used expressions is the following, given by Nishi (1981):

$$M_n + R + C + E + J = 0 \tag{6.12}$$

where M_n is the rate of metabolic heat loss from the skin, R is the rate of heat loss by radiation exchange, C is the rate of heat loss by convection transfer and J the rate of loss of heat stored by the body.

The human body can adapt its thermal response within a wide range of temperature and humidity conditions. However, the range in which humans perceive comfort is narrower and differs substantially

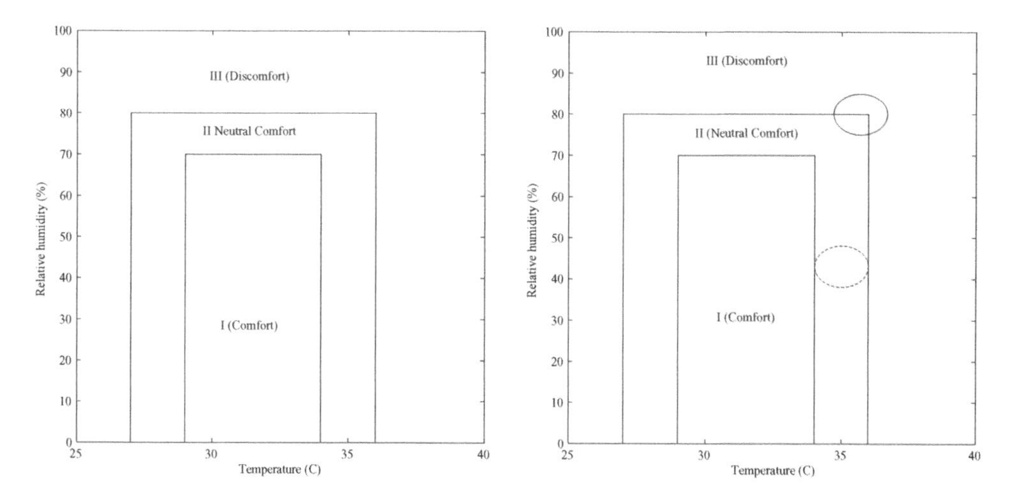

Figure 6.23 Areas of comfort according to González (2007) (left) and comfort in the forearm (right, dashed ellipse) and the hand (right, full line ellipse)

from open environments to microenvironmental conditions in which air circulation and air velocity are reduced. According to González (2007), three different regions can be defined with respect to comfort (see Figure 6.23 (left)): *comfort*, between 29 and 34°C and below 70 % humidity; *neutral comfort*, between 27 and 36°C and below 80 % humidity; and *discomfort*, outside these values.

6.6.3 Climate conditions in clothing and wearable devices

It has been seen before that when a body part is covered with any kind of material, a microenvironment is developed between the skin and the cover. The characteristics of the microenvironment will depend on the skin temperature (hence the importance of all metabolic and thermal control aspects), the vapour conditions (humidity) of the environment and the thermal characteristics of the materials that constitute the cover.

The principal effect of the cover is to isolate from the environment. This isolation implies mechanisms of dried heat loss and damped heat loss; therefore, the comfort experienced by the user can be modified by altering the isolation parameters.

The comfort experience will depend ultimately on the thermal resistance

$$E = \frac{T_{sk} - T_e}{R_t} \tag{6.13}$$

and vapour resistance.

$$E = \frac{p_{sk} - p_e}{R_v} \tag{6.14}$$

Thermal resistance, R_t, is the heat loss from the skin, E, based on the temperature difference between the skin T_{sk} and the environment T_e. Vapour resistance, R_v, is the heat loss from the skin, E, based on the vapour pressure difference between the skin p_{sk} and the environment p_e (Havenith, Holmer and Parsons, 2002).

Generally speaking, the vapour resistance of the materials that constitute the cover should be as low as possible in order to enhance comfort and avoid skin maceration. The optimum thermal resistance will depend on the characteristics of the environment. In a cold environment, high thermal resistance

Table 6.8 Results from the questionnaires

	Thermal comfort	Hygroscopic comfort	Overall comfort
Start (0 min)	8.3	9.2	8.8
20 min	7.4	7.7	7.6
End (40 min)	6.9	7.5	7.2

will be the optimum, but in a hot environment low thermal resistance will be preferred. Nowadays, with the emergence of phase change materials (PCMs), covers can be designed with tunable thermal resistance.

6.6.4 Measurement of thermal comfort

Thermal comfort can be measured by combining registers of microclimate sensor capsules and surface temperature sensors and checking the results against user questionnaires. Following are the results of an experiment that was conducted with five young subjects wearing a wrist orthosis.

Three sensors were used in the test: sensor 1 (microclimate) is placed on the palmar side of the hand under the palm valve of the orthosis (in the central area of the palm); sensor 2 (microclimate) is placed on the dorsal side of the forearm under the area of the orthosis fixation; and sensor 3 (surface temperature) is placed in the inner region of the forearm under the orthosis.

After the trials the users completed a questionnaire about perceived thermal comfort. Figure 6.23 (right) shows a projection of the evolution of microclimate sensors in the areas defined for comfort. As can be seen from these projections, climate comfort in the forearm is better than climate comfort in the hand (which lies outside the neutral comfort area). These results are consistent with the answers to the questionnaires (see Table 6.8).

6.7 CASE STUDY: BIOMIMETIC DESIGN OF A CONTROLLABLE KNEE ACTUATOR

J. C. Moreno, L. Bueno and J. L. Pons

Bioengineering group, Instituto de Automática Industrial, CSIC, Madrid, Spain

Knee–ankle–foot orthoses (KAFOs) are prescribed as a partial solution for joint disorders to provide stability and keep joints in their functional positions. Conventional systems provide stability during walking by maintaining the knee in a fixed position, but this produces unnatural gait patterns. While the user walks with a locked knee, he/she cannot flex the knee during the pre-swing and swing phases. As a result the user has to swing the body to the side of the nonaffected leg in order to clear the swinging leg. Besides, having the knee locked throughout the gait cycle imposes excessive energy consumption. There is also no loading response during knee flexion.

A knee actuator system based on energy storage release has been designed for a KAFO to apply functional compensation to the knee throughout the gait cycle. This is done by means of two elastic actuators whose elastic constants adapt to the different phases of the cycle so as to approach a normal profile. The approach adopted is a biological one, and the principle underlying the design consists in using biomechanical data from the leg to determine the configuration of the actuators and actions that are applied at joint level.

The absence of the necessary muscle control in the leg segments can affect locomotion in a variety of ways, from an undesired gait pattern to bodily collapse. For designers of actuator systems it is helpful to analyse the possible situation in each joint when these problems occur.

6.7.1 Quadriceps weakness

Knee extensors are active in both the stance and swing phases. Their role is to control the rate of knee flexion, which is induced by the ground reaction force in stance and by the inertia of leg movement in swing. In cases of quadriceps weakness, at the end of the stance phase the joint knee flexes in an uncontrolled manner, putting the patient at risk of falling. Moreover, extension of the leg at the end of the swing phase is also compromised, resulting in undesired movements. The hamstrings or calf muscles are less able to accomplish the necessary functional knee control.

One strategy to overcome the absence of knee extensor power and avoid uncontrolled knee flexion during the stance is to prevent the ground reaction force from passing behind the knee joint. This can be achieved by enhancing hip extensor activity and anterior trunk bending (which displaces the centre of gravity of the body forwards). A possible long-term consequence of this condition is hyperextension of the knee due to the continued thrusting of the joint into full extension for security. If hyperextension develops, the knee will become more inherently stable due to its posterior displacement relative to the line of the ground reaction force during the stance phase. However, the increased extension moment will induce a further hyperextension deformity, to the extent that this secondary problem may become a major clinical concern.

6.7.2 Functional analysis of gait as inspiration

The requirements for an actuator system can be established in terms of the intended function that the knee and ankle orthotic joints should perform in order to approach a normal gait. If it is proposed to design an external actuator system for this purpose, an analysis of the functional actions of the musculoskeletal system during the gait cycle from a mechanical point of view could help to decide how a portable system should act on the lower limb joints, with a view to imitating this functionality.

On the basis of biomechanical data from normal subjects at natural cadence (Winter, 1990) (see Figure 6.24), the relationships between gait variables and energy transformations during the cycle have been studied in order to analyse the mechanical basis of lower limb joint performance. All data are normalized to body weight.

It can be seen that between approximately 5 and 15 % of the gait cycle (at stance), when the joint is absorbing the impact (through the action of the quadriceps), the performance of the joint, by comparison of the angular position with the torque at the knee (load–displacement relationship), resembles elastic performance in that the relationship is approximately linear. During this interval, the musculoskeletal activity is dedicated to power absorption; then immediately afterwards, corresponding to extension recovery, this power is recovered until approximately 30 % through the phase. This roughly elastic relationship starts near the beginning of the gait cycle at 1.4 % and is maintained past the energy generation zone, until approximately 50 % through the cycle when the knee extension is complete and the knee is ready to swing.

The evolution of the ratio between knee joint torque and angle is illustrated in Figure 6.24 (normalized data). Various different modes of operation can be identified by analysing how the rotational displacement of the joint relates to its torque in the gait cycle. According to the intervals defined, the shock absorption area and the recovery of extension just before the swing phase are characterized by the segment A–B–C. The flexion phase during swing is characterized by the segment C–D and the extension phase during swing is characterized by the segment D–E.

Based on the functions of the leg muscles during the gait cycle, the functions of the knee actuator system are defined as:

- Stance phase. Shock absorption and knee assistance back to full extension.
- Swing phase. Free knee flexion in early swing and assisted knee extension in late swing to prepare for the next foot contact.

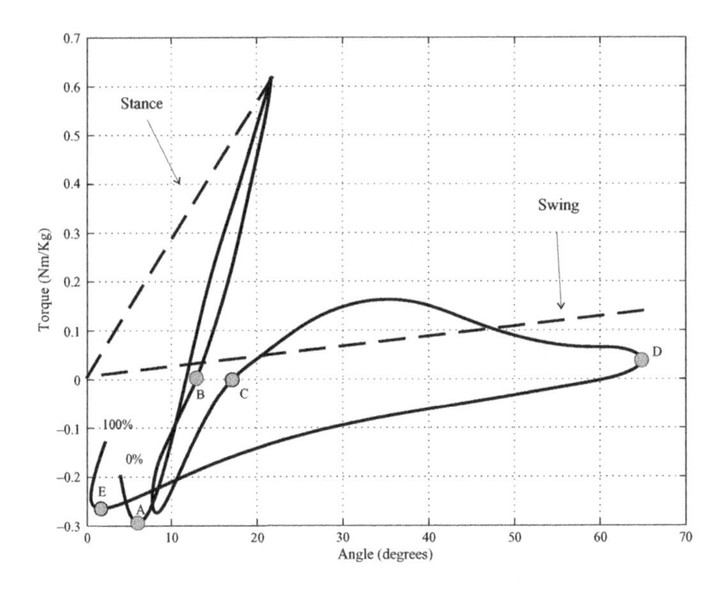

Figure 6.24 Average biomechanical data for normal gait. Knee angle versus. torque in the sagittal plane throughout a gait cycle. Identification of elastic constants (dashed lines)

The design considers the most critical scenario, where there is no residual function around the knee (no muscle power). The switch from one spring to another will be made with the leg completely extended. Two elastic springs are included in the attachable knee actuator. Therefore, the theoretical torsional elastic constant of an action based on the joint angle needs to be calculated for a given body weight. Consider the intervals defined in the previous section:

- Interval 1. shock absorption–recovery of extension. The interval may be considered to be from the beginning to 50 % of the stance phase as an approximately linear relationship between applied torque and the flexion angle. This zone corresponds to the interval between points A–C in the torque–gait percentage diagram in Figure 6.24. Interval 1 may be approximated as a constant ratio similar to the elastic constant defined by Hooke's law, which establishes proportionality between mechanical stress and strain, and so a line equation can fit this interval.

  The solution chosen (dashed line in Figure 6.24) was to fit a line from point 0–0 to the maximum torque value, point 22–0.62 in the angle–torque diagram. From the equation for this line, a torsional elastic constant can be derived:

$$T = K_1 \alpha_{\text{stance}} \tag{6.15}$$

  where T is the joint torque in N m/kg, α_{stance} is the joint angle for stance phase in degrees and K_1 is the torsional elastic constant in N m/kg degree.

- Intervals 2–3: flexion and extension during the swing phase. Interval 2 presents a nonlinearity in the torque–displacement curve which may be called pseudoelasticity. This behaviour is present when the deformation strain augments considerably with minimum applied stress after a given loading stress is reached. Superelasticity phenomena are known for their nonlinearity during unloading. In the case of the knee joint, the torque–angle relationship only sustains pseudoelastic behaviour during loading. A second interval featuring an approximately linear relationship between torque and angular displacement, and hence elastic working behaviour, can be approximated between points D and E (interval 3). The optimal fit for both intervals is a line between A and D, but the

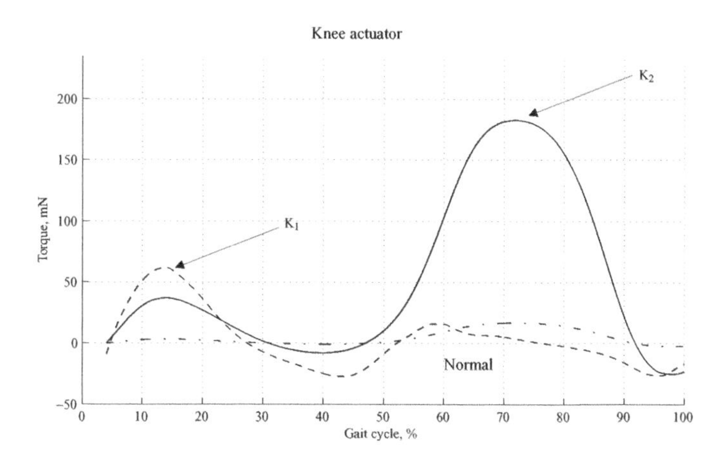

Figure 6.25 Knee actuator theoretical action: piecewise functioning to the stable stance phase (K_1) and free swing (K_2)

need to start from the same neutral position as in stance imposes point 0–0 in the torque–angle diagram as the starting point. The final fit can be seen in Figure 6.24. A torsional elastic constant can be estimated for the swing phase from the expression

$$T = K_2 \alpha_{\text{swing}} \tag{6.16}$$

where T is the joint torque in N m/kg, α_{swing} is the joint angle for swing phase in degrees and K_2 is the torsional elastic constant in N m/kg degree.

The theoretical action of the actuator with the elastic constants thus calculated is depicted in Figure 6.25 versus average data of healthy subjects.

6.7.3 Actuator prototype

An actuator prototype for a KAFO has been developed to test the concept. It was decided to use a linear actuator placed in the sagittal plane. This actuator is able to apply force to the orthosis. The design requirements are a stance range of movement of 0–22°, a swing range of movement of 0–65°, a maximum range of flexion of 95° and minimum torque values for 90° of knee flexion (sitting position).

The solution adopted aligns the force applied by the actuator with the centre of rotation of the orthotic joint, minimizing the applied moment. The knee actuator is composed of two telescopic cylinders, one containing the stance spring and the other containing the swing spring. The concept is illustrated in Figure 6.25.

Two types of springs are used in the prototype: swing springs are right-hand-helix compression springs made of stainless steel and stance springs are composed of small discs, which can be grouped according to the required constant. Stance and swing springs in the knee actuator are selected on the assumption that there are no residual actions on the joints. Equations for fitting spring rates with the patient's body weight (BW) were used to derive the maximum torque values (stance and swing):

$$T = K_{1/2} \alpha_{\text{stance/swing}} \text{ BW} \tag{6.17}$$

The torque provided by the actuator along the sagittal plane is estimated by the expression

$$T = F_{\text{actuator}} (x \cos \alpha + y \sin \alpha) \tag{6.18}$$

For an orthotic knee joint based on a four-bar mechanism (see Case Study 3.5), it is necessary to calculate the instant centre of rotation at those maxima in order to derive x and y distances. The expression for a linear spring gives the linear elastic constant of the springs:

$$F_{\text{actuator}} = K_{\text{spring}}\ \Delta l_{\text{actuator}} \tag{6.19}$$

where F_{actuator} is the force provided by the actuator. The action performed by the actuator with the constants derived for both springs is different from the action calculated in the elastic fit due to relative displacement of the actuator in the orthosis, but these differences are small and not significant.

Preliminary tests of the design hypothesis for selection of the springs with healthy subjects have shown a good approximation to the ranges of knee joint motion during gait. Experimental tests with controllable systems are described in Case Study 9.1.

REFERENCES

Bader, D., Chase, A., 1993, 'The patient–orthosis interface', in *Biomechanical Basis of Orthotic Management*(eds P. Bowkder, D.N., Condie, D.L.L, Bader and D.J. Pratt), Butterworth-Heinemann Ltd.

Bar-Cohen, Y., (ed.), 2001, *Electroactive Polymer (EAP) Actuators as Artificial Muscles – Reality, Potential and Challenges*, SPIE Press, Bellingham, Washington.

Bilas, V., Santic, A., Lackovic, I., Ambrus, D., 2001, 'A low-power wireless interface for human gait assessment', in *Proceedings of the 18th IEEE Instrumentation and Measurement Technology Conference*, vol. 1, pp. 614–618.

Blaya, J., Herr, H., 2004, 'Adaptive control of a variable-impedance ankle–foot orthosis to assist drop foot gait. *IEEE Transactions on Neural Systems Rehabilitation Engineering* **12**: 24–31.

Brunetti, F., Moreno, J.C., Ruiz, A.F., Rocon, E., Pons, J.L., 2006, 'A new platform based on IEEE802.15.4 wireless inertial sensors for motion caption and assessment', in *Proceedings of the 28th Annual Conference of the IEEE Engineering in Medicine and Biology Society*, pp. 6497–6500.

Caparros-Lefebvre, D., Blond, S., Vermesch, P., 1993, 'Chronic thalamic stimulation improves tremor and l-dopa induced dyskinesias in Parkinson's disease', *Journal of Neurology, Neurosurgery and Psychiatry* **56**: 268–273.

Carrozza, M.C., Micera, S., Massa, B., Zecca, M., Lazzarini, R., Canelli, N., Dario, P., 2001, 'The development of a novel biomechatronic hand-ongoing research and preliminary results', in *Proceedings of the 2001 IEEE/ASME International Conference on Advanced Intelligent Mechatronics* vol. 1, pp. 249–254.

Covalin, A., Feshali, A., Judy, J., 2005, 'Deep brain stimulation for obesity control: analyzing stimulation parameters to modulate energy expenditure', in *Proceedings of the 2nd International IEEE EMBS Conference on Neural Engineering*, pp. 482–485.

Davis, S., Tsagarakis, N., Canderle, J., Caldwell, D.G., 2003, 'Enhanced modelling and performance in braided pneumatic muscle actuators', *International Journal of Robotics Research* **22**: 213–227.

Ferrario, C., 2006, 'Functional electrical stimulation (FES) leg cycling exercise in paraplegia – a pilot study for the definition and assessment of exercise testing protocols and efficacy of exercise', PhD Thesis University of Glasgow.

Ferrazzin, D., di Domizio, G., Salsedo, F., Avizzano, C.A., Tecchia, F., Bergamasco, M., 1999, 'Hall effect sensor-based linear transducer', in *Proceedings of the 8th IEEE International Workshop on Robot and Human Interaction*, pp. 219–224.

Fletcher, R., 1996, 'Force transduction materials for human–technology interfaces', *IBM Systems Journal* **35**(3–4).

Gharooni, S., Heller, B., Tokhi, M., 2000, 'A new hybrid spring brake orthosis for controlling hip and knee flexion in the swing phase', *IEEE Transactions on Neural Systems Rehabilitation Engineering* **9**: 106–107.

González, J.C., 2007, 'Modelización de la influencia del calzado y las condiciones ambientales en la respuestas termofisiológica del pie y el comfort térmico', PhD Thesis, Universidad Politécnica de Valencia.

González, S.L., de Peralta, R.G., Thut, G., Millán, J., Morier, P., Landis, T., 2006, 'Very high frequency oscillations (VHFO) as a predictor of movement intentions', *NeuroImage* **23**(1): 170–179.

Hanafusa, A., Isomura, T., Sekiguchi, Y., Takahashi, H., Dohi, T., 2002, 'Contact force evaluation of orthoses for the treatment of malformed ears', in *Proceedings of the MICCAI 2002 5th International Conference, Part I*, pp. 224–231.

Havenith, G., Holmér, I., Parsons, K., 2002, 'Personal factors in thermal comfort assessment: clothing properties and metabolic heat production', *Energy and Buildings* **34**: 581–591.

Hermens, H., Freriks, B., Merletti, R., Stegeman, D., Blok, J., Rau, G., Disselhorst-Klug, C., Hägg, G., 1999, 'European recommendations for surface electromyography', *Roessingh Research and Development* **8**.

Hollerbach, J.M., 1982, 'Workshop on the design and control of dexterous hands', MIT-AI Memo No. 661, Massachusetts Institute of Technology, Cambridge, Massaechusetts.

Jansen, J., Richardson, B., Pin, F., Lind, R., Birdwell, J., 2000, 'Exoskeleton for soldier enhancement systems feasibility study', ORNL.

Kazerooni, H., 1998, 'Human power extender: an example of human–machine interaction via the transfer of power and information signals', in *Proceedings of the International Workshop on Advanced Motion Control*, pp. 565–572.

Kyberd, P.J., Pons, J.L., 2003, 'A comparison of the Oxford and Manus intelligent hand protheses', in *Proceedings of the IEEE International Conference on Robotics and Automation*, pp. 3231–3236.

Liu, J., Shkel, A.M., Niel, K., Zeng, F.G., 2003, 'System design and experimental evaluation of a MEMS-based semicircular canal prosthesis', in *Proceedings of the First International Conference on IEEE EMBS Neural Engineering*, pp. 177–180.

Lotters, J., 1998, 'Design, fabrication and characterization of a highly symmetrical capacitive triaxial accelerometers', *Sensors and Actuators* **66**: 205–212.

Lou, E., Chan, C., Raso, V., Hill, D., Moreau, M., Mahood, J., Donauer, A., 2005, 'A smart orthosis for the treatment of scoliosis', in *Proceedings of the IEEE International Conference on Engineering in Medicine and Biology Society*, pp. 1008–1011.

Luca, C.J.D., 2002, 'Surface electromyography: detection and recording', delsys incorporated, from http://www.delsys.com/
Attachments_pdf/WP_SEMGintro.pdf, last visit: 11 May, 2007.

Luinge, H., 2002, 'Inertial sensing of human movement', PhD Thesis, University of Twente, The Netherlands.

Mak, A.F., Zhang, M., Boone, D., 2001, 'State-of-the-art research in lower-limb prosthetic biomechanics socket interface: a review', *Journal of Rehabilitation Research and Development* **38**(2): 161–174.

Malmivuo, J., Plonsey, R., 1995, *Bioelectromagnetism: Principles and Applications of Bioelectric and Biomagnetic Fields*, Oxford University Press.

Marins, J., Yun, X., Bachmann, E., McGhee, R., Zyda, M., 2001, 'Extended Kalman filter for quaternion-based orientation estimation using MARG sensors', in *Proceedings of the IEEE/RSJ International Conference on Intelligent Robots and Systems*, vol. 4, pp. 2003–2011.

Mascaro, S., Asada, H.H., 2001, 'Finger posture and shear force measurement using fingernail sensors: initial experimentation', in *Proceedings of the IEEE International Conference on Robotics and Automation*, vol. 2, pp. 1857–1862.

Mayol, W.W., Tordoff, B., Murray, D.W., 2004, 'Towards wearable active vision platforms', in *Proceedings of the IEEE International Conference on Systems, Man, and Cybernetics*, vol. 3, pp. 1627–1632.

Micera, S., Carpaneto, J., Scoglio, A., Zaccone, F., Freschi, C., Guglielmelli, E., Dario, P, 2004, 'On the analysis of knee biomechanics using a wearable biomechatronic device', in *Proceedings of the IEEE/RSJ EMBS Conference on Intelligent Robots and Systems*, vol. 2, pp. 1674–1679.

Miyoshi, Y., Mitsubayashi, K., Sawada, T., Ogawa, M., Otsuka, K., Takeuchi, T., 2005, 'Wearable humidity sensor with porous membrane by soft-MEMS techniques', in *13th International Conference on Solid-State Sensors, Actuators and Microsystems, 2005* vol. 2, pp. 1290–1291.

Moreno, J.C., Rocon, E., Ruiz, A., Brunetti, F., Pons, J.L., 2006, 'Design and implementation of an inertial measurement unit for control of artificial limbs: application on leg orthoses', *Sensors and Actuators B* **118**: 333–337.

Nicolopoulos, C., Anderson, E., Solomonidis, S., Giannoudis, P., 2000, 'Evaluation of the gait analysis FSCAN pressure system: clinical tool or toy?', *The Foot* **10**(3): 124–130.

Nishi, Y., 1981, 'Measurement of thermal balance of man', in *Bioengineering: Thermal Physiology and Comfort*, (eds K. Cena and J.A. Clark), Elsevier.

Omega Engineering, 2006, Omega Engineering Technical Reference, from http://www.omega.com/temperature/Z/pdf/z021-032.pdf, last visit: 11 May, 2007.

Pappas, I., Popovic, M., Keller, T., Dietz, V., Morari, M., 2001, 'A reliable gait phase detection system', *IEEE Transactions on Neural Systems and Rehabilitation Engineering* **9**(2): 113–125.

Pons, J.L., 2005, *Emerging Actuator Technologies. A Micromechatronic Approach*, John Wiley & Sons, Ltd.

Pratt, J., Krupp, B., Morse, C., 2002, 'Series elastic actuators for high fidelity force control', *Industrial Robot Journal* **29**(3): 234–241.

Reed, K.B., Peshkin, M., Hartmann, M.J., Colgate, J.E., Patton, J., 2005, 'Kinesthetic interaction', in *Proceedings of the 9th International Conference on Rehabilitation Robotics*, pp. 569–574.

Richards, G., 2003, 'Triaxial force pin sensor array', United States Patent 6,536,292, 25 March 2003.

Riviere, C., Ang, W.T., Khosla, P., 2003, 'Toward active tremor canceling in handheld microsurgical instruments', *IEEE Transactions on Robotics and Automation* **19**: 793–800.

Rocon, E., Ruiz, A.F., Pons, J.L., Belda-Lois, J.M., Sánchez-Lacuesta, J.J., 2005, 'Rehabilitation robotics: a wearable exo-skeleton for tremor assessment and suppression', in *Proceedings of the International Conference on Robotics and Automation 2005 – ICRA05*, pp. 241–246.

Schrand, D., 2007, 'Crosstalk compensation using matrix methods', *Sensors and Actuators B, Industrial Control Designline*, 20 April 2007.

Searle, A., Kirkup, L., 2000, 'A direct comparison of wet, dry and insulating bioelectric recording electrodes', *Physiological Measurement* **21**(2): 271–283.

Sitti, M., 2003, 'Teleoperated and automatic nanomanipulation systems using atomic force microscope probes', in *Proceedings of the IEEE Conference on Decision and Control* **3**(9–12): 2118–2123.

Smith, B., Johnston, T., Betz, R., Mulcahey, M., 2005, 'Implanted FES for upright mobility in paediatric spinal cord injury: a follow-up report', in *Proceedings of the 2nd International IEEE EMBS Conference on Neural Engineering*, pp. 372–373.

Sooyong, L., Jangwook, L., Woojin, C., Munsang, K., Chong-Won, L., Mignon, P., 1999, 'A new exoskeleton type masterarm with force reflection: controller and integration', in *Proceedings of the IEEE/RSJ International Conference on Intelligent Robots and Systems*, vol. 3, 1438–1443.

Veneman, J.F., Ekkelenkamp, R., Kruidhof, R., van der Helm, F.C.T., van der Kooij, H., 2006, 'A series elastic- and Bowden-cable-based actuation system for use as torque actuator in exoskeleton-type robots', *International Journal of Robotics Research* **25**: 261–281.

Winter, D.A., 1990, *Biomechanics and Motor Control of Human Movement*, John Wiley & Sons, Ltd.

Yun, S., Kang, S., Kwon, D., Choi, H., 2006, 'Tactile sensing to display for tangible interface', in *Proceedings of the IEEE/RSJ International Conference on Intelligent Robots and Systems, 2006*, pp. 3593–3598.

7

Communication networks for wearable robots

F. Brunetti and J. L. Pons

Bioengineering Group, Instituto de Automática Industrial, CSIC, Madrid, Spain

7.1 INTRODUCTION

Everyone is surrounded by communication networks in their daily activities. The latest technological advances have provided the means for these networks to proliferate. Devices are connected within the concept of the global village. Moreover, the communication concept is not restricted to interconnection of distant nodes. Nowadays, close interaction among all devices close at hand can be achieved to improve access to new technologies and automate processes, or simply to connect people.

The *ambient intelligence* concept aims to create such a local interaction driven by short-range networks. It envisages many different devices gathering and processing information from many sources to control physical processes and interact with human users. It is part of the global hierarchy of networks defined by the now-familiar wireless area network (WAN) and local area network (LAN).

Research on ambient intelligence shares many goals with mobile ad hoc networks (MANETs), personal area networks (PANs) and body area networks (BANs). Wireless technologies have strongly impacted these fields, inspiring new concepts and research objectives, e.g. wireless PANs, wireless sensor networks (WSNs), wearable BANs and clothing area networks (CLANs). Popular protocols like Bluetooth and ZigBee are products of these innovative efforts. The wireless sensor network is another concept that promises to have a strong impact on ambient intelligence.

Many of the networks mentioned above look similar, but they differ in essential features that require specific solutions. Various different networks can be built depending on the complexity of the nodes. For instance, in WSNs the nodes are fairly simple, while Bluetooth devices are normally more complex, e.g. computer peripherals or mobile phones. The type of connection also affects the network design and architecture. In some scenarios, point-to-point connections are required to establish a bidirectional link. In data centric networks, on the other hand, the data flow to a specific location,

Wearable Robots: Biomechatronic Exoskeletons Edited by José L. Pons

usually referred to as the sink. Another key issue is the purpose of the network. Sometimes this is to monitor and control a process continuously, while in other cases it is simply to transmit sporadic events.

In a smart environment, nodes are highly heterogeneous. In addition, some of them are included in a periodic maintenance service and some are not. The flow of data can be really high, but meaningful information has to be retrieved for end-users in a comprehensible way.

These developments involve many research fields, e.g. signal processing and cognitive sciences. Moreover, there are still many unsolved problems regarding these networks: scalability and inclusion in mature networks, cost, size and dense deployment, security, context awareness and ethical issues.

The applications for all these networks are numerous. Healthcare, wellness systems, habitat monitoring, consumer electronics, tracking and any kind of telematic application are all potential use scenarios.

The field of robotics is not closed to all these possibilities based on new technologies, but robotic networks have to define their own profiles. Experience in the networks mentioned above will be helpful in their development; it is important to bear in mind that a new concept relating to robotic networks does not imply a complete new paradigm. Moreover, robotic networks will be embedded in this smart ambience and in the global village.

The new development will include key features like fault tolerance schemes and prioritized protocols to support the control tasks that arise in the field of robotics. Based on this approach, researchers will be able to develop modular devices, sensors and actuators that will help to accelerate the innovation in the field. Figure 7.1 depicts the concept of communication networks for WRs.

The wireless sensors and actors networks (WSANs) project being run by the BWN Lab at Georgia Institute of Technology is one of the first steps in the field. This project proposes a network of sensors and actors, rather than actuators, to accomplish cooperative goals. An actor is defined as a node capable of controlling several actuators and sensors in order to perform a specific task based on high-level primitives or commands.

This chapter presents useful concepts, technologies, protocols and experiences in the field. It also includes case studies illustrating them and the current trends and applications, e.g. smart textiles.

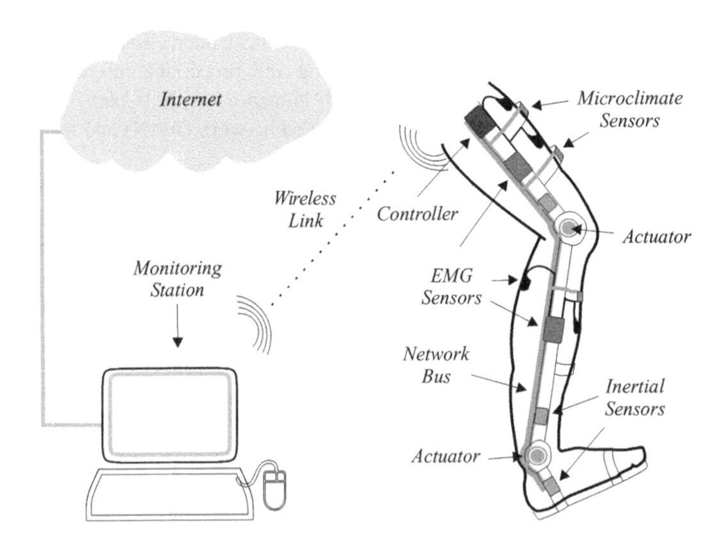

Figure 7.1 A network architecture to monitor/control a lower limb exoskeleton

7.2 WEARABLE ROBOTIC NETWORKS, FROM WIRED TO WIRELESS

A network for wearable robotic applications is a tradeoff between industrial communication networks, personal area networks and sensor networks. Each network is designed to fulfil the requirements of a specific application. Robustness is a major requirement in industrial applications, and therefore they are usually wired. Modern standards for industrial networks are fieldbus based. The term 'fieldbus' is applied to new digital industrial networks that replace the current 4–20 mA analogue signal. The new standards include bidirectional links and multihopping mechanisms based on a serial bus. These networks have been designed to link sensors, actuators, controllers and other devices such as data loggers. Because they have a specific role in the network, they are designed and optimized with cost and efficiency in mind.

In WSNs the freedom associated with a wireless link is a highly appreciated attribute. These usually use low-power radios and microcontrollers to save as much energy as possible, since nodes are designed for practically unattended supervision and low maintenance. Different developments are influenced by the nature of these different networks, and often solutions for one specific network are not suitable for another. Finally, the designer will have to choose the appropriate technology to implement the right network based on the requirements and the desired performance. One of the special and novel features of networks for wearable robotics is the coexistence of portable and low-cost sensors and actuators. This makes such networks unique, and therefore new solutions have to be developed.

7.2.1 Requirements

There are many common features describing a network. These features or parameters are used to compare different available technological solutions to implement a network.

7.2.1.1 Network throughput

The network throughput is usually referred to as the amount of information transmitted during a period of time. This is generally expressed in kbps (kilobits per second) or higher multiples like Mbps (mega bits per second) or Gbps (giga bits per second). On the one hand, industrial networks are generally intended to control and monitor processes. Although the nodes are intelligent, there is a master controller or unit working as the main node or controller. The user can control the whole process by means of such a main controller. This interaction demands continuous signalling between the main controller and the nodes, producing *constant and deterministic* network traffic.

On the other hand, WSNs use very simple nodes subject to cost and energy constraints. They are oriented towards event detection and transmission. Within this framework, the traffic is not constant, and moreover may be random. This difference restricts the design and use of protocols to access the transmission medium. In wearable robotics there are various different scenarios. The range of the network is short, so most of the physical layer protocols for industrial networks are adequate.

7.2.1.2 Clustering and data fusion mechanisms

Clustering concepts make sense in large networks with many nodes. In many WSN applications, nodes monitor environmental variables such as temperature and humidity. In near-range nodes, these variables are highly correlated. Data fusion is the process of combining the information acquired by many different sensors or systems into a compact and coherent descriptor of the observed phenomena. Data fusion mechanisms can be applied in order to save power and optimize the network traffic, e.g.

the mean value of the readings or detection of an event after processing the values from many sensors and nodes. In this case, the traffic is data centric. Every cluster has an identified sink that collects the data from the different sensors and, after processing it, the sink transmits the information to the root node. Similar mechanisms can be used in industrial communication networks, but the control functions performed by the main controller may require the raw data gathered by the sensors.

7.2.1.3 *Quality of service (QoS)*

In communication networks the term 'quality of service' (QoS) refers to the probability of delivering a service successfully. High QoS does not mean high performance of the network in terms of latency or bit rate, but good referring and protocols to control the network. Mechanisms to improve the robustness of the network can be added on every layer (Karl and Willig, 2005). For instance, wiring topologies, acknowledgement schemes, retransmission mechanisms, scheduled or referred media access, priority management and other mechanisms help to improve the QoS of a network. Industrial communication networks demand a high QoS. The first consequence of this requirement is the preference for wired physical links with referred or scheduled medium access, like the ones used by a fieldbus. In WSNs, the QoS is improved by means of retransmission mechanisms, acknowledgement schemes and dynamic routing algorithms. The first approach (fieldbus) also guarantees low latency, which is critical for control applications, while the second (WSN) does not address this issue. Evidently, there is a tradeoff between controlling and monitoring, robustness and cost. Both monitoring and controlling applications can be found for wearable robots; what system is most appropriate will depend on the requirements.

7.2.1.4 *Latency*

Latency can be defined as the elapsed time between an event or action and its effects. In wearable robotics a high latency can cause many problems. Moreover, the human–robot interaction can be affected, thus reducing the dependability of the robot and hence the overall performance of this symbiotic system. Again, control tasks demand a lower latency than monitoring tasks. The best way to reduce latency is to implement a deterministic protocol to access the communication link. Mechanisms like medium access based on priority of a device address can also help to achieve lower latencies for critical tasks.

7.2.1.5 *Cost, size and weight*

In every wearable application, cost, size and weight should all be as small as possible. The use of wires is also a drawback in this field, since many robotic structures or interfaces cannot easily be wired. Wires may also restrict movement and alter normal movement patterns. Bulky sensors, actuators and energy storage devices have the same effect. Finally, the use of wearable technological solutions is directly related to cost. All these features are common to WSNs and affect the QoS of the network.

7.2.1.6 *Power consumption*

Naturally, wearable robots tend to be portable. One of the major problems with portable solutions is power consumption, i.e. the autonomy of the system. WSNs devote a great deal of effort to reducing power consumption, as reflected in the use of low-power components, short-range radios and optimized media access control (MAC) protocols. This is not so in the case of industrial networks since these always use mains power.

7.2.2 Network components: configuration of a wearable robotic network

In the sphere of communication networks for wearable robotic applications, robot components can be developed modularly. Modular development permits easier hardware upgrades, using a set of different configurations while spreading the range for potential applications. A network can comprise different components; each component has a specific function and this function defines the technological requirements and the subcomponents.

7.2.2.1 Controllers

The main node of the network is the controller, and all networks must have one. Its functions are as follows:

- to implement the control strategies governing the process;
- to manage the overall network and the nodes, including discovery of services, addresses and priority assignment, application of QoS mechanisms, retrieval and processing of data from sensors, and generation of commands or signals to control the actuators.

Depending on the application, the architecture of this node can be based on microcontrollers or more powerful systems like digital signal processing (DSP), PDA or industrial PCs. This node acts as a master in the network. All the other components should be able to communicate with this node during the setup of the network. For some applications the controller has to be able to offer specific services to network nodes, such as dynamic communication slot assignment for new nodes in the network, interruption handlers for specific events generated by nodes and gateway services to access higher level networks in order to enable interaction within the two networks.

7.2.2.2 Sensors

The type and number of sensors in a wearable robotic network can be large. Further details of these sensors can be found in Chapter 6. Sensors can be controlled internally by a smart device or they can be embedded in dummy subsystems. The resolution, the generated information and the nature of the measured phenomena can have a major impact on network traffic. Sensors may require the following services: communication slots to transmit continuous data, flexible packet length to support discrete values, different resolution and event notifications, and interruption capabilities to notify the occurrence of events.

The sensor can also provide some services to the controller: a description of sensor node services, adjustment mechanisms including calibration of the sensor, sampling frequency, antialiasing filter settings and resolution, notification to the controller of whether the command execution was successful or not and any other problem related to the sensing system.

7.2.2.3 Actuators

One of the special features of this wearable network is the presence of actuators. Actuators, like sensors, can be embedded in either smart or simple devices. In the case of smart devices, the device itself can generate the control signals for the actuator on the basis of high level-commands received from the controller and from sensors also attached to the device. When they are embedded in a simple device, the controller has to generate the control signals and transmit them over the network. These two approaches affect network traffic in different ways. In WR networks, an actuator has well-defined functions: to provide the network controller with a description of the actuation system, to implement the control signals or commands sent by the network controller, to advise the controller whether the command execution was successful or not and any other problem related to the actuation system.

7.2.2.4 Other components

Many other different nodes can be attached to a network. In portable robotics for rehabilitation, it is useful to save the data generated by the sensors during daily activities in order to analyse the evolution of a therapy. This can be done by means of data loggers. Thanks to new advances in data storage technologies, engineers can use commercial memory cards to store any kind of information related to the system. The writing speed can differ among different types and models of cards, as can the writing protocol. These issues need to be considered along with the effect that the data logger will have on the network traffic. The controller may treat the data logger as an additional actuator, hence providing the same services, which are applicable to both functions.

In some applications, the gateway functions can be performed by another network component in order to simplify the design of the controller. The advantage of this modular design is that the gateway device can implement a particular network protocol depending on the network to be accessed. This approach also has a positive impact on the final cost of the device.

7.2.3 Topology

Network topology means the way in which physical interconnection between nodes is implemented. Figure 7.2 depicts the most common topologies. In wired networks the meaning of the topology is clearer, as the connection is based on the physical links between nodes. However, in wireless networks, the topology is determined more by the hierarchical and logical structure of the network, including the kind of MAC protocol that the network implements.

Different topologies perform in different ways under certain conditions (Middaugh, 1993). Most protocols are designed to work with a specific topology.

7.2.3.1 Star

This topology can be used in wired and wireless networks. In wired networks it implies that there is a point-to-point link between each node and a central one. This central node acts as the master of the network and controls access to it. In wearable robotics, the central node is usually the controller. Transmission medium access is simpler in wired networks since each link is only used by two nodes. In wireless networks, nodes often share a single medium, which has to be controlled by the master. The advantages of this topology are limited communication afforded by nodes connected to a central node, simplicity of communication and routing since each node communicates point-to-point, especially in wired networks line faults only affect one node in wired networks and easier network management, since it is done by a single node.

It can be seen that this topology is quite simple, but it does have some disadvantages: overloading of the central node, the number of nodes is limited by the capacities of the central node, wiring costs in the case of wired networks, full dependence on central node and complexity of network scalability.

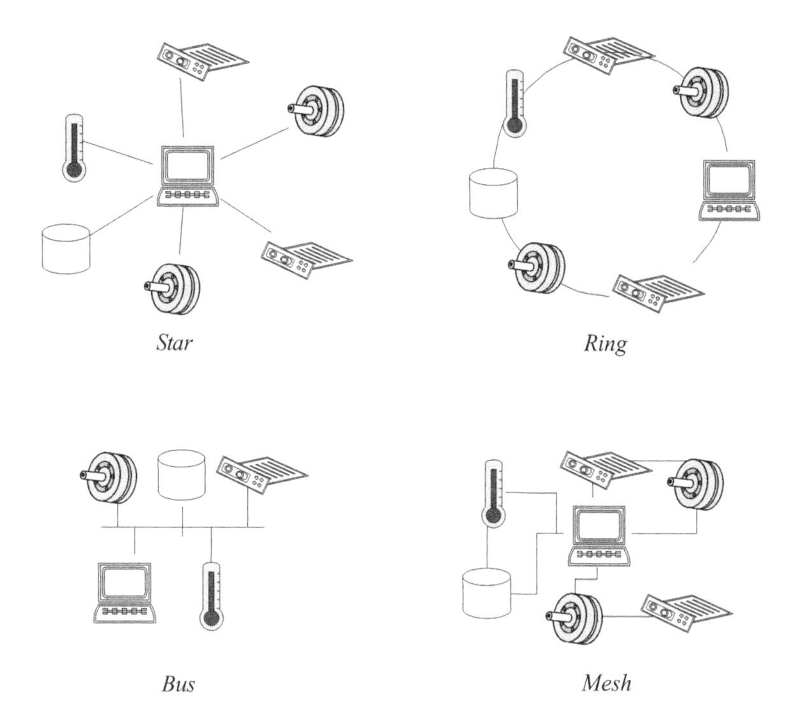

Figure 7.2 Network topologies hosting different WR devices

There are more complex topologies based on this one that are usually applied in wireless networks. One of these is the tree topology. The difference is that the central node can also act as a node of a higher-hierarchy star network at the same time. This approach eliminates one of the drawbacks of star topology, since now the number of nodes can be higher. However, it presents scalability problems, since it is not easy to manage all the communication links. Nevertheless, it makes sense in some applications where clustering and data fusion or aggregation mechanisms are used. More complex than this is graph topology. Here, every node can be connected to one or more other nodes. Depending on the number of nodes and established links, this topology can become very complex. More complex topologies produce a mesh network, common in wireless applications. In a mesh network the links between nodes are highly dynamic, so that there is continuous connecting and reconfiguration of the network. In this scenario hierarchical control ceases to make sense.

7.2.3.2 Ring

In this topology there is a common network link for all the nodes. A message travels around the link in such a way that every node has access to it. In general, protocols use addresses to send messages to a specific device, which is preidentified with a unique address. In this scenario, the throughput of the network can be optimized since every node has a specific time slot to transmit data. When a node receives the token, it can use the channel. When it finishes, it will pass the token on to the next node of the ring. The advantages of the ring topology are optimized channel use, there is no central node and simple network management protocol.

The disadvantages are network vulnerability; e.g. if the ring is broken, the network fails and wiring is complex and network expansion costly.

The ring topology is suitable for static networks that are not intended to grow and are kept in a safe environment, or for networks built with high-quality components to minimize the probability of failure. It is therefore not suitable for wireless applications since these are dynamic and the channel is exposed to noise and other problems that can affect the performance of the link.

7.2.3.3 Bus

This is the most common topology in current networks. Nodes are connected to a common bus. Each message transmitted by a node is broadcast and received by all nodes. In the ring topology, the nodes generally act as message repeaters since each node is directly connected only to its two neighbours in the ring. This is not necessary in the bus topology due to the characteristics of the channel. Any node can access the channel to transmit data.

The advantages of this topology are there is no central device, a node failure does not compromise the network function, good support for broadcasting messages, easy network growth and reconfiguration. Disadvantages include collision of transmitted data can occur, vulnerability of the bus and more complex protocols for media access.

7.2.4 Wearable robotic network goals and profiles

The field of wearable robotic applications has many uncommon features. Analysis of the requirements of this application and the various different topologies suggest that the most suitable solution for wired networks is a bus topology including smart devices. Smart devices mean devices that can execute more complex tasks from a high-level primitive. For instance, the controller may send a basic command to the device in order to execute a more complex task, a task that could entail interaction of several sensors and actuators with processing algorithms. This concept also allows for a modular system setup, which has numerous advantages during development. Finally, it is important to emphasize that in WRs the organization is always hierarchical, which means that there is a master node controlling the network.

In this scenario, there are many possible different goals that strongly influence the network design. These goals are:

1. *System monitoring.* The object of the system is to monitor the wearable robot. This is a common application since actuating technologies are not suitable for wearable robotics. Some robots include mechanically controlled actuators. The function of the wearable robotic network is to monitor the performance of the overall system. The protocols are designed to cope with data heterogeneity, sampling frequency of the sensors and data logging. The control functions are simpler, including on/off actuators. The network is data centric. WSN technologies are the first choice; however, as continuous monitoring systems, wearable robotic networks require different protocols to execute this task.

2. *System controlling.* Complex wearable robots may include several DoFs and active actuating systems, and the network has to be very robust in these robots. Several protocols may be implemented between the network establishment phase and the stable operating phase. The controller has to be able to handle priorities and discover services. Commands and services of network devices have to be defined within the protocols. Fieldbus network designs typically include the control function.

A network-based WR can be used for different purposes depending on what devices are connected. A profile means a description of the types and roles of connected devices. A wearable robotic network can use only one profile at a given time. Protocols should be developed to support different profiles, but the final in-device implementation may be partial to optimize the cost of the protocol.

Common profiles for wearable robotic networks are:

- *Rehabilitation.* In this profile the controller includes basic routines to control the wearable robots. The system includes several smart nodes (combination of sensors, actuators and small controllers). Basic routines can be implemented in the firmware of the devices; for instance, the controller may command a task using high-level primitives. The downlink (controller to node) has a higher priority than the uplink (node to controller).
- *Robotic control.* This profile is designed to support control of the cHRI by means of EMGs, EEGs or biomechanical monitoring devices. The profile demands robust protocols. Smart monitoring devices should provide muscle activation detection, mental states, joint positions or other services. All this information may be supplied on demand, continuously or event-triggered. Based on these services, the controller can generate the required actuation commands and other additional tasks.
- *Evaluation.* This profile is designed to provide real-time and offline monitoring of the robot and the HRi; it is the simplest profile. The controller has to supply the gateway or storage device with the data generated by different sensors.
- *Universal control.* The aim of this profile is to render the wearable robotic network capable of interacting with the environment. It is a hybrid between robotic control and evaluation controls. In this profile a predefined protocol for interaction between the external world and the WR has to be managed by the controller and the gateway. Smart monitoring devices in the wearable robotic network have to provide similar services to those provided by the robotic control profile.

7.3 WIRED WEARABLE ROBOTIC NETWORKS

After defining the purpose and requirements of the network and its expected overall performance and priorities, the next step is to choose appropriate technologies and suitable protocols for the application. This section describes wired technologies. The description focuses on lower-layer protocols, more specifically the physical layer and data link layer protocols.

These two layers provide the network structure for communication with different network nodes. However, additional protocols may be developed for upper layers depending on the application. Usually, these are not part of the standards that describe the lower-layer protocols. Figure 7.3 depicts the protocol stack of the candidate technologies described hereafter.

7.3.1 Enabling technologies

Wired technologies applicable to robotics come from the field of industrial networks. Industrial networks implement different protocols. This section gives a brief description of the most relevant protocols and identifies potential wearable robotic applications and useful mechanisms.

7.3.1.1 Interchip connection protocols

The transmission of digital data using parallel protocols is virtually impossible due to the large number of lines needed. This was established years ago by integrated circuit (IC) manufacturers, who developed advanced serial protocols to support the interconnection of multiple ICs.

Philips Semiconductor developed I2C as a standard for connecting networked integrated circuits. The main purpose is to interconnect microcontrollers or connect microcontrollers with other peripherals like sensors or analogue-to-digital converters (ADC).

The protocol defines a two-wire serial interface (I2C-Bus Specification, 2000). The first is a bidirectional data wire. The channel is half duplex and allows rates up to 3.4 Mbps in the high-speed

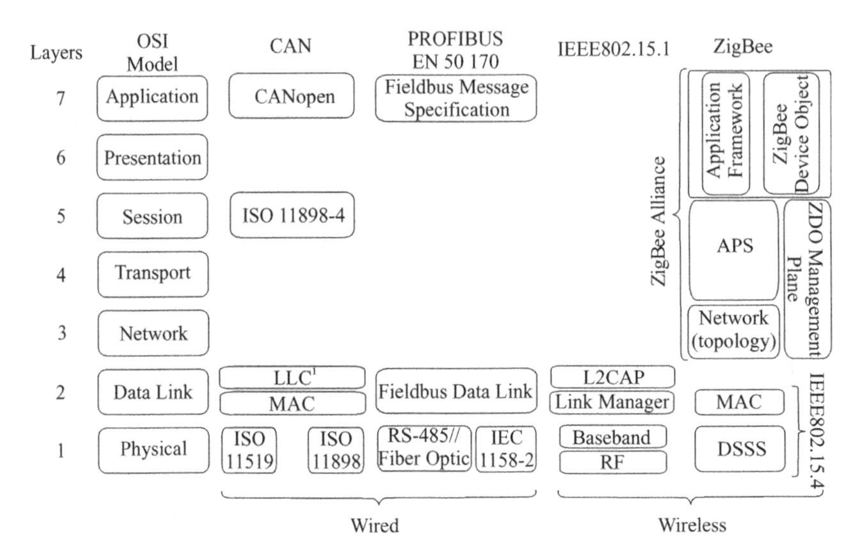

Figure 7.3 Basic protocol stack for wearable robotic applications, and the candidate technologies

mode. The second line is a clock signal given by the master. The master initiates the communications and sends the control signals to the devices. Each node has a 7-bit or 10-bit address, which is used to identify a node in the network. The standard limits the capacitance of the bus to 400 PF. Up to 1024 nodes can be connected to the bus without exceeding its capacitance limit.

Another protocol to interconnect ICs is the serial peripheral interface (SPI) developed by Motorola. This is a three-wire serial bus for 8-bit data transfer applications. The standard uses two wires to transfer data and one to transmit a clock signal. Like I2C, the network has one master and the other devices are slaves. The master generates the clock signal and controls the network. If many slaves are attached to the bus, an extra line is needed for each device. This signal is called Chip Select, and basically controls the impedance of the device interface to the bus. This is a disadvantage when SPI is compared to I2C. Data on the SPI bus can be transferred at up to 1 Mbps. SPI is an attractive bus for networks with a small number of nodes (Freescale Semiconductor Manual, 2001).

I2C has a further advantage over SPI. In I2C, new sensors or modules can be connected to the network without affecting it. The master finds the address of the new node in order to establish a normal networked communication. This is not so easy in SPI because it requires hardware and wiring changes.

Both these protocols and other similar ones are often included in popular microcontrollers developed by a number of manufacturers. Depending on the application, these protocols have two major disadvantages. The first is that none of these protocols supports interruptions, and that can be critical for event detection networks. The master has to execute polling mechanisms to retrieve the data from the nodes, and a continuous poll of the nodes entails higher power consumption. The second disadvantage is that they have no protection or mechanisms to withstand noisy environments.

7.3.1.2 RS-232, RS-422, RS-485

RS-232 is one of the most widely used communication standards. It was introduced in 1962, and after 45 years its simple interface still makes it a good communication option to interconnect devices in almost every field. In recent years, RS-232 has been replaced in many applications by the universal serial bus (USB), but it is still present in many others. The RS-232 standard allows bidirectional short-range communication between two nodes.

RS-232 signals are represented by voltage levels with respect to a system common (power and logic) ground. It has many handshaking lines and a MAC communication protocol. Both are defined in the standard.

RS-232 is not suitable for noisy environments. Differential data transmission was therefore developed to get around the vulnerability of RS-232 to noise. The outcome was the protocol RS-422, which supports a data rate of up to 100 kbps over distances up to 1200 m. In addition, RS-422 allows up to 10 receivers to be connected to a bus with only one driver.

The next step was RS-485, which allows up to 32 drivers and 32 receivers on a single (two-wire) bus. With this protocol, a ready-built network can be used. A master must control access to the medium to avoid collisions.

These protocols only describe the physical layer and some details of the link layer or MAC protocol. Upper-layer protocols for wearable robotics have to be developed separately. However, the simplicity of this physical interface makes it attractive for simple applications.

7.3.1.3 IEEE P1451.3

IEEE P1451.3 is a standard for transducer networks developed by the IEEE P1451 project (Lee, 2000). This recent protocol uses a single line to power the transducers and transmit data. The transmission medium can be a coaxial cable or a twisted pair cable. It implements synchronization and interruption mechanisms.

One controller commands the bus. It is embedded in the network capable application processor (NCAP). The other components of the network are the transducers, which use a transducer bus interface module (TBIM) to connect to the bus. The NCAP also acts as a gateway for communication within broader networks.

TBIMs have five communication functions. These are network communication, TBIM communication to transducer controller, data transfer, synchronization and triggering to start specific tasks. These functions are multiplexed into five communication channels using one physical medium.

Another useful feature of this standard is the transducer electronic data sheet (TEDS) concept, which describes the type, attributes, operation and calibration of the transducer. This information is stored in a nonvolatile memory. Using the TEDS, an NCAP can ascertain the sensor's features exactly and recognize it in the network.

The future of IEEE 1451.3 is very promising. It should enable a universal connection and self-deployment of network transducers. However, the protocol has not been widely adopted as yet.

7.3.1.4 Controller area network (CAN)

Controller area network (CAN) is a network standard developed in the 1980s for the automotive industry. The protocol is described in ISO standard 11892-1. The CAN protocol defines the physical layer and the data link layer of the OSI reference model.

The CAN is a fieldbus protocol (Corrigan, 2002). A broadcasting mechanism is implemented so that every node of the network can access the data transmitted. All the nodes have their own unique address. Access to the bus is based on priorities, but here is where the CAN presents an interesting solution: the priorities are not for specific nodes or addresses; the CAN uses a priority scheme based on message contents. In this way, the network will define the set of messages with their priorities *a priori*. This approach makes the protocol especially suitable for event detection networks. Low latency is guaranteed and the master does not have to poll different nodes. With the CAN also, a multimaster hierarchy increases the redundancy and robustness of the networks. It also supports Plug&Play devices, since no hardware changes have to be made to add new devices to the network.

As an industrial network protocol, it includes many bit error detecting mechanisms. The CAN is not an open bus protocol; terminators are needed at the end of the bus. ISO 11898-2 is the most commonly used physical layer standard for the CAN. According to this standard, bitrates up to 1 Mbps are possible using a bus length of 40 m. The standard specifies a two-wire differential bus. The number of nodes is limited by the electrical load of the bus.

There are many other standards describing CAN-related protocols. One that deserves mention is the CANopen protocol, (CAN-in-Automation, 1996). This is a higher-layer protocol for CAN networks. The standard covers the application and communication profile by means of defined objects for real-time data, configuration data, network management data and special functions. Among these functions, the protocol describes time stamps, synchronization and emergency messages. For instance, a master may get data from the node on demand, synchronously or upon detection of a new event alert by the node. Another useful feature of CANopen is the Object Dictionary. This is a kind of table where all the information about the device is stored. The protocol defines services to provide remote access to the Object Dictionary of a device.

New microcontrollers and DSPs integrate CAN support on a chip. They can be attached to a CAN bus using a simple driver. Such easy integration enhances the usefulness of the CAN in multiple applications.

7.3.1.5 PROFIBUS

PROFIBUS is one of the most popular industrial network standards. It is an open protocol that can use five different transmission technologies. RS485 is one of them. The bus is a twisted pair copper cable (PROFIBUS System Description, 2002). Transmission rates of 12 Mbps can be reached. Up to two nodes can be attached to a single segment. If more that 32 nodes are needed, the network has to include repeaters. Plug&Play is also possible.

The protocol provides cyclic functions, including cyclic data exchange, station, module and channel diagnostic and interruption mechanisms. It also provides acyclic functions for parameter configuration, operation and visualization and functions to allow external connections to the fieldbus network. The maximum number of nodes allowed in the network is 126. It permits the coexistence of more than one master in the network. One interesting feature of PROFIBUS is the implementation of interslave communication. The protocol takes advantage of the fact that a fieldbus is a broadcasting network, so that a slave can read data from another slave (the publisher) without the intervention of a master. This scheme reduces the response time of the bus by up to 90 %.

The protocol is very comprehensive. It covers almost all the needs of the sector and can address a wide range of applications with considerable flexibility. Its drawback is overall complexity. It is designed for use with commercial products for industry, but not for prototypes or research activities.

7.3.1.6 IEEE 802.4 token bus

This protocol was defined to get the best out of token ring protocols and bus topologies. Bus networks governed by contention-based MAC protocols were not suitable for industrial networks because of their nondeterministic behaviour and latency. The token bus uses a 75 ohm coaxial cable. It allows a single or double bus without terminators. The standard defines three analogue modulation schemes: two frequency shift keys (FSK) and one phase shift key (PSK). It supports bit rates of 1.5 or 10 Mbps.

Every node of the network knows the address of both its right and left neighbours. In order to transmit data the node must have possession of the token (TOKENBUS, 1985, 1990). It passes on

Preamble (1 byte)	Start Delimiter (1 byte)	Control Frame (1 byte)	Destiny Address (6 bytes)	Source Address (6 bytes)	Data (up to 8192 bytes)	CRC (4 bytes)	End Delimiter (1 byte)

Figure 7.4 IEEE 802.4 frame

the token when the transmission is complete or when the token holding time runs out. The token is passed from higher to lower address devices, following the numeric sequence. In this way, the protocol defines a logical ring. The protocol also defines different internal priorities, which are applied in the device. If a node has many data packets, it verifies which one has the highest priority so that it can be transmitted first when it has the token.

Periodically, a node sends the Solicit Successor message, which starts up the procedure for aggregation of a new node to the network. This procedure envisages collisions when two new nodes try to answer this packet. The protocol describes simple commands for management of the ring.

The frame defined by the protocol is depicted in Figure 7.4. Many network protocols use similar frames. Top-layer protocols have to be developed separately since they are not included in the standard.

7.3.2 Network establishment, maintenance, QoS and robustness

As in any other network, there are two operating phases in wearable networks for robotic applications. The first phase comprises the network setup process when all the components are powered up. In this phase, the network's master node has to take the lead. First it announces its presence in the network and starts the service discovery procedures. According to the master's goal, it will use the services provided by the different nodes available in the network; for instance, it can use continuous monitoring of a joint angle provided by a device to control this joint on the basis of high-level commands. The number of possible scenarios is large.

Once all the components have been recognized, the second phase starts. During this phase, the controller executes the task with the help of different nodes. The network tends to be static, but in certain cases a new node can be connected to the network, e.g. a wireless gateway for real-time monitoring in a clinic during rehabilitation therapies. Exoskeletons can work with different profiles, as noted in Section 7.2.4. There are numerous benefits to be gained from a modular approach. Wearable robotic networks may consist of several smart devices with specific functions, and a new profile can be configured depending on which ones are connected.

The controller can receive the order to execute tasks related to the selected profile. Given this high-level primitive, it has to apply all the mechanisms for network formation in order to address the overall objective. Many of the above introduced protocols allow Plug&Play mechanisms. However, service discovery mechanisms are not specific to wearable robotic applications, and they have to be defined separately. Tools and schemes like TEDS, which help to identify sensors univocally and describe their parameters, are very useful for wearable robotic networks. Services provided by the smart device can also be included in a structure of this kind. This scheme is applicable to every component of the network.

Such a flexible network configuration does not compromise the performance of the network in terms of safety and robustness. However, if a network device is unplugged or fails during the second phase when the robot is working normally, the human–robot physical and cognitive interaction and working conditions will be affected and the consequences may be unacceptable. In order to avoid this, or to minimize the impact, several mechanisms can be implemented at different levels. First, in the physical layer, wired technologies are more robust than wireless ones. Some protocols, e.g.

PROFIBUS, allow multimaster configuration in a network. A redundant scheme like this will make the system more robust but will increase the cost of the solution. If a sensor set fails, application-layer algorithms have to adopt emergency procedures, e.g. blocking or releasing joints of the exoskeleton, and stop system actuators that may make the system unstable. These actions can also be included in smart devices in case of control signal loss. Another good way to improve the robustness of the H–R system is to monitor whether the actuator tasks are executed correctly by means of internal sensors.

As regards the QoS of the network, some protocols allow different task priorities in the network, so that transmission slots are guaranteed for critical devices. CAN uses a message-based scheme to guarantee low latency for the transmission of data with high priority. This scheme can be very useful in wearable robotic scenarios. Again, service and message priorities can be included in the device description table. During the network formation phase, dynamic message priority schemes can also help to assign different priorities to the devices, taking into account the currently selected working profile.

7.4 WIRELESS WEARABLE ROBOTIC NETWORKS

Wiring a robot is not a trivial task. Moreover, in portable devices wires can affect robot function. Depending on the application, numerous profiles can be defined in WRs. For the system to be able to have this multiprofile feature, the network must meet two major requirements. The first, already discussed, concerns the network's Plug&Play capabilities. The second concerns the level of complexity to don&doff the system. Wiring problems and don&doff complexity can be minimized by using small wireless systems.

Wireless technologies are often used to interconnect wearable robotic networks with wider networks. A typical example is the use of a wireless link to monitor remotely the overall performance of the network from a base station. In this scenario the gateway functions are not too critical, since their failure will not affect the overall system performance. Wireless technologies provide mobility, but they cannot compete in robustness with wired technologies.

7.4.1 Enabling technologies

Use of wireless technologies has become widespread over the last decade, in particular for computer networks and MANETS. The reduction of device cost and size, the adoption of mature low-power microelectronic technologies and advances in energy storing have all contributed to make wireless technologies one of the first alternatives considered for many applications. Many important telecommunication and high-tech companies have invested a great deal of effort in developing standardized wireless protocols, and one of the most successful fields has been short range communication networks. The wireless personal area network (WPAN) describes this type of interaction between near devices. The massive proliferation of computer peripherals and mobile phones has encouraged the research community to develop new WPAN solutions. IEEE 802.15 is a working group devoted to developing standard protocols and recommending practices and guidelines for interconnection and coexistence of different WPAN standards.

7.4.1.1 IEEE 802.15.1 and Bluetooth

Bluetooth is a standard for wireless communications developed by Toshiba, Sony, Intel, Nokia, Motorola and others in 1998. The first aim of this protocol was to replace cables for interconnection of devices, and so Bluetooth is oriented towards point-to-point communications. However, it can also be used for small wireless networks. The standard is developed and licensed by the Bluetooth Special Interest Group (SIG).

The standard defines three power class-dependent ranges. The communication ranges are 1, 10 and 100 metres respectively. Bluetooth is based on the IEEE 802.15.1 standard, which defines the lower transport layers including the MAC and physical protocols (Bluetooth Specification, 2001).

Bluetooth uses the ISM band (2.4 GHz). It avoids interference by using a frequency hopping spread spectrum (FHSS) transmission scheme. In order to establish a link, two or more Bluetooth devices define *a priori* a pseudo-random sequence of spaced channels in a wide band. Frequency hopping (FH) is a technique used in low-power and low-cost radios. The Bluetooth protocol defines 16 000 frequency hops per second. It also defines the concept of a piconet, which is a network of up to eight Bluetooth active devices. All of them share the same frequency hopping sequence. In every piconet there is one master and the other devices are slaves. Up to 255 devices can be inactive in a piconet. The master can activate them at any time.

There are two types of links in Bluetooth. The first is the asynchronous connectionless link (ACL), which is designed for data transmission and broadcasting. It supports a maximum bit rate of 721 kbps. The second type is the synchronous connection oriented (SCO) link, which is mainly used for voice traffic.

The protocol stack of Bluetooth is very comprehensive. Currently, it is used as a model reference for new WPAN developments. The Bluetooth SIG defined a set of profiles for usage and implementation of the technology. These profiles may be seen as vertical sections of the layered protocol stack. Partial implementations of the protocol are optimized for specific applications in terms of size and functions. In addition, the compatibility of different devices is assured if they use the same Bluetooth profile. These profiles include the service discovery application profile (SDAP), the generic access profile (GAP), the serial port profile (SPP), the generic object exchange profile (GOEP) and others that are even more device-specific.

Bluetooth nodes have a unique 48-bit address that is used to identify a node over the network. To set up a connection any device can broadcast an inquiry. Any node can respond to this inquiry with an information set including the device name, the device class and the list of services. If a node knows the address of the target node, it can establish the link directly. The protocol also defines pairing schemes and some other mechanisms to prevent unwanted network data access.

Bluetooth is hardly used in WSNs due to the complexity of the protocol and the high power consumption required for applications of this kind. However, networks for WRs can use similar mechanisms, e.g. the Service Discovery.

7.4.1.2 IEEE 802.15.4 and ZigBee

ZigBee is a standard developed by companies in the ZigBee Alliance. The standard was published recently (ZigBee Standards Organization, 2004). The MAC and physical layers of the standard are based on IEEE 802.15.4 (IEEE Computer Society, 2003). Developed as a Bluetooth complementary protocol, it uses the direct sequence spread spectrum (DSSS) instead of the FHSS. The standard also uses the ISM band, where it defines 16 channels, each one with a bandwidth of 5 MHz. The bit rate is up to 250 kbps.

The MAC uses carrier sense multiple access with collision avoidance (CSMA-CA). This allows nodes to sleep when they are not active in order to save energy. The MAC protocol in IEEE 802.15.4 can operate in both beacon-enabled and beaconless modes. In the beaconless mode, the protocol is essentially a simple CSMA-CA protocol. Since most of the unique features of IEEE 802.15.4 are in the beacon-enabled mode, here attention will be focused on that.

In the beacon mode, IEEE 802.15.4 uses a superframe structure. A superframe begins with beacon frames sent periodically by the coordinator at an interval that can range from 15 ms to 245 s. There are both active and inactive portions in the superframe. Devices communicate with their WPAN only during the active period and enter a low-power mode during the inactive period. Parameters establish the length of the beacon interval and the length of the active portion of the superframe. The active portion of each superframe is further divided into 16 equal time slots and comprises three parts: the beacon, a contention access period (CAP) and a collision-free period (CFP), which is only present if guaranteed time slots (GTSs) are allocated by the WPAN coordinator to some of the devices. Each GTS consists of an integer multiple of CFP slots and up to seven GTSs are allowed in a CFP. To save energy, the nodes first ask the access point before starting a communication using the CAP.

The protocol was designed to be implemented in small, cheap microcontrollers with memory constraints. The main advantages as compared to Bluetooth are that the protocol is much simpler and consumes less power. The complete ZigBee solution can be embedded in a small system in package (SiP) component.

7.4.1.3 IEEE 802.15.3a and Certified Wireless USB

A certified wireless USB is intended to support high-speed wireless links. It uses a multi-band orthogonal frequency division multiplexing (MB-OFDM) ultra-wideband (UWB) radio platform developed by the WiMedia Alliance (WUSB, 2007). The physical layer is described by standard IEEE 802.15.3a (IEEE Computer Society, 2007). It was developed to replace the current universal serial bus (USB) connections. This standard defines a physical layer that uses an ultra-wideband (UWB) scheme to transmit data and it defines the physical and MAC layer protocols, as in IEEE 802.15.1 and 802.15.4. The protocol supports bit rates up to 55 Mbps. For a range of 10 metres the maximum speed is up to 1 Mbps, while for 50 metres it is reduced to 62.5 kbps.

7.4.2 Wireless sensor network platforms

Novel low-power and low-cost microelectronics components enable promising wireless sensor network (WSN) solutions for a wide range of applications. A WSN comprises a large number of homogenous low-cost nodes. Typical applications are habitat monitoring, surveillance, intelligent buildings and wellness systems, among others. The WSN research community is growing rapidly. Even though these networks are designed to cover wide areas, platforms and protocols can also be used in short-range applications. Considering that nodes use low-power radios and therefore have short-range communication capabilities, the topology hardly respects hierarchical structures in wide area applications. In addition, the nodes are prone to failure due to environmental conditions and infrequent maintenance of the nodes. Consequently, network topologies for WSN can be complex and protocols for dynamic configurations are the most appropriate.

The scenario in wearable robotic is quite different. Short-range communications take place and nodes work under more controlled conditions, so that more static configurations are suitable. The controller may act as the network master and control access by the different devices to the transmission medium.

The main component of a WSN is the node. A brief description of popular WSN nodes in presented in Table 7.1. In general, all the nodes in the network are similar. However, the information generated by a set of nodes all flows in the same direction, i.e. towards a common central node or access point to a higher hierarchy network. Central nodes are usually more powerful and provide access to the WSN information.

A WSN node typically comprises the following components:

- *Microcontroller.* This runs a code stored in a nonvolatile memory to capture the data from the sensors. It also controls the other network peripherals. A WSN takes advantage of low-power microcontroller architectures to reduce the energy consumption of the node. Most platforms use simple microcontrollers like the TMS430 or Atmel AVR. Both microcontrollers are based on RISC

Table 7.1 WSN nodes

	MicaZ	Telos	TinyNode	NMRC node
Microcontroller	Atmega128	MSP430	MSP430	Atmega128
Radio	CC2420	CC2420	XE1205	nRF2401
Frequency	2.4 GHz	2.4 GHz	869 MHz	2.4 GHz
Maximum TX power	0 dBM	0 dBM	15 dBM	0 dBM
Range[a]	100 m	100 m	600 m	100 m
Data rate	250 kbps	250 kbps	76.8 kbps	1 Mbps

[a] These are approximate ranges for outdoor conditions.

architectures. The clock frequency is usually low, ranging from 3 to 16 MHz approximately. They include several peripherals, e.g. analogue-to-digital converter (ADC) and universal synchronous asynchronous receiver and transmitter (USART) ports, among others. They also support hardware interruptions.

- *Memory.* This is used to store the program code, the acquired data and the communication protocols. Memory is frequently included in the microcontrollers. Flash memory is a good alternative for storing code, while random access memory (RAM) is still the first choice for volatile data. For massive data storage, current memory cards, e.g. secure digital (SD), are a good option.
- *Communication.* The most common way to transmit data is using radio frequency (RF) links. Other choices are optical and ultrasound links. Most popular platforms use RF components, which are usually compliant with specific standards; e.g. the Chipcon 2420 radio is compliant with 802.15.4 but does not implement all the protocol in the chip. Many radio chips include sleep modes in order to save power. When selecting a component, the designer has to consider all the radio parameters that will affect the application, e.g. included services to upper layers, data rates, carrier frequency and channels, range and carrier sense capabilities, etc.
- *Sensors and actuators.* The range of sensors and actuators is extremely wide. For more details on types of sensors and actuators, especially those used in WRs, see Chapter 6.
- *Power supply.* Small batteries are the most common solution. The latest research is oriented towards developing energy scavenging or harvesting technologies. These are not yet mature, but they can at least help to prolong battery life, and in some cases are useful for recharging.

7.4.2.1 Operating systems

Although WSN nodes are quite simple, the research community has developed tiny operating systems (OSs) for them. An operating system allows a modular approach to be used in the development of protocols and hardware drivers that can be reused by other applications. There are two important OSs currently available:

- *TinyOS.* This is an open-source operating system. Today, it is supported and maintained by the TinyOS Alliance, which includes several universities and companies. TinyOS, which is used by more than 500 research groups and companies, proposes a modular approach for programming nodes. It is based on components (e.g. *timer.nc*, a component that implements timers using TinyOS) that use common interfaces to interconnect with other components.

  The OS is event-based. A component supports tasks, and it provides and can execute commands and implement event handlers. Version 2.0 was released recently, with improved timing mechanisms among other new functionalities. TinyOS is supported by many available platforms. In addition, a new compliant platform can be built using libraries for common microcontrollers, e.g. Texas and Atmel, and radios, e.g. Nordic, Chipcon and RFM.

- *Ecos.* This is a real-time OS for embedded applications. It can be executed in 16-, 32- and 64-bit architectures and is intended for more powerful processors, e.g. ARM. The real-time kernel can handle interruptions, exceptions, different schedulers and threads. It can also implement several timers, counters and alarms. The OS can host USB slave devices; it also supports serial and Ethernet communications and includes a TCP/IP stack.

7.4.2.2 MAC protocols

Organized access to a shared transmission medium by numerous mobile nodes using low-power radios and simple microcontrollers is a huge challenge. MAC protocols describe the way nodes access the transmission channel and the control mechanisms. As noted earlier, WSN nodes can be deployed over a wide area. Given the short range of the radio link, not all the nodes can directly reach the central node. The network has to apply multihopping and dynamic link establishment protocols in order to provide access to distant nodes deployed under nonstable conditions.

Basically there are two different approaches for MAC protocols:

- *Deterministic.* These techniques are more suitable for stable networks where the traffic load is highly deterministic. Static frequency channels (frequency division multiple access, or FDMA) or time slots (time devision multiple access, or TDMA) are common features (Rajendran, Obraczka and Garcia-Luna-Aceves, 2006). This approach is suitable for most of the working profiles of wearable robotic networks. They usually adopt a star topology, where each node can connect directly to the central one. The number of connections is limited by the master node and the channel bandwidth. In addition, in homogenous wireless networks, clustering and scalability issues pose a major problem. Deterministic protocols are rarely used given the nondeterministic nature of classic WSNs.
- *Contention-based.* These protocols use mechanisms, e.g. CSMA-CD (CSMA-collision detection), that enable shared use of the transmission channel by many nodes without a 'big brother' controlling the network. They are more suitable for dynamic scenarios such as mesh and multihopping networks. They have to cope with some problems associated with these types of access: collision of packets, overhearing, hidden-neighbour problems, priority management, complex routing protocols and optimized channel use. S-MAC (Ye, Heidemann and Estrin, 2002) and B-MAC (Pollastre, Hill and Culler, 2004) are good examples of this type of protocol for WSNs. Both are implemented using TinyOS architecture.

7.5 CASE STUDY: SMART TEXTILES TO MEASURE COMFORT AND PERFORMANCE

J. Vanhala

Tampere University of Technology, Finland

7.5.1 Introduction

New innovations are accepted by customers only when there is a genuine need – when they solve an existing problem. Good applications for smart clothing can be found in fire and rescue services. Smart clothing techniques can be used to enhance the safety and functionality of firefighting suits.

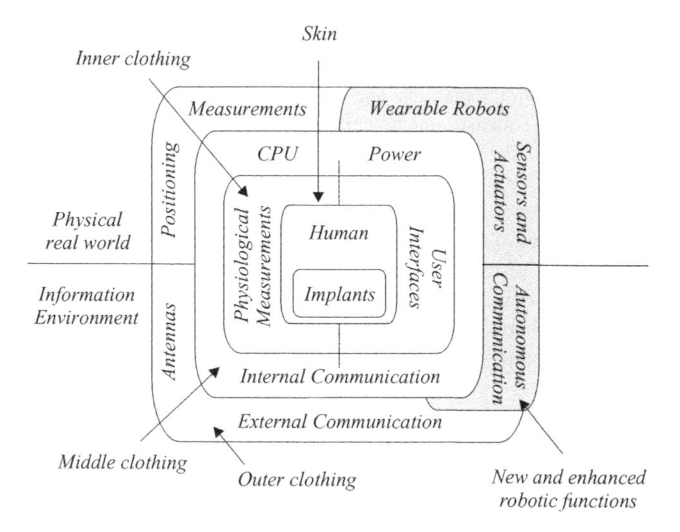

Figure 7.5 Layers of clothing (Rantanen and Hännikäinen, 2005). Reproduced from Rantanen and Hännikäinen, 2005

Clothing forms a protective layer against the environment. Ordinary clothing is developed for ordinary conditions and the most important design criteria are wearing comfort, looks and low cost. Protection is against normal conditions of use: winter clothes protect against cold, outdoor clothes against wind and rain, and indoor clothing mostly only covers the body.

Many work tasks are performed in environments containing potentially hazardous conditions such as extremely high or low temperatures, corrosive and toxic agents or explosives. These can damage unprotected skin, eyes, respiratory organs or other body parts, or cause poisoning or suffocation. Such working conditions are found, for example, in disaster rescue work, in military environments and in some industrial tasks.

In most cases normal clothing consists of several layers with different properties. Underwear and hosiery are close to the body, middle layers provide insulation and outer layers are for appearance and protection. All of these layers will be modified in smart clothing as electronics and sensors are embedded in the textiles. Conceptual layers of clothing are shown in Figure 7.5 (Rantanen and Hännikäinen, 2005). As connected information appliances, smart clothing systems exist in two environments: the real physical world (top) and the information environment (bottom).

Because clothing is continuously worn, it provides an ideal platform to incorporate sensors. Sensors embedded in underwear are in direct contact with the skin and can be used to measure human physiological signals such as body temperature, skin conductivity, perspiration, ECG and EMG.

The middle layer is a good place for siting electronics, which have to be protected from the environment. A central processing unit, batteries, internal communication links and sensors are well protected under the outer layer of clothing without being in direct contact with the skin, which would be uncomfortable. Sensors in the middle layers can register factors connected with wearing comfort, e.g. temperature and humidity of the microclimate inside the garment and pressure induced on the body (see Case Study 6.6). Heating or cooling elements would also be embedded in the middle layer.

The outer layer of clothing protects the system. Sensors on the outer layer can measure environmental factors such as temperature, radiation, chemical substances and noise. User interfaces to electronic functions are on the outer layer, where they are readily accessible. They are also in direct contact with the user while in use. Communication to and from the clothing, together with antennas, are on the outer layer. This causes no attenuation of the signal passing through the textile layers, and the gap between the body surface and the antenna generally makes it easier to design and implement the antenna structure.

The smart clothing concept can be extended further to take in actuators to assist the user in various tasks. Wearable robotic systems will be connected physically to the outer layer of clothing and will utilize the sensors, communication capability and user interfaces in the smart garment; for instance, heart rate and body temperature measurements can provide information on the physical status of the user and indicate whether he or she is capable of performing a given task using the enhanced robotic functions.

There is a direct correspondence between conceptual layers and physical layers of clothing. Layers are separate, disconnected pieces of textile, and that poses the technical problem of how to interconnect electronics on different layers of clothing. This study describes the development of smart clothing for firefighters. The work was done in the CLAN (clothing area network) project studying body area networks, i.e. communication between electronic modules embedded in clothing. Protective clothing for firemen was chosen for a prototype implementation because it can potentially benefit from added intelligence.

The study further describes the user requirements, the design and implementation of the system and the results from user testing. The goal was to develop general architecture for a smart work garment that could be readily adapted to different applications and environments.

The project was conducted jointly by Tampere University of Technology, VTT and Helsinki University of Technology. It was funded by TEKES, the Finnish Work Environment Fund and industrial partners.

7.5.2 Application description

Modern firefighting suits are developed to fulfil the stringent requirements of rescue work. They have to withstand heat, flames and mechanical wear. They must also be waterproof and breathable. Standards set requirements and lay down rules for firefighting suits. They regulate the structural and material properties: resistance to heat, flame and heat radiation; permeation of water, water vapour and other chemicals; mechanical strength; visibility; ergonomic tests (European Union, 2005). For instance, the materials have to withstand 5 minutes at 180°C and the material must not ignite, melt or shrink by more than 5 %. The material must also be able to sustain 10 seconds of direct flame exposure.

Rescue work, and especially fire extinction, is very demanding both physically and mentally. An operation may last for hours in rapidly changing conditions. In Finland a basic unit for all smoke diving and other extinction tasks is composed of one or more diving pairs and an engine operator. The task of the engine operator is to communicate with the diving pair using a VHF or TETRA phone and operate the firefighting equipment at the fire engine end. The operator also takes care to ensure that the diving pair exit the building in time, when either the maximum time for the dive is reached or the situation demands a quick exit. The normal duration of a dive is up to 15 minutes, the limiting factors being the supply of pressurized oxygen and the physical endurance, especially the heat balance, of the firemen.

The most common injuries to firemen are distensions, distortions, scratches and especially minor burns. Most burns occur in the area between the mask and the protective hood and at the shoulders. A common problem is thermal strain. Long-term tolerance of heat exposure will be reduced by severe symptoms of heat strain. These are sunstroke (overheating of the CNS), collapse due to heat strain (circulatory disorder in the brain), heat exhaustion (dehydration and loss of salt) and heat stroke (collapse of temperature control of the system). Although severe conditions in themselves, these symptoms may actually cause fatal injuries if they occur during the operation. Symptoms of heat strain are hard to perceive in oneself while active in an extremely stressful situation.

Heat exposure as such is not a danger, as the protective clothing is able to insulate from the heat for as long as the diving pair are allowed to operate, thus preventing burns. A more important factor is the slow increase of body temperature. The work is physically very demanding and internal heat generation adds to the temperature rise (Rantanen *et al.*, 2001). Low visibility, hazardous chemicals, falling and collapsing structures are additional risks.

A definition of the requirements for the proof of concept implementation of the CLAN architecture was initiated with a brainstorming session at the Emergency Services College, which represents the highest level of expertise in fire service in Finland. Also, five firemen on active duty were interviewed. The brainstorming session produced a shortlist of development targets that could be made feasible by adding intelligence to fire equipment. This shortlist included environmental temperature measurement, measurement of performance, work load and stress, information on the environment (maps, known hazards, etc.), positioning inside a building, communication between the diving pair and engine operator, and oxygen resources remaining.

The list was discussed in the interviews. The ability to measure the temperature of the environment was seen as the topmost target by all interviewees. A sudden increase in temperature and an unnoticed increase in body temperature were felt to be the most significant risks. Other areas where new development was felt necessary were surviving in an unknown environment with heavy equipment, help for locating victims and hot spots, help for taking care of one's own safety and that of the diving pair, extinguishing techniques and operation management.

7.5.3 Platform description

The goal was to define architecture for a system that will enable information to be collected from application-specific sensors embedded in different parts of work clothing (underwear, jacket, trousers, gloves, boots, helmet, etc.) and from devices used by the worker (mobile phone, external sensor devices, camera, etc.). The collected information is processed locally and communicated to an external operator. The central component of the architecture is an internal communication network, which also gives the system its name, CLAN (clothing area network). The architecture is used to create a general modular platform design for a class of applications (Kärki *et al.*, 2005). A proof of the concept implementation was built for a firefighting suit (see Figure 7.6).

CLAN architecture defines the internal communication network and protocols at different levels, connection interfaces to sensor devices, resources for local processing, and communication with external devices and an external operator. The architecture also sets requirements for tolerance of environmental conditions.

Sensors to be connected to the CLAN system will require only a low communication bandwidth. Such sensors measure, for example, temperature, radiation, pressure, ECG, location or gas composition. Voice and video communication will be handled by other means. Sensors will be embedded in the clothing and therefore have to be small, lightweight, durable and flexible. All sensors are smart so that they can hook up to the communication network.

The CLAN is a centralized system in the sense that all information is gathered into a central unit, and except for local signal processing at the smart sensors, all information processing is done in the central unit. The central unit implements the communication networks and protocols, has processing capability for sensor fusion and handles the user interface functions.

The CLAN network uses three different methods for communication. Wired communication is used whenever possible because it is reliable, it is low cost and power can be distributed over the same channel. Local communication between disconnected pieces of clothing, e.g. a jacket and a glove, is handled by inductive communication. An inductive link was chosen because it is simple, has very low power consumption and is compatible with the external sensor devices used. The communication link between a CLAN system and an external system is implemented via a low bandwidth radio link. The reason for the choice of a radio modem instead of using, for instance, the TETRA network is that the CLAN is a standalone system and does not rely on the availability of extra communication devices. In addition to the three communication methods defined, the architecture is open to include

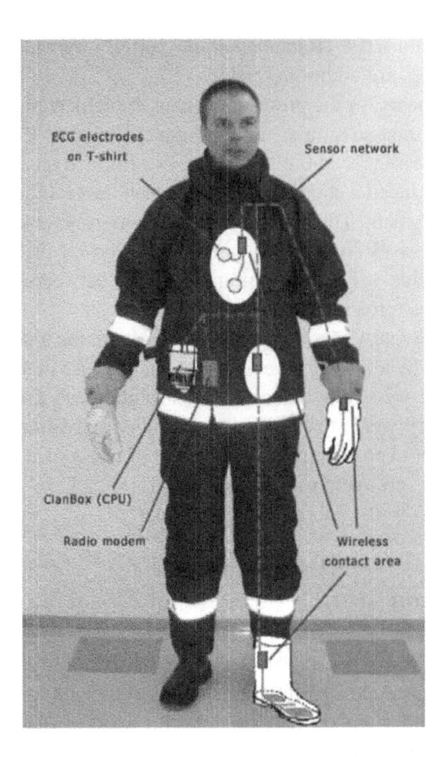

Figure 7.6 The structure and components of the CLAN prototype system

any other communication link as long as it has the same interface functions for both the sensors and the central unit; e.g. Bluetooth or ZigBee could be used.

7.5.4 Implementation of concepts

The CLAN prototype suit was designed on the basis of user requirements, the general architecture and the available technologies. There are five Dallas semiconductor one-wire temperature sensors, two measuring the inside temperature and three the outside temperature. Heartbeat is measured with a standard heart rate monitor chest belt. The pressure in the oxygen tank is transmitted by a commercial scuba diver's pressure transmitter. The wireless network linking up the boots, gloves, trousers, jacket, pressure sensor and heart rate monitor was implemented as an inductive link. The central unit design is based on an Atmel ATMega128 microcontroller and also incorporates a receiver for the inductive signal from the belt and the pressure sensor, a one-wire master and connections for inductive links. External communication is implemented using a radio modem from Satel. Except for the power switch there is no user interface in the suit itself. As soon as the system is powered-up, it starts sending measurements via the radio modem, which are shown in real time.

7.5.5 Results

The suit was tested in four exercises at the Tampere Fire Department training site at Sulkavuori. The system was tested by three firemen all around 30 years old. In a basic exercise the fireman

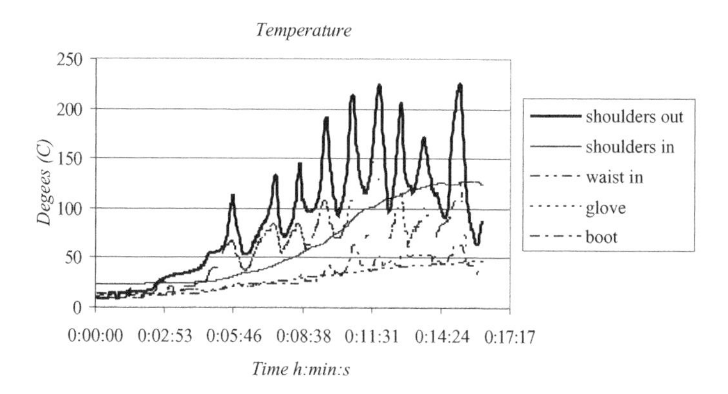

Figure 7.7 The graph shows the measurement results from the smoke diving experiment

is stationary on his knees, as close as possible to the fire, and sprinkles the fire with short pulses of water. The idea is that the water droplets will evaporate and cool down the hot air. During the exercise the measurements were monitored with a laptop computer outside the building using the engine operator's application CLANWare. Figure 7.7 shows the data recorded for later analysis. From the graph the effect of the water pulse extinguishing technique can be observed. The temperature sensor on the shoulder gives readings that correspond to the air temperature at a height of about 1 metre. The temperature decreases rapidly after each water pulse. Once extinction was completed, the firemen cleared the site and returned to the fire station. One hour after the exercise, the firemen were interviewed once again.

7.5.6 Discussion

There are clear advantages in developing a smart firefighting suit. Real-time data on the physiological status of the fireman and measurements from the working environment increase safety and make it possible to manage the operation with less uncertainty. The chosen measurements – heartbeat, oxygen and various temperature readings – proved useful. In fact, implementing a prototype and testing it in a training situation simulating a real operation was in itself also seen as very important. Where previously the feedback from the situation was based only on intermittent voice communication with the engine operator, now the operation managers could have a clear real-time view of the fire extinguishing personnel.

The prototype presented some technical problems, which would be intolerable in a commercial product. Nevertheless, the problems were easily identified and provided valuable information on how to improve the design. The electronics, sensors and interconnections must tolerate heat, shocks and immersion.

The system must not require any special attention during the operation, and regular maintenance at the fire station must be simple. In particular, the system must not prolong response times. The software at the monitoring computer in the lead car must be easy to use so that it actually helps in the management of the operation rather than introducing new tasks to be taken care of. Automatic alarms based on the known condition of individual firemen must be included. Individual differences in fitness and skills can be readily perceived from the data and can be used to personalize training programmes. The system can be utilized in practising even while still not authorized for operational use.

Since personal endurance is one of the limiting factors in rescue and fire extinguishing work, the wearing comfort and usability of the system directly affect the safety and efficiency of the operation.

7.6 CASE STUDY: EXONET

F. Brunetti and J. L. Pons

Bioengineering group, Instituto de automática, CSIC, Madrid, Spain

The Berkeley lower limb extremity exoskeleton (BLEEX) is a WR that uses a network to implement the control architecture. The aim of this exoskeleton is to empower load-carrying human capability. The exoskeleton sensor set comprises sixteen accelerometers, ten encoders, six force sensors, eight foot sensors, one load cell and one inclinometer. The control architecture uses the signals from these sensors to control six hydraulic servo valves. The overall complexity of the wiring persuaded the researchers to develop a networked control architecture. In this way they developed the ExoNET, which provides a physical medium to interconnect all the distributed sensors and actuation systems with the main controller of the exoskeleton.

7.6.1 Application description

The BLEEX consists of two lower limb exoskeletons and one large backpack containing the power supply and controller. This backpack is also used to locate the load to be transported. The hydraulic actuators are located on the hip, knee and ankle joints. There are no sensors located on the user's body, so the exoskeleton structure has been used to carry the network wiring. In view of the number of sensors and actuators and the centralized control architecture, a high-speed protocol based on Cypress HotLink technologies was selected.

7.6.2 Network structure

The ExoNET is based on four ExoRings and one GuiRing. Each ExoRing is a ring-like network. It does not use token mechanisms to access the channel. An actuation packet sent by the controller always has the highest priority; e.g. it grants permission to access the medium. Signals for all actuators are included in this packet. There are no collisions or congestion. This approach guarantees a low latency for control signals. The main controller manages four subnetworks using a special board. This board comprises two different physical interfaces, called SIOMs, connecting to the different remote I/O modules (RIOMs) of the exoskeleton, and an extra interface, called GuiSIOM, connecting to an external computer. Each SIOM has two different channels, A and B. SIOM1 uses channel A to connect to RIOMs 0, 1 and 2, which are located on the right part of the exoskeleton, while channel B is used to connect to RIOMs 3 and 4, which are located on the back. SIOM2 uses a similar scheme to connect to RIOMs 5, 6, 7, 8 and 9, located on the left part of the exoskeleton (see Figure 7.8).

7.6.3 Network components

The ExoNET includes 12 network components:

- *ExoBrain.* This comprises three modules. The first module is the ExoCPU, which implements the control algorithms using the data from the sensors located in the system memory. The results of the algorithms are also located in this memory at each control loop. The second module is the

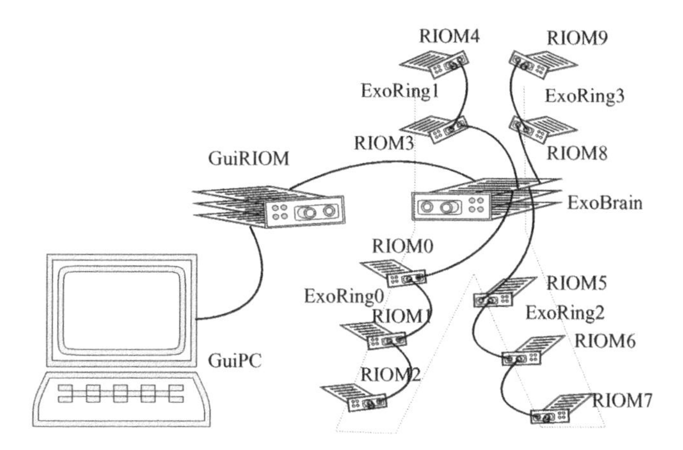

Figure 7.8 ExoNET, the network architecture of the BLEEX project

ExoPCI, which acts as an interface to transport the data from the system memory to the SIOMs and vice versa. The last component is the SIOM board, built with three SIOMs whose task is to transfer the control data to the physical communication medium and return the sensor data to the ExoPCI.

- *RIOM*. There are ten RIOMs distributed strategically about the exoskeleton structure. Each one has three ADCs, one DAC, one quadrature decoder and an 8-bit digital input port.
- *GUI computer*. This is equipped with one network PCI communication block to connect to one SIOM of the ExoBrain. The purpose of the machine is to monitor the overall working of the system.

Each component has a communication block that enables access to the network. The communication blocks were built using a Cypress CY7C924ADX transceiver. This is a point-to-point link that enables high-speed serial data transfer. The ExoNET uses firewire specifications for the physical layer. There is a six-wire cable between each network node. Two wires are used for power transmission and the other four for differential TX/RX lines. The transmission scheme is sequential. The controller transmits data to RIOM0 using SIOM0. RIOM0 receives the data and transmits it to RIOM1. The process is then repeated until the data packet reaches the last RIOM. The TX line of the last node is connected to a loop-back terminator.

7.6.4 Network protocol

The protocol is a simple one based on a message-passing scheme. A control loop is started by the master, which sends a message to RIOM0. This message contains three packets and message control bits. Each packet contains the actuation signals for RIOM0, RIOM1 and RIOM2 respectively. When RIOM0 receives the message, it reads the data related to its actuator and appends its sensor data packets to the message before passing it on to the next RIOM. Each RIOM repeats the same procedure. Thus, when the message is transmitted back, ExoBrain knows if an error has occurred in any of the RIOMs. The sensor packet, marked with the RIOM generator address, acts like a node's proof-of-life. The same scheme is repeated in every ExoRing.

7.7 CASE STUDY: NEUROLAB, A MULTIMODAL NETWORKED EXOSKELETON FOR NEUROMOTOR AND BIOMECHANICAL RESEARCH

A. F. Ruiz and J. L. Pons

Bioengineering Group, Instituto de Automática Industrial, CSIC, Madrid, Spain

NeuroLab refers to an experimental platform designed to enhance studies in human movement and neuromotor control. The platform comprises a robotic exoskeleton and some other standalone devices. All of these components include communication capabilities integrated in the hardware. Thus they can work cooperatively taking advantage of a networked architecture. This multiprofile wearable robotic platform caters for research in the following fields: biomechatronic design and control, BCI algorithms, movement science, research and development of portable electronics and communication networks engineering.

7.7.1 Application description

Human movement and neuromotor control is a very complex research field, due to the complexity of the mechanisms involved and the difficulty of access to the components of the overall system. For this reason, the research community tries to exploit all kinds of valid information (EMG, EEG, kinetics and kinematics) relating to movement planning and execution in order to understand this system.

One common and generally accepted approach to understanding and modelling the human motor system is to monitor and analyse movement-related data during different motor tasks. A common approach to understanding the dynamics of the motor control system is to manipulate the mechanical conditions of each joint independently while acquiring the biomechanical signals and the generated biopotentials while the human motor system adapts to those new applied conditions.

In NeuroLab, there are independent devices or modules that communicate with each other, based on a personal area network (PAN) concept. Each device has a specific function and helps to address the overall goal of the platform.

The robot is an upper limb exoskeleton that allows the mechanical conditions of each limb joint to be manipulated independently (Ruiz *et al.*, 2006). The networked platform enables combined measurement of biomechanical variables (kinematics and kinetics variables) and biopotentials, such as EMG and EEG. Goals of NeuroLab include:

1. Study of human movement in subjects with motor disorders such as pathological tremor or spasticity. The information provided by the platform during execution of specific motor tasks can be used as a tool to diagnose and assess motor disorders.

2. Study of neuro-adaptative strategies for learning and training of specific motor patterns through the application of selected force fields to the upper limb. This application could potentially be of considerable impact in patients suffering from cerebral injuries (Krebs *et al.*, 1998).

3. Validation of neurophysiological models of human motor control in upper and lower limbs. This will help to gain a better understanding of the integration of sensory information and the underlying mechanisms for generation of motor commands (Mistry, Mohajerian and Schaal, 2005).

4. Study of human body behaviour under external loads. The load application is the basis for several technical aids to compensate functional disability.

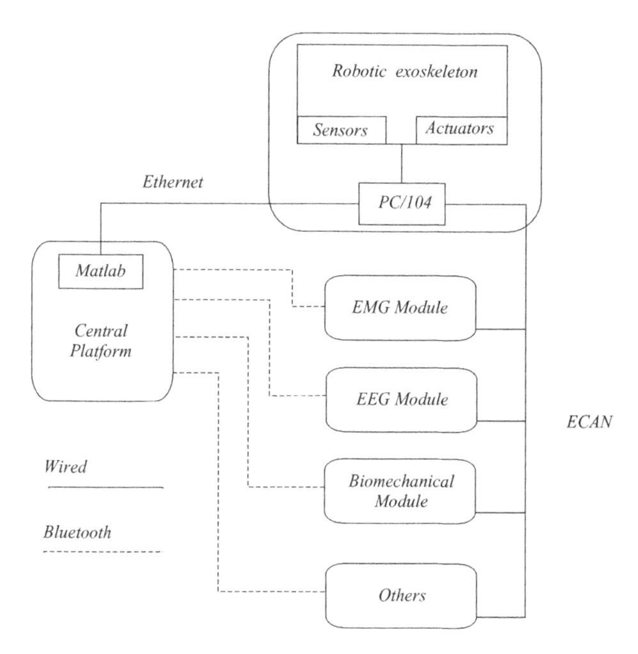

Figure 7.9 Layout of the NeuroLab system. Different modules can connect to a base station using Bluetooth. At the same time, all the modules are part of the wearable robotic network, called BioNET

5. Exploration of new communication channels in human–robot interfaces. This is potentially feasible through the use of EMG and EEG information to control the exoskeleton.
6. Assessment and quantification of upper limb parameters, e.g. mechanical impedance. These parameters are now considered important for understanding of the control mechanisms of the human joints, the generation of control signals and the execution of movements under changing conditions.

7.7.2 Platform description

NeuroLab integrates several devices in a global architecture. The platform, depicted in Figure 7.9, is composed of modules and devices that provide several capabilities: an upper limb robotic exoskeleton, an EMG module, a Motion module and an EEG module. It can further be expanded with other peripherals. A software platform is defined to manage the system, e.g. plan the experiments and acquire data. Safety and reliability were priority considerations in the development.

The powered exoskeleton and the devices can communicate with each other using a CAN-based network and specific protocols. Each element of the platform provides several services that can be requested by other devices. There are therefore different primitives in the upper layers of the protocol, for instance to retrieve the data acquired by a module or to control a joint of the exoskeleton.

7.7.3 Implementation of concepts and algorithms

The upper limb robotic exoskeleton in NeuroLab spans the human elbow and wrist joints (Rocon *et al.*, 2005). The sensors (gyroscopes, potentiometers and force sensors) measure the biomechanics of the arm. Using these data, limb movements, motor tasks and several postures can be assessed under different mechanical conditions.

The exoskeleton is controlled following an impedance control strategy which includes a position feedback loop. The goal of the controller is to modify the apparent HRi impedance.

NeuroLab has a real-time target computer system (xPC Target) to control the exoskeleton. Control is implemented using the MatLab Real-Time suite by MathWorks, Inc. This environment provides mathematical libraries making it easy to implement control strategies. The algorithm can be coded in C-language and compiled in an executable application.

7.7.3.1 EMG module

Measurements supplied by electromyography provide valuable information regarding physiology and muscle activation patterns. This information describes the forces that will be generated by the muscles and the timing patterns of the motor commands. It can also be used to assess the response of the human motor system to external dynamic conditions or perturbations.

The EMG module allows for acquisition of data on four muscle groups. Since the EMG signal is very small (50 μV–5 mV), it may be affected by interference from other biological and environmental noise sources, e.g. movement artefacts, electrical noise and muscle noise, among others. In order to minimize the effects of noise, the EMG module amplifies and filters the raw EMG signals before they are digitalized.

Additionally, a battery is used to power the EMG acquisition module in order to reduce 50 Hz harmonics (powerline noise). In the light of international safety regulations regarding electronic devices connected to human beings, several topics were addressed in connection with electrical isolation of the EMG module. In particular, galvanic isolation using a wide-band, unity-gain isolation amplifier was implemented in the EMG module. For more details on similar solutions, see Chapter 6.

7.7.3.2 Biomechanical monitoring module

This module uses inertial sensors to acquire kinematic and kinetic information on the wearable robotic system. Thanks to the modular approach of the NeuroLab project, different devices can be used in many different applications.

The biomechanical monitoring module comprises the following logical components:

- *The controller.* This uses a TMS320F2812 DSP from Texas Instruments, which is powerful enough to run all the signal processing algorithms. The clock frequency is up to 150 MHz.
- *The sensor set.* Two inertial sensors (see Section 6.2) can be connected to the controller using an SPI interface.
- *The data logging block.* An ATMega32 microcontroller is used to manage an SD card. The microcontroller implements a FAT16 file system.
- *Communication block.* The communication block includes four different communication interfaces for networks. The first is the SPI, which is embedded in the DSP and is used to communicate with the sensors and the data logger. The second block comprises a Bluetooth module for wireless communication with a base station for real-time monitoring. The third interface is a CAN port provided by the DSP. It can be attached to the NeuroLab BioNET using simple CAN drivers. The last interface is a USB port for data transfer and real-time monitoring.
- *Power supply.* This is based on an ion–lithium battery with a capacity of 900 mA h. The module uses the USB connection to charge this battery.

7.7.3.3 EEG monitoring module

As presented in Chapter 4, EEG can be used to study movement planning and to control WRs. The development of a portable EEG module for research purposes is not a trivial task. Noisy environments and movement artefacts affect the quality of the EEG signals. Moreover, processing techniques are usually complex and require a powerful platform to execute these algorithms.

Early developments have been based on a PC/104 computer platform. A special acquisition board has been designed; this board and a CAN network board are attached to the computer platform.

7.7.3.4 BioNET

The purpose of NeuroLab is to integrate several different devices in order to study HR interaction (both cognitive and physical) and neuromotor systems using noninvasive techniques. In view of the wide range of profiles and applications of the system, a distributed modular approach was selected to implement the proposed concepts.

A network of smart devices was identified as the optimum solution to achieve the goal. The network is called BioNET and is CAN-based. Several network protocols, e.g. service discovery, synchronization and priority management mechanisms, are included. A table describing the device, its services and its parameters, is stored in the device itself. This concept is similar to TEDS, used in IEEE P1451.3.

Researchers are currently working to develop monitoring and rehabilitation profiles for the network.

7.8 CASE STUDY: COMMUNICATION TECHNOLOGIES FOR THE INTEGRATION OF ROBOTIC SYSTEMS AND SENSOR NETWORKS AT HOME: HELPING ELDERLY PEOPLE

J. V. Martí, R. Marín, J. Fernández, M. Nuñez, O. Rajadell, L. Nomdedeu, J. Sales, P. Agustí, A. Fabregat and A. P. del Pobil

Computer Engineering and Science, University Jaume I (UJI), Castellón, Spain

The Jaume I Robotics Intelligence Laboratory is very concerned with the integration of sensor networks and networked robotic platforms in order to achieve an advanced ambience intelligence system that is able not only to monitor and extract data from reality but also manipulate the environment in order to give assistance to people in different application domains. One very recent project in that field, which is still under development, is Jaume, the UJI Librarian Robot. The challenge becomes even more interesting when an attempt is made to place a robot in a domestic environment. In this case the design of a multipurpose networked mobile manipulator is fundamental in order to give assistance to handicapped or elderly people at home (Prats *et al.*, 2007). In this case study a survey of the communication technologies is provided that can be used to integrate a networked assistance robot with sensor networks (i.e. including cameras) and can go to make up an entire ambience intelligence system for helping elderly people at home. Technologies like Pico Radio, Zigbee, 1-Wire and EIB are briefly analysed. Some IP communication protocols are also studied from the standpoint of the congestion protocol.

7.8.1 Introduction

Technological advances in the electronics industry mean than new processors, microcontrollers, sensors and actuators can be offered with improved size, performance and power consumption. As a result, these new devices can be embedded in other types of devices, endowing them with a degree of intelligence. Electronic calendars, mobile telephones, intelligent tags, electrical appliances, networked cameras, wearable devices and even robotic manipulators are some of the components that will make up future networks of connectable intelligent devices. Large numbers of such devices can be connected in ad hoc networks to make up new types of applications, thus marking the birth of a new discipline: ubiquitous computation.

By adding intelligent user interfaces to ubiquitous computation, an ambient intelligence (or simply AmI) entity can be produced. Ambient intelligence is a new multidisciplinary paradigm born out of the ideas of Norman (1998) and ubiquitous computing. The basic idea is to have a multilayer distributed architecture that allows ubiquitous communication and supports advanced person–machine communication protocols.

From the standpoint of user interfacing, the main approach adopted to make the smart environment interact with living people is to try and understand what they need without their explicit intervention. By learning normal personal behaviour patterns and detecting deviations from them, abnormalities can be detected and the requisite assistance provided.

In 2001, the European Commission launched the AmI challenge, proposing ideas for the Sixth EU Framework Programme (2002–2006). Some of the ambient intelligence projects specifically address the problems of elderly people; they are divided between those which apply the benefits of techniques to improving everyday life and those focused on medical implants. Although Senior Citizens are not targeted exclusively, they are clearly favoured. None of the projects studied concerns the design of a multipurpose manipulator for helping elderly people at home. These projects include the following: IPCA, HEALTHY AIMS, ASK-IT, NETCARITY, CAALYX, PERSONA, SHARE-it, AHRI and SHARP.

The present case study gives an overview of the communication technologies that will make it possible to integrate a multipurpose manipulator in the home and a set of networked sensors like cameras, presence sensors, etc. Systems like Pico Radio, Zigbee, 1-Wire and EIB are analysed. Moreover, the communication systems listed above are specifically designed for ready integration of sensor networks that do not consume very much energy, and they do not provide very much bandwidth. The case of the manipulator and the networked cameras is different in that in some situations (e.g. an external operator that needs to interact with the system) the available bandwidth would need to be very high. For these situations the problem is analysed in terms of software (i.e. the IP protocols), considering that the manipulator is connected to the network using at least a WiFi 54 Mpbs connection.

7.8.2 Communication Systems

7.8.2.1 Pico Radio

The main target of the Pico Radio project at the Berkeley Wireless Research Center, University of California, has been to make ubiquitous wireless sensor networks possible. While increasing network intelligence and performance it is necessary to reduce sensor size, cost and power consumption. The Pico Radio group (Rabaey *et al.*, 2000) specifies the acceptable values for size (1 cubic centimetre), weight (less than 100 grams), cost (less than 1 dollar) and power consumption (substantially less than 100 microwatts) to ensure that these devices perform as expected.

Reducing the effective range to a couple of metres at most will dramatically cut down power consumption, while a self-configuring scalable network infrastructure will allow distant nodes to communicate with one another. In addition, by giving high priority to energy-saving directives in all layers of the design process, it should be possible to achieve an energy reduction of many orders of magnitude. With such consumption levels it is possible to build self-powered nodes that extract energy from the environment (energy scavenging or harvesting). Some of the sources consulted use tiny variable capacitors to extract energy from vibrations, noise or even human movement.

The Pico Radio philosophy envisages extremely low data rates (even less than 1 hertz), nodes that are not continuously awake (duty cycle about 1%), need-to-know sensor location in order to achieve network efficiency and addressing techniques that include node locations instead of only their identification codes.

Transmission from node to node may require packet forwarding by intermediate nodes. This is called multihop networking. A further study of data radio transmission leads to the conclusion that transmission using intermediate near-to-each-other nodes is more energy efficient than between end-to-end distant nodes. In this way a densely populated Pico Radio network can achieve cost efficiency.

7.8.2.2 ZigBee

ZigBee has been developed to manage wireless sensor networks in three main scenarios: residential home control, commercial building control and industrial plant management. The technology is analysed in detail in Section 7.4. In this kind of system, nodes communicate frequently, transmitting small amounts of information. It is not necessary to have high bandwidth, but it is desirable to achieve low power dissipation so as to prolong the battery life of self-powered nodes. Network protocols should consider the possibility of adding, removing or repositioning nodes and should automatically update related node location and readiness information (Callaway *et al.*, 2002).

ZigBee is based on Standard IEEE 802.15.4 (IEEE Computer Society, 2003), using both 2.4 GHz and 868 MHz/900 MHz bands featuring low data rates, very low power consumption (multimonth to multiyear battery life) and very low complexity. Standard IEEE 802.15.4 deals with the physical (PHY) and media access controller (MAC) layers while the ZigBee specifications cover network, security and application framework layers, offering a platform to end-user applications. The network uses 64-bit IEEE and 16-bit short addressing, allowing up to 65 000 nodes per network. The maximum packet size is 128 bytes with a variable data field of up to 104 bytes.

7.8.2.3 EIB

The EIB (European installation bus) is a domotic communications standard, whose main purpose is to link every device in a typical electric installation (counters, HVAC, alarms, security systems, appliances), bringing together a large number of manufacturers, installers and users. The EIB may use many different physical layers to achieve device communication, the most widely implemented being the twisted pair version (i.e. EIB.TP at 9600 bps). EIB is now converging with Batibus and EHS to achieve a single European home and office automation standard.

7.8.2.4 1-Wire

By using a single wire (plus ground), Dallas Semiconductor/Maxim provides both communication and power supply to devices attached to the net. This technology, called 1-Wire, allows a master (PC or microcontroller) to communicate and control multiple slaves through a single twisted pair up to 300 metres long.

Each 1-Wire device has a global unique 64-bit identifier containing information about the family to which it belongs. In addition, the 1-Wire protocol allows devices to be connected or disconnected at any time. These two characteristics make 1-Wire devices suitable for access control.

1-Wire devices may be presented in two different packages:

- Single chip. This is a standard semiconductor package (SOIC, TSOC, TO92, etc.) depending on the number of pins and pin density. This package allows 1-Wire devices to be installed in electronic circuit boards.
- iButton. This is a button cell-like package with a data contact (called a lid) and a ground contact (called a base) matching the two 1-Wire contacts. This package can be mounted on personal items, like watches, rings, keyrings, etc., and can be used for applications such as control of access to buildings and computers, asset management and various data logging tasks.

There are several families of 1-Wire devices such as RAM, EEPROM and ROM memories, ADCs and DACs, digital switches, thermometers, data loggers, protocol converters (RS232, USB, etc.) and id-only, which can be used to implement a wide variety of automation applications.

On the Dallas/Maxim 1-Wire overview webpage it can be seen that there are many possible applications, but all are restricted to fixed nodes, where some may be iButton readers, allowing identification, access control, etc., from mobile devices but not from wireless devices. The main application focus is on centralized data acquisition and access control in fixed environments with very low bandwidth and easily expanded wiring structures.

7.8.3 IP-based protocols

Having looked at the communication technologies available for integration of domestic sensor networks with a networked manipulator, it can be seen that this can be readily achieved by using an IP (i.e. Internet protocol). Each node (i.e. robots and sensors) would have a unique IP identifier in the network, allowing the use of already available protocols like HTTP, FTP, Telnet, etc.

However, the protocols already available for the Internet are not specifically designed for real-time and remote control. The basic transport protocols available on the Internet for implementing remote control applications are the UDP (user Datagram protocol) and the TCP (Transmission Control protocol) (Park and Khatib, 2005). The UDP is a protocol that does not maintain a connection with the Server side; moreover, it does not provide retransmission of lost packets, it does not control network congestion, nor does it provide confirmation of the packets that have reached the destination. The advantage of the UDP for remote control of devices via the Internet is that given good network conditions, communication is accomplished without significant delays and without major fluctuations (i.e. delay jitter). However, the UDP does not ensure that the packets have reached the destination in the same order as they were sent; in fact, the UDP does not even report whether packets have even been received or not. In addition, the UDP does not run a congestion control mechanism, which means that the sending rate is not adjusted according to the actual bandwidth available. This means that another protocol is needed for remote device control via the Internet.

On the other hand, the TCP is a highly sophisticated protocol that establishes a virtual connection between the sender and the receiver. Moreover, as the TCP handles confirmation of properly received packets, it is possible to sure that the communication will be reliable. However, when the TCP was designed, the designers had in mind reliable communication for applications like e-mails and files (ftp), and not control of devices like robots. The congestion control and connection establishment mechanisms entail considerable delay jitter (fluctuation), which will not do for applications such as Internet teleoperation of a robot manipulator using a haptic device. Figure 7.10 shows the results when a robot is controlled using the TCP and UDP.

Most current networked applications using the Internet use the TCP or UDP. For this purpose, the variable time-delay and bandwidth effect is resolved at the application level by using intelligent sensors, predictive displays and high-level commands. In teleoperation, on the other hand, there is a need to find applications that are closer to real time (Park and Khatib, 2005).

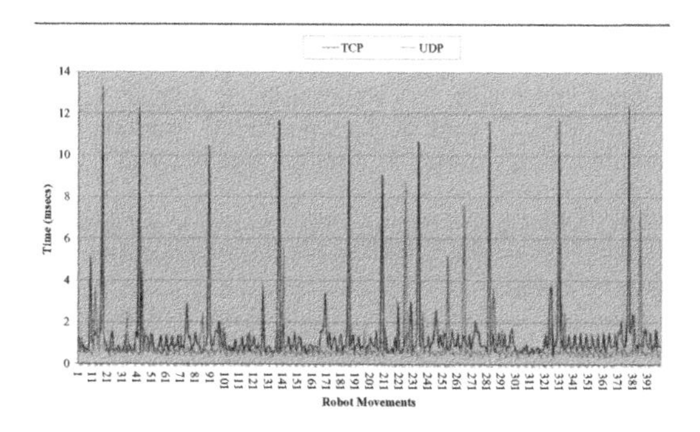

Figure 7.10 Delay response when controlling an industrial Motoman robot via the Internet using the UDP and TCP

As this is an emerging field of research, the scientific literature contains few articles describing specific protocols for controlling networked devices (e.g. robots) via the Internet. On the other hand, many protocols can be found for designing networked applications that require the transmission of multimedia content via the Internet: (1) TFRC (TCP-Friendly Rate Control protocol), RAP (Rate-Based Adaptation protocol), LDA (Loss-Delay Adjustment protocol), SIMD (Square-Increase/Multiplicative-Decrease protocol) and RTP (Real-Time protocol) (Jin *et al.*, 2001; Rejaie, Handley and Estrin, 1999). These protocols are not very suitable for networked systems, as they use an intermediate buffer to compensate for the delay jitter when receiving video and audio. In network robotics, using buffers produces a greater overall delay that strongly affects the immediate control of robots. The few protocols for teleoperation include the following: Trinomial method, Real-Time Network protocol (RTNP) and Interactive Real-Time protocol (IRTP) (Liu *et al.*, 2005; Uchimura and Yakoh, 2004).

Acknowledgements

This work has been partially funded by the Spanish Ministry of Education and Science (MEC) under Grants DPI2005-08203-C02-01, DPI2004-01920 and TSI2004-05165-C02-01; by the Fundació Caixa Castelló under Grants P1-1B2003-15 and P1-1A2003-10; and by the EU-VI Framework Programme under Grant IST-045269 GUARDIANS of the EC Cognitive Systems Initiative.

REFERENCES

Bluetooth Specification, 2001, Volume 1, *Specification of the Bluetooth System*, Core Version 1.1, Bluetooth SIG.
Callaway, E., Gorday, P., Hester, L., Gutierrez, J.A., Neave, M.H., Bahl, V., 2002, 'A developing standard for low-rate wireless personal area networks', *IEEE Communication Magazine*, **40**: 8.
CAN-in-Automation, 1996, CANopen device profile for I/O modules, CiA DSP-401, Version 1.4, 1996.
Corrigan, S., 2002, *Introduction to the Controller Area Network (CAN)*, Texas Instruments.
European Union, 2005, Preliminary Standards prEN 469, prEN 659, prEN 190, prEN 191.
Freescale Semiconductor Manual, 2001, SPI Block Guide V03.06, *Freescale Semiconductor*.
IEEE Computer Society, 2003 IEEE Std 802.15.4-2003, Part 15.4: *Wireless Medium Access Control (MAC) and Physical Layer (PHY) Specifications for Low-Rate Wireless Personal Area Networks (LR-WPANs)*, IEEE.
IEEE Computer Society, 2007, IEEE Std 802.15.3a website, http://www.ieee802.org/15/pub/TG3a, last visit: 3 May 2001.
I2C-Bus Specification, 2000, Version 2.1, *Philips Semiconductor*.

Jin, S., Guo, L., Matta, I., Bestavros, A., 2001, 'TCP-friendly SIMD congestion control and its convergence behaviour', in *9th IEEE International Conference on Network Protocols*, Riverside, California, pp. 156–164.

Kärki, S., Honkala, M., Cluitmans, L., Cömert, A., Hännikäinen, J., Mattila, J., Peltonen, P., Piipponen, K., Tieranta, T., Vätäänen, A., Hyttinen, J., Lekkala, J., Sepponen, R., Vanhala, J., Mattila, H., 2005, Smart clothing for firefighters, in *Ambience 05, Intelligent Ambience and Well-Being* (ed. P. Talvenmaa), Tampere, 19–20 September 2005.

Karl, H., Willig, A., 2005, *Protocols and Architectures for Wireless Sensor Network*, John Wiley & Sons, Inc.

Krebs, H.I., Hogan, N., Aisen, M.L., Volpe, B.T., 1998, 'Robot-aided neurorehabilitation', *IEEE Transactions on Rehabilitation Engineering* **6**(1): 75–87.

Lee, K., 2000, 'IEEE 1451: a standard in support of smart transducer networking', in *Proceedings of the 17th IEEE Conference on Instrumentation and Measurement Technology*, pp. 525–528.

Liu, P.X., Meng, M.Q.H., Liu, P.R., Yang, S.X., 2005, 'An end-to-end transmission architecture for the remote control of robots over IP networks', *IEEE Transactions on Mechatronics*, **10**.

Middaugh, K.M., 1993, 'A comparison of industrial communications networks', *IEEE Transactions on Industry Applications* **29**: 846–853.

Mistry, M., Mohajerian, P., Schaal, S., 2005, 'Arm movement experiments with joint space force fields using an exoskeleton robot', in *Proceedings of the International Conference on Robotics and Automation*, pp. 2350–2355.

Norman, D.A., 1998, *The Invisible Computer*, The MIT Press, Cambridge, Massachusetts.

Park, J., Khatib, O., 2005, 'Robust haptic teleoperation of a mobile manipulation platform', in *Experimental Robotics IX, Star*, Springer Tracts in Advanced Robotics, Springer.

Pollastre, J., Hill, J., Culler, D., 2004, 'Versatile low power media access for wireless sensor networks', in *Proceedings of the 2nd International Conference on Embedded Networked Sensor Systems*, ACM Press.

Prats, M., Sanz, P.J., del Pobil, A.P., Martinez, E., Marín, R., 2007, 'Towards multipurpose autonomous manipulation with the UJI service robot', *Robotica International Journal* **25**(2): 245–256.

PROFIBUS System Description, 2002, Version October 2002, *PROFIBUS Nutzerorganisation e.V. PNO.*

Rabaey, J.M., Ammer, M.J., da Silva Jr, J.L., Patel, D., Roundy, S., 2000, 'PicoRadio Supports ad hoc ultra-low power wireless networking', *Computer* **33**(7): 42–48.

Rajendran, V., Obraczka, K., Garcia-Luna-Aceves, J.J., 2006, 'Energy-efficient, collision-free medium access control for wireless sensor networks', *Wireless Networks* **12**: 63–78.

Rantanen, J., Hännikäinen, M., 2005, 'Data transfer for smart clothing: requirements and potential solutions', in *Wearables and Photonics*, (ed. X. Tao) Woodhead Publishing, England, pp. 198–222.

Rantanen, J., Vuorela, T., Kukkonen, K., Ryynänen, O., Siili, A., Vanhala, J., 2001, 'Improving human thermal comfort with smart clothing, in *IEEE Transactions on Systems, Man, and Cybernetics Conference*, pp. 795–800.

Rejaie, R., Handley, M., Estrin, D., 1999, 'RAP: an end-to-end rate-based congestion control mechanism for realtime streams in the Internet', in *IEEE Infocom*, pp. 1337–1345.

Rocon, E., Ruiz, A.F., Pons, J.L., Belda-Lois, J.M., Sánchez-Lacuesta, J.J., 2005, 'Rehabilitation robotics: a wearable exo-skeleton for tremor assessment and suppression', in *Proceedings of the International Conference on Robotics and Automation 2005*, pp. 241–246.

Ruiz, A.F., Forner-Cordero, A., Rocon, E., Pons, J.L., 2006, 'Exoskeletons for rehabilitation and motor control', in *Proceedings of the 2006 IEEE International Conference on Biomedical Robotics and Biomechatronics* (BIOROB06).

TOKENBUS, 1985, *IEEE Standard for Local Area Networks: Token-passing BUS, Access Method and Physical Layer Specification*, IEEE, Inc.

TOKENBUS, 1990, ISO/IEC 8802-4, *Information Processing Systems – Local Area Networks – Part 4: Token-Passing Bus Access Method and Physical Layer Specification*, IEEE, Inc.

Uchimura, Y., Yakoh, T., 2004, 'Bilateral robot system on the real-time network structure', *IEEE Transactions on Industrial Electronics*, **51**.

WUSB, 2007, Universal Serial Bus website, http://www.usb.org/developers/wusb/, last visit: May 2007.

Ye, W., Heidemann, J., Estrin, D., 2002, 'An energy-efficient MAC protocol for wireless sensor networks', in *Proceedings of the 21st Annual IEEE Conference on Computer Communications, INFOCOM 2002.*

ZigBee Standards Organization, 2004, *ZigBee Specification ZigBee Alliance.*

8

Wearable upper limb robots

E. Rocon, A. F. Ruiz and J. L. Pons

Bioengineering Group, Instituto de Automática Industrial, CSIC, Madrid, Spain

This chapter presents a selection of case studies to illustrate the applications, designs and technologies of some of the most relevant and outstanding upper limb exoskeletons identified in the literature. The main function of the arm is to position the hand for functional activities. The hand must be able to reach any point in space, especially any point on the human body, in such a way that the person can manipulate, draw on and move objects towards or away from the body. Therefore, the kinematic chain consisting in the shoulder, elbow and wrist articulations together with the upper arm, forearm and hand segments has a high degree of mobility; see Section 3.3 for a detailed analysis of upper limb kinematics. The upper limb is one of the most anatomically and physiologically complex parts of the body.

The upper limb is very important because it is able to execute cognition-driven, expression-driven and manipulation activities. Furthermore, it intervenes in the exploration of the environment and in all reflex motor acts. For this reason, any alteration or pathology that affects upper limb motion range, muscle power, sensitivity or skin integrity will disturb its operation.

The primary applications of upper limb exoskeletons were originally teleoperation and power amplification. Exoskeletons have since come to be considered for rehabilitation and assistance of disabled or elderly people, e.g. upper and lower limb orthoses. Lastly, because robotic exoskeletons are able to apply independent dynamic forces to human joints and segments, such devices are providing the basis for technologies in experimentation and study of motor control, motor adaptation and learning, and for neuromotor research in general.

Each of the possible fields of application of WRs generates specific requirements. This chapter presents case studies to illustrate the different topics discussed throughout the book from the particular perspective of portable exoskeletons interacting with human upper limbs. The discussion is approached from the point of view of implementation, addressing aspects such as structural design, the technologies involved and examples of control strategies.

Wearable Robots: Biomechatronic Exoskeletons Edited by José L. Pons

8.1 CASE STUDY: THE WEARABLE ORTHOSIS FOR TREMOR ASSESSMENT AND SUPPRESSION (WOTAS)

E. Rocon and J. L. Pons

Bioengineering Group, Instituto de Automática Industrial, CSIC, Madrid, Spain

8.1.1 Introduction

Tremor is a movement disorder that has a considerable impact on the quality of life of people who suffer from it (Rocon *et al.*, 2004). It can affect the head, face, jaw, voice or upper and lower extremities. Tremor affecting the upper limbs is of particular interest, since it can be very disabling as regards leading an independent life. It is a symptom associated with some abnormal neurological conditions or cerebral lesions and degenerative diseases, including Parkinson's disease, essential tremor, orthostatic tremor, cerebellar diseases, ethylic intoxication and others. As well as medication, rehabilitation programmes and surgical interventions, it has been shown that the application of biomechanical loading to tremor movement can suppress the effects of tremor on the human body (Rocon *et al.*, 2004).

Starting from this principle, the wearable orthesis for tremor assessment and suppression (WOTAS) device was presented, within the framework of the DRIFTS project, as a promising solution for patients who cannot use medication to suppress the tremor (Manto *et al.*, 2003). This case study describes the general concept of WOTAS, highlighting its special features, the design and the selection of system components. The results of the exoskeleton clinical trials are presented at the end.

8.1.2 Wearable orthosis for tremor assessment and suppression (WOTAS)

WOTAS was developed to provide a means of testing nongrounded tremor reduction strategies. It follows the kinematic structure of the human upper limb and spans the elbow and wrist joints (see Figure 8.1). It exhibits three degrees of freedom corresponding to elbow flexion–extension, forearm pronation–supination and wrist flexion–extension. WOTAS constrains the adduction–abduction movement of the wrist; see Section 3.3 for a description of anatomical movements.

The mechanical design of the elbow and wrist joints is similar to other orthotic solutions and is based upon hinges, as they model the anatomical elbow and wrist joints reasonably well. The axis of rotation for the elbow joint is located on the line between the two epicondyles. The axis of rotation for the wrist joint is located on the line between the capitate and lunate bones of the carpus. The mechanical design for the pronation–supination movement is more complex and is explained in Case Study 3.6.

There are no passive orthoses capable of achieving tremor suppression, because tremor is intrinsically dynamic. Passive orthoses used as tremor suppression mechanisms tend to lose their alignment instead of suppressing tremor. To determine the points of the upper limb where the dynamic forces should be applied, i.e. the points where the arm supports should be placed for the physical interface between the actuators and the arm, a number of biomechanical and physiological factors have to be considered. These are described in Case Study 5.6. The actual design of the supports is described in Section 5.3.

8.1.2.1 Sensors

The system aims to allow both monitoring of tremor data and implementation of tremor suppression strategies. It is therefore equipped with kinematic (angular velocity) and kinetic (interaction force between the limb and orthosis) sensors.

Tremor force, position, velocity and acceleration are the data required to implement the two control strategies. MEMS gyroscopes were selected to measure kinematics information (see Section 6.2). Since gyroscopes provide absolute angular velocity on their active axis, two independent gyroscopes are required, placed distal and proximal to the joint of interest. Some electronic circuits have been developed to integrate the sensor in the WOTAS architecture: a bandpass filter with a low cutoff frequency of 0.3 Hz in the sensor output and a higher cutoff frequency of 25 Hz.

Strain gauges were selected as sensors for extracting the kinetic characteristics from the tremor movement (see Section 6.2). The gauges are responsible for measuring the torque applied by the motors to the WOTAS structure and are therefore mounted on the structure. This means that they only measure the force perpendicular to the motor axis, and so their measurement is not affected by forces acting in undesired directions. The strain gauges are connected to a Wheatstone bridge circuit in a combination of four active gauges (full bridge).

8.1.2.2 Actuators

Before selection of specific actuators to suppress tremor, the torque and power requirements were estimated. This was done by analysing kinematic tremor data as presented in Case Study 3.7. An aluminium alloy was selected for the exoskeleton structure so that it would be lightweight but have sufficient rigidity to support the strains.

A number of actuators were selected as possible candidate technologies for tremor suppression, bearing in mind the application requirements. The evaluation determined that ultrasonic and DC motors were the best solutions for activating the exoskeleton. A number of experiments were done to evaluate the performance of the ultrasonic motors. They concluded that although ultrasonic motors offer a number of advantages for the field of rehabilitation robotics, e.g. small size and silent operation, they are unsuitable due to their poor response at low speed; i.e. they cannot be used to track slow movements like human motions (Rocon, Ruiz and Pons, 2004).

In view of the problems encountered with ultrasonic motors, a new exoskeleton was constructed using DC motors for actuation. The DC motor selected to activate WOTAS articulations was a Maxon EC45 flat brushless DC motor. A harmonic transmission drive was used to match the speed and torque of the DC motor to the application requirements. The particular drive selected for this application was the HDF-014-100-2A. This configuration can deliver a maximum torque of 8 N m; however, the maximum torque was limited electronically to 3 N m to ensure user safety. Figure 8.1 illustrates the final configuration of a WOTAS system activated by DC motors.

The total weight of the final system is roughly 850 g. A testing protocol was followed to assess its usability and workspace it affords to a normal user. The system was used in the laboratory. These preliminary tests successfully demonstrated that the system worked correctly and was able to access the human workspace without affecting the wearer's normal range of motion (Rocon *et al.*, 2007).

8.1.2.3 Control architecture

The WOTAS control architecture basically consists of three components: (1) *the orthosis*, with its structure, sensors and actuators; (2) *a control* unit, responsible for executing the algorithms in real time to suppress the tremor (including an acquisition card for interfacing the sensors and actuators with the controller); and (3) *a remote computer*, which in the present case executes an application

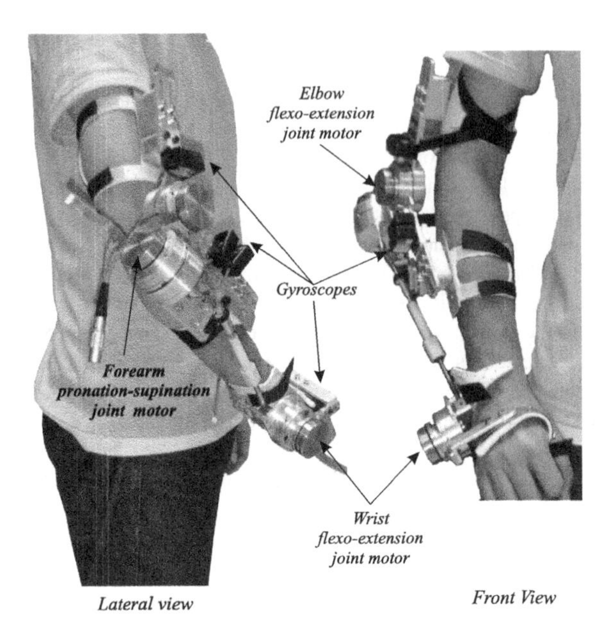

Figure 8.1 Final version of WOTAS for control of three human upper limb movements: flexion–extension of the elbow, flexion–extension of the wrist and pronation–supination of the forearm

developed to establish an interface between the entire system and the person in charge of the therapy, be it a doctor, therapist or orthotist. WOTAS operates in three different modes:

1. *Monitoring*. In this mode, WOTAS measures and characterizes the tremor, both qualitatively and quantitatively. It therefore presents no opposition to the patient's voluntary or tremor movement. To implement this mode, the lower loop of the control strategy presented in Case Study 5.7 was programmed in the control unit.
2. *Passive intervention*. In this mode, WOTAS is a system that can mechanically damp out the tremor movements. This is done by simulating the application of viscosity or inertia to the upper limb to dissipate vibrations caused by the tremor and enhance the user's voluntary movement. The control loop implemented in this mode is presented in Case Study 5.7.
3. *Active intervention*. In this mode, WOTAS is able to apply forces opposed to the tremorous movement on the basis of a real-time estimation of the involuntary component of motion. This produces active compensation and effective suppression of tremor. The control loop implemented in this mode is presented in Section 5.4.

The control of the entire active orthosis was implemented in the MATLAB real-time environment, which provides mathematically complex control strategies in real time. The interface between the MATLAB environment and the active orthosis is based on a standard data acquisition board.

A computer tool was developed to manage the system. The application is installed in an external computer and provides communication with the device, data storage, signal acquisition, information analysis and display, and report generation. One of its basic functions is to evaluate the control algorithms, and on that basis it controls their execution, monitors algorithm variables and adjusts or tunes algorithm parameters in real time. Moreover, it implements the specified measurement protocol and the clinical trials, and displays quantitative information on the performance of the algorithms being evaluated.

8.1.3 Experimental protocol

A number of experiments were conducted to assess the capacity of the exoskeleton to suppress pathological upper limb tremor. To that end a protocol was defined for the experiments. The trials were carried out in two different countries. In the first evaluation stage, pre-clinical trials were conducted at the Department of Neurology of Hôpital Erasme in Brussels, Belgium. The second clinical trial stage was conducted at the Department of Neurology at the Hospital General in Valencia, Spain. The experimental protocol was approved by the ethics committees of both hospitals.

The performance of the WOTAS exoskeleton was evaluated in 10 patients with tremor-related diseases. The pathology of each patient was first diagnosed by a neurologist at the hospital using the quantification functional scale proposed by Fahn, tolosa and Marin (1998). Ten users participated in these experiments (three women, mean age 52.3 years). Users presented different pathologies, but the majority were affected by essential tremor (ET). ET was moderate in users 1, 3, 4 and 7 and severe in users 2, 5 and 6. User 8 suffers from multiple sclerosis, user 9 from post-traumatic tremor and user 10 is affected by a mixed tremor. All users gave their informed consent. All the experiments were recorded. The users still exhibited tremor despite regular intake of the drugs conventionally administered for tremor, at high doses in some cases.

During the experiments the only person who knew what particular mode of operation was being applied each time was the operator; in other words, neither the patient, the therapist nor the doctor knew whether the orthosis was applying an active or a passive strategy to suppress the tremor, or whether it was working in free or monitoring mode. This approach was adopted to offset the placebo effect (Belda-Lois *et al.*, 2004).

8.1.3.1 Evaluation tasks

Within the framework of the project, a set of tasks was selected for the evaluation trials, based on medical considerations. The tasks selected were (1) keeping the arms outstretched, (2) touching the nose with the finger, (3) keeping the upper limb in a resting position and (4) drawing a spiral. All the tasks performed by the patients were clinical and functional tasks that neurologists use to diagnose tremor-related diseases. These trials provide relevant amplitude and frequency information that can be used to classify the pathological tremor. Three different measurements had to be taken for each task and operation mode to guarantee the repeatability of the data. Similarly, for each task the time taken, the time between measurements and a specific file code to be stored on the disk were defined. To avoid patient fatigue during the trials, the total number of task repetitions was selected so that the total time of the measurement session did not exceed 1 hour. During the experiments, the different WOTAS operating modes and tasks were generated randomly. This approach reduced the learning effects and the subsequent effect on the data analysis (Belda-Lois *et al.*, 2004).

8.1.3.2 Data analysis

The data analysed were the output voltages from the gyroscopes fitted to the active orthosis. These output voltages were sampled at a rate of 2000 Hz. The data were filtered using a kernel smoothing algorithm and a 51-point-wide Gaussian window. The figure of merit adopted to quantify the reduction achieved by the exoskeleton is the ratio between the tremor power in monitoring mode (P_{mm}) and the tremor power in suppression mode (P_{sm}) in either passive or active modes:

$$R = \frac{P_{sm}}{P_{mm}} 100 \tag{8.1}$$

Thus, tremor reduction was measured under the same user conditions: with the orthosis placed on the upper limb. The estimated reduction, then, was the remaining tremor in suppression mode with respect to the tremor in monitoring mode.

The parameter selected to compare the tremor level was the power contained in the 2–8 Hz frequency band (Rocon *et al.*, 2004).

8.1.4 Results

Figure 8.2 illustrates the performance of WOTAS for all subjects in this experiment when operating in suppression mode. Note that the efficiency of the exoskeleton improves with tremor power. A statistical analysis was run to characterize tremor suppression. In this way it is possible to identify a lower limit for efficient tremor suppression – that limit is roughly 0.15 rad^2/s^3. In Figure 8.2 it can be seen that the robotic exoskeleton has a minimum tremor suppression limit; i.e. if the spectral density of tremor movement is below this lower limit, WOTAS will not effectively suppress tremor.

The results also indicated that the range of reduction in tremor energy for signals above this orthosis operational limit is from 3.4 % (percentile 5) to 95.2 % (percentile 95) with respect to energy in monitoring mode. These reductions can be appreciated in Figure 8.3, which illustrates the effects of WOTAS on tremorous movement using both strategies. Figure 8.3 (left) illustrates the time series corresponding to the tremorous movement of the elbow joint of user 2 while the arm is outstretched. The top part of the figure shows the time signal with WOTAS in *monitoring mode*. Note that in both *passive* and *active modes* the amplitude of tremor is clearly lower than in monitoring mode.

Figure 8.3 (right) illustrates the same reduction in the frequency domain. The power spectrum densities (PSDs) were obtained from the part of the signal with tremor. The top part of the figure illustrates the PSD of the tremorous movement with WOTAS operating in *monitoring mode*. There is a peak of tremor activity clearly visible close to 4 Hz. The middle part shows the PSD while WOTAS was operating in active mode. Note that the energy associated with tremor activity has been substantially reduced. In the bottom part of the figure there is also a clear reduction in the energy peak corresponding to the tremorous activity when WOTAS is in passive mode. These results indicate that WOTAS is able to suppress tremor, and they validate both the active and the passive control

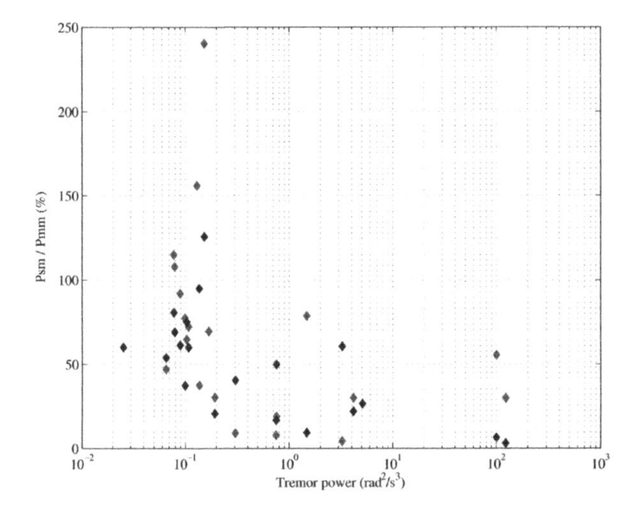

Figure 8.2 Tremor reduction (y axis) achieved by WOTAS operating in active (black markers) and passive (grey markers) suppression mode; x axis represents the user's tremor energy with WOTAS in monitoring mode

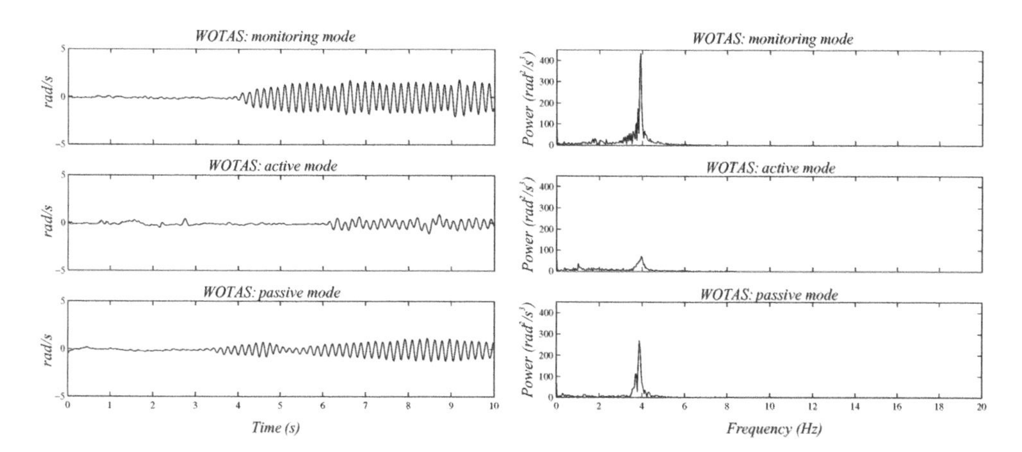

Figure 8.3 This figure illustrates: (left) the oscillations of elbow tremor with WOTAS in monitoring and suppressing modes and (right) the associated power spectral density (PSD) in monitoring and suppressing modes in user 2. Note the sharp reduction in amplitude and power of tremor when suppressing actions are applied

strategies. Note that the frequency of tremor does not change when the exoskeleton is working in suppression modes.

A detailed analysis of the data showed that the active suppression strategy achieved higher levels of tremor suppression (81.2 % mean power reduction) than the passive suppression strategy (70 % mean power reduction). This suggests that performance in tremor suppression is better in the active mode.

8.1.5 Discussion and conclusions

Analysis of the videos recorded during the experiments showed that in most of the users there was no tremor displacement to proximal joints of the upper limb. The results of the experiments indicated that the device could consistently achieve 40 % tremor power reduction for all users and was capable of attaining a reduction ratio in the region of 80 % tremor power in specific joints of users with severe tremor. In addition, the users reported that the exoskeleton did not affect their voluntary motion. These results indicate the feasibility of tremor suppression through biomechanical loading. Nevertheless, the users reported that the exoskeleton was not a practical solution to their problem because it is bulky and heavy. They also thought that the use of such a device could cause social exclusion. This was not unexpected since the exoskeleton was developed as a platform to evaluate the concept of mechanical tremor suppression and not as a final orthotic solution. The principal wish expressed by the potential users was that the exoskeleton could be hidden under clothing (Rocon *et al.*, 2007).

The results also indicated that active tremor suppression performed better than passive tremor suppression. However, the authors believe that this could be because the same viscosity value was added to the movement of all users. Customizing of the viscosity or inertia added to the upper limb to suit the biomechanical characteristics of each user should improve the efficiency of the passive tremor suppression strategy.

It was noticed that the degree of tremor reduction was dependent upon the tremor power. There are lower limits for robotic tremor suppression closely related to the mechanical contact impedance at the human–exoskeleton interface.

During the trials two users spontaneously reported that they noticed a reduction in the amplitude of their tremorous movement and consequently felt more confident about executing the task. This indicates that visual feedback from a smooth movement has a positive impact on the user.

8.2 CASE STUDY: THE CYBERHAND

L. Beccai, S. Micera, C. Cipriani, J. Carpaneto and M. C. Carrozza

ARTS Lab, Scuola Superiore Sant'Anna, Pisa, Italy

8.2.1 Introduction

Functional substitution of the hand is necessary for patients suffering from upper limb amputation. The hand, from a functional point of view, is the end-effector of the human arm and is necessary to perform reaching, grasping and manipulation tasks. The hand is also the means whereby humans perform important exploratory functions in the environment. Hand contact with an object provides important information about the object, such as shape, surface texture, size and orientation.

Upper limb amputations induce a poor self-body image in the patient and have serious psychological implications (Gaine, Smart and Branably-Zachary, 1997). The loss of fine, coordinated hand movements and of proprioceptive and exteroceptive sensation and feedback, as well as the aesthetic implications, can be compensated for to a very small extent by the solutions offered by state-of-the-art prosthetic technology. The principal reasons why amputees do not regularly use their prosthetic hands are low functionality or controllability, in addition to poor cosmetic appearance (Carrozza *et al.*, 2002). In view of the gravity of the clinical problem and of the lack of suitable technological solutions, robotics scientists and neuroscientists are motivated to focus their research on the complete functional substitution of the human hand. The cybernetic prosthesis approach is therefore being adopted in order to arrive at a solution to this complex issue.

There are two fundamental characteristics required for implementation of a cybernetic prosthesis:

- It must feel to the user like the lost limb thanks to sensory feedback delivered by stimulation of the sensory nerve fibres (afferent system).
- It must be controlled in a natural way by extracting motor commands from the (peripheral) nervous system (efferent system).

Within this framework, the connection with the PNS will occur through a neural interface whose basic function is to transmit the user's intentions as motor commands to the prosthetic device and to restore the patient's natural sensory perception; see Chapter 4 for a discussion on cHRI.

This case study describes the CyberHand (Carrozza *et al.*, 2006; Micera *et al.*, 2006) as an example of the efforts involved in implementing a dexterous artificial hand, its sensory system and the control system. At the same time it presents the issues raised by the neural interface (Navarro *et al.*, 2005) that will be dedicated to the exchange of sensory (afferent) and motor (efferent) information between the nervous system and the artificial hand.

8.2.2 The multi-DoF bioinspired hand prosthesis

The biological model of the hand has been studied and analysed to derive the specifications for a bio-inspired artificial hand; see Chapter 2 for a detailed analysis of bioinspiration. Of all the features of human hand performance, the following parameters are considered the most important from the standpoint of biomechatronics (Carrozza *et al.*, 2006):

- the number of degrees of freedom (DoFs), i.e. 22;

- the force range, which goes from a few newtons in fine manipulation to at least 500 N for power grasp;
- the number and the type, distribution and location of natural receptors encoding proprioceptive and exteroceptive information (see Section 5.1.1 for a brief description of these natural receptors).

Robotic technology is striving to fulfil as many 'natural' requirements as possible. Nevertheless, there are still a number of limitations imposed by the current technology; thus some tradeoffs are necessary to provide effective solutions to the architecture of the artificial hand.

One such issue is the technology available for the actuation system; at present this is unable to offer a high power density (ratio between the power delivered by the actuators and their volume) and high efficiency (Carrozza *et al.*, 2004); see Section 6.3 on actuator technologies. Consequently, the possible number of actuators and DoFs is limited, but augmented dexterity demands a larger number of actuators. The solution adopted in CyberHand is that of an underactuated system where the number of DoFs is greater than the number of the actuators. An underactuated mechanism is based on a differential mechanical principle and is capable of automatically distributing the forces and torques supplied by the actuators among the joints (Hirose, 1985). In this way, an adaptive grasp is achieved (Hirose and Umetani, 1978), the artificial phalanges wrap around the grasped object and the three phalanges are actuated by a single motor.

The advantage of this kind of mechanism is that, with only a small number of actuators, it is possible to achieve a more natural and anthropomorphic kind of grasp without demanding particular control capabilities of the user. Very importantly, in this way only a few control signals are needed to achieve multiple degrees of motion, and most of the grasping patterns of the human hand can be imitated without augmenting mechanical and control complexity.

The biomechatronic CyberHand, which is designed for direct interfacing with the human peripheral nervous system, consists of five fingers with cylindrical aluminium alloy phalanges attached to a central palm. Flexion (agonistic action) of all the phalanges of an underactuated finger is achieved by a cable pulled by a DC motor. Finger extension (antagonistic action) is implemented by a torsion spring embedded in each finger joint. The hand has a total of 16 DoFs: 3 DoFs for each finger plus 1 DoF for thumb opposition movement. Thumb movements are accomplished by a 2 DoF joint corresponding to the human trapezo-metacarpal joint at the base of the palm. The motor for thumb adduction and abduction is located inside the palm, while the motors for finger movement are all external. The overall volume of the CyberHand is 50 cm^3 (only the actual hand) like the natural hand, and the overall weight is 360 g (without actuators) as compared to 400 g in the case of the human hand (without extrinsic muscles). The resulting allowed force for a power grasp is about 70 N and the joint speed is 45 deg/s. Figure 8.4 depicts the CyberHand and the performance of four basic grasps.

8.2.2.1 The artificial sensory system

The information from the sensory system of a cybernetic artificial hand is used for two main purposes: (1) as input to an afferent feedback system in order to restore the natural perception of the lost limb as far as possible; (2) to close the low level control loop of the artificial hand.

One important issue that has to be addressed when choosing the type of sensors to be integrated in the final version of the cybernetic hand is how to understand and identify which types of sensors are designated for each of the two main areas described above. The biomechatronic approach adopted here was to consider the biological model of the sensory system while selecting technology and materials that allowed mechatronic integration in the prosthetic hand.

One feature of the CyberHand is that, since the hand mechanism is based on underactuation, complex sensors for grasp optimization are not required. A redundant sensory, open platform has

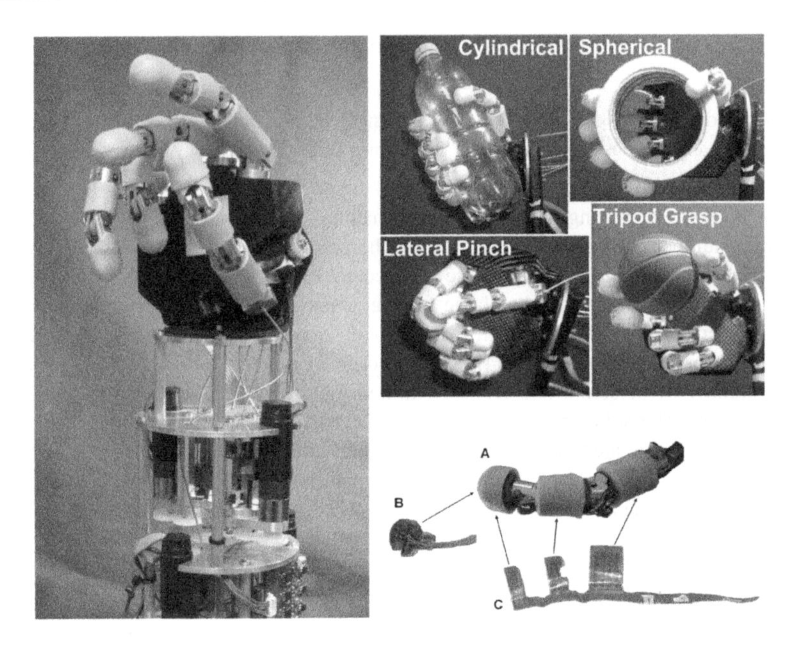

Figure 8.4 The five-fingered biomechatronic hand, CyberHand, of the ARTS Lab, Scuola Superiore Sant'Anna, Italy(left). Four basic grasps performed by the CyberHand (top right): cylindrical, spherical and tripod grasps, and lateral pinch. Image showing (below right): (A) the artificial finger of the hand with; (B) triaxial three-dimensional force sensor based on aluminium structure and semiconductor strain gauge technology integrated at the fingertip; (C) one prototype of the polyimide based on–off contact sensor consisting of two flexible polyimide layers with an array of Cu electrodes (1 mm × 1 mm) spaced by 2 mm polyurethane foam layers. Matrices of 8 × 3 contact points are provided at the fingertip and at the middle phalanx, and an 8 × 4 matrix at the proximal phalanx. The contact sensor array is covered with a 1 mm thick compliant silicone skin when integrated in the finger (A). Reproduced from Carrozza *et al.*, 2006

been developed, i.e. one in which sensors and actuators can be easily added or reduced in the hardware system. Its flexibility is crucial in testing and validating both the sensory and the control system.

The CyberHand artificial sensory system can provide proprioceptive and exteroceptive sensory feedback (Carrozza *et al.*, 2006; Edin *et al.*, 2006, 2007). The proprioceptive sensors supply information on the joint positions of the hand and the forces exerted by the actuators, in a similar way to the joint receptors and Golgi tendon organs of the human hand that furnish information about tendon stretches. Joint position sensors based on the Hall effect (see Section 6.2), integrated at the finger joints, and incremental magnetic encoders integrated at each hand motor, are used to provide information about the position of all the phalanges during grasping tasks. Cable tension sensors are used to detect the tension of the cables that run through each finger. These are strain gauge sensors delivering an output voltage proportional to the tension force (up to 120 N with a resolution of about 20 mN), and they are used by the low level controller (described in the next section) (Cipriani *et al.*, 2006).

The exteroceptive sensors provide information about the interaction of the hand with the environment. As this is a bio-inspired approach, it is important to provide basic information of this kind, such as contact of the hand with the object, of the object with the environment and slippage (Edin *et al.*, 2006). For this purpose, tactile arrays of on–off sensors and aluminium-based strain gauge triaxial fingertip sensors (see Figure 8.4) were built into the hand. The design goal was to emulate the sensitivity (with the contact sensors) and the dynamic behaviour (with the triaxial force sensors) of the mechanoreceptors of the human hand. Pressure thresholds of the artificial on–off sensors are

<15 mN/mm^2, and the triaxial force sensor has a bandwidth of 0–700 Hz. The two types of sensors are complementary and could be used to supply the patient with information about the mechanical events that occur during typical activities of daily living (ADL) task, such as grasping, lifting and replacing an object (Edin *et al.*, 2006, 2007). Currently, MEMS technologies are being applied to develop tactile sensors integrated in the artificial hand, towards a sensory system that will mimic the spatial resolution, sensitivity and dynamics of the human mechanoreceptors. To this aim, it has been demonstrated that a silicon based triaxial force sensor having a maximum dimension of 1.5 mm (Beccai *et al.*, 2005) can provide slippage information to the artificial hand when a proper packaging is used to develop a soft and compliant tactile sensor (Beccai *et al.*, 2006).

8.2.2.2 The control system

The CyberHand control architecture has been developed as an open platform, so that the number of sensors and actuators can be modified while comparing different control strategies (Cipriani *et al.*, 2006). The underactuation principle (see Section 6.3) on which the artificial hand is based does not allow for manipulation. This is because 3 DoFs are driven with only 1 DoM (degree of mobility) for each finger. It is not possible to implement a control architecture that can control each phalanx independently, as the final configuration of each finger is dictated by the shape of the grasped object. The use of these novel interfaces is analysed in Case Study 4.7.

Following a bio-inspired approach, there is more than one level of control of the CyberHand (Carrozza *et al.*, 2006; Cipriani *et al.*, 2006): a high-level controller interacts directly with the operator, who specifies the type of grasp and the force needed; a low-level controller performs the specified grasp while assuring its stability. The high and low levels of the control system depend on the artificial sensory system, which can detect key events during the grasping task and let the controller identify the transitions between different states of operation. The grasp types are triggered by a higher-level unit that can identify the user's intentions and use appropriate grasping primitives.

The control algorithm for the grasping task is composed of two separate consecutive phases. The first is pre-shaping, which is based on the grasp type. In this phase the hand is pre-shaped by means of a PID position control algorithm and the desired finger tendon force is selected according to the grasping primitives. During the second phase, the fingers involved rapidly close around the object, seeking a balanced distribution of the forces within the hand until the desired global force is reached. The control system, which consists of several subsystems such as PCI input/output cards and stand alone motion modules, is capable of performing many stable grasps, e.g. cylindrical, spherical, tripod and lateral pinch, with a high degree of reliability and robustness.

8.2.3 The neural interface

In the last thirty years, several neural interfaces have been developed with different characteristics (see Figure 8.5). For example, cuff electrodes have proven very reliable and robust; they have the advantage of reduced invasiveness but suffer from limited selectivity, but albeit some interesting results have been achieved using multisite cuff electrodes (Tarler and Mortimer, 2004). On the other hand, sieve electrodes could present a very interesting solution for the development of neuroprosthetic and hybrid bionic systems, but there are still a number of problems limiting their usability (Lago *et al.*, 2005), and they are only applicable to sectioned nerves.

For this reason, in the recent past a number of research groups started investigating the possibility of developing and using neural interfaces featuring needles that are inserted longitudinally – LIFE electrodes (Lawrence *et al.*, 2004) – or transversally – USEA electrodes (McDonnall, Clark and Normann, 2004) – in the PNS. This approach looks very promising in that it combines reduced invasiveness with good selectivity.

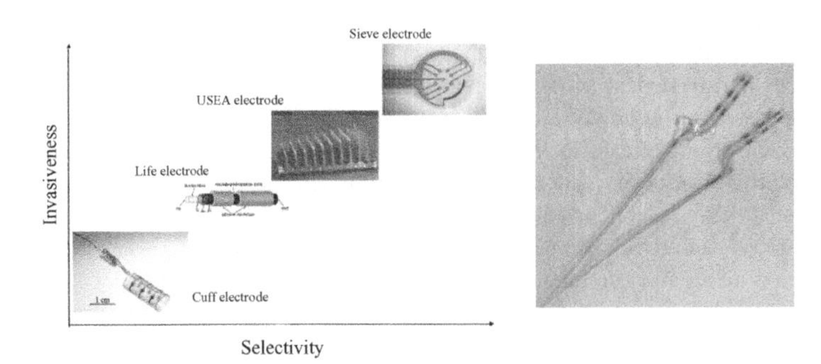

Figure 8.5 The characteristics of the different neural interfaces in terms of invasiveness and selectivity (left) (Navarro *et al.*, 2005). Image showing a tf-LIFE (right). Each half of the polyimide ribbon structure consists of a ground electrode, an indifferent electrode and four recording sites, and ends in a U-shaped portion containing the pads for the connections with ceramic connectors (Lago *et al.*, 2007)

Even though intraneural PNS interfaces look very interesting – good results have been achieved during preliminary experiments in bidirectional control of hand prostheses (Dhillon and Horch, 2005) – they suffer from a number of drawbacks that limit their usability: (1) the biomechanical/mechanical properties of the PNS have not been studied in detail to gather information applicable to the design of effective but not too invasive interfaces; (2) the implantation procedure is carried out 'blind', with no possibility of selecting the final position; (3) advanced algorithms need to be developed to extract useful neural information.

The above issues are currently being addressed, with the long-term aim of developing a new generation of intraneural PNS interfaces. Some preliminary results are presented in the following sections.

8.2.3.1 Actuation of intraneural electrodes

As noted earlier, LIFEs look promising for their good selectivity and limited invasiveness, although it has been shown that – in many cases – they can record only during a short period of time. Also, they have been improved in the recent past by using flexible materials as insulation on which conductive metals can be deposited. This new version, called tf-LIFEs (Hoffmann *et al.*, 2006), is composed of thin-film polyimide on which eight active sites have been placed. The flexibility of the innovative electrode combined with a larger number of active sites reduces drifts and enhances signal-to-noise ratios, making tf-LIFE a good solution for long-term implantations.

Another drawback of conventional LIFEs is that chronic intrafascicular electrodes cannot be moved once implanted. This means that electrodes may not be positioned near selected cells and there is no flexibility for targeting specific cell types or receptive field positions. It would therefore be desirable to be able to control electrode positions once they are implanted, adjusting them in the tissue to improve the longevity of cell recordings. To that end, the tf-LIFEs were integrated with microactuators to provide movable contacts. In particular, shape memory alloys (SMAs) (see Section 6.3) were used as smart actuators to move the contact points of the tf-LIFEs, selectively altering their shape. A 'serpentine' shape was memorized into SMA thin films, which were covered by polyimide thin films to simulate the tf-LIFE structure. At the same time, the SMA was also glued directly to the electrode (see Figure 8.6 for the concept).

The results of the characterization (Bossi *et al.*, 2007) showed that flexible intrafascicular electrode actuation by the SMA could be a promising new technique for controlling the position of the active sites of the tf-LIFE inside the nerve, thus improving their performance for long-term applications.

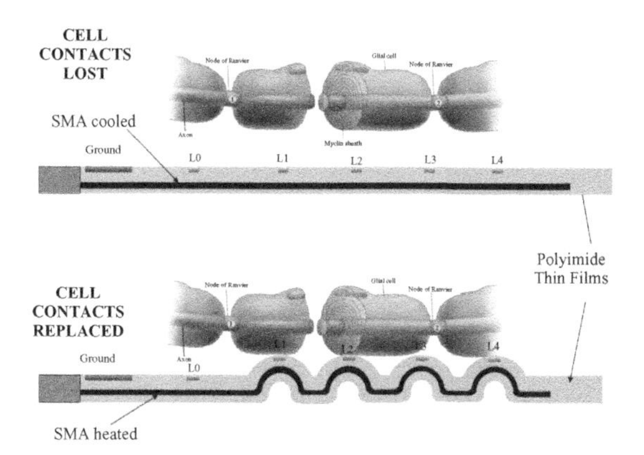

Figure 8.6 A basic scheme depicting a tf-LIFE and how its lost contacts may be replaced by SMA actuation (Bossi *et al.*, 2007). Reproduced from Bossi *et al.*, 2007

8.2.3.2 Extraction of neural information using advanced processing algorithms

Although many experiments have been carried out in the past on the processing of ENG signals recorded by means of cuff electrodes (e.g. Cavallaro *et al.* 2003; Micera *et al.* 2001) in only a few cases (e.g. McNaughton and Horch 1994; Mirfakhraei and Horch 1997) has an attempt been made to understand the potentials of intraneural electrodes (and in particular tf-LIFEs) for extraction of neural information. For example, it is possible to extract spikes from the recorded ENG signals using intraneural electrodes, whereas using cuff electrodes only compound signals can be recorded. Thanks to this feature, it is possible to develop algorithms (Citi *et al.*, 2006) with the potential capacity to detect different spikes (related to different neural stimuli), thus enhancing the selectivity of the electrode. The first step taken was to implant tf-LIFEs in the sciatic nerve of rabbits. Different kinds of stimuli were applied to the paw of the rabbit: (1) squeezing the foot with the knee at 90°; (2) squeezing the foot with the knee released; (3) ankle flexion; (4) toe extension; (5) toe extension combined with ankle flexion. Each was repeated four to six times.

These signals were processed to determine whether the different modes of information could be decoded. Signals were Kalman filtered, wavelet denoised and spike sorted. The classes of spikes found were then used to infer the stimulus applied to the rabbit. Although the signals acquired from a single tf-LIFE gave poor stimulus recognition, the combination of the signals from multiple sites led to better results (see Figure 8.7). The spike sorting algorithm is also helped by temporal correlation among the channels. The results point to the possibility of extracting different neural information by exploiting the potentials of multisite neural interfaces.

With the results of these experiments it could be possible in the near future to develop a new class of intraneural interfaces that can reduce invasiveness, e.g. thanks to the use of advanced materials and of the information gathered using biomechanical models (Sergi *et al.*, 2006), to extract various kinds of information (using advanced algorithms), to deliver sensory feedback (by increasing the number of contacts in the interface) and to be able to place the electrodes in the optimum way for the desired signal-to-noise ratio.

8.2.4 Conclusions

The CyberHand system (Carrozza *et al.*, 2006; Micera *et al.*, 2006) is an interesting example of parallel design of the different modules necessary to implement a cybernetic hand: the mechanism, the sensory system, the control systems and the neural interface.

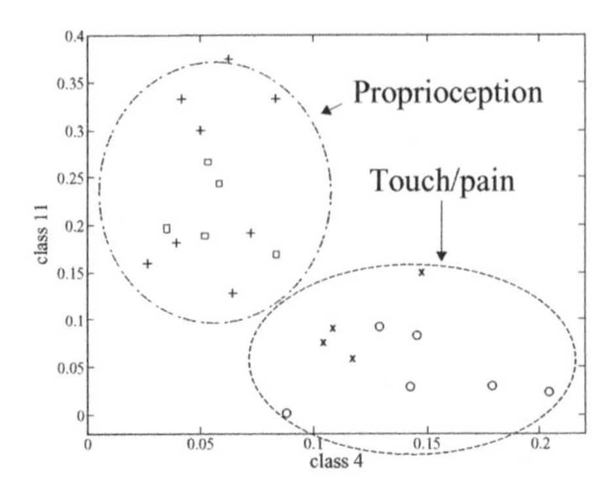

Figure 8.7 Scatter plot of a fraction of spikes in class 4 versus a fraction of spikes in class 11 for different stimuli: squeezing foot with knees at 90° (○) and released (×), ankle flexion (+), toe extension (∗), toe extension combined with ankle flexion (□). The different kinds of stimuli are approximately separable. Two categories are well separated: touch/pain (○,×) and proprioception sensations (+,∗,□) (Citi *et al.*, 2006). Reproduced from Citi *et al.* 2006

At present the neural connection is the bottleneck in the cybernetic hand system, as it is not yet possible to have an implantable neuroelectronic system capable of restoring the large number of sophisticated functional connections between the natural hand and the nervous system (Micera *et al.*, 2006). The design approach of the CyberHand was adopted in view of the fact that currently available neural interfaces can provide only a few channels for the bidirectional connection to allow exchange of information for efferent and afferent pathways. However, the ultimate purpose of the CyberHand is to investigate methodologies whereby it will be possible to connect the hand with the PNS; its modular design makes it possible to work towards that goal in incremental steps. At the same time, the CyberHand system can be used to test neuroscientific hypotheses regarding sensorimotor control and the functional roles of the biological sensory systems during manipulation tasks. It is therefore of the utmost importance to this research to carefully merge robotic science with neuroscience, as it is from this fusion that suitable guidelines can be drawn for experimentation and for clinical adaptation of the CyberHand for the prosthetic application.

8.3 CASE STUDY: THE ERGONOMIC EXARM EXOSKELETON

A. Schiele[1,2]

[1]*Mechanical Engineering Department, Automation & Robotics Section, European Space Agency (ESA), Noordwijk, The Netherlands*
[2]*Mechanical Engineering Faculty, Biomechanical Engineering Department, Delft University of Technology (DUT), Delft, The Netherlands*

8.3.1 Introduction

The ESA human arm exoskeleton, EXARM (Schiele and van der Helm, 2006; Schiele and Visentin, 2003), is being developed as a human–machine interface for master–slave robotic teleoperation with force feedback. The EXARM will allow astronauts inside the International Space Station (ISS) to remote-control EUROBOT (Schoonejans *et al.*, 2004), a humanoid space robot, on the outside.

EUROBOT will assist crew during maintenance on the ISS and will support future manned or unmanned exploration missions to other celestial bodies in the solar system, e.g. the Moon or Mars. In the first instance, however, EUROBOT will be devoted to supporting extravehicular activities (EVA) on the ISS. Depending on the application, the robot can be equipped with two or three redundant, 7 DoF robotic arms. With respect to other space manipulators, such as the SRMS or SSRMS, each arm of EUROBOT is similar in function and size to a human arm. For ISS applications the third arm is used as a leg for fixing the robot to the outer structure of the station. For planetary use, EUROBOT will have two arms only and will be supported by a movable, centaur-like, base. In both cases, a versatile set of end-effectors will be provided, which the robot can exchange for different tasks, e.g. assembly, inspection, transportation, etc. Two different modes of operation are envisaged for control of EUROBOT: autonomous and manual.

In autonomous mode, arm movement can be pre-planned and programmed offline, in which case on-board hazard detection and obstacle avoidance will be used. This mode is used to operate in well-defined, structured environments. Typically, EUROBOT will then handle orbital replaceable units (ORUs), e.g. on ISS, that are equipped with robotically compatible interfaces. In manual control mode, a more reactive control based on master–slave teleoperation is envisaged. Force feedback with a human operator in the loop allows operations in unstructured environments that are difficult to model *a priori* from the ground. An example of an unstructured environment would be a regolithic or sandy planetary surface that is scattered with rocks. On ISS, for instance, unstructured environments are those that contain soft structures, such as multilayer insulation (MLI) panels, fluid lines or similar. Those environments exist mostly on an active worksite, on which astronauts carry out maintenance, installation or repair tasks. However, during close cooperation of the remote-controlled robot with an astronaut during an EVA, a sort of reactivity is required that can best be provided by a human that joins as a master into the control loop of the slave robot. In manual control mode, the human operator, who is hooked into the robot control loop, can adapt more quickly to the situation and react with faster recovery actions than any autonomous path-planning and re-planning algorithm today could provide. This mode is therefore envisaged for crew assistance or for rapid intervention in emergencies. The EXARM exoskeleton was developed to make the master–slave control mode as intuitive as possible for the astronauts.

The exoskeleton senses the motion of the astronaut's arms and translates it to the robot arm. At the same time, the device feeds forces and torques back to the operator's arms. Such forces can, for instance, be measured from the robot's contacts with the real environment, or more probably from collisions of the robot with imposed virtual constraints. Thanks to those constraints, the robot end-effector and limbs can be kept in a safe work envelope at all times. This is important in order to restrict robot motion in the near vicinity of life-critical and sensitive hardware infrastructure.

An exoskeleton offers significant advantages over other input devices such as force-feedback joysticks. A key advantage is the possibility of controlling the robot end-effector in Cartesian space while also controlling its joint-space motion or the geometrical pose of the robot. This is an important asset for free navigation in complex environments. Control of the pose of dexterous 7 DoF robots is otherwise difficult to achieve. In purely Cartesian control, such robots exhibit a self-motion that is induced by inverse kinematics algorithms. Seven degree of freedom robots have an infinite number of solutions for joint-space trajectories to reach a specific Cartesian location. In order to choose an appropriate solution, their iterative inverse kinematic algorithms optimize for a specific constraint that is chosen by the developer. One constraint typically used is to avoid singular positions of the manipulator. Thus, during movement of the arm, the pose of the arm is altered in order to avoid the manipulator's singularities. This pose adjustment induces the so-called self-motion, which can drive the robot limbs into unwanted collisions with the environment. With an exoskeleton that controls the pose of the robot, forces and torques on each robot segment can be translated to the corresponding joints of the human arm as torque feedback. This gives the robot operator an intuitive understanding and feel of the robot configuration at all times and can thus be regarded as illustrative of a physically triggered cHRI (see Chapter 4).

The challenges to the designers of the EXARM exoskeleton came from the application scenario. The device must be able to sustain prolonged command of the slave robot. It is not unusual for an EVA-supported operation on the ISS to last in the region of 6 hours. Furthermore, EXARM needs to be handled by users with a wide range of different statures and masses. Therefore, the exoskeleton has to be as 'ergonomic' and 'comfortable' as possible. Another desired feature for the exoskeleton was ready adaptability to varying user statures without requiring mechanical adjustments and software calibration. The design of EXARM adopted a novel approach for ergonomic exoskeletons, presented in Schiele and van der Helm (2006). A new actuation principle was also developed (Schiele *et al.*, 2006) to make implementation of the ergonomic structure practically feasible. The ergonomic properties of the exoskeleton are detailed in the following paragraphs.

8.3.2 Ergonomic exoskeleton: challenges and innovation

When using force-feedback devices inside a low-gravity (μ-G) environment, any force or torque fed back to the user must be counteracted by the user's body. When force- or torque-feedback devices are used inside a reduced gravity (μ-G) environment, only body-grounded feedback should be used. Then, force or torque applied to the operator via the device will create an appropriate reaction force in the operator muscles. Such forces will always apply between body segments only. In non-body-grounded, fixed-base force-feedback devices, the forces or torques stemming from the slave can only be counteracted by the inertia of the entire operator body. As a consequence, when force feedback is applied, e.g. to the operator arm, the entire body will be set into motion. Such a force or torque feedback would be highly counterintuitive; it will push an operator away rather than helping to interpret correctly the contact situation of the remotely located robot in relation to its environment. Therefore, in contrast to most currently existing exoskeletons, the EXARM has been implemented as a wearable, body-grounded device. The EXARM has also been designed as a compact and portable system because this provides more flexibility during operational use. For instance, after a typical operation with the exoskeleton inside the ISS, the device can be easily stowed away or transported to another station segment.

A portable device will also not disturb the floating sensation that astronauts become used to during their time in μ-G. However, the EXARM also needed to be compatible for 1-G use. This is crucial for development and testing on the ground, and therefore EXARM had to be as lightweight as possible. Lightweight structural materials such as carbon fibre reinforced plastics (CRP) were employed. As mentioned above, another key challenge was to develop an exoskeleton that can fit a wide range of user statures without requiring complex adjustment and calibration procedures. Astronaut crew statures fall within the 5th percentile of Japanese female and the 95th percentile of the US male population (1.49–1.90 m). Some previous exoskeleton developments have mechanically adjustable limbs whose length can be changed so as to align the exoskeleton's main axes to the human joint axes. However, alignment of an exoskeleton to the human arm joints is difficult, as explained in detail in Schiele and van der Helm (2006) and in Chapters 3 and 5. It is difficult because the positions of the human joint axes are not exactly known and because they change during movement of the joint. Furthermore, the axis positions vary between subjects. Should an exoskeleton be misaligned, interaction forces will be created between the device and the human operator. These interaction forces can be relatively large (up to over 160 N), as shown in Section 5.2 and Case Study 5.5, and can contribute significantly to the discomfort of the device. Moreover, these interaction forces can restrict the natural range of motion–the workspace–of the human arm. In order to avoid the disadvantages of requiring alignment, the EXARM was designed to be independent of individual statures and biomechanics, within the range typical of an astronaut crew. One particular benefit of this design feature is that the EXARM can be donned quickly by different operators. Donning and doffing the exoskeleton takes no more than about 30 seconds for an untrained person. This answers the requirement of a rapid intervention capability in emergency situations.

8.3.3 The EXARM implementation

The EXARM exoskeleton is a serial manipulator. It has a total of 16 DoFs, of which only 8 DoFs are to be actuated. The remaining 8 DoFs are passive. The passive joints are for alignment compensation and are permanently free to move with the serial chain. Despite these noncontrolled joints, the EXARM can be actuated fully when worn by an operator. Then, it resembles a closed parallel loop robot combined with the human arm. This tolerates the existence of passive joints in the structure of the exoskeleton. EXARM can thus be controlled and transmit torque to all human joints on the right arm, from the shoulder to the wrist. The base of the exoskeleton is attached to a chest-vest. The structure is fixed to metal inserts implemented on the CRP sheet of the vest (see Figure 8.8). Next, the exoskeleton is attached to the operator's upper arm and forearm by means of inflatable air-cushions. At the most distal tip, the device can be secured to the palm with an orthopedic glove. The operator's fingers are free to move, which allows for implementation of a simplified grasping interface later on. The upper arm and forearm attachments allow rotation of the limbs. The EXARM incorporates 6 DoFs for interacting with the operator's shoulder movement.

The surfaces depicted in Figure 8.9 show the optically measured and averaged shoulder workspace of five different test subjects. Data were measured with optical markers attached to bony landmarks on the human arm. Four Optotrack 3020 camera systems were used. The less transparent surface in dark grey shows the workspace boundary that the MPII bony landmark (base of the index finger) can draw when the EXARM is worn. The more transparent, light grey surface depicts the boundary of the naturally available shoulder workspace when the exoskeleton is not worn. For visual clarity, a virtual skeleton model is projected on to the data. The model has a stature of 1.80 m. Note that the wearer can reach almost the entire human shoulder workspace while wearing the EXARM. In addition, the exoskeleton wearer has full access to the natural-limb elbow and wrist workspace, without restrictions.

Figure 8.8 The ESA ergonomic exoskeleton worn by ESA Astronaut Frank de Winne during an evaluation session. Reproduced from ESA

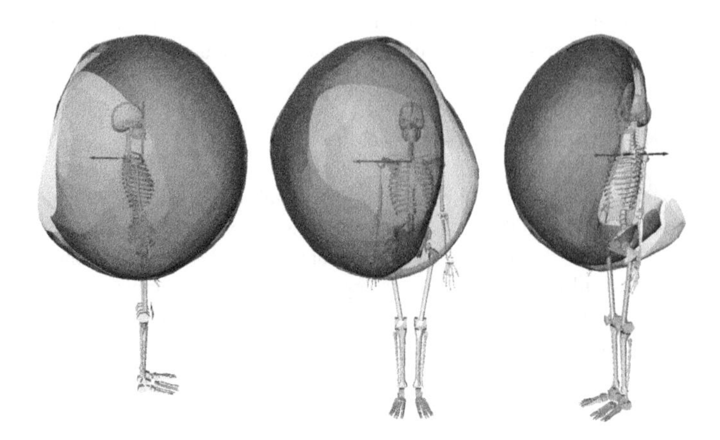

Figure 8.9 Measured shoulder workspace of the right arm. The naturally available workspace is shown in light grey; the workspace available while wearing the EXARM exoskeleton is shown in dark grey. Data were measured with optical markers attached to the human arm, averaged over five test subjects and wrapped with a surface

All joints of the EXARM are equipped with high-precision conductive plastic potentiometers to sense human arm motion. Four joints of the shoulder articulation have additional provisions for actuation. These joints, like all other actuated joints in EXARM, are equipped with cable-pulleys, which convert the linear motion of externally guided cable tendons into joint motion. All pulleys are equipped with integrated torque sensors for motor control. Between the upper arm and the forearm, the exoskeleton incorporates 3 DoFs to enable natural elbow flexion and forearm pro-supination. Only two of these joints are equipped with pulleys for actuation. To enable natural movement of the operator's wrist, six joints are implemented in the EXARM structure, only two of which are actuated for torque feedback to wrist abduction–adduction and wrist flexion–extension. To summarize, the kinematic structure of the EXARM may be described as 'bridging over the human joints'. Bridging means that the exoskeleton provides a different type of kinematic structure from the human arm, while offering the same type of motions. The structure does not explicitly imitate the limb kinematics and contains motorized links that are not tightly coupled to each other. Adjustment of the limb sizes of the current EXARM prototype to individuals is therefore not necessary for persons of 1.75 ± 0.09 m stature and 68.7 ± 12.8 kg mass. As explained in Chapter 5, the EXARM does not create large constraint forces on the human joints during movement, even when there are misalignments with the principal human joints. As a result, wearing the exoskeleton feels natural and very comfortable for the operator. For more detail on the ergonomic design and the rationale thereof see Schiele and van der Helm (2006).

The EXARM weighs only about 5 kg overall, of which the mass of the movable system along the human arm is less than 2.5 kg. For testing and operating in a laboratory environment, the device is suspended on a counterbalancing system. Most structural components of the exoskeleton are made in laminated CRP. Functional components carrying motors, joint sensors or bearings are implemented in aluminium.

The serial exoskeleton structure is fixed to the operator's chest, but the motor units are fixed to an external plate. In future versions this will be a wearable backplate. Bowden cable transmissions transfer the actuator motion from the motor units to the EXARM joints. With this joint actuation of EXARM, the mass of the device can be kept to a minimum. The only provisions for actuation required in the moving exoskeleton system are the pulleys that hold the cable tendons. Before deciding on which actuation technology to use, an extensive survey and prototyping campaign was carried out to compare various different actuator technologies. Numerous actuators were prototyped, including

DC motors in various configurations (e.g. direct drive, low reduction, highly reduced), ultrasonic motors and passive devices such as magnetorheologic fluidbrakes. The main purpose of the prototyping was to determine performance differences between the technologies in a haptic master–slave control loop. The next step was a campaign to find the most suitable candidates for a lightweight portable exoskeleton implementation. The selection criteria were, among others, actuator torque-to-mass ratio, dynamic range, maximum peak torque, torque rise time, power consumption and so forth. Controllability was analysed experimentally by quantifying metrics such as transparency in free-air motion and stability of slave contact with the environment. For the experimental analysis, a variety of controller implementations were also tested, from simple position-error controllers to complex multichannel controllers. A summary of the motor prototyping is provided in Letier *et al.* (2006). The best haptic performance was still produced by slightly reduced DC motors, but their torque-to-volume and torque-to-mass ratios were poor. It was therefore decided to relocate the motors off the structure and use Bowden cable transmissions, as mentioned above.

The cable actuation gear was successfully tested in a 1 DoF haptic loop with the exoskeleton. Performance was determined within a typical master–slave control scheme (Schiele *et al.*, 2006). The Bowden cable actuator can transmit a torque of approximately 1.5 Nm to the EXARM joints, which is sufficient in magnitude to create haptic feedback. The free movement friction is only about 0.1 Nm, which can hardly be felt on the operator arm. By relocating the motors off the exoskeleton joint, it was possible to increase the power density on the exoskeleton joints more than sixfold, from 2.5 to 16.0 mNm/cm^3. The mass spent on each exoskeleton joint for actuation was thereby reduced from about 1 kg per joint to only about 150 g. Figure 8.10 shows the final 1 DoF actuator prototype on the test bench, before mounting on EXARM. On the left, an output bar emulates a joint of the EXARM exoskeleton. The relocated DC motor unit with a cable capstan reducer is on the right. A reduction ratio of 10:1 was implemented in the drivetrain. This was a good compromise, trading off the maximum output torque the motor can deliver at acceptable mass and size against the increase in the apparent inertia. With a low reduction ratio, the contribution of the motor inertia to the felt end-effector inertia is smaller. Together with the relocation of the motors, this helps to increase the transparency of the exoskeleton significantly. On either side of the transmission, strain gauge-based torque sensors are fitted into spokes (not visible) of the cable pulleys. On the motor side of the joint, a 500 pulse-per-revolution incremental encoder is used for speed estimation. The control scheme that was found to perform best for haptic interaction with the real slave robot joint is based on a four-channel controller, as proposed in Lawrence (1993).

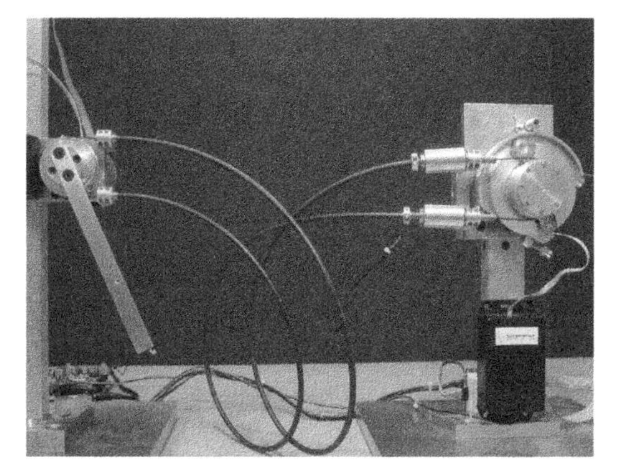

Figure 8.10 The EXARM's Bowden cable actuator on the test bench. Relocation of the motor unit off the exoskeleton joint made it possible to increase the joint power density more than sixfold

The EXARM is not kinematically equivalent to the slave robot it is to control (16 DoFs versus 7 DoFs), which makes multidegree of freedom bilateral control challenging.

The robot arm used for EUROBOT will be one custom-developed by ESA, but for the moment an industrial Mitsubishi PA-10 is used for controller development and testing (Schiele *et al.*, 2006b). Several strategies are followed for the forward link between the exoskeleton and the slave robot, ranging from end-point control with inverse kinematics optimization to joint cluster to joint mapping. In the first method, an additional constraint is imposed on the solution of the slave robot inverse kinematics that controls the geometric pose of the slave. In the present case, this is the angle of the elbow plane with respect to the horizontal plane. In the second method, EXARM joint clusters are mapped on to single joints of the slave robot. This is a modified form of explicit, joint-to-joint mapping. The joint clusters are selected so that the robot joint motion imitates the human joint motion. For force-feedback telemanipulation, the joint torques are inversely mapped. Real-time control of the exoskeleton is run by a QNX Neutrino operating system on the exoskeleton controller computer. This computer communicates with the slave robot controller, or alternatively with a computer running a virtual model of the slave in a contact environment. The environment makes use of the open dynamics engine (ODE) for collision detection as well as for contact dynamics computations. All computers communicate via an Ethernet point-to-point link using TCP/IP. For the current 1 DoF implementation, set-points are exchanged between the master and the slave at 500 Hz sampling frequency. Control set-points of the four-channel controller include the joint velocities and joint torques. Data from the motor units of the exoskeleton are transferred to the exoskeleton controller via a CAN bus network. This limits the data rate for the full force-feedback implementation of EXARM and so will eventually be replaced by a real-time SpaceWire link that is currently under development at ESA.

8.3.4 Summary and conclusion

So far, various upper arm exoskeletons have been proposed for force-feedback teleoperation and haptic interaction with virtual environments, as well as for rehabilitation and physical training (Bergamasco *et al.*, 1994; Frisoli *et al.*, 2005; Sánchez *et al.*, 2006; Tsagarakis and Caldwell, 2003a; Williams *et al.*, 1998). Most such exoskeletons were designed like classical serial manipulator robots, with little attention to compatibility with the natural kinematics or comfort of the operators. However, poor user compatibility was reported especially in the case of rehabilitation exoskeletons, indicating a need for better physical human–robot interaction (Hidler and Wall, 2005). More recent developments of exoskeletons seek to approach those problems with innovative solutions. The Georgetown University exoskeleton (Carignan, Liszka and Roderick, 2005), for instance, aims to solve the alignment problem for the shoulder with an additional joint for scapula rotation.

The ESA EXARM exoskeleton (see Figure 8.11) has some distinctive innovative features for user ergonomics and improved physical human–robot interaction:

- The EXARM has an ergonomic kinematic structure that is inherently prone to misalignment between the operator's physiological joints and the robotic exoskeleton joints. Therefore, the EXARM does not create interaction forces that disturb or even harm the operator. Also, the EXARM does not restrict the natural motion of the human limbs.
- The EXARM actuation has been entirely relocated off the device. In this way the EXARM can be made lightweight, compact, wearable and portable. Thanks to these principal features, the EXARM (1) can operate smoothly with a wide range of user statures without requiring adjustments, (2) is highly comfortable during prolonged tasks, (3) is easy and quick to don and doff (less than 30 seconds) and (4) interacts naturally with the full range of shoulder, elbow and wrist motions.

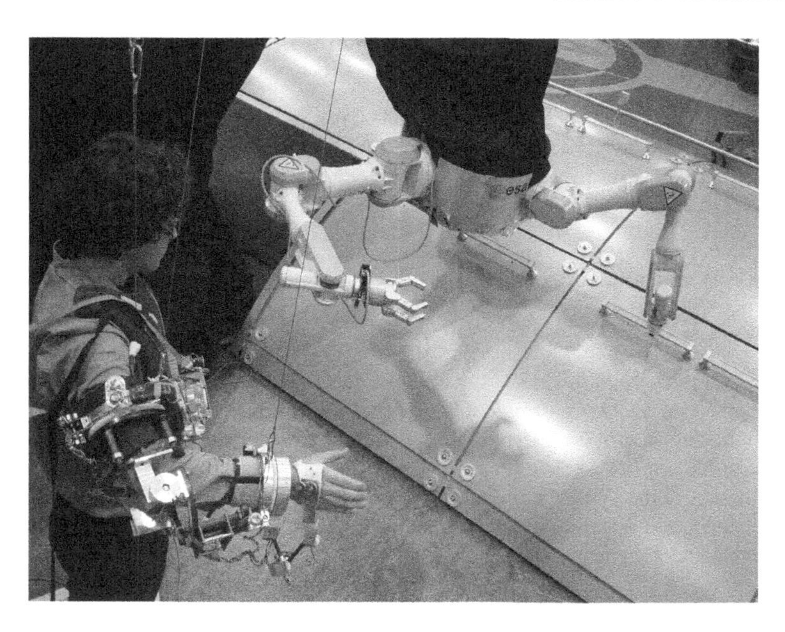

Figure 8.11 The EXARM exoskeleton used during teleoperation experiments with the EUROBOT prototype at the ESA automation and robotics laboratory

8.4 CASE STUDY: THE NEUROBOTICS EXOSKELETON (NEUROExos)

S. Roccella, E. Cattin, N. Vitiello, F. Vecchi and M. C. Carrozza

ARTS Lab, Scuola Superiore Sant'Anna, Pisa, Italy

A number of research groups have developed several complex and advanced robotic solutions to support upper limb motor activities by integrating advanced technologies in terms of actuation, sensors (see Chapter 6), control (see Chapters 4 and 5) and mechanisms (see Chapter 3), in particular Caldwell and Tsagarakis (2003); Carignan, Liszka and Roderick (2005); Cavallaro *et al.* (2005); Kiguchi, Tanaka and Fukuda (2004). However, unlike the examples cited above, in which the interaction between the human upper limb and exoskeleton is localized at a fixed number of points, a novel *totally wearable robotic system* will be physically coupled to the human arm. This means that the human upper limb will have a distributed interface with the exoskeleton. The exoskeleton design will therefore be based on human biomechanics, control and learning strategies, and hence the development of innovative exoskeletons for the upper limb demands a close alliance of robotics and neuroscience.

The novel robotic platform named NEUROExos is being developed within the framework of the NEUROBOTICS Integrated Project (fusion of NEUROscience and roBOTICS, IST-FET Project 2003-001917). The design and the development of NEUROExos will combine neuroscience and robotics. The objectives of the project are:

- To determine how humans can control the robotic system with a noninvasive interface and an easy, simple and fast intention decoder. The objective is to develop a *natural* control interface based on natural user behaviour.

- To determine how to couple the external manipulator (exos) to the internal human arm (internal manipulator) by monitoring the mechanical interface between them (inner exos surface versus human skin) and the forces.
- To control the external artificial actuator system acting parallel to the internal natural musculoskeletal system. The idea is to investigate the human control strategy and replicate it in the external actuator system. The user's natural movement will be enhanced (and not hampered) by the external actuated system.
- To continuously match the exos impedance to the human arm impedance by agonist/antagonist control of the NEUROExos joints.

In addition, as discussed in this book, the following principal requirements need to be considered for the development of a *totally wearable robotic exoskeleton*: kinematic coupling of the mechanical structure with the human arm (see Section 5.2); mechanical structure able to support the arm and to transmit force at the arm–exoskeleton interface (see Section 5.3); actuation system able to generate the required torques at the joints; high wearability, comfort and safety; and a control system able to predict/drive human arm motion with support/augmentation functions (see Section 5.4).

The design methodology of the NEUROExos is summarized in the following steps:

1. *Definition and development of an experimental setup to be carried out under controlled reference experiments.* Catching a moving object was selected as a benchmarking task because it is well known in neuroscience and is challenging in terms of kinematics and dynamics. An ad hoc catching apparatus has been developed for experiments with human subjects.
2. *Experimental and analytical definition of a biomechanical model of the human arm.* This will be used to determine the typical kinematic and dynamic performances of the human arm during execution of the reference task. ADAMS software (LIFEMOD plug-in) has been used to derive the analytical model of the human arm on the basis of experimental kinematic data acquired using the specific experimental setup (see step 1). The biomechanical model of the arm has also been defined. It can be used to estimate the kinematic and dynamic parameters of the joints and body segments of the upper limb, the synergies of all the muscle activations and the mechanical impedance of the arm joints, and to simulate the dynamic and kinematic coupling between the human arm and exoskeleton. Some examples of application of the resulting biomechanical model are depicted in Figure 8.12.
3. *Definition of technical specifications of the actuation system.* This step is based on a biomechanical model of the human arm. Two actuators will act on each joint using two cables to implement a bio-inspired agonist/antagonist configuration. Joint stiffness will be controlled by coactivation of two actuators and transmission by two cables and two nonlinear springs.
4. *Design and development of a two link–two joint robotic arm, called NEUROARM.* This platform will be actuated with four actuators in an agonist/antagonist configuration (two actuators for each joint) to replicate the human arm in terms of: physical constraints (ranges of motion, mass, inertia, stiffness, weight); an agonist/antagonist scheme of actuation system; high planar motion performance during the reference task; and sensory functions. The impedance and other control strategies of the robotic platform are implemented to replicate the cinematic and dynamic performances of the human arm during execution of catching and reaching tasks.
5. *Integration of NEUROARM in NEUROExos.* The NEURARM will replicate the human arm during the execution of reaching tasks and will be coupled with an actuated exoskeleton in order to test the coupled control system in totally safe conditions. The NEUROExos will be made up of active shells, rigid joints, remote actuators, Bowden cables embedded in the structure, load cells, linear

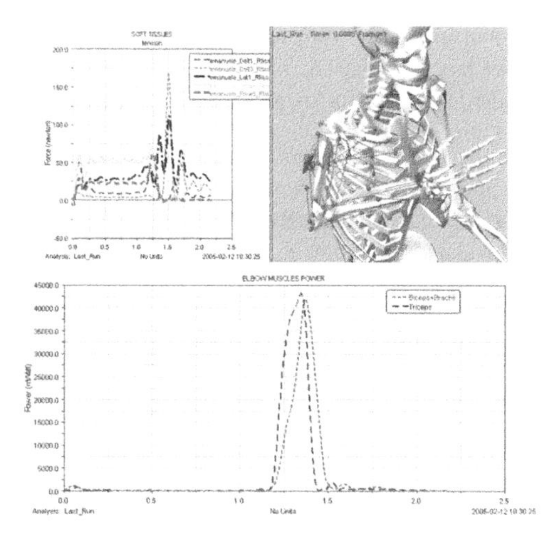

Figure 8.12 The biomechanical model may be used to estimate muscle activation during execution of the task, see the shoulder agonistic/antagonistic muscle activations (top view), and to estimate the dynamics of the joints, the elbow agonistic/antagonistic muscle power (bottom view)

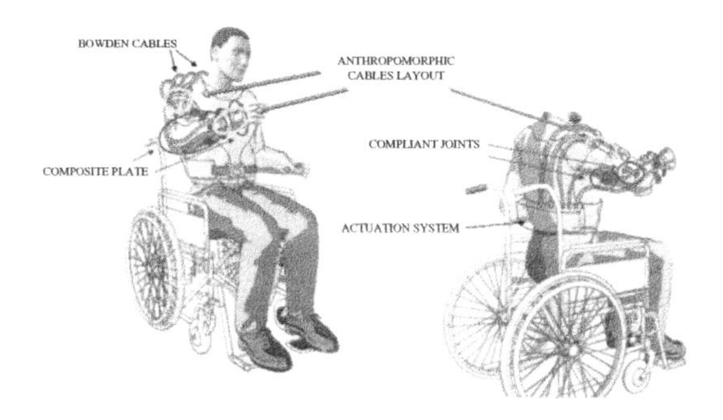

Figure 8.13 The concept of the final NEUROExos prototype

potentiometers, and mechatronic internal and external skin. Figure 8.13 depicts the concept of the final NEUROExos system.

8.4.1 Exoskeleton control approach

The NEUROExos system will be a hybrid bionic system integrating advanced control and learning models based on a complex sensory system. The following list of ongoing experiments is intended to define the human–machine interfaces and the control strategies that will be integrated in the final NEUROExos:

1. *To investigate the gaze and the artificial vestibular system as anticipatory triggers.* Normally, human gaze and human head movements anticipate the movements of the human body. It is proposed to investigate the application of this anticipatory behaviour as a trigger (Salvini, Laschi and Dario, 2006). This is an innovative solution intended to cut down the remapping normally required when

using traditional human–robot interfaces (for instance, to use a joystick to reach an end position with a manipulator, the user has to remap the movements of the hand to drive the manipulator) and is expected to reduce the learning effort required and improve system acceptability.

2. *To evaluate sensory fusion and real-time control.* A number of sensory data are processed and a number of control signals are generated by the NEUROExos controller, all in real time. These processes should not be noticeable to the NEUROExos user: the robot will be physically coupled to the user's arm and should adapt in real time to its movements, to gravity and to external events.

3. *To investigate human/robot force/torque/equilibrium position interaction.* The human will physically interact with NEUROExos. It is proposed to research three models of interaction using the NEURARM, and also shared control of the robot's end-point stiffness. Strategies for planning the NEUROExos end-point stiffness will likewise be investigated.

4. *To investigate the EMG interfaces for control of the force and equilibrium position of the NEURARM end-point.* The sEMG signals may be used in conjunction with kinematic signals to estimate the kinematics and torques of the joints, to predict the joint torque and stiffness, or to predict the kinematics of a joint using the kinematics of another joint (e.g. prediction of elbow kinematics through shoulder kinematics).

5. *To investigate the hybrid mechatronic–EMG approach to control of the exoskeleton.* The sEMG signals may also be used in conjunction with other sensory information from the mechatronic system to achieve anticipatory triggering between preselected trajectories or to implement more complex algorithms combining EMG–kinematics information in order to transform EMG-based force fields into joint space.

6. *To develop a mechatronic sensory skin that will be applied to the inner surface of the exoskeleton.* This will be used as a control interface (Beccai *et al.*, 2005, 2006). The mechatronic sensory skin, provided with contact and proximity sensors, will also be applied to the outer surface of the exoskeleton to implement automatic obstacle avoidance behaviours.

7. *To develop novel wearable control and cognitive feedback interfaces.* The ACHILLE interface is a wireless sensory insole that has been developed at the ARTS Lab and has been used to control multijoint hand prostheses and the PC pointer (Carrozza *et al.*, 2007). The same concept may be used to control some features of the NEUROExos and to feed cognitive information back to the user.

8.4.2 Application domains for the NEUROExos exoskeleton

The NEUROExos has been specifically designed as an orthotic device that is composed of two shells and can passively adapt to the kinematics of the human arm, ensuring comfort and safety for the user by means of innovative kinematic and transmission solutions (see Figure 8.14). Five patents are currently being prepared for the mechanical solutions adopted to design the arm–hand exoskeleton; also, novel bio-inspired control and learning strategies are being researched and will be integrated in the final system. The NEUROExos will be used in two main rehabilitation fields: to assist the physiotherapist during execution of exercises for neuro rehabilitation of the upper limb and to assist the disabled person during execution of functional motor tasks.

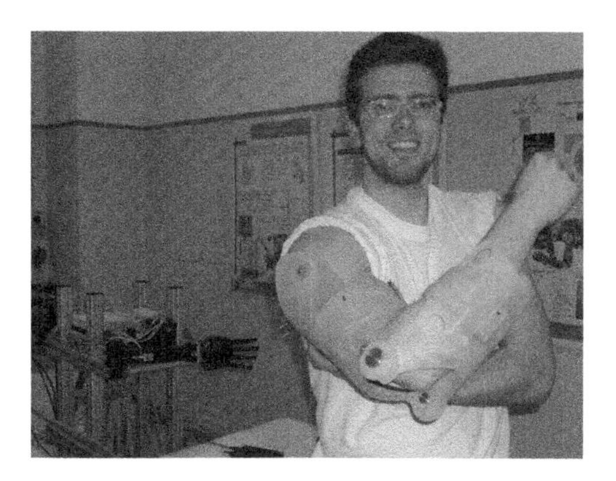

Figure 8.14 The first version of the NEUROExos

8.5 CASE STUDY: AN UPPER LIMB POWERED EXOSKELETON

J. C. Perry and J. Rosen

Department of Electrical Engineering, University of Washington, Seattle, Washington, USA

The system described in this case study is the third generation of an upper limb powered exoskeleton. Previous generations included a 1 DoF system (elbow joint, Figures 8.15(a) and (b)) and a 3 DoF system, with 2 DoF at the shoulder joint and 1 DoF at the elbow joint (Figure 8.15(c)). These systems were previously used to develop neural control algorithms for the upper limb (Rosen, Fuchs and Arcan, 1999; Rosen *et al.*, 2001). The third exoskeleton generation is a 7 DoF system which is the end result of a research effort reported in part by this case study. The system underwent several design iterations, depicted in Figures 8.15(d), (e) and (f). The final two-arm exoskeleton system is depicted in Figure 8.15(g).

8.5.1 Exoskeleton design

The design and development of a high-performance robotic device is a process with numerous competing factors. The mechanism weight and stiffness exist at opposite ends of the spectrum, the goal being to achieve the highest structural rigidity while maintaining the lowest segmental inertias. Contributing to these underlying requirements are factors such as the operational workspace, desired joint torques, motor placement, link design and cable selection. Since the device will operate in direct contact with humans, additional requirements emerge regarding comfort and safety of operation.

8.5.1.1 Design requirements

Kinematics and dynamics. In order to promote high performance while ensuring safe operation, the requirements must be realized and understood both from their technical as well as their functional

Figure 8.15 The first two prototypes of an upper limb powered exoskeleton. (a) A 1 DoF (elbow joint) powered exoskeleton was developed as a proof of the concept using myosignals as the primary command signals. (b) The 1 DoF exoskeleton system tested with a disabled person suffering from Tay-Sachs. (c) The 3 DoF (two shoulder joints, one elbow joint) powered exoskeleton was developed to study joint dependency during manipulation. (d) A wooden mockup including 7 DoF similar to the final design – note the singular configuration of the shoulder joint due to the joints orientation. (e) A conceptual CAD model of the 7 DoF exoskeleton arm – note how the two exoskeleton shoulder joints were reoriented to position the singular configuration on the periphery of the human arm workspace. (f) A detailed CAD design of the 7 DoF Exoskeleton arm (g) Two 7 DoF Exoskeleton arms

aspects. To better understand the kinematic and dynamic requirements of an exoskeleton arm for functional use a subject study ($n = 6$) analysing the kinematics and the dynamics of the human arm was first performed. In the study, upper limb kinematics were acquired with a motion capture system while subjects performed 24 activities of daily living. Based on previous surveys of the disabled community, the 24 activities were divided among the following four activity categories: general reaching, functional actions, eating and drinking, and hygiene. Utilizing a 7 DoF computational model of the human arm, the equations of motion were used to calculate joint torques from measured kinematics, resulting in a database that may provide a fundamental basis towards the development of assistive technologies for the human arm. Further details of the study may be found in Perry (2006) and Rosen *et al.* (2005).

Results of joint position and joint torque about each axis were condensed to a single set of histograms (Perry, 2006) (see Figure 8.16). While some position distributions appear quite normal in shape, others possess a bimodal or even trimodal form, where the centres of modes correspond to key anthropomorphic configurations. These configurations are positions of the arm that occur commonly throughout daily activities, often where joint velocities remain near zero at the initial or final periods of motion trajectories. In joint torque calculations, velocital and accelerational components were

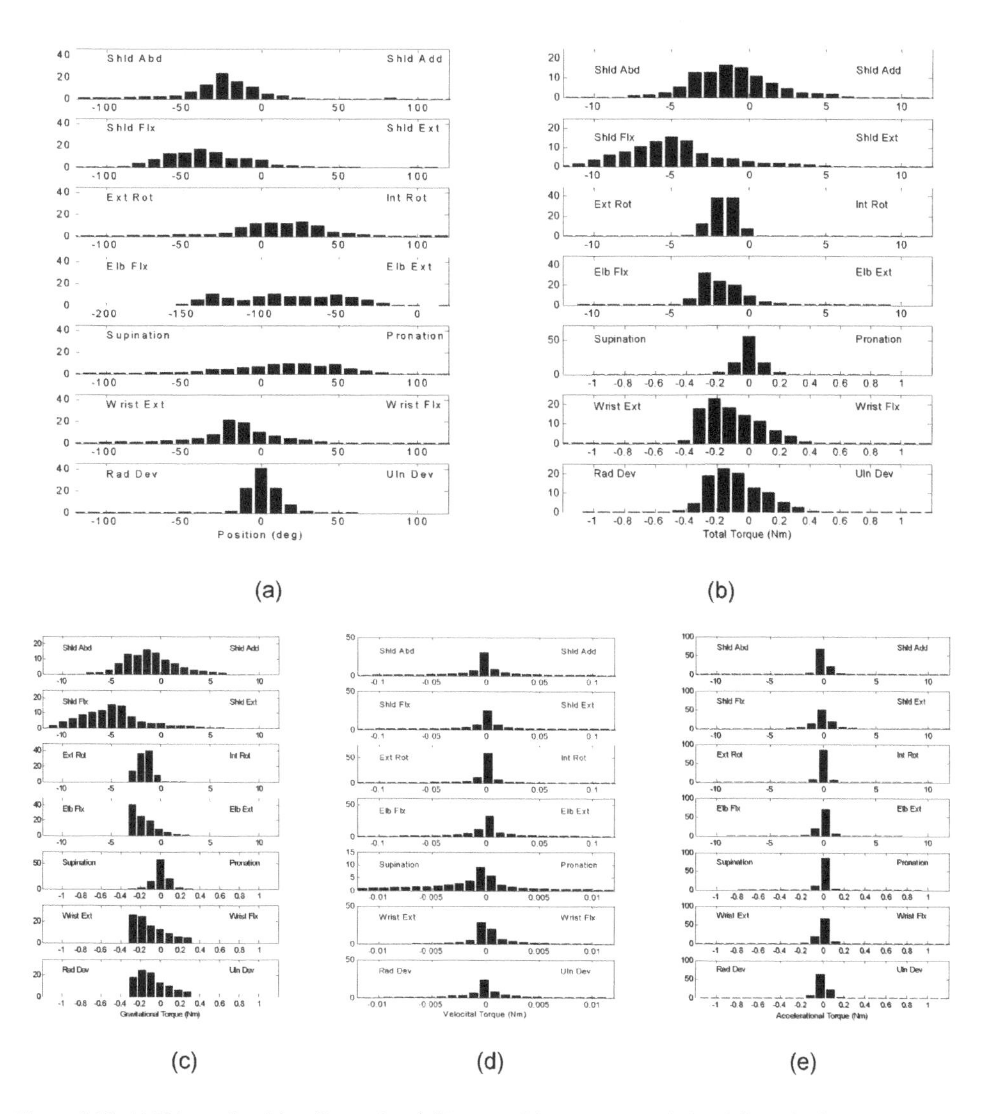

Figure 8.16 (a) Values of position (degrees) and (b) torque (N m) generated during daily activities at each of the 7 DoFs. Torque is expressed in terms of gravitational torque (c), velocital torque (d), and accelerational torque (e)

normally distributed about 0 N m, whereas gravitational component distributions varied with the joint. Additionally, velocital effects were found to contribute only one-hundredth of the total joint torque, whereas accelerational components contribute one-tenth of the total torque at the shoulder and elbow, and nearly half of the total torque at the wrist. The results of the study led directly towards the definition of mechanical and functional requirements for the design, and also provided insight regarding dominant aspects of dynamic motion that can be exploited in the implementation of a controller.

Mechanical human machine interfaces (mHMI). The mHMI, known throughout this book as pHRi, are the physical components that mechanically couple the human arm and the exoskeleton structure and enable force transmission between them. With awareness that one intended population of users will possess varying levels of muscular and functional impairment, an emphasis was placed on designing an interface that can easily be attached to the user.

To achieve axial rotation of exoskeleton limbs, three primary exoskeletal configurations are conceivable, and are illustrated in Figure 8.17. The first two configurations (Figures 8.17(a) and (b)) involve a single DoF bearing with its axis of rotation aligned collinearly with the approximate anatomical axis of rotation of the segment, while the third configuration (Figure 8.17(c)) involves a first axis that is displaced from the anatomical axis and a minimum of two additional noncollinear axes. In the first two configurations, the exoskeleton joint can be placed at either end of the long axis of the segment (Figure 8.17(a)), or axially between the ends of the segment (Figure 8.17(b)) using a bearing of minimum radius, r_b, greater than the maximum anthropometrical radius, r_a, about the corresponding segment axis. The additional axes of the third configuration are required to correct for noncollinearity of the first axis with respect to the rotating segment.

The configuration in Figure 8.17(a) offers a simple solution that allows for proximal placement of heavy components such as bearings and actuators, reducing inertial effects on power consumption, but the placement is undesirable due to human–machine interferences during shoulder abduction. The configuration in Figure 8.17(c) can avoid the interferences by displacing the joint axis laterally from the segment axis of rotation. However, the two additional joints, adding undesired weight and complexity to the design, are necessary to maintain proper rotation, as was achieved in previous configurations through the use of a single joint. The configuration in Figure 8.17(b) offers an alternative single DoF solution where the human–machine interferences associated with configuration 3a can

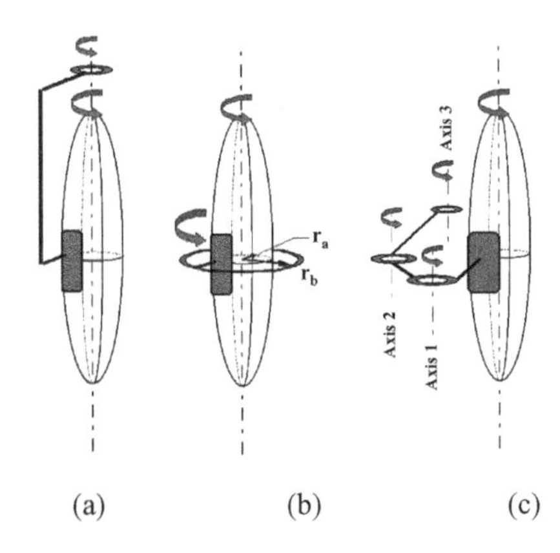

Figure 8.17 Three exoskeleton configurations that achieve rotation about the long axis of a limb segment, involving (a) a proximally placed single DoF, (b) a circumferencially placed single DoF and (c) three parallel, noncollinear DoFs

be removed. Full 360° bearings in this arrangement interfere with the torso when the arm is at rest or during motions that place distal arm joints near the body. Alternatively, these interferences can be removed through substitution of the full bearing with a partial bearing where the bearing track is affixed to the proximal exoskeleton link.

Current strength-to-weight limitations of available hardware necessitate nonmobile platforms for immediate upper limb exoskeleton technologies and, consequently, more user-friendly mHMIs. Strength-to-weight ratios of existing materials and electric motors, as well as energy-to-weight ratios of power supplies are not yet at the level necessary to support development of mobile platforms for partial-body upper limb exoskeletons. As a result, a full-body exoskeleton is required to support the existing weight of state-of-the-art power supplies, onboard controllers, and other upper limb hardware.

Modelling the human arm. Anthropomorphic joint approximations can be modelled at varying degrees of accuracy and complexity (Kapandji, 1982). The level of complexity needed for a suitable representation greatly depends on the desired tasks to be performed and replicated using the model. Shoulder motion, for example, composed of glenohumeral (GH), acromioclavicular and sternoclavicular articulations, can be represented by the GH joint for a variety of arm activities involving up to 90° of arm elevation. With minimal activity exceeding this range, a simplified model of the shoulder was deemed appropriate for the study. The GH movement can further be simplified to a ball-and-socket joint composed of three orthogonal axes intersecting at the centre of the humeral head, although the true centre of rotation is known to vary with arm orientation (Kapandji, 1982). Rotations about these orthogonal axes may be treated as Euler rotation. The order of flexion–extension and abduction–adduction about the first two axes is arbitrary but should be noted, while the third rotation corresponds to internal–external rotation.

Pronosupination of the forearm has been treated interchangeably in literature as a freedom of the elbow and as a freedom of the wrist. In either case, it should be considered directly adjacent to the forearm, occurring after elbow flexion and before either wrist flexion or deviation, with the axis of rotation running approximately through the 5th metacarpal–phalangeal joint (Kapandji, 1982).

The wrist can be modelled as two orthogonal axes with a fixed offset between them (Kapandji, 1982). The proximal and distal axes of the wrist correspond to wrist flexion–extension and wrist radial–ulnar deviation respectively.

Performance. A widely used quantitative measure to evaluate system performance is bandwidth. Systems having a higher bandwidth are controllable under higher frequency command signals. Limited by the system's lowest natural frequency, the bandwidth is a measure of how successfully tradeoffs between weight and stiffness are made. A target bandwidth of 10 Hz was selected based on the achievable frequency range of the human arm, which resides between 2 and 5 Hz (Kazerooni, 1990). Additional target values for the design include: weight (moving links) of 6.8 kg, maximum static payload of 2.5 kg, maximum angular deflection of 2° per joint, and bandwidth of 0–10 Hz. The actual weight was 3.5 kg and 6.3 kg for link 1 and links 2–7 respectively, where links are numbered sequentially between joints, and link 1 corresponds to the segment between joints 1 and 2 (Figure 8.17(e)).

Safety Requirements. Paramount to all HMIs is the guarantee of safe operation. Safety precautions have been implemented on three levels, built into the mechanical, electrical and software designs. In the mechanical design, physical stops prevent segments from excessive excursions that could hyperextend or hyperflex individual joints. Electrical brakes attached to the actuators allow the system to freeze the exoskeleton arm configuration mechanically in response to an emergency stop (e-stop). The electrical system is equipped with three emergency shutoff switches: an enable button that terminates the motor command signal upon release, a large e-stop button for complete power shutoff by the observer, and a similar e-stop foot switch for the user.

Ideally, the above safety measures would go unused as a result of adequate safeguards at the software level. Redundant position sensors, one at either end of the powertrain, monitor both joint motion as well as motor position. Differentiation of position provides knowledge of velocity and acceleration, both of which are incorporated into the control structure to prevent undesirable effects when approaching the joint limits. Redundancy of position sensing also enables software to monitor power transmission integrity. Any slip occurring between the motors and the end-effector will result in a position discrepancy and lead to immediate system shut-down. Software limits are implemented on commanded motor currents, i.e. motor torques.

8.5.1.2 Exoskeleton design

Exoskeletal joint design. Articulation of the exoskeleton is achieved about seven single-axis revolute joints: one for each shoulder abduction–adduction, shoulder flexion–extension, shoulder internal–external rotation, elbow flexion–extension, forearm pronation–supination, wrist flexion–extension and wrist radial–ulnar deviation. The exoskeletal joints are labelled 1 to 7 from proximal to distal in the order shown in Figure 8.18(e).

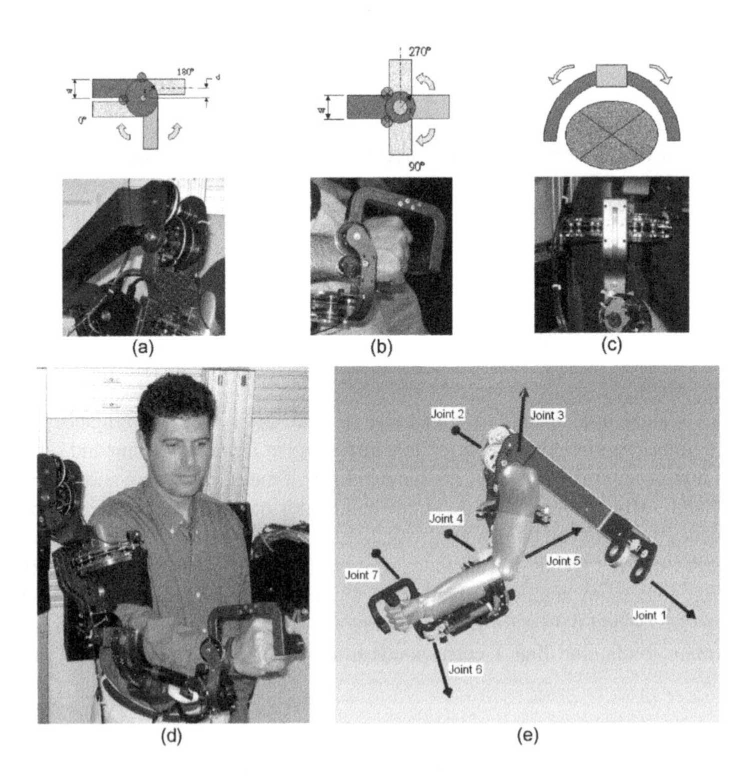

Figure 8.18 The exoskeleton is composed of three joint configurations: (a) 90° joints, (b) 180° joints and (c) axial joints. Together the joints produce an exoskeleton structure that achieves full glenohumeral, elbow and wrist functionality (d). (e) Exoskeletal axes assignment in relation to the human arm. Positive rotations about each joint produce the following motions: (1) combined flexion/abduction, (2) combined flexion/adduction, (3) internal rotation, (4) elbow flexion, (5) forearm pronation, (6) wrist extension and (7) wrist radial deviation

In the design of the current exoskeleton, three joint configurations emerged. The configurations can be classified as one of the following: (a) 90°, (b) 180°, or (c) axial. The distinction pertains to the relative alignment of adjoining links when the joint is approximately centred within its range of motion. While some joints of the body articulate about their mid-RoM when adjoining links are near orthogonal (Figure 8.18(a)), others do so when the links are near parallel (Figure 8.18(b)). A third configuration emerges in axial rotation of both the upper and lower arm segments (Figure 8.18(c)). As shown in Figure 8.18(d), exoskeleton joints 1 and 7 are modelled as 180° joints (Figure 8.18(b)), joints 2, 4 and 6 are 90° joints (Figure 8.18(a)), and joints 3 and 5 are axial joints (Figure 8.18(c)). Joint RoM in 90° and 180° configurations can be increased either by increasing the central radius, r, or decreasing the link width, w (top of figures 8.18(a) and (b)). Adjusting the link offset distance, d, shifts the joint limits, illustrated with small transparent circles, and effectively tunes the joint's mid-RoM.

The shoulder complex is reduced to the glenohumeral (GH) joint articulation and the GH joint is considered a spherical joint composed of three individual axes intersecting at its centre. The elbow is modelled by a single axis orthogonal to the third shoulder axis. A joint stop prevents the joint from hyperextension. Exoskeletal pronation–supination takes place midway between the elbow and wrist joints as it does in the physiological mechanism. Finally, two intersecting orthogonal axes represent the wrist. Although anthropometrically it would be more accurate to incorporate a slight offset between the flexion–extension and radial–ulnar deviation axes, this offset has been neglected for simplicity. The RoM achievable with the exoskeleton arm is as follows: 180° for shoulder flexion–extension and abduction–adduction, 166° for shoulder internal–external rotation, 150° for elbow flexion–extension and, at the wrist, 120° of flexion–extension, 60° of radio–ulnar deviation and 155° of pronation–supination. The current exoskeleton mHMI uses a semicircular bearing design to allow users to don the device without strain or discomfort. The semicircular guides are composed of three 60° curved rail–bearing segments (THK, Tokyo Japan).

Singularity placement. A singularity is a device configuration where a degree of freedom is lost or compromised as a result of an alignment of two rotational axes. In the development of a 7 DoF exoskeleton, the existence of singularities will depend on the desired reachable workspace. For devices that require large ranges of motion, i.e. motions greater than or equal to 180° in at least one joint, singularities cannot be eliminated. In this case, the challenge is to place the singularity in an unreachable or near unreachable location, such as the edge of the workspace.

For the exoskeleton arm, singularities occur when joints 1 and 3 or joints 3 and 5 align. To minimize the potential for this occurrence, the axis of joint 1 was positioned such that singularities with joint 3 take place only at locations that are anthropometrically hard to reach. To allow some user-specific flexibility in the design, the singular position is movable in 15° increments. For the placement shown in Figure 8.19, the singularity can be reached through simultaneous extension and abduction by 47.5° and 53.6° respectively (Figure 8.19(c)). Similarly, the same singularity can be reached through flexion and adduction of the upper arm by 132.5 and 53.6° respectively (Figure 8.19(d)). The singularity between joints 3 and 5 naturally occurs only in the full elbow extension (Figure 8.19(e)), i.e. on the edge of the forearm workspace. With each of these singularity vectors at or near the edge of the human workspace, the median of the workspace is free of singularities.

Another aspect to consider when placing singularities is mechanical isotropy. For optimal ease of movement in any direction, singular axes should be placed orthogonal to directions where isotropy is of the highest importance. For the singularity placement shown, isotropy will be maximized in 42.5° of shoulder flexion and 26.4° of shoulder abduction, values that lie in the median of shoulder RoM from the ADL study.

Power actuation and transmission. To date, the transmission of power from one location to another is achieved through a variety of means such as shafts, cables, fluid lines, and geartrains. Each method

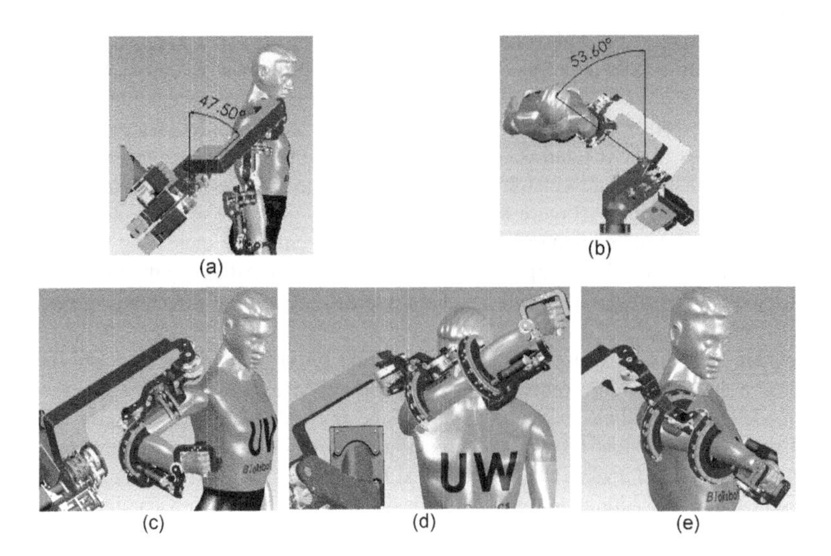

Figure 8.19 Mechanical singularities between axes 1 and 3 occur around the shoulder internal–external rotation axis in configurations (c) and (d). A singularity between axes 3 and 5 also occurs in the full elbow extension (e)

has specific applications where its characteristics are best suited. In the field of wearable robotics, weight is a critical factor that frequently must be sacrificed for the sake of strength or rigidity. However, development of a rigid structure that lacks adequate bandwidth is as ineffective of a tool as one that is lightweight but lacks structural rigidity. To achieve both rigidity and bandwidth, critical decisions were made regarding transmission type and placement of actuators.

Cable-drive systems. Cable-drive or tendon-driven systems have been in use on larger-scale devices long before their introduction into the world of biorobotics and microsurgery. In robotic haptics and wearable robotics applications (Salisbury *et al.*, 1988), cable drives are used due to their ability to transmit loads over long distances from an actuator located on a stationary base without the friction or backlash inherent to gears. The low friction associated with cable drives make them back-drivable – a characteristic that is essential for applications in haptics. Moreover, the absence of backlash is achieved through the structural continuity of the cable, enabling a direct link between the driving shaft and the shaft or link being driven. For these reasons, a cable-driven design was selected.

Selection and placement of actuators. As the heaviest components in the design, placement of the motors was a crucial decision. Motors for joints 1 to 4 were mounted on the stationary base, achieving a 60 % reduction in overall weight of the moving parts. The remaining three motors, whose torque requirements are substantially less, were positioned on the forearm. As each motor carries the weight and inertia of the more distally placed motors, the importance of high power-to-weight ratios increases from shoulder to wrist. Shoulder and elbow joints are each driven by a high-torque and low power-to-weight motor (6.2 N m, 2.2 N m/kg), while wrist joints are driven by a lower-torque and high power-to-weight motor (1.0 N m, 4.2 N m/kg). Motors are rare earth (RE) brushed motors (Maxon Motor, Switzerland).

Two-stage pulley reduction. Pulley arrangements can be used to create speed reductions in cable transmissions. Neglecting frictional losses, power throughout the transmission remains constant while tradeoffs between torque and angular velocity can be made. At the motor, the required torque is low while angular velocity is high, whereas at the joint, the torque is high and angular velocity is low. Lower torque corresponds to lower cable tension in stage 1, resulting in less strain and, therefore, less stretch per unit length of cable. Minimizing the length of stage 2 and routing the cable in stage 1 through the majority of the exoskeleton structure maximizes the overall transmission stiffness. Two-stage pulley reductions have been implemented in joints 1 to 4, whereas reductions at the wrist are composed of a single-stage pulley reduction following a single-stage planetary gear reduction. Total reductions for each joint are approximately as follows: 10:1 (joints 1–3), 15:1 (joint 4), 30:1 (joints 5–7).

Cable selection. Steel cables, also referred to as wire rope, are available in a variety of strengths, constructions and coatings. Although cable strength generally increases with diameter, the effective minimum bend radius is decreased. Cable compliance, cost, and construction stretch generally increase with strand count. A 7×19 cable, composed of 133 individual strands, offers moderate strength and flexibility and is recommended for use with pulleys as small as 25 times the cable diameter (SAVA Industries, Riverdale). Applications requiring high-strength cables and small-diameter pulleys, less than 1/25th the cable diameter, should utilize a higher-count cable construction. The exoskeleton has been developed with both 7×19 and 7×49 cable constructions.

8.5.1.3 System integration

The system is controlled by two PCs (the servo PC and virtual reality (VR) PC). The servo PC is responsible for maintaining low-level servo control. The VR PC runs VR applications and projects a visual view of the virtual environment into the VR goggles (Figure 8.20). The VR PC maintains the current state of the VR environment and calculates via the inherent physics engine the force feedback that needs to be rendered and applied by the exoskeleton arms. The UDP protocol is used as the communication protocol between the servo PC and the VR PC. The servo PC acquires the joint positions of the exoskeleton, which is physically coupled with the human arm, as well as the interaction forces and torques between the operator and the exoskeleton device. Based on the required force feedback that is calculated by the VR PC, the servo PC provides servo command to the actuators

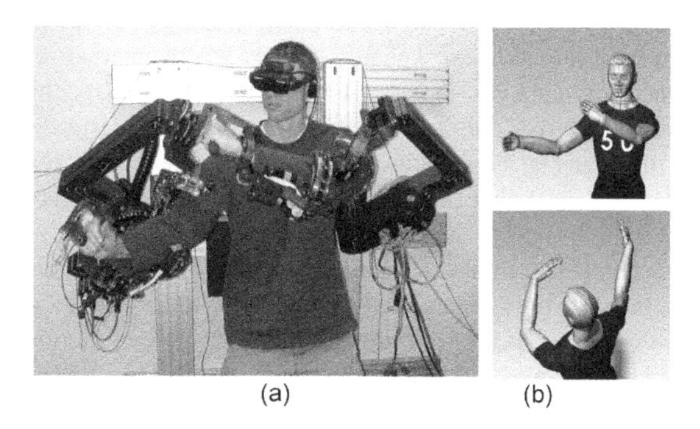

(a) (b)

Figure 8.20 The exoskeleton system operating in a VR environment mode: (a) a user wearing the arm and a head-mounted display for viewing a virtual environment and (b) the virtual environment representation as seen through the head-mounted display

to generate the appropriate joint torques that simulate physical interaction with the virtual object. The current joint position is transmitted by the servo PC to the VR PC for keeping the VR environment up to date.

Three multiaxis force/torque sensors are located at the exoskeleton mHMI (upper arm, forearm, and hand) measuring the interaction between the human arm and the exoskeleton system. Redundant position sensing capabilities were also incorporated into the mechanism (encoders located on the servo DC motors and potentiometers located on the joints themselves). Analogue-to-digital converters (ADCs) as well as encoder counters are used to acquire all the analogue and digital signals. A digital-to analogue converter (DAC) is used to control the DC motor through their linear amplifiers.

8.5.2 Conclusions and discussion

The integration of a human and wearable robot into a single system offers remarkable opportunities for creating a new generation of assistive technology and human–computer interface that may benefit members of both healthy and disabled populations. The same device with different control algorithms may be used in four fundamental modes of operation, although existing devices, see below, are typically limited to one or two. The exoskeleton system developed in this research effort has been designed to operate under the following four modalities:

1. Physiotherapy. The patient wearing an exoskeleton performs task-based occupational or physical therapy in an active or passive mode (Fasoli *et al.*, 2003; Hogan *et al.*, 1992; Krebs *et al.*, 2002).
2. Assistive device (human amplifier). The operator feels scaled-down loads while interacting with objects in the environment, most of the load being carried by the exoskeleton (Kazerooni, 1996).
3. Haptic device. The subject physically interacts with virtual objects while the forces generated through the interactions are fed back to the user through the exoskeleton, conveying shape, stiffness, texture or other characteristics of the virtual objects (Frisoli *et al.*, 2005).
4. Master device. Replacing the virtual environment with a real robot, the operator uses the exoskeleton to control a robotic system in a teleoperation (master–slave) mode, where the exoskeleton reflects back to the user the forces generated as the slave robot interacts with the environment (Jau, 1988).

The aim of this case study was to describe the development of a 7 DoF upper limb exoskeleton that is based on anthropometric data as well as on the kinematics and dynamics of the arm in activities of daily living. In contrast to previous exoskeleton devices where internal–external rotation joints and pronosupination joints fully enclosed the arm (Frisoli *et al.*, 2005; Kiguchi, Tanaka and Fukuda, 2004; Repperger, Remis and Merrill, 1990), the current exoskeleton design uses open mHMIs for both upper and lower arm segments. This feature greatly reduces challenges associated with donning and doffing by impaired users, a task that can be difficult and even uncomfortable with closed bearing configurations.

Although some studies report that joints, particularly at the shoulder, can achieve ranges of motion exceeding 180°, most joints can only reach such excursions with contributions from neighbouring joints. Despite the GH joint appearance of providing more than 180° of motion about all three axes, this is due largely to scapular motion. As a result, joints capable of providing 180° of motion, or less, using the three configurations described previously are sufficient to develop an arm exoskeleton with full GH, elbow, and wrist joint functionality.

Due to the unique placement of the shoulder singularity in this device, pure shoulder flexion is achieved through a combination of rotations about the first two joints of the shoulder (exoskeleton joints 1 and 2). Additionally, this unique placement moves the region of highest shoulder joint isotropy into the area of the workspace most often utilized during functional tasks. This combined effect of placing the singularity in the periphery of the workspace while maintaining high isotropy in the central workspace leads to a device configuration that is highly suited for exoskeleton applications.

As a final remark, it is worth noting several aspects of transmission integrity with regard to cable-driven systems. Depending on the length of the transmission, even small changes in cable length may result in excessively high or excessively low cable tensions, both sources of undesirable effects. For this reason, care must be taken at all cable termination sites to ensure constant cable lengths, as cables wrap around either reduction pulleys, drive pulleys, or capstans. The tension that results, and therefore the amount of length change allowable, will vary with the length of cable in the particular stage. Cable runs that are long can endure higher amounts of cable stretch without undergoing significant increases in tension. This, combined with large gear reductions, can result in significant lateral travel of cables as they wrap around motor capstans. The nearest idler pulley should be located sufficiently far from the capstan to maintain proper alignment (less than $\approx 2.5^\circ$ offset) between the cable and helical grooves in the capstan.

The research effort described in this case study represents not only a contribution towards the advancement of haptics and human–computer interfaces but also towards a more general understanding of the human upper limb. The exoskeleton is a unique but versatile high-performance two-way interface, designed, fabricated and integrated to the highest industry standards, and will support further research along a number of academic pathways towards a deeper understanding of the human body, the neuromuscular system, and the optimal modalities for neuromuscular rehabilitation.

8.6 CASE STUDY: SOFT EXOSKELETON FOR USE IN PHYSIOTHERAPY AND TRAINING

N. G. Tsagarakis[1], D. G. Caldwell[1] and S. Kousidou[2]

[1]*Italian Institute of Technology, Genova, Italy*
[2]*Centre of Robotics and Automation, University of Salford, Salford, UK*

Full or partial loss of function in the shoulder, elbow or wrist is an increasingly common ailment associated with a wide range of injuries, disease processes and other conditions including sports injuries, occupational injuries, spinal cord injuries (SCIs) and strokes. These impairments can be of varying degrees of severity. Hemiplegia, the most common impairment resulting from a stroke, leaves the survivor with a stronger unimpaired arm and a weaker impaired one. Impairments such as muscle weakness, loss of range of motion, reduced reaction times and disorderly movement organization create deficits in motor control that affect patients' ability to live independently (Parker, Wade and hangton, 1986).

In most such cases intensive and repetitive physiotherapy may be necessary to modify neural organization and recover functional motor skills (Carr and Shepherd, 1987). However:

1. upper-limb disability rates low on the priority list for urgent medical assistance because it is seldom considered life-threatening. Therefore, physiotherapy tends to follow only days or even weeks after admission.

2. Treatment for these conditions typically relies to some extent on manipulative physiotherapy procedures, which by their very nature are highly labour intensive, requiring intensive one-to-one attention from highly skilled medical personnel.

Therefore, reducing the task load for these professionals through the use of assistive orthotics could produce major benefits in terms of overall healthcare provision and the cost of that provision. At the same time, providing greater access to effective rehabilitation regimes could be of significance to the healing process of the patient (Nakayama *et al.*, 1994).

This article presents the construction and testing of a soft-exoskeleton system for upper limb training/physiotherapy. The system makes use of a new range of pneumatic muscle actuators (pMA) as a power source for the system. This type of actuator addresses the need for safety, simplicity and lightness and also has an excellent power-to-weight ratio. The article shows how the system takes advantage of the inherent controllable compliance to produce a device that can provide a wide range of functionality with a high assurance of safety for the patient. The general layout of the arm orthosis is presented. This includes the mechanical design, the actuation and the control description. Results of preliminary experiments demonstrate the potential of the device as an upper limb training, physiotherapy and power assist (soft exoskeleton) system.

8.6.1 Soft arm–exoskeleton design

8.6.1.1 Mechanical structure

The mechanical structure of the soft exoskeleton has 7 DoFs corresponding to the natural motion of the human arm from the shoulder to the wrist but excluding the hand (see Sections 3.2 and 3.3). The structure, which is built primarily of aluminium and composite materials with high stress joint sections fabricated in steel, has 3 DoFs in the shoulder (flexion–extension, abduction–adduction and lateral–medial rotation), 2 DoFs at the elbow (flexion–extension and pronation-supination of the forearm) and 2 DoFs at the wrist (flexion–extension and abduction–adduction) (Figure 8.21).

The exoskeleton arm is built for use by a ‘typical adult’ with only minor changes to the setup. Arm link lengths can be readily and quickly changed if necessary, making it easy to accommodate a range of users. This is an important aspect of the design. The arm is attached to the user’s arm at the elbow and the wrist using two broad Velcro(r) bands. This was found to be adequate during operation, combining secure attachment with ease of system donning and removal (Tsagarakis and Caldwell, 2003a).

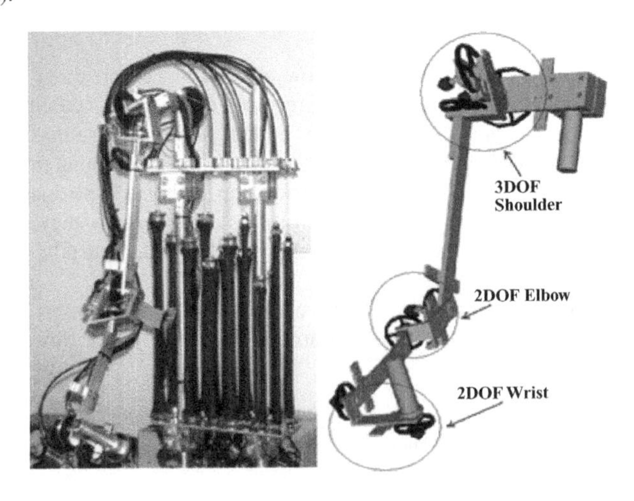

Figure 8.21 Soft Exoskeleton mechanical structure

8.6.1.2 Actuation

Actuators and actuation systems are essential, possibly defining, parts of all the exoskeleton structures that provide the forces, torques and mechanical motions needed to move the joints, limbs or body (see Section 6.3). Their performance is usually characterized by parameters such as power (particularly power-to-weight and power-to-volume ratios), strength, response rate, physical size, speed of motion, reliability, controllability, compliance and cost. The nature of the drive source proposed in this work forms a key subsystem making use of 'soft' compliant actuator technology. This is fundamentally different from any methods developed previously and is a key to the success of the technique (Tsagarakis and Caldwell, 2003b). The system uses braided pneumatic muscle actuators (pMA) to provide a clean, low-cost actuation source with a high power-to-weight ratio and good safety thanks to the inherent compliance.

These pneumatic muscle actuators (pMA) are built in the form of a two-layered cylinder with an inner rubber liner and outer containment layer of braided material, together with two end-caps that seal the open ends of the muscle. The braided material in this instance was nylon thread woven into a double helix, although other thread materials can also be used. The nature of this weave is such that when the actuator expands in the radial direction (following the introduction of air), the outer shell contracts in length while retaining its cylindrical shape, thus providing the required force through a 'tendon' attached at the ends of the muscle. The complete unit can safely withstand pressures up to 700 kPa (7 bar), although 600 kPa (6 bar) is the operating pressure for this system. Details of the construction, operation and mathematical analysis of these actuators can be found in Caldwell, Medrano-corda and goodwin (1995).

The structure of the muscles lends the actuator a number of desirable characteristics: muscles can be produced in a range of lengths and diameters, with size increments to produce increased contractile force; actuators have exceptionally high power-to-force to weight-to-volume ratios >1 kW/kg; the actual achievable displacement (contraction) depends on the construction and loading but is typically 30–35 % of the dilated length, which is comparable to the contraction achievable with natural muscle; when compared directly with human muscle the contractile force for a given cross-sectional area of actuator can be over 300 N/cm^2 for the pMA as compared to 20–40 N/cm^2 for natural muscle; soft construction and finite maximum contraction make pMA safe for human–machine interaction; force control using antagonistic pairs for compliance regulation is possible in the same way as for natural muscle; and the actuators can operate safely in dusty, aquatic or other liquid environments and are highly tolerant of mechanical (rotational and translational) misalignment, thus reducing engineering complexity and cost.

Joint motion and torque control in the soft-exoskeleton is achieved by producing appropriate antagonistic torques through cables and pulleys driven by the pneumatic actuators. Since the pneumatic muscle actuator is a one-way-acting element (contraction only), two elements are needed for bidirectional motion/force operation. These two acting elements work together in an antagonistic scheme, simulating a biceps–triceps system to provide the bidirectional motion/force (see Figure 8.22). In this setup, $L_{\min}$ denotes the length of the muscle when it is fully contracted; $L_0 = L_{\min} + (L_{\max} - L_{\min})/2$ is the initial dilated length of the muscle, which is equal to half of the maximum muscle displacement in order to maximize the range of motion of the joint; $L_{\max}$ is the maximum dilated length; r is the radius of the pulley; and P_1, P_2 are the gauge pressures inside the two muscles, monitored by miniature strain-gauge-based pressure sensors incorporated in each muscle (Tsagarakis, Tsachouridis and Coldwell, 2003). As the system rotates anticlockwise, the muscle on the top shortens while the muscle on the bottom lengthens, and vice versa.

Flexible steel cables are used for the coupling between the muscles and the pulley. Since most of the joints require a range of rotation in excess of 90°, double-groove pulleys are used. The pulleys are three-spoke structures made with solid, internally machined aluminium pieces. High-linearity sensors provide position sensing on each joint while joint torque sensing is achieved by strain gauges

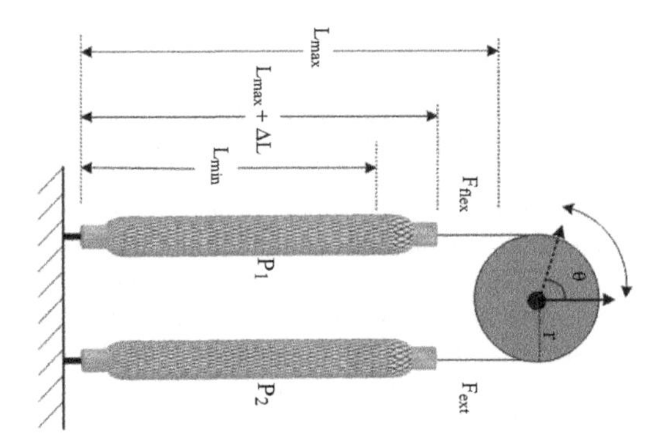

Figure 8.22 Antagonistic muscle configuration

(mounted on the internal spokes) built into each joint pulley. The muscles used in the soft exoskeleton have a diameter of 2–4 cm and an 'at rest' length varying from 15 to 45 cm.

8.6.2 System control

8.6.2.1 Low level joint control

Low-level joint torque or position control has been implemented on each joint (see Figure 8.23). Using the torque/position feedback signal, a high-bandwidth torque or position control loop can be formed around each individual joint depending on the operation. The torque or position control loop uses the torque/position error to calculate the required amount of pressure change in the two muscles of the antagonistic pair. The command pressures for the muscles at each cycle are given by

$$P_1 = P_0 - \Delta P \tag{8.2}$$

$$P_2 = P_0 + \Delta P \tag{8.3}$$

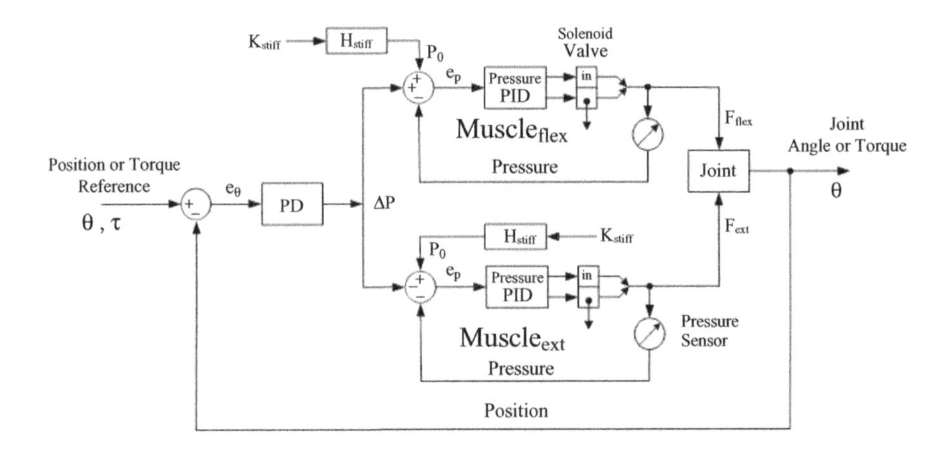

Figure 8.23 Block diagram of a single joint torque control scheme

where ΔP is computed using a PD control law

$$\Delta P = K_{pr}e + T_d\dot{e} \tag{8.4}$$

where

$$e = \tau_d - \tau_s \quad \text{or} \quad e = \theta_d - \theta_s \tag{8.5}$$

is the joint torque or position error.

The two command pressures, P_1 and P_2, form the input for the two inner pressure control loops. The pressure feedback signal in the pressure loops is supplied by means of the pressure sensors contained in each muscle of the antagonistic pair. The output of these inner pressure control loops are the times t_1 and t_2 corresponding to the duty cycle of the pulse width modulation (PWM) signal that drives the solenoid valves. Positive values for t_1 and t_2 activate the filling valves while negative values switch on the venting sequences. The above control loop makes the actuator/joint system behave as a pure torque or position source and provides improved torque or position response. With this control method, open-loop joint stiffness control can also be achieved by varying the amount of pressure, P_0, that is added to the output of the outer control loop (Tsagarakis and Caldwell, 2003a).

8.6.2.2 Soft exoskeleton assistive/training control scheme

An impedance control scheme operating in an assistive/augmentation (or resistive for exercise) mode is employed for the overall exoskeleton system to enable execution of complex assistive/resistive exercises (Tsagarakis and Caldwell, 2003a). The dynamic behaviour of the exoskeleton is described by the following equation:

$$\mathrm{M}(\boldsymbol{q})\ddot{q} + \boldsymbol{V}(\boldsymbol{q}, \dot{\boldsymbol{q}}) + \boldsymbol{F}(\dot{\boldsymbol{q}}) + \boldsymbol{G}(\boldsymbol{q}) + \mathrm{J}^{\mathrm{T}} F_{\mathrm{R}} = \tau_{\mathrm{joint}} \tag{8.6}$$

where $\boldsymbol{q}$ is the joint variable $\boldsymbol{n}$-vector, τ_{joint} is the joint torque vector, $\mathrm{M}(q)$ is the inertia matrix, $V(q, \dot{q})$ is the coriolis/centripetal vector, $F(\dot{q})$ is the friction vector, $G(q)$ is the gravity vector, F_{R} is the force that the arm generates at the end-tip and J^{T} is the transpose Jacobian of the exoskeleton (see Section 3.2.2). Whenever the user's limb is attached to the exoskeleton, the above equation can be used to describe the interaction between the operator and the exoskeleton.

Let F_{H} denote the force that the human exerts on the exoskeleton arm end-tip (which is actually the force felt by the user), F_{R} be the force that the exoskeleton applies to the operator and $Z_{\mathrm{E}}(s)$ be the system-simulated mechanical impedance. To make the operator feel the simulated augmentation/resistive dynamics, the following equation must be applied:

$$Z_{\mathrm{E}}(s)\,(x - x_{\mathrm{E}}) = M_{\mathrm{E}}\ddot{x} + B_{\mathrm{E}}\dot{x} + K_{\mathrm{E}}\,(x - x_{\mathrm{E}}) = F_{\mathrm{H}} \tag{8.7}$$

where M_{E}, B_{E} and K_{E} are the inertia, damping and stiffness coefficients respectively. The above equation defines the desired characteristics of the motion of the pair (operator, exoskeleton). Having specified the desired behaviour of the system, the control law can now be derived by eliminating $\ddot{x}$ and $\ddot{q}$ from Equations (8.15) and (8.17). To do this the following equations are introduced, which relate the velocities and accelerations of the exoskeletal trainer end-point to the velocities and accelerations in joint space:

$$\dot{\boldsymbol{x}} = \mathbf{J}\dot{\boldsymbol{q}} \tag{8.8}$$

$$\ddot{\boldsymbol{x}} = \mathbf{J}\ddot{\boldsymbol{q}} + \dot{\mathbf{J}}\dot{\boldsymbol{q}} \tag{8.9}$$

Solving Equations (8.7) and (8.9) for $\ddot{x}$ and $\ddot{q}$ respectively gives

$$\ddot{\boldsymbol{x}} = \mathbf{M}_{\mathrm{E}}^{-1}\,(\mathbf{F}_{\mathrm{H}} - \mathbf{B}_{\mathrm{E}}\dot{\boldsymbol{x}} - \mathbf{K}_{\mathrm{E}}\,(x - x_{\mathrm{E}})) \tag{8.10}$$

$$\ddot{\boldsymbol{q}} = \mathbf{J}^{-1}\left(\ddot{\boldsymbol{x}} - \dot{\mathbf{J}}\dot{\boldsymbol{q}}\right) \tag{8.11}$$

Combining Equations (8.6), (8.10) and (8.11), $\ddot{q}$ can be eliminated to give

$$\mathbf{M}(q)\mathbf{J}^{-1}\left((F_{\mathrm{H}} - \mathbf{B}_{\mathrm{E}}\dot{x} - K_{\mathrm{E}}(x - x_{E})) - \dot{\mathbf{J}}\dot{q}\right) + G(q) = \boldsymbol{\tau}_{\mathrm{joint}} - \mathbf{J}^{\mathrm{T}} F_{\mathrm{R}} \tag{8.12}$$

To keep the Cartesian inertia of the human arm/exoskeleton unchanged:

$$\mathbf{M}_{\mathrm{E}} = \mathbf{J}^{-1}\mathbf{M}\mathbf{J}^{-\mathrm{T}} \tag{8.13}$$

Considering slow motions typical in rehabilitation/physiotherapy applications, Equation (8.12) gives

$$\boldsymbol{\tau}_{\mathrm{joint}} = -\mathbf{J}^{\mathrm{T}}\left(\mathbf{B}_{\mathrm{E}}\dot{x} - K_{\mathrm{E}}(x - x_{\mathrm{E}})\right) + G(q) \tag{8.14}$$

The above equation describes the impedance control law for the overall soft exoskeleton. The damping and the stiffness matrices $\mathbf{B}_{\mathrm{E}}$ and K_{E} are 6×6 diagonal and depend on the dynamics that are to be modelled. To enable simulation of effects such as static force, the control equation (8.14) can be modified by including a bias force matrix F_{bias} as follows:

$$\boldsymbol{\tau}_{\mathrm{joint}} = -\mathbf{J}^{\mathrm{T}}\left(\mathbf{B}_{\mathrm{E}}\dot{x} - K_{\mathrm{E}}(x - x_{\mathrm{E}}) + \mathbf{F}_{\mathrm{bias}}\right) + G(q) \tag{8.15}$$

where $\mathbf{F}_{\mathrm{bias}}$ is a 6×1 bias force/torque matrix that can be used for simulation of special effects like virtual weight lifting.

In the assistive control mode the exoskeleton system applies assistive force signals using Equation (8.14), depending on the motion desired by the operator. To enable detection of the user's desired motion, a force/torque (F/T) sensor is mounted on the distal section (end-tip) of the exoskeleton. This sensor monitors the force signals applied by the user. The desired position of the system is updated on the basis of these force signals in Equation (8.14).

The following formula is used to derive the new desired position using the sensed force signal:

$$\boldsymbol{x}_{\mathrm{E}}^{i} = \boldsymbol{x}_{\mathrm{E}-1}^{i} + \int \boldsymbol{x}'^{i}_{f}\,\mathrm{d}t, \quad i = 1, \ldots, 6 \tag{8.16}$$

$$\boldsymbol{x}'^{i}_{f} = \left\{ \begin{array}{cc} k_a\left(\boldsymbol{F}_s^i - a\right), & \boldsymbol{F}_s^i > a \\ 0, & -a < \boldsymbol{F}_s^i < a \\ k_a\left(\boldsymbol{F}_s^i + a\right), & \boldsymbol{F}_s^i < -a \end{array} \right\} \tag{8.17}$$

where $x_{\mathrm{E}}^{i} = \left[x_{\mathrm{E}}^{1}, \ldots, x_{\mathrm{E}}^{6}\right]^{\mathrm{T}}$ is the desired position vector, $F_{\mathrm{s}}^{i} = \left[F_{\mathrm{s}}^{1}, \ldots, F_{\mathrm{s}}^{6}\right]^{\mathrm{T}}$ is the force vector from the F/T sensor, k_{a} is a sensitivity coefficient that can be adjusted according to the user's physical state and a is the deadband parameter. By injecting the desired position vector derived from Equations (8.16) and (8.17) into Equation (8.14), assistive forces for the user's desired actions/motions can be generated.

8.6.2.3 Soft exoskeleton overall control architecture for physiotherapy and rehabilitation

In addition to the above low–medium level control schemes, a highly modular object-oriented control framework has been developed to enable the execution of rehabilitation/physiotherapy procedures with the soft exoskeleton. The main framework components, grouped by functionality and the flow of data, are depicted in Figure 8.24.

Through the graphical user interface (GUI), the therapist can select an existing protocol or create a new one. A protocol can be synthesized by combining the basic tasks that comprise its building blocks. A basic task can be very simple, e.g. elbow flexion or shoulder extension, or more complex, e.g. reaching for a mug. Other attributes of a protocol are the number of repetitions and the rest period

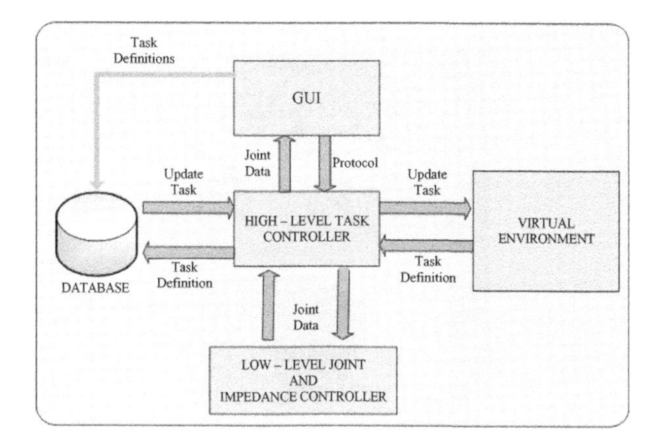

Figure 8.24 Overall control framework

between two consecutive tasks. The simplest tasks that can comprise a protocol, called primitive tasks, are the possible movements around all the arm joints. Primitive tasks have attributes such as start and stop angles and speed. Nonprimitive tasks can also have their speed adjusted.

In order to synthesize a protocol, the therapist combines a number of tasks (primitive or nonprimitive) with information about the number of repetitions, the interval after each repetition and the rest period between the different tasks. After a protocol is synthesized, the therapist can save it in the database, where it can be globally accessed. Alternatively, the therapist can select a protocol from the database and customize its parameters for a patient. In order for the protocol to be executed, the GUI module passes it on to the high-level task controller (HLTC), where it is resolved to low-level joint signals that are fed to the low-level joint controller (LLJC) (Kousidou *et al.*, 2006).

8.6.3 Experimental results

A number of preliminary experiments were carried out to evaluate the performance of the system as a physiotherapy and augmentation device for the upper limb. The results are presented in the following sections.

8.6.3.1 Shoulder strengthening with weight training

There is one group of rehabilitation exercises, often used for shoulder training or treatment after injury, that are based on consistent repetitive motions using small weights. In this experiment the training exoskeleton was configured to simulate the forces generated by a virtual constant load located at the elbow joint. The arm exoskeleton was securely attached to the operator's arm and the control matrices in Equation (8.15) were set up to simulate a 2 kg load. During the experiment the operators repeated a shoulder abduction–adduction exercise, as shown in Figure 8.25. During these motions the position and the output torque of the shoulder abduction–adduction joint were recorded.

The results presented in Figure 8.26(a) show the abduction–adduction motions for a typical test subject. Graph (b) introduces the desired and the output torque of the shoulder abduction–adduction joint as recorded during the experiment. The next graphs illustrate the output load as a function of (c) time and (d) joint position. The maximum external load error is less than 2.5 % for the whole range of motion. In terms of actual sensation, the test subjects (10 male subjects aged 22–35)

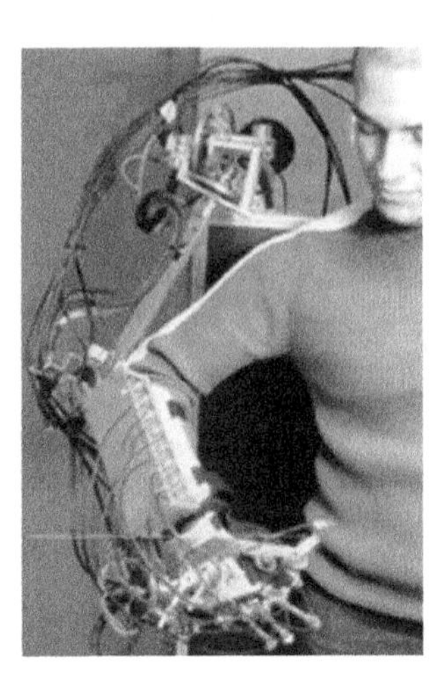

Figure 8.25 Shoulder training experiment

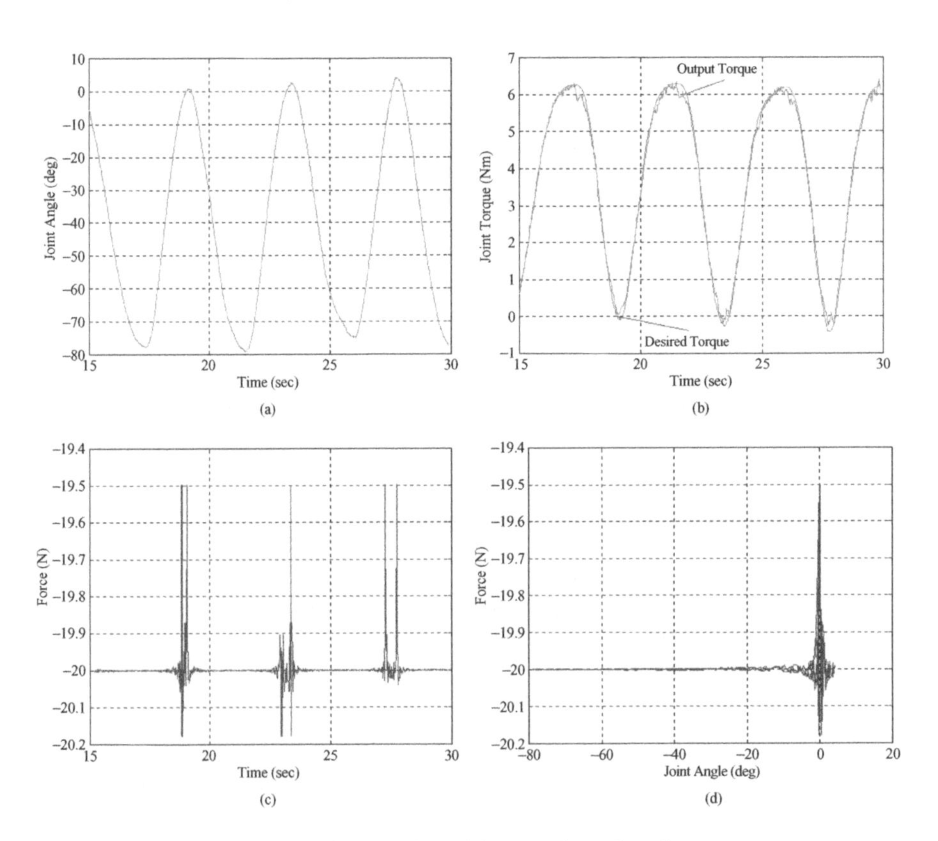

Figure 8.26 Shoulder training experimental results

reported that the sensation produced by the external load was very similar to the natural sensation. The feedback that they gave was very encouraging as regards the possibility of using the system as a training/rehabilitation device and applying this to a more extensive range of physiotherapy training regimes in addition to applications as a power assist system.

8.6.3.2 Assistive mode experiment

In this experiment, the assistive capacity was assessed by evaluating the performance of the soft exoskeleton in supplying assistive force signals to healthy subjects. The device was securely attached to the operator's arm and was configured to work in assistive mode. A 2 kg load was attached at the exoskeleton end-tip and the users were prompted to manipulate the exoskeleton device and try to lift the attached load vertically. Meanwhile, the force sensed at the F/T sensor was recorded. Figure 8.27(a) shows the force profile detected by the sensor, which is the actual force felt by the user. The force initially increases as the user starts trying to move the exoskeleton system in order to lift the load. The increased force levels cause changes in the desired vertical coordinate Y (see Figure 8.27(b)), which subsequently causes assistive forces governed by Equation (8.17) to be generated at the tip of the system. As a result, the force applied by the user (see Figure 8.27(a)), decreases. This is indicative of the capability of the system and the proposed assistive control scheme to supply assistive forces under user motion control.

8.6.4 Conclusions

In this section an upper limb multipurpose exoskeleton device was presented and the mechanical and actuation system was described. The device has 7 DoFs corresponding to the natural motion of the human arm. Particular attention is paid to the use of pneumatic muscle actuators; these have an extremely good power-to-weight ratio and thanks to their highly flexible and soft nature have beneficial attributes in applications where a power device is in close proximity to a user. This is especially appropriate in medical applications where the users' conditions may mean that they are more at risk than an entirely healthy operator in an industrial type of environment. This soft design concept offers a basis for an inherently safer system, thanks to the definite limits set on actuator contraction (35 % maximum). In addition, antagonistic action permits compliance control, which again has advantages of enhanced safety and more human 'soft' interaction, providing a facility that is reminiscent of the compliance-controlled feel of human manipulation. It has been shown

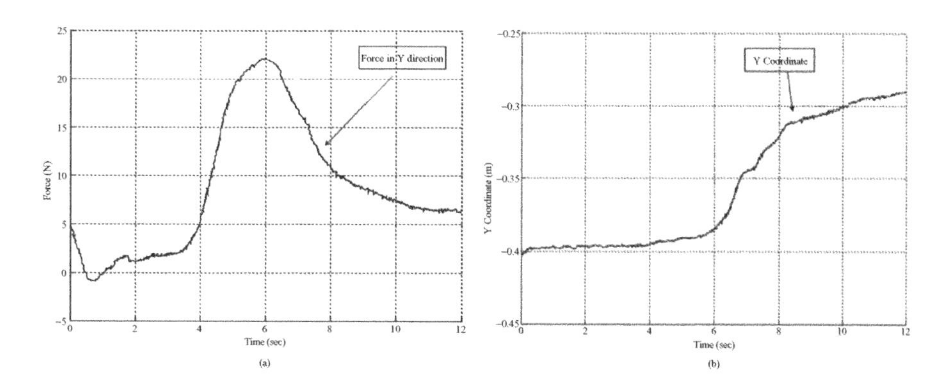

Figure 8.27 Assistive mode experimental results

that the device can be used as an exercise facility for the joints of the upper limb as well as a rehabilitation/power assist orthosis for persons with loss of/reduced power in the limb.

REFERENCES

Beccai, L., Roccella, S., Arena, A., Valvo, F., Menciassi, A., Carrozza, M.C., Dario, P., 2005, 'Design and fabrication of a hybrid silicon three-axial force sensor for biomechanical applications', *Sensors and Actuators A* **120**(2): 370–382.

Beccai, L., Roccella, S., Ascari, L., Valdastri, P., Sieber, A., Carrozza, M.C., Dario, P., 2006, 'Experimental analysis of a soft compliant tactile microsensor to be integrated in an anthropomorphic artificial hand', in *Proceedings of ESDA2006 8th Biennial ASME Conference on Engineering Systems Design and Analysis*.

Belda-Lois, J., Vivas, M., Castillo, A., Peydro, F., Garrido, J., Sanchez-Lacuesta, J., Poveda, R., Prat, J., 2004, 'Functional assessment of tremor in the upper limb', in *Proceedings of the 8th Congress of European Federation for Research in Rehabilitation*, MP-30.

Bergamasco, M., Allotta, B., Bosio, L., Ferretti, L., Parrini, G., Prisco, G.M., Salesdo, F., Sartini, G., 1994, 'An arm exoskeleton system for teleoperation and virtual environments applications', in *Proceedings of the IEEE International Conference on Robotics and Automation* vol.2, pp. 1449–1454.

Bossi, S., Menciassi, A., Koch, K.P., Hoffmann, K.-P., Yoshida, K., Dario, P., Micera, S., 2007, 'Shape memory alloy microactuation of tf-LIFEs: preliminary results', *IEEE Transactions on Biomedical Engineering* **54**(6): 1115–1120.

Caldwell, G.D., Tsagarakis, G.N., 2003, 'Development and control of a 'soft-actuated' exoskeleton for use in physiotherapy and training', *Autonomous Robots* **15**: 21–33.

Caldwell, D.G., Medrano-Cerda, G.A., Goodwin, M.J., 1995, 'Control of pneumatic muscle actuators', *IEEE Control Systems Journal* **15**(1): 40–48.

Carignan, C., Liszka, M., Roderick, S., 2005, 'Design of an arm exoskeleton with scapula motion for shoulder rehabilitation', in *Proceedings of the 12th IEEE International Conference on Advanced Robotics*, pp. 524–531.

Carr, J.H., Shepherd, R.B., 1987, *A motor relearning programme for stroke*, Butterworth-Heinemann, Oxford.

Carrozza, M.C., Massa, B., Micera, S., Lazzarini, R., Zecca, M., Dario, P., 2002, 'The development of a novel prosthetic hand – ongoing research and preliminary results', *IEEE/ASME Transactions on Mechatronics* **7**(2): 108–114.

Carrozza, M.C., Suppo, C., Sebastiani, F., Massa, B., Vecchi, F., Lazzarini, R., Cutkosky, M.R., Dario, P., 2004, 'The SPRING hand: development of a self-adaptive prosthesis for restoring natural grasping', *Autonomous Robots* **16**(2): 125–141.

Carrozza, M.C., Cappiello, G., Micera, S., Edin, B.B., Beccai, L., Cipriani, C., 2006, 'Design of a cybernetic hand for perception and action', *Biological Cybernetics* **95**(6): 629–644.

Carrozza, M.C., Persichetti, A., Laschi, C., Vecchi, F., Lazzarini, R., Vacalebri, P., Dario, P., 2007, 'A wearable biomechatronic interface for controlling robots with voluntary foot movements', *IEEE/ASME Transactions on Mechatronics* **12**: 1–11.

Cavallaro, E., Micera, S., Dario, P., Jensen, W., Sinkjaer, T., 2003, 'On the intersubject generalization ability in extracting kinematic information from afferent nervous signals', *IEEE Transactions Biomedical Engineering* **50**: 1063–1073.

Cavallaro, E., Rosen, J., Perry, J.C., Burns, S., Hannaford, B., 2005, 'Hill-based model as a myoprocessor for a neural controlled powered exoskeleton arm – parameters optimization', in *Proceedings of the 2005 IEEE International Conference on Robotics and Automation*, pp. 4514–4519.

Cipriani, C., Zaccone, F., Stellin, G., Beccai, L., Cappiello, G., Carrozza, M.C., Dario, P., 2006, 'Closed-loop controller for a bio-inspired multi-fingered underactuated prosthesis', in *Proceedings of the 2006 IEEE International Conference on Robotics and Automation*, pp. 2111–2116.

Citi, L., Carpaneto, J., Yoshida, K., Hoffmann, K.P., Koch, K.P., Dario, P., Micera, S., 2006, 'Characterization of tfLIFE neural response for the control of a cybernetic hand', in *Proceedings of the 2006 IEEE International Conference on Biomedical Robotics and Biomechatronics (* BIOROB06), pp. 477–482.

Dhillon, G., Horch, K., 2005, 'Direct neural sensory feedback and control of a prosthetic arm', *IEEE Transactions on Neural Systems and Rehabilitation Engineering* **13**: 468–472.

Edin, B.B., Beccai, L., Ascari, L., Roccella, S., Cabibihan, J.J., Carrozza, M.C., 2006, 'A bio-inspired approach for the design and characterization of a tactile sensory system for a cybernetic prosthetic hand', in *Proceedings of the 2006 IEEE International Conference on Robotics and Automation*, pp. 1354–1358.

Edin, B.B., Ascari, L., Beccai, L., Roccella, S., Cabibihan, J.J., Carrozza, M.C., 2007, 'Bio-inspired sensorization of a biomechatronic robot hand for the grasp-and-lift tasks', *Brain Research Bulletin* (accepted).

Fahn, S., Tolosa, E., Marin, C., 1998, 'Clinical rating scale for tremor', in *Parkinson's Disease and Movement Disorders* (eds E. Tolosa and J. Jankovic), Urban & Schwarzenberg, Baltimore, Maryland.

Fasoli, S.E., Krebs, H.I., Stein, J., Frontera, W.R., Hogan, N., 2003, 'Effects of robotic therapy on motor impairment and recovery in chronic stroke', *Archives of Physical Medicine and Rehabilitation* **84**: 477–482.

Frisoli, A., Rocchi, F., Marcheschi, S., Dettori, A., Salsedo, F., Bergamasco, M., 2005, 'A new force-feedback arm exoskeleton for haptic interaction in virtual environments', in *First Joint Eurohaptics Conference and Symposium on Haptic Interfaces for Virtual Environments*, pp.195–201.

Gaine, W.J., Smart, C., Bransby-Zachary, M., 1997, 'Upper limb traumatic amputees Review of prosthetic use', *Journal of Hand Surgery British* **22**: 73–76.

Hidler, M., Wall, A.E., 2005, 'Alterations in muscle activation patterns during robotic-assisted walking', *Clinical Biomechanics* **20**: 184–193.

Hirose, S., 1985, 'Connected differential mechanism and its applications', in *Proceedings of the International Conference on Advanced Robotics*, pp. 319–326.

Hirose, S., Umetani, Y., 1978, 'The development of soft gripper far the versatile robot hand', *Mechanism and Machine Theory* **13**: 351–358.

Hoffmann, K.P., Koch, K.P., Dörge, D., Micera, S., 2006, 'New technologies in manufacturing of different implantable microelectrodes as an interface to the peripheral nervous system', in *Proceedings of the 2006 IEEE International Conference on Biomedical Robotics and Biomechatronics*, pp. 414–419.

Hogan, N., Krebs, H.I., Charnnarong, J., Srikrishna, P., Sharon, A., 1992, 'MIT-MANUS: a workstation for manual therapy and training. I', *IEEE Proceedings of the International Workshop on Robot and Human Communication*, pp.161–165.

Jau, B.M., 1988, 'Anthropomorphic exoskeleton dual arm/hand telerobot controller', in *IEEE International Workshop on Intelligent Robots*, pp. 715–718.

Kapandji, I.A., 1982, *The Physiology of the Joints: Annotated Diagrams of the Mechanics of the Human Joints*, 5th revised edition, Churchill Livingstone; Edinburgh.

Kazerooni, H., 1990, 'Human–robot interaction via the transfer of power and information signals', *IEEE Transactions on Systems, Man, and Cybernetics* **20**(2): 450–463.

Kazerooni, H., 1996, 'The human amplifier technology at the University of California, Berkeley', *Robotics and Autonomous Systems*, **19**: 179–187.

Kiguchi, K., Tanaka, T., Fukuda, T., 2004, 'Neuro-fuzzy control of a robotic exoskeleton with EMG signals', *IEEE Transactions on Fuzzy Systems* **12**: 481–490.

Kousidou, S., Tsagarakis, N.G., Caldwell, D.G., Smith, C., 2006, 'Assistive exoskeleton for task based physiotherapy in 3-dimensional space', in *Proceedings of the 2006 IEEE International Conference on Biomedical Robotics and Biomechatronics*, pp.266–271.

Krebs, H.I., Volpe, B.T., Ferraro, M., Fasoli, S., Palazzolo, J., Rohrer, B., Edelstein, L., Hogan, N., 2002, 'Robot-aided neuro-rehabilitation: from evidence-based to science-based rehabilitation', *Topics in Stroke Rehabilitation* **8**(4): 54–68.

Lago, N., Ceballos, D., Rodríguez, F., Stieglitz, T., Navarro, X., 2005, 'Long term assessment of axonal regeneration through polyimide regenerative electrodes to interface the peripheral nerve', *Biomaterials* **26**(14): 2021–2031.

Lago, N., Yoshida, K., Koch, K.P., Navarro, X., 2007, 'Assessment of biocompatibility of chronically implanted polyimide and platinum intrafascicular electrodes', *IEEE Transactions on Biomedical Engineering* **54**(2): 281–290.

Lawrence, D., 1993, 'Stability and transparency in bilateral teleoperation', *IEEE Transactions on Robotics and Automation* **9**(5): 624–637.

Lawrence, S.M., Dhillon, G.S., Jensen, W., Yoshida, K., Horch, K.W., 2004, 'Acute peripheral nerve recording characteristics of polymer-based longitudinal intrafascicular electrodes', *IEEE Transations on Neural Systems and Rehabilitation Engineering* **12**(3): 345–348.

Letier, P., Avraam, M., Horodinca, M., Schiele, A., Preumont, A., 2006, 'Survey of actuation technologies for body-grounded exoskeletons', in *Proceedings of the EuroHaptics Conference*.

Manto, M., Topping, M., Soede, M., Sanchez-Lacuesta, J.J., Harwin, W., Pons, J.L., Willimas, J., Skaarup, S., Normie, L., 2003, 'Dynamically responsive intervention for tremor suppression', *IEEE Engineering in Medicine and Biology Magazine* **22**(3): 120–132.

McDonnall, D., Clark, G.A., Normann, R.A., 2004, 'Interleaved, multisite electrical stimulation of cat sciatic nerve produces fatigue-resistant, ripple-free motor responses', *IEEE Transactions on Neural Systems and Rehabilitation Engineering* **12**(2): 208–215.

McNaughton, T.G., Horch, K.W., 1994, 'Action potential classification with dual channel intrafascicular electrodes', *IEEE Transactions on Biomedical Engineering* **41**(7): 609–616.

Micera, S., Carrozza, M.C., Beccai, L., Vecchi, F., Dario, P., 2006, 'Hybrid bionic systems for the replacement of hand function', *Proceedings of the IEEE* **94**(9): 1752–1762.

Micera, S., Jensen, W., Sepulveda, F., Riso, R.R., Sinkjaer, T., 2001, 'Neuro-fuzzy extraction of angular information from muscle afferents for ankle control during standing in paraplegic subjects: an animal model', *IEEE Transactions on Biomedical Engineering* **48**: 787–794.

Mirfakhraei, K., Horch, K., 1997, 'Recognition of temporally changing action potentials in multiunit neural recordings', *IEEE Transactions on Biomedical Engineering* **44**(2): 123–131.

Nakayama, H., Jorgensen, H.S., Raaschou, H.O., Olsen, T.S., 1994, 'Recovery of upper extremity function in stroke patients: the Copenhagen Stroke Study', *Archives of Physical Medicine and Rehabilitation* **75**: 394–398.

Navarro, N., Krueger, T.B., Lago, N., Micera, S., Dario, P., Stieglitz, T., 2005, 'A critical review of interfaces with the peripheral nervous system for the control of neuroprostheses and hybrid bionic systems', *Journal of the Peripheral Nervous System* **10**(3): 229–258.

Parker, V.M., Wade, D.T., Langton, H.R., 1986, 'Loss of arm function after stroke: measurement, frequency, and recovery', *International Rehabilitation Medicine* **8**: 69–73.

Perry, J.C., 2006, 'Design and development of a 7 degree-of-freedom powered exoskeleton for the upper limb', PhD Dissertation, University of Washington, Seattle, Washington.

Repperger, D.W., Remis, S.J., Merrill, G., 1990, 'Performance measures of teleoperation using an exoskeleton device', in *Proceedings of the IEEE International Conference on Robotics and Automation* vol.1. pp. 552–557.

Rocon, E., Ruiz, A.F., Pons, J.L., 2004, 'On the use of ultrasonic motors in orthotic rehabilitation of pathologic tremor', in *Proceedings Actuator 2004*, pp. 387–390.

Rocon, E., Belda-Lois, J.M., Sánchez-Lacuesta, J.J., Pons, J.L., 2004, 'Pathological tremor management: modelling, compensatory technology and evaluation', *Technology and Disability* **3**: 3–18.

Rocon, E., Manto, M., Pons, J.L., Camut, S., Belda-Lois, J.M., 2007, 'Mechanical suppression of essential tremor', *The Cerebellum* **6**: 73–78.

Rosen, J., Fuchs, M.B., Arcan, M., 1999, 'Performances of hill-type and neural network muscle models: Toward a myosignal based exoskeleton', *Computers and Biomedical Research* **32**(5): 415–439.

Rosen, J., Brand, M., Fuchs, M.B., Arcan, M., 2001, 'A myosignal-based powered exoskeleton system', *IEEE Transactions on Systems, Man and Cybernetics – Part A: Systems and Humans* **31**(3): 210–222.

Rosen, J., Perry, J.C., Manning, N., Burns, S., Hannaford, B., 2005, 'The human arm kinematics and dynamics during daily activities – toward a 7 DOF upper limb powered exoskeleton', in *EEE Proceedings of the 12th International Conference on Advanced Robotics*, 18-20 July 2005, pp.532–539.

Salisbury, K., Townsend, W., Eberman, B., DiPietro, D., 1988, 'Preliminary design of a whole-arm manipulation system (WAMS)', in *Proceedings of the IEEE International Conference on Robotics and Automation*, vol.1, 254–260.

Salvini, P., Laschi, C., Dario, P., 2006, 'From robotic tele-operation to tele-presence: A bio-inspired approach to presence and natural interface', in *Proceedings of the 1st IEEE/RAS-EMBS International Conference on Biomedical Robotics and Biomechatronics*, pp. 408–413.

Sánchez, R.J., *et al.*, 2006, 'Automating arm movement training following severe stroke: functional exercises with quantitative feedback in a gravity-reduced environment', *IEEE Transactions on Neural Systems and Rehabilitation Engineering* **14**(3): 378–389.

Schiele, A., van der Helm, F.C.T., 2006, 'Kinematic design to improve ergonomics in human machine interaction', *IEEE Transactions on Neural Systems and Rehabilitation Engineering* **14**(4): 456–469.

Schiele, A., Visentin, G., 2003, 'The ESA human arm exoskeleton for space robotics telepresence', in *7th International Symposium on Artificial Intelligence, Robotics and Automation in Space.*

Schiele, A., D.e., Bartolomei, M., van der Helm, F.C.T., 2006, 'Towards intuitive control of space robots: a ground development facility with exoskeleton', in *Proceedings of the 2006 IEEE/RSJ International Conference on Intelligent Robots and Systems*, pp. 1396–1401.

Schiele, A., *et al.*, 2006, 'Bowden cable actuator for force-feedback exoskeletons', in *Proceedings of the IEEE/RSJ International. Conference on Intelligent Robots and Systems*, pp. 3599–3604.

Schoonejans, P., *et al.*, 2004, 'Eurobot: EVA-assistant robot for ISS, Moon and Mars', in *Proceedings of the 8th ESA Workshop on Advanced Space Technologies for Robotics and Automation*, ASTRA, Noordwijk.

Sergi, P.N., Carrozza, M.C., Dario, P., Micera, S., 2006, 'Biomechanical characterization of needle piercing into peripheral nervous tissue', *IEEE Transactions on Biomedical Engineering* **53**(11): 2373–2386.

Tarler, M.D., Mortimer, J.T., 2004, 'Selective and independent activation of four motor fascicles using a four contact nerve-cuff electrode', *IEEE Transactions on Neural Systems and Rehabilitation Engineering* **12**(2): 251–257.

Tsagarakis, N.G., Caldwell, D.G., 2003a, 'Development and control of a soft-actuated exoskeleton for use in physiotherapy and training', *Journal of Autonomous Robots, Special Issue on Rehabilitation Robotics* **15**: 21–33.

Tsagarakis, N.G., Caldwell, D.G., 2003b, 'Development and control of a physiotherapy and training exercise facility for the upper limb using soft actuators', in *Proceedings of the IEEE International Conference on Advanced Robotics*, pp.1092–1097.

Tsagarakis, N.G., Tsachouridis, V., Caldwell, D.G., 2003, 'Modeling and control of a pneumatic muscle actuated joint using on/off solenoid valves', in *Proceedings of the IEEE International Conference on Advanced Robotics*, pp.929–934.

Williams, I.I., R.L., *et al.*, 1998, 'Kinesthetic force/moment feedback via active exoskeleton', in *Proceedings of the IMAGE Conference*, Scottsdale, Arizona.

9

Wearable lower limb and full-body robots

J. C. Moreno, E. Turowska and J. L. Pons

Bioengineering Group, Instituto de Automática Industrial, CSIC, Madrid, Spain

Lower limb and full-body exoskeletons are wearable devices that can be classified according to their particular application as assistive devices: for human impaired movement or for human power augmentation. This chapter provides a wide range of case studies of lower limb and full-body exoskeletons, including functional descriptions of the different approaches followed in designing the robot and how the H–R interaction was conceived in a particular development. The collection of case studies seeks to describe the different possible applications, with details on the design of sensors, actuators and control aspects. These selected case studies represent leading-edge research and development of rehabilitation (orthotics and prosthetics) and human power augmentation exoskeletons.

9.1 CASE STUDY: GAIT–ESBIRRO: LOWER LIMB EXOSKELETONS FOR FUNCTIONAL COMPENSATION OF PATHOLOGICAL GAIT

J. C. Moreno and J. L. Pons

Bioengineering Group, Instituto de Automática Industrial, CSIC, Madrid, Spain

9.1.1 Introduction

Assistive systems to restore gait have been proposed as an alternative to functional electrical stimulation (FES) for muscle activation and can be broadly classified as: (a) improved mechanisms such as orthotic hinges with braking or clutching functionalities (Gharooni, Heller and Tokhi, 2000; Irby *et al.*, 1999); (b) exoskeletal robots for training limbs in stroke and incomplete paraplegics (Colombo *et al.*, 2000; Colombo, Wirz and Dietz, 2001); and (c) control systems acting on a single joint to correct a specific dysfunction (Blaya and Herr, 2004).

Wearable Robots: Biomechatronic Exoskeletons Edited by José L. Pons

In the Bioengineering Group at the IAI-CSIC, the GAIT exoskeleton has been developed as a KAFO device for young, active poliomyelitis and cerebral palsy subjects, who may suffer from quadriceps, tibialis anterior, soleus and/or gastrocnemius weaknesses (Moreno *et al.*, 2004, 2005). The system is conceived as a monitoring tool for collecting real life data (Moreno *et al.*, 2006). This case study presents the biomechanical aspects considered in the design of the exoskeleton, the sensor, actuator and control systems, and the results of its application in clinical cases. It introduces the ESBIRRO exoskeleton, an extension of GAIT consisting of a bilateral leg exoskeleton device aimed at improving current hip–knee–ankle–foot orthoses (HKAFOs) and incorporating the limit-cycle walking strategy.

9.1.2 Pathological gait and biomechanical aspects

Absence of the necessary muscle activity on body segments can lead to bodily collapse. To overcome this problem, the robotic orthosis must compensate for the missing moment of force around the joint with the aim of stabilizing and compensating for the lack of muscle strength. The exoslekeleton with the actuators should apply the external joint moment to the body segments in an appropriate way. Here the most suitable external force systems for the two joints are considered:

9.1.2.1 Compensation for knee instability

The system is intended to compensate for loss of control of knee extensors, especially quadriceps, which poses the risk of the knee collapsing during flexion when bearing weight. In walking, the effect of paralysis of these muscles will be greatest just after heel-strike. During this interval the combination of gravity and the forward momentum of the body will cause the knee to bend. Consequently, the knee will collapse unless stability is maintained by means of compensation (e.g. use of a hip extensor or forward bending of the trunk in early stance) or an orthosis. The exoskeleton is required to perform the following actions:

- Stance phase: to restrain knee flexion and assist the knee back to full extension.
- Swing phase: to allow the knee to flex in early swing and assist knee extension in late swing to prepare for the next foot contact.

The knee can be stabilized by means of either a three- or a four-point force system. In the case of four-point support, the interface forces are smaller, as are the maximum internal moments in the sagittal plane. A four-point support system also enables the knee to be stabilized through the application of an external moment by means of an actuator placed on the orthotic joint. This is depicted schematically in Figure 9.1.

9.1.2.2 Compensation for ankle instability

Compensation requirements for the ankle are determined by the compensatory strategies adopted at the knee level. Patients with unilateral weakness of the knee extensors will commonly be successful in overcoming this problem if they adopt an interiorly flexed posture of the trunk to bring the line of action of the ground reaction force in front of their knee, thus creating a stabilizing knee moment. The success of this manoeuvre requires sufficient plantar flexion power to prevent ankle dorsiflexion as the patient leans forward. The robotic orthosis should provide sufficient dorsal flexion power to prevent involuntary plantar flexion. The dorsiflexion is achieved by means of an actuator set up at the ankle level.

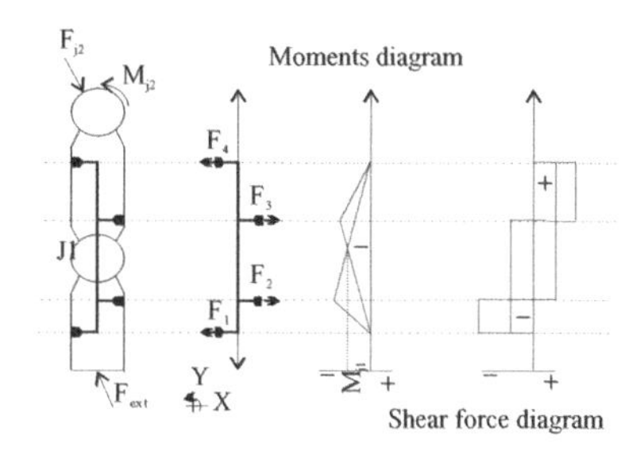

Figure 9.1 Exoskeleton four-point support; moment and shear force diagrams

9.1.2.3 User acceptance considerations

An appropriate exoskeleton design must take into account comfort considerations for the human–robot interface (HRI). Comfort is defined to a very great extent by the biomechanical interaction between user and robot. In addition, the biomechanical interaction will be defined by the combined interaction between the compliant, soft body tissues and the support surface through which forces are transmitted. The factors that most affect the comfort felt by users of orthotic devices are pressure and shear forces.

Pressure is an important factor to consider in the development of any device that is in contact with the human body. Bodily response to loads is still not fully understood. There are differences regarding pressure pain sensitivity in different zones in the same body region, an issue analysed in Section 5.3. An analysis of lower limb sensitivity, performed to determine the best areas on the lower limb for application of external loading, is presented in Case Study 5.6.

9.1.3 The GAIT concept

The wearable robotic orthosis takes the form of a unilateral exoskeleton fitted with two semi-active actuators for knee and ankle joints, as well as a portable control system based on a redundant sensor set to coordinate joint performance during walking, as depicted in Figure 9.2 (left).

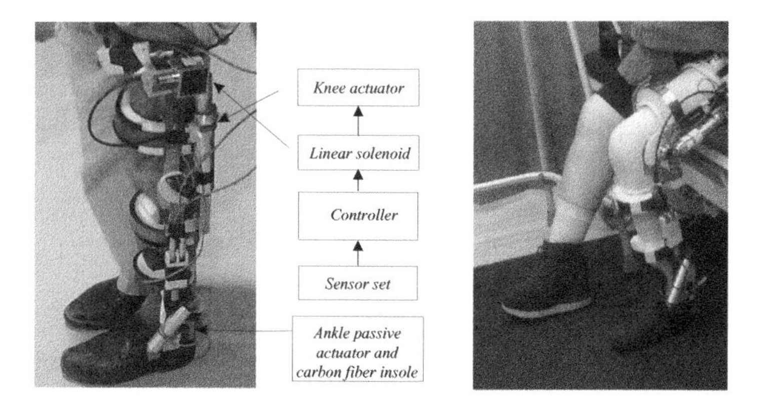

Figure 9.2 System components (left): interface between the impaired human motor control and the robotic orthosis formed by the exoskeleton, a set of sensors, knee and ankle actuators, and an embedded controller. System fitted to a patient (right)

9.1.4 Actuation

No current actuation technology separately fulfils the desired functionalities and properties for the wearable robot, largely because of power, size and weight limitations (see Section 6.3). For example, to meet the peak power demand during walking (3.3 W per body weight kilogram) for a 100 kg subject the actuator must deliver a torque of around 120 N m at push-off to achieve plantar flexion and swing the leg. A purely active solution requires continuous control and activation of the actuator during the entire gait cycle in order to approach a natural pattern. What the wearable robotic orthosis needs is a suitable compromise between volume, weight and functionality. The actuation system is designed to provide the required amount of stiffness at joint level in order to approach a more natural gait profile. This design (presented in detail in Case Study 6.7) considers parametric identification from functional analysis of the musculoskeletal system, and also the construction of knee and ankle actuators according to anthropometrical data to compensate for the loss of functionality.

9.1.5 Sensor system

An interface between the human body and an assistive technology such as an exoskeleton can be achieved with kinetic and kinematic information from the HRI (see Chapter 4). In functional electrical stimulation (FES) systems, the failure of a single sensor usually has catastrophic consequences for control, since only a few sensors are utilized and all are vital. The sensor setup consists of an *inertial measurement unit* (IMU) on the foot element inside the shoe (below the orthotic ankle joint) and a second unit for the lower bar of the exoskeleton. Each IMU is composed of: (a) a single miniature MEMS rate gyroscope which senses the Coriolis force during an angular rate by measuring capacitance (Analog Devices ADXRS300, volume of 0.15 cm^3, weight of 0.5 g) with maximum sensitivity $\pm$ 300/s and (b) a complete dual-axis 200 mV/g accelerometer (ADXL202 5 mm $\times$ 4.5 mm $\times$ 1.78 mm). Units housed in boxes attached to foot and shank orthotic bars sense rotational motion, tilt, and tangential and radial segment accelerations in orthogonal directions (X and Y), while most orthotic rotations at the level of joints and bars take place on the locomotion progression plane (sagittal) due to mechanical constraints imposed by the structure. Movements of interest occur at normal (2.6 km/h) and low (2 km/h) gait speeds, and therefore signals outside the band frequency associated with gait kinematics (0.3–20 Hz) are screened out of the sensor outputs by -3 dB lowpass filters, while the noise floor is lowered by bandwidth restricting. A precision angular position sensor with an effective range of 340° is fixed to one rotation axis of the four-bar mechanism of the knee joint for continuous tracking of the knee joint angle on the sagittal plane. A resistive pressure sensor (active area diameter of 5 mm, thickness of 0.30 mm) is used to monitor the status of the knee-locking mechanism.

It is important that electrical connections be robust, as each device will have to bear a large number of movements, impacts and efforts. The IMU on the ankle bar measures rotational movements, inclination and tangential and radial accelerations of the limb in the orthogonal direction on the sagittal plane. The same rotations and movements are measured in the leg by a second IMU, mounted on the lateral side of the bar. An IMU on the foot bar is inserted in the subject's shoe (below the ankle joint); there is a second IMU on the foot bar (on the ankle joint).

9.1.6 Control system

The portable control architecture is based on a microcontroller and is equipped with embedded memory for data logging and a Bluetooth node for monitoring. The microcontroller generates signals to command the knee actuator. The selective knee actuator mechanism features two discrete states, R0 for stance and R1 for swing, during cyclical walking. Opportune cyclical transition of the system between the states permits locomotion with a safe stance phase and a free swinging leg and allows for functional compensation. The transition from R0 to R1 is achieved by the action, at a given *onset*

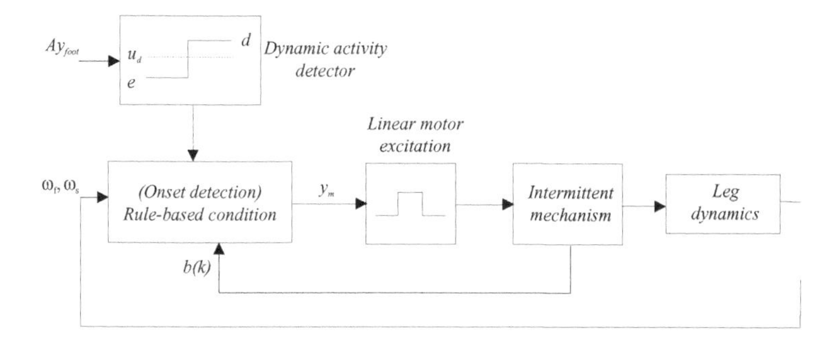

Figure 9.3 Gait cyclical controller. The dynamic activity detector tracks foot vertical acceleration and defines gait initiation. The rule-based condition is established with respect to the rate velocity of the foot and shank segments

time, of a linear pulling solenoid applying the required pull force to effect an exchange between springs and release the knee joint, during a given activation period. The actuator is battery-powered and digitally controlled by means of a switching circuit in the intermittent duty mode (intermittent action, variable time cycles).

Adequate onset timings for the switching mechanism in the knee are defined with the goal of ensuring stability of the joint during the stance phase. In patients with anterior knee instability, a safety criterion states that secure release of the joint at the start of the swing phase can only be achieved if there is an extension moment at the knee, i.e. the force vector of the ground reaction force elapses ventral of the knee hinge. The appropriate *activation onset* can be determined from the kinematics of the lower leg (joints angles, segment rate of rotation) and tracked cyclically by spatio-temporal parameters such as step length, stance and swing phase periods.

A reactive controller has been designed for gait in patients with knee instability (see Figure 9.3). The aim of the discrete controller is to detect the transition instant (TI) using real-time evaluation, at a sampling frequency of 100 Hz, of the information obtained from the inertial measurement units (IMUs) of the exoskeleton. System activation onsets during each stride are calculated by rule-based conditions. The criteria for adapting the activation period (pulse width) of the actuator cycle-to-cycle is defined by considering stance-phase-related temporal parameters of the current $S(k)$ and previous $S(k-1)$ strides. Initial conditions $S(0)$ are given by average expected values. The inertial sensors attached to the frame measure ω_f and ω_s, foot and shank angular velocities respectively, and a_f, foot tangential acceleration along the sagittal plane. Offset values in sensor signals are corrected during static conditions with the foot flat on the ground, when ω_f and ω_s are approximately zero and total accelerometer output corresponds to 1 g.

9.1.7 Evaluation

The system was tested by two patients with post-polio syndrome in need of technical assistance to be able to walk and lead an active life. The patients were affected by weakness in the left quadriceps group.

A left leg unilateral exoskeleton was customized to each patient (see Figure 9.2 (right)). The construction of the joints restricted movement on the sagittal plane. Special attention was paid to achieving the right mechanical adjustment and adaptation of the exoskeleton in order to guarantee comfort and the appropriate transmission of forces, with the constant assistance of an expert in orthopaedics. The securing pieces were shaped and material was added to adjust to the anatomical form in such a way that no marks were left on the skin after using the system for 20 minute periods. Ankle and toe heights were adjusted by adding material to the shoe insole as necessary. The actuators in each prototype were made of springs built to deliver compensations calculated from the moment polynomial adjustment coefficients versus angle, according to the subject's weight.

A solenoid (12 V DC, 10 W, Belling Lee) was anchored as the intermittent mechanism of the knee actuator; this transmits the force necessary (up to 700 g, 3 mm trajectory) to effect the switch between springs and unlock the joint during the stance phase. The solenoid was fed electrically from a lead-sealed battery (Dryfit 1.2 A h, 12 V, Sonnenschein) and controlled digitally using a switching circuit in the pulsated mode to reduce power consumption. The circuit implements the discharge via a capacitor to allow rapid unlocking.

9.1.7.1 Results

The GAIT exoskeleton made it possible for patient S1 to walk for the first time without the need of a compensating hip movement, which was necessary with other knee-locking orthoses. The knee actuator using energy recovery obviously assists knee extension so that it is effectively completed before contact with the ground and a low and constant gait speed is maintained. It is uncertain what percentage of assistance to the knee extension is due to the actuator action and what percentage is determined by movement inertia. The hypothesis is that the extension recovery spring counteracts movement at the onset of extension.

Patient S2 uses a knee-and-ankle orthosis to be able to walk. The orthosis has the knee joint locked while the ankle joint restricts plantar flexion and dorsal flexion mobility. The gait pattern with her orthosis is such that the knee is locked during the stance and swing phases. Immediately after the stance response, there is a dorsal flexion movement greater than normal in the ankle. During the swing phase the foot is protected from plantar drop with the restriction imposed by the exoskeleton. To put the foot forward, the patient compensates with her body on the transversal and frontal plane.

Using the controlled exoskeleton, subject S2 required 30 minutes of training to walk with the crutches instead of standing on the parallel bars used in the first tuning assays of the cable-driven mechanism. After 30 minutes, the patient was able to walk with free swing of the knee (mean maximum flexion of 50°) using the two crutches, and although the speed adopted was low, the knee clearly reached extension at the end of the swing phase (see Figure 9.4). After a short time, the patient learnt to manage joint locking at the beginning of stance, used only one crutch and felt confident without any risk of falling.

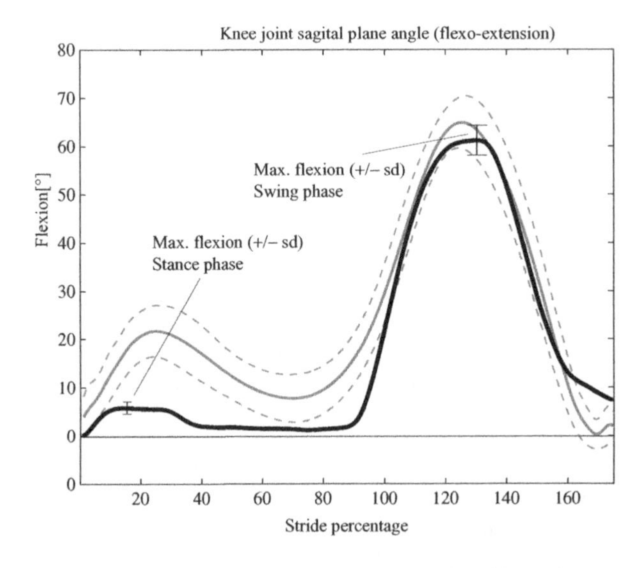

Figure 9.4 Effects (average values) on knee joint kinematics

Examination of the ankle angle showed that dorsal flexion, which was excessive with the subject's habitual orthosis, was adequate during the stance phase thanks to the ankle actuator. Plantar drop was restricted by at most 5°(mean of 4.5°).

The differences found in the patients' kinematic gait patterns during the application of functional compensation to the lower limb reflected significant differences in the subjects' usual gait. Both patients exhibited rapid adaptation, adopting new motor commands to manage the exoskeleton with the constraints imposed on the limb. The benefits of the correct release of the knee in both instances are clearly evident in the improved gait patterns, which were much closer to normal patterns, as depicted in Figure 9.4.

The patterns with ambulatory assistance using the robotic exoskeleton were significantly better than those offered by traditional orthoses or basic aids. Each pathological case has its own intrinsic characteristics, and the system of mechanical adaptation and control therefore requires a customized robotic solution. The evidence gathered from the two post-polio subjects demonstrates the feasibility of gait compensation using wearable robotic exoskeletons as a means of improving the quality of daily life in subjects with lower limb joint disorders.

9.1.8 Next generation of lower limb exoskeletons: the ESBiRRo project

The GAIT exoskeleton demonstrates the functional compensation concept as applied to quadriceps weakness. The ESBiRRo exoskeleton is suitable for potential GAIT users, and its user groups extend to patients with a bilateral handicap affecting the complete lower limb (hip, knee and ankle). The patient groups considered for this exoskeleton are early multiple sclerosis (MS) or geriatric patients and patients with no leg control due to spinal cord injury (SCI), advanced MS or one-leg hemiparetic cerebrovascular accident (CVA). For the ESBiRRo exoskeleton, a limit-cycle walking strategy was investigated to enable the system to walk in an energy-efficient manner and to recover from disturbances. To that end, situations of those kinds (e.g. major disturbances, leg elevating/lowering recovery strategies, etc.) were analysed and simulations and tests were run with the ESBiRRo biped robot (see Case Study 2.5).

Within this framework, trajectory generation methods and gait stabilization strategies were incorporated in an exoskeleton. The ESBiRRo exoskeleton is thus conceived as a lightweight design having a control system structured as a state machine triggered by sensor signals (joint angles, trunk tilt or contact switches), which implements biped limit-cycle control and has bioinspired reflexes for large-disturbance recovery.

9.2 CASE STUDY: AN ANKLE–FOOT ORTHOSIS POWERED BY ARTIFICIAL PNEUMATIC MUSCLES

D. P. Ferris

Division of Kinesiology, Department of Biomedical Engineering and Department of Physical Medicine and Rehabilitation, The University of Michigan, Ann Arbor, Michigan, USA

9.2.1 Introduction

In the Human Neuromechanics Laboratory at the University of Michigan, pneumatically powered orthoses were built for assisting ankle motion during human walking (Ferris *et al.*, 2006; Gordon and Ferris, 2007; Sawicki, Domingo and Ferris, 2006). There was particular interest in powered plantar flexion because the plantar flexor muscles are critical to the generation of forward velocity and support of the centre of mass during human walking (Kepple, Siegel and Stanhope, 1997; Neptune, Kautz and Zajac, 2001; Winter, 1983). The idea was that powered ankle–foot orthoses could be used both for basic science studies and for gait rehabilitation after neurological injury. Because only laboratory or clinical testing and training are required for these purposes, there was specifically no focus on making the orthosis portable.

From a basic science perspective, a powered orthosis can be used to perturb the relationships between neural control, movement biomechanics and metabolic energy expenditure. By inputting mechanical energy into limb dynamics it should be possible to reduce muscle mechanical work and/or force. This would in turn presumably reduce the metabolic cost of locomotion. By connecting the orthosis to the nervous system via neural signals (e.g. electromyography, electroencephalography, cortical electrodes), researchers could investigate how the nervous system learns to control novel limb dynamics (Gordon and Ferris, 2007) (see Case Study 4.7 for a detailed discussion on this topic). In both cases, studies are highly dependent on a comfortable biomechanical human–machine interface.

For purposes of rehabilitation, a powered orthosis could help individuals re-learn how to walk after neurological injury (Ferris, Sawicki and Domingo, 2005). Breakthroughs in clinical neuroscience have revealed that humans with spinal cord injury or stroke can enhance their motor capabilities through intense task-specific practice (Dietz *et al.*, 1998; Harkema, 2001; Hesse *et al.*, 1995). Therapy often requires manual assistance from several physical therapists (Behrman and Harkema, 2000). Robotic devices could substantially reduce manual labour costs. Several groups are working on robotic devices to aid gait rehabilitation after neurological injury (Colombo, Wirz and Dietz, 2001; Jezernik *et al.*, 2003; Reinkensmeyer *et al.*, 2006). However, none of these devices provide plantar flexor assistance during walking. Given the amount of mechanical energy supplied by the plantar flexors during push-off at the end of stance, a powered ankle–foot orthosis would presumably be a valuable contribution to the field of rehabilitation robotics.

9.2.2 Orthosis construction

The orthosis was custom fabricated from carbon fibre, polypropylene and a metal hinge based on a cast of the subjects' lower limbs (see Figure 9.5). A bivalve design was used for the shank section to increase ease of donning and doffing. Two pegs located on the posterior shank piece fit snugly into two slots on the anterior shank piece. Two plastic cycling shoe buckles (Sidi Ultra SL, Veltec Sports, Inc., Monterey, Califorinia) secured the shank pieces together tightly. The univalve foot section was fabricated from polypropylene (3 mm thick) for comfort, previous carbon fibre foot sections having sometimes resulted in areas of high pressure on the foot. A Velcro(r) strap was attached to the foot section to help prevent the foot from sliding in the orthosis. The metal camber axis hinge (Model 750, Becker Orthopedics, Troy, Michigan) provided a rigid connection between the shank and foot sections. Angle brackets were attached to the anterior and posterior shank sections to connect artificial pneumatic muscles. The brackets were made from stainless steel ribbon (19.0 mm × 4.5 mm) (Becker Orthopedic, Troy, Michigan) and bent to an angle of 90°. A stainless steel ribbon support welded to the inside of the bracket angle attenuated bending when the artificial muscles produced tension. A flat stainless steel ribbon was affixed to the bottom of the foot section, with machine screws for the distal insertion of the artificial plantar flexor muscle. The artificial muscles were attached to the orthosis with interchangeable brackets for easy adjustment of the length and moment arm of the artificial muscles. A stainless steel D-ring (Westhaven Buckle, Addison, Illinois) on the dorsum of the foot section provided an attachment for the artificial dorsiflexor insertion.

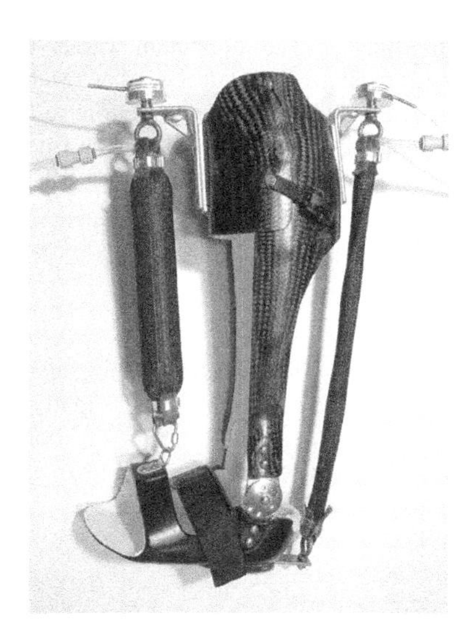

Figure 9.5 Ankle–foot orthosis. The shank section is made from carbon fibre and the foot section is made from polypropylene. The artificial dorsiflexor muscle is fully inflated and the artificial plantar flexor muscle is relaxed for comparison. Artificial muscle tension is monitored by load sensors in series with the artificial muscles

9.2.3 Artificial pneumatic muscles

The orthosis is actuated by artificial pneumatic muscles (sometimes called McKibben muscles or flexible pneumatic actuators). The artificial pneumatic muscles consist of an expandable internal bladder surrounded by a braided shell. When the internal bladder is pressurized, it expands like a balloon. The braided shell constrains the expansion. As the volume of the internal bladder augments with increasing pressure, the pneumatic muscle shortens and produces tension if coupled to a mechanical load. Artificial pneumatic muscles are good for this kind of application because they are lightweight, capable of delivering large forces and inherently compliant (Davis *et al.*, 2003; Reynolds *et al.*, 2003; Tondu and López, 2000). Their mechanical properties (force–length, force–velocity, force–pressure ratios) have been analysed in detail during benchtop testing (Klute, Czerniecki and Hannaford, 1999; Klute and Hannaford, 1998, 2000). During isometric benchtop testing, artificial pneumatic muscles demonstrated a linear force–length relationship (Gordon, Sawicki and Ferris, 2006). Each artificial muscle of the size used produced a peak force of 1700 N when fully activated at maximal length. Increasing the cross-sectional area of the muscle with large-diameter tubing and sheathing produced greater maximum muscle force. The muscle lost all its force when contracted to 71% of its maximum length. The force bandwidth of this artificial pneumatic muscle as determined from benchtop testing was 2.4 ± 0.1 Hz (mean ± s.d.) (Gordon, Sawicki and Ferris, 2006), which is similar to the force bandwidth of human muscle of 2.2 Hz (Aaron and Stein, 1976).

9.2.4 Muscle mounting

Forged eyebolts connected the artificial muscles to the orthosis through holes in the stainless steel brackets. The eyebolt shank went through the hole of an aluminium washer and tension/compression force transducer (Model LC8150-375-1K 0-1000 LBS, Omega Engineering, Inc., Stamford,

Connecticut) placed on top of the bracket. The aluminium washer was custom machined to provide a stable base with distributed loading. A spherical washer (McMaster-Carr, Cleveland, Ohio) was added between the bracket top and aluminium washer bottom so that the eyebolt could rotate in relation to the bracket as the ankle angle changed. This spherical washer kept the load transducer perpendicular to the artificial muscle's line of action. Nuts and lock washers (McMaster-Carr, Cleveland, Ohio) secured the eyebolt shank ends.

9.2.5 Orthosis mass

The ankle–foot orthosis has an approximate total mass, including artificial muscles and force transducers, of $\approx$1.1 kg. Previous versions from the laboratory were slightly heavier, but there was success in reducing the mass of the orthosis shell by shifting to prepreg carbon fibre in current versions. Orthosis mass is important because distally located added mass can greatly increase the metabolic cost of walking (Browning *et al.*, 2007).

9.2.6 Orthosis control

Proportional myoelectric control was implemented by means of a desktop computer and real-time control board (ACE Kit 1103, dSPACE, Inc., Northville, Michigan). The software program was written in Simulink (The Mathworks, Inc., Natick, Massachusetts) and converted to ControlDesk (dSPACE, Inc., Northville, Michigan) (see Section 4.3 for additional details on EMG as the basis for cHRI). The program regulated air pressure in the artificial pneumatic muscles proportionately to the processed muscle activation pattern. The artificial plantar flexor muscle and tibialis anterior were activated by soleus EMG and the artificial dorsiflexor muscle by EMG (for a description of EMG systems see Chapter 6). EMG signals from the soleus and tibialis anterior were highpass filtered with a second-order Butterworth filter (cutoff frequency of 20 Hz) to remove movement artefact, full-wave rectified and lowpass filtered with a second-order Butterworth filter (cutoff frequency of 10 Hz) to smooth the signal. Threshold cutoffs eliminated background noise and adjustable gains scaled the control signals. When the lowpass filtered soleus EMG signal was above threshold, the software inhibited all activation of the artificial dorsiflexor muscle. The software sent 0–10 V analogue signals to the proportional pressure regulators and solenoid valves to control activation and deactivation of the artificial pneumatic muscles.

9.2.7 Performance data

Figure 9.6 shows muscle activation patterns, joint kinematics and joint net muscle moments from one healthy subject walking while not wearing the orthosis and walking while wearing the orthosis passively. The subject had never worn the orthosis before this test session. The mean rectified EMG was similar for the two conditions. There were small differences in ankle joint kinematics and net muscle moment at the knee joint between the no-orthosis and passive orthosis conditions. In general, however, the gait pattern was relatively similar for the two conditions. These findings demonstrate that the added mass and restrictions on ankle degrees of freedom had relatively little effect on the walking dynamics.

The subject walked overground immediately after turning on the proportional myoelectric control (see Figure 9.7). The dorsiflexor inhibition from plantar flexor activation prevented coactivation of the two artificial muscles, facilitating gait in the naive subject. The largest changes in muscle activation between the passive orthosis condition and the active orthosis condition were greatly increased recruitment of the tibialis anterior and medial hamstrings (see Figure 9.7). A comparison of joint kinematics for the two conditions reveals reduced dorsiflexion during stance and increased plantar

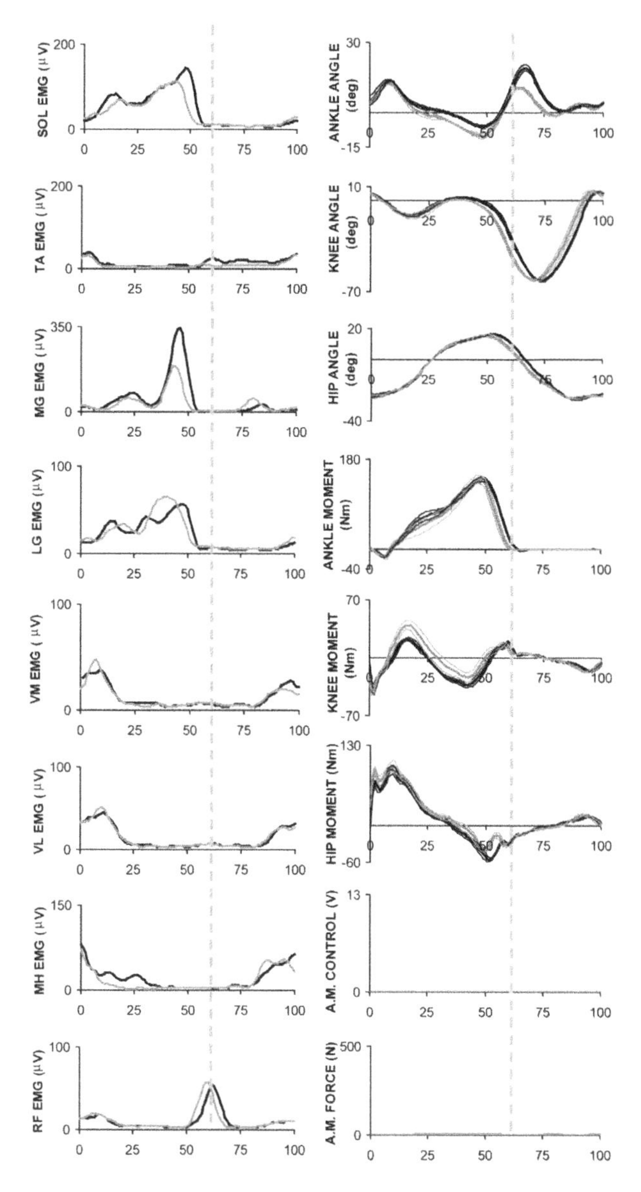

Figure 9.6 Comparison of no-orthosis and passive orthosis conditions. Mean lowpass filtered EMG patterns were very similar for the no-orthosis (black) and passive orthosis (grey) conditions. The largest difference in joint kinematics was at the ankle joint. In the passive orthosis condition (grey) there was slightly more dorsiflexion during stance and less plantar flexion during swing compared to the no-orthosis condition (black). The largest difference in net muscle moment about the joints was at the knee joint. There was a slightly greater knee extension moment magnitude and slightly lower knee flexion moment magnitude during stance in the passive orthosis condition (grey) as compared to the no-orthosis condition (black). Joint angle and moment graphs include the mean standard deviation one. Standing posture was defined as zero degrees for joint angles. Joint extension and plantar flexion are positive for joint angle and moment graphs. There were no artificial muscle (AM) control signals during either condition. AM forces during the passive orthosis condition were less than 30 N (data are means of three trials). Toe-off was at 61% of the gait cycle for the passive orthosis condition and 64% of the gait cycle for the no-orthosis condition. The light grey dashed vertical line represents 61% of the gait cycle

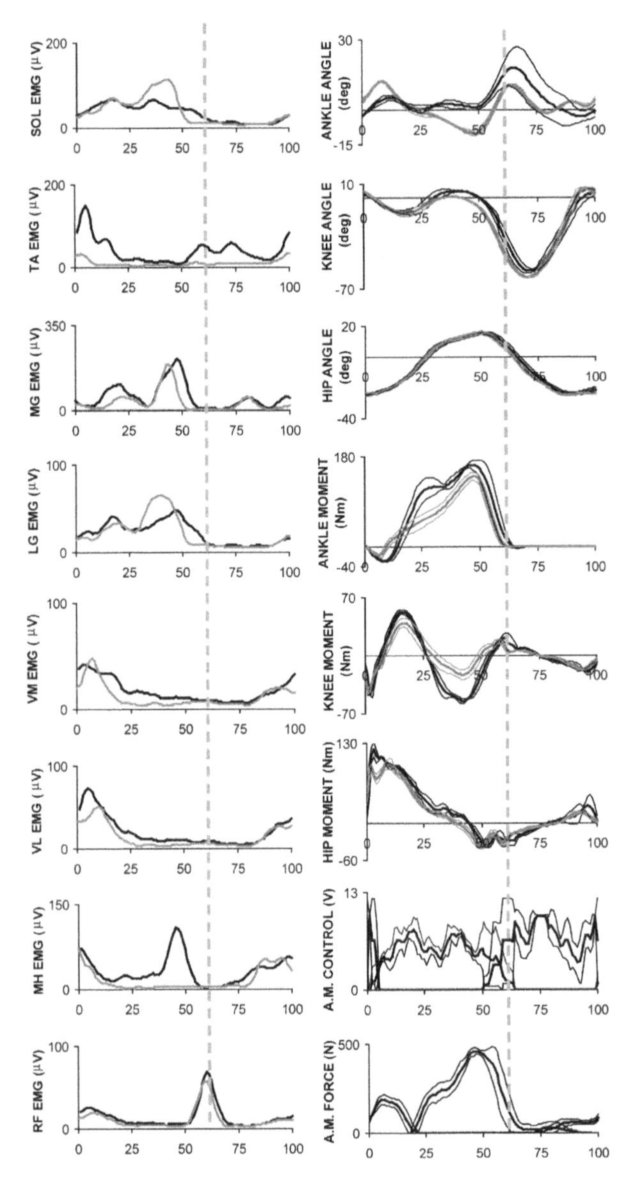

Figure 9.7 Comparison of active orthosis and passive orthosis conditions. Mean lowpass filtered EMG patterns of most muscles were similar for the active orthosis (black) and passive orthosis (grey) conditions. In the active orthosis condition (black) there was much less dorsiflexion during stance and more plantar flexion during swing compared to the passive orthosis condition (grey). There was a greater magnitude of plantar flexion torque and knee flexor torque during stance when walking with the active orthosis. Only the active condition control signals and artificial muscle (AM) force are evident in the graphs because they were zero and virtually zero respectively in the passive condition. The AM control signals appear to show coactivation because they reflect the mean of three separate trials and not an individual trial. Joint angle, joint moment, AM control and AM force graphs are the mean standard deviation one. Standing posture was defined as zero degrees for joint angles. Joint extension and plantar flexion are positive for joint angle and moment graphs (data are means of three trials). Toe-off was at 61% of the gait cycle for the passive orthosis condition and 63% of the gait cycle for the active orthosis condition. The light grey dashed vertical line represents 61% of the gait cycle

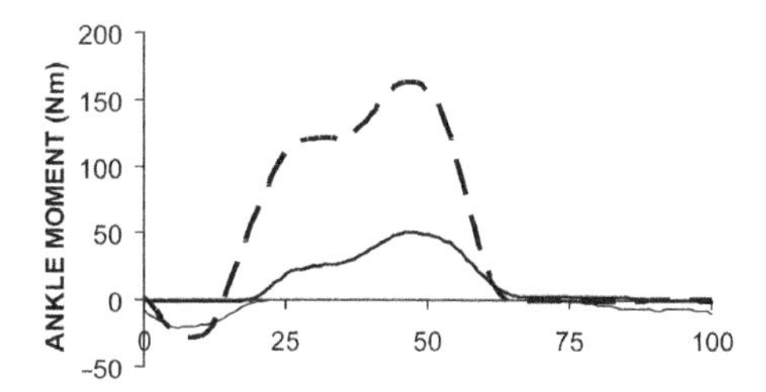

Figure 9.8 Moment contributions at the ankle joint. The dashed black line represents the total net muscle moment at the ankle joint during the active orthosis condition. The thick solid line is the contribution of the artificial plantar flexor muscle to the total net muscle moment. The thin solid line is the contribution of the artificial dorsiflexor muscle to the total net muscle moment

flexion during swing in the active orthosis condition (see Figure 9.7). There was also a greater magnitude of plantar flexion torque and knee flexor torque during stance when walking with the active orthosis. The artificial pneumatic muscles produced large forces during stance and performed substantial positive work. The artificial plantar flexor muscle delivered a peak torque (50.7 N m) that was 36 % of the peak plantar flexor torque in the passive orthosis condition (140 N m). The artificial dorsiflexor muscle delivered a peak torque (20.7 N m) that was 123 % of the peak dorsiflexor torque in the passive orthosis condition (16.8 N m). The contribution of the artificial muscles to the overall net muscle moment at the ankle joint is shown in Figure 9.8.

9.2.8 Major conclusions

In other studies that were conducted with the ankle–foot orthosis design presented here, it was found that the orthoses can substantially assist ankle motion during walking. Depending on the amount of practice and the type of control implemented, the orthosis can produce around 50 % of the normal ankle plantar flexor torque during stance. The orthosis peak torque does not change with walking speed (Gordon and Ferris, 2007). When testing at different speeds, the orthosis produced 0.28 J/kg of positive work at 1.0 m/s and 0.31 J/kg of positive work at 1.5 m/s. These values are about 70 % of the positive plantar flexor work done during normal walking (Eng and Winter, 1995). Artificial muscle work also remained fairly constant across speeds.

Although most of the previous work has focused on healthy, neurologically intact subjects wearing the orthoses, currently their use as rehabilitation aids after spinal cord injury and stroke is being investigated. In future studies an investigation will be carried out to see whether practice walking with added power at the ankle has any long-term beneficial effects on walking dynamics after the orthoses are removed.

9.3 CASE STUDY: INTELLIGENT AND POWERED LEG PROSTHESIS

K. De Roy

Össur, Reykjavik, Iceland

Revelations produced by modern science contribute to the constant development of new materials and technologies. This case study describes the integration of such materials and technologies in systems that can operate in synergy with the disabled human body.

The functionality of the human body has been investigated and quantified by means of gait analysis and neuroresearch, to the extent that little or nothing remains to be discovered about how the intact human body supports different levels of functionality during activities of daily living (ADL). Researchers have used advanced imaging technologies and oxygen intake measuring equipment to investigate and reveal how the functionality and performance of the trans-femoral amputee differ from those of nonamputees. Discrepancies have been reported in the prevalence of degenerative conditions, reduced biomechanical efficiency and abnormal gait dynamics among amputee subjects. Technical improvements of the conventional leg prosthesis have focused on passive mechanics, and findings indicate some limitations on the capacity of those mechanisms to replace the functionality lost due to the amputation of a leg at the thigh.

During design of the first intelligent powered prosthesis (also known as Power Knee™), control engineers focused on the integration of a multilevel human–system interface through advanced sensory control, artificial intelligence and actuation techniques. The system provides active lifting power, the ability to regenerate kinematics and control motion trajectories, and a sensing module connecting the prosthetic device to the human body. By integrating these intelligent structures into the design of prosthetic legs, it is hoped to meet more of the functional requirements of above-knee amputees. Reduction of physical and mental effort and restoration of normal gait dynamics are expected to reduce the risk of secondary complications and cases of comorbidity.

Continuous investigation is constantly being combined with additional practical experience to help identify and objectify how these advanced prosthetic solutions meet the medical requirements of the amputee population. At this stage, however, clinical investigation only allows for identification of the functional limitations experienced by above-knee amputees as compared to nonamputees when wearing a passive mechanical prosthetic knee design. In addition, theory combined with practical experience in the application of the first intelligent and powered leg prosthesis indicates that this prosthetic technology can augment the functional level of above-knee amputees.

In conclusion, it is fair to say that complete simulation and substitution of complex human locomotion and activity will require further development in sensor technology, artificial intelligence and actuator technology. Clinicians currently confirm the positive effect of these structures on the locomotive characteristics of the amputee and on the variety of daily living activities performed by the amputee population.

9.3.1 Introduction

Historically prosthetic design consists of the application of passive mechanical structures. Esquenazi and DiGiacomo (2001) quantified the added oxygen consumption requirement in different types of amputees. It was found that above-knee amputees wearing certain types of passive prosthesis use up to 50 % more energy than nonamputees while walking. For hemipelvectomy amputees the energy consumption for level ground walking is double that of nonamputees. In fact it has been found that increased effort requirements lead to a limitation on user activity or even to avoidance of certain activities of daily living (ADL). According to the National Institutes of Health, obesity and overweight together are the second leading cause of preventable death in the United States. Along with food and genetic predisposition, physical activity is an important causal factor.

Kinematic analysis on the locomotion of active amputees objectifies compensatory strategies used to overcome the shortcomings of mechanical prosthetic design. Deviations from normal motion and locomotion during different activities produce alternative strain on the human body. Studies indicate increased prevalence of secondary pathologies among the population of amputees (Burke, Roman and wright, 1978) found that there was a significant increase in osteoarthritis in the knee of the unamputated leg compared with the amputated side. According to Norvell *et al.* (2005), stresses on the contralateral knee of amputees may contribute to secondary disability such as osteoarthritis. Kulkarni *et al.* (2005) indicated that above-knee amputees were more likely to suffer from back pain.

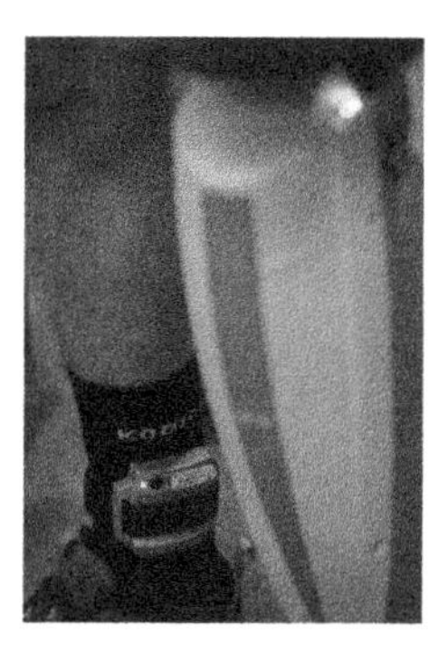

Figure 9.9 Human–system interface

The investigation of the biomechanics of passive mechanical prosthetic design in the above-mentioned studies is less complete than investigation of the human body, and therefore all functionalities cannot be restored in a human-like manner. Modern prosthetic design therefore aims to implement intelligent structures that will improve control of the leg prosthesis and regain more of the functionalities lost due to amputation, helping the amputee to overcome obstacles by means of improved kinematics and reduced effort.

The intelligent structures incorporated in modern leg prosthesis design are examples of advanced sensor technology, artificial intelligence and actuator technology. The actions of the leg prosthesis are controlled by sensory data collected at the sound leg as well as at the level of the artificial limb. The sensors combine to provide information on the positioning and loading of the sound leg and the prosthetic leg. This user–system sensory interface (see Figure 9.9) provides the basis for control of a motorized leg prosthesis, in synergy with the human body and equipped to respond appropriately to the activities to which the amputee is confronted.

The artificial intelligence operates on high- and low-level layers to monitor continuously the overall state of the human–system interface. The high-level layer is responsible for management of biomechanical events in a variety of daily life activities and of amputee–prosthesis interaction. The low level artificial intelligence manages the artificial knowledge, the system's response and the interaction between units.

The motorized actuator module generates power according to the user's need to negotiate different portions of locomotion adequately. Portions requiring specific power management are level ground walking, ascending or descending stairs and inclines, sitting down and standing up. The motorized knee unit substitutes for the eccentric and concentric muscle-work required during this type of action, allowing effortless, natural performance.

The leg prosthesis is intended to be emblematic of individual independence, equipping the above-knee amputee to face any challenge. The concept's great strength is its freshness and novelty; future developments will ensure that the product lives up to the high expectations it elicits.

9.3.2 Functional analysis of the prosthetic leg

The ability to achieve normal walking dynamics depends on the integrity of the human locomotion system. Synergy between the different structures of this system is imperative to achieve the dynamics that constitute normal gait. The most common way of describing the walking activity is to identify the functional significance of the events within the gait cycle and designate the intervals as the functional phases of gait.

An amputation inherently affects that integrity and will therefore affect the functionality of the human locomotion system. This chapter describes a selection of important functional phases of gait where discrepancies are observed between above-knee amputees and nonamputees. It highlights the difference between impaired dynamics, which is the gait pattern recognized for amputees using traditional mechanical prosthesis, and the restored dynamics provided by the intelligent powered prosthetic leg.

9.3.2.1 Normal walking dynamics

Normal walking dynamics describes the synergetic motion of the joints and muscle activity during the different stages of gait. These dynamics are the result of kinematic, kinetic and electromyographic analysis. The following synopsis of relevant aspects of the gait cycle is based upon the findings described in Perry (1992).

Initial swing. Initial swing is the part of the gait cycle where the foot is lifted and the limb is advanced by hip flexion, increased knee flexion and slight lifting of the foot. The objective during this phase is to provide foot clearance from the ground to avoid stubbing the toes while the trailing limb is advancing. The gracilis and sartorius muscles control the three-dimensional path of the limb and induce knee flexion as they act at the hip. The most important muscles for an understanding of this article are the two single joint muscles, popliteus and the short head of the biceps (BFSH), which provide direct knee flexion up to 40°, and the pre-tibial muscles, which perform a toe-lift at this stage.

Mid swing. Mid-swing is the part of the gait cycle where knee flexion reaches a maximum of 60° at an angular velocity of 350 deg/s. The activity at the hip is reduced at this time, and the advancement of the limb is a result of the continuation of the action occurring in initial swing. At this stage the popliteus and BFSH continue to provide direct knee flexion, contributing to the foot clearance required to avoid toe stubbing. At the ankle there is continued ankle dorsiflexion.

Terminal swing. Terminal swing is the final preparation in the transition phase between the swing and stance. During this phase the muscle activity is intended to prepare the advancing limb for load acceptance. At the level of the hip and the knee the two joint hamstring muscles control the deceleration of the hamstring muscle. At the hip the gluteus maximus and adductor longus also contribute. Some produce slight hyperextension of the knee while others maintain a minor degree of flexion, but in general the final knee posture at the end of terminal swing averages 5° of flexion in preparation for initial contact.

Initial contact and loading response. At initial contact the knee is flexed about 5°. Following the impact, the knee rapidly flexes in response to the loading phase and reaches a maximum of 18°. While the main objective at initial stance is stable weight bearing, shock absorption is the principal function during the loading response. The quadriceps has functioned eccentrically to restrain the degree of flexion.

9.3.2.2 Impaired walking dynamics

Losing a lower extremity above the knee greatly affects the synergetic motion of the joints and muscle activity during the different stages of gait. Different types of prosthetic design have attempted to restore those synergies by means of mechanical prosthetic solutions. The findings described in this paragraph are partly objective and partly based upon clinical empirical findings.

Initial swing. The gracilis and the sartorius muscle function mentioned above is partly affected by amputation above the knee. The length will be compromised to the extent of the amputation, influencing the effectiveness of the muscle function. In general, amputees are thought to overcome this limitation by increasing the hip-flexor muscle activity in order to lift the prosthetic foot off the ground and initiate the swing phase. In addition to hip flexion, amputees will also need to rotate the pelvis and the lower trunk while applying load on the toes of the prosthetic foot. Since there is no muscle activity at the level of the knee or the foot to initiate the toe-off of the foot, ground clearance is often compromised.

One of the most important shortcomings is the loss of the two single joint muscles, popliteus and the short head of the biceps (BFSH), which provide direct knee flexion, and the pre-tibial muscles, which perform a toe-lift at this stage. Since the foot is not lifted, this will affect foot positioning during the entire swing phase and compromise user safety during the initial swing phase.

Mid-swing. The shortcomings mentioned in the initial swing phase persist through the advancement of the leg during the mid-swing phase. Ground clearance is compromised, and it is during this part of the swing phase that the risk of toe-stubbing and falling is greatest.

The limitation of knee flexion becomes more noticeable at this stage. The absence or reduction of knee flexion will lead to recognizable gait deviations such as lifting of the prosthesis up from the hip (hip-hiking), jumping on the contralateral side at mid-swing of the prosthetic leg (vaulting) or circumduction gait to compensate for the reduced ground clearance and avoid stubbing the toes on the ground. These gait deviations result in asymmetric loading of the body; they also put strain on the lower back and require additional energy from the amputee. Some of these limitations of passive prosthetics are inherent to the knee design, which may require toe-load to initiate flexion; others are related to the biomechanical interaction between the human body and the prosthetic device.

When walking on uneven terrain, at low walking speeds, taking turns or walking in confined spaces, the trans-femoral will confirm the difficulty or inability to perform the required combination of movements (hip flexion, pelvic rotation and loading the prosthetic toe). Gait deviations such as hip hiking, circumducting, vaulting and walking with a stiff knee as a compensatory strategy become even more apparent and efficiency and safety are compromised.

Terminal swing. Any mechanical prosthetic knee design has an inherent low-end friction, which reduces the acceleration of the prosthetic knee joint during advancement of the lower leg in swing phase. The difference in anthropometrical characteristics between the mechanical prosthetic knee design and the human leg has an impact on the efficiency of the pendulum motion. The user is required to overcome this residual friction and the abnormal anthropometric characteristics by means of additional muscle activity and abnormal motion patterns in order to complete the swing phase. Following are some of the recognized deviations during this phase of gait.

Typically the user kicks the knee into extension, usually by contraction of the gluteus muscles, which reverses the heel-rise motion of the knee and initiates the forward swing. The activity of the gluteus is abnormal during this stage of swing and affects the quality (efficiency) of gait as well as the stability. The kicking motion which is required to move the passive prosthesis forward produces a subtle posterior positioning of the amputee's centre of mass. This is counter to the walking direction and negatively affects balance and walking effectiveness.

The passive swing extension limits the possibility of stumble recovery. In the event that an obstacle (high grass, wind, sand, stone, etc.) interferes with the trajectory of the knee, it can obstruct the motion of the knee and the knee will not be able to complete the full range of motion. In such a case, if the knee does not attain extension in time and so cannot accept the load of the next step, there is considerable likelihood of the user falling.

Initial contact and loading response. As noted earlier, this phase of the gait cycle is responsible for stance stability and shock absorption. The prosthetic leg of the transfemoral amputee does not produce the knee flexion motion intended to absorb the shock, due to the lack of active joint control subsequent to loss of quadriceps muscle function (Suzuki *et al.*, 1983). During weight bearing, the knee must be fully extended, and often hyperextended, to prevent it from buckling into flexion due to loss of quadriceps control (Perry, 1992).

In daily practice it is observed that the amputee seeks stability at terminal swing by kicking the prosthesis into extension; this affects the slightly flexed prepositioning of the knee, making initiation of flexion at initial contact more difficult and producing a sensation of instability. Hay (1985) demonstrated the negative influence of a greater vertical displacement of the CoM on the energy cost, again drawing attention to the lack of biomechanical efficacy.

9.3.2.3 Restored walking dynamics

A prosthesis that combines intelligence and motorized actuation has the ability to regenerate kinematics. Regeneration of kinematics means that the knee is pre-programmed to execute normal gait dynamics during all phases of the gait cycle, and is able to meet more of the functional requirements with which the amputee is challenged in activities of daily living, such as negotiating inclines or stairs or simply sitting down and standing up. Through advanced human–system interfaces at the level of the sensing technology and the artificial intelligence, the system is able to act in synergy and symmetry with the user, which contributes to the smoothness of the prosthetic behaviour.

This section explains how the intelligent powered prosthesis manages to replace the walking dynamics that are lost when the amputee loses the leg, and with it the muscles responsible for execution of the different stages of gait.

Initial swing. The intelligent powered prosthetic knee joint is the only prosthetic solution that actively lifts the foot off the ground at initial swing to provide the required ground clearance for safe gait, irrespective of the walking speed or type of activity.

The normal human leg moves at an angular velocity of 350 deg/s during this phase. It is no coincidence that the intelligent powered knee prosthesis contains gyrometer technology capable of monitoring angular displacement over 300 deg/s and a motor capable of reproducing the angular displacement needed to support this phase of the gait cycle. In this way the system is able to overcome the limitations of passive prosthetics mentioned above.

Since the system has active lifting capacity, the user no longer requires abnormal movement or muscle activity to initiate the heel rise. On level ground this contributes to safety at initial swing and restores normal heel rise, which is furnished by the BFSH and the popliteus muscle in nonamputees. Gait deviations such as hip hiking, vaulting and circumduction are visibly reduced; users report a feeling of reduced abnormal muscle activity, and the overall effort required to walk becomes more comparable to that of a nonamputee.

The additional ground clearance provided by the power knee and the fact that the knee actively lifts the foot off the ground enhances user safety when walking on uneven terrain. When walking on inclined walkways the improved gait detection mechanism and the active heel rise of the power knee facilitate this type of walking effort, allowing the user to walk without requiring additional effort, without gait deviations and without causing additional strain on the remaining body structures.

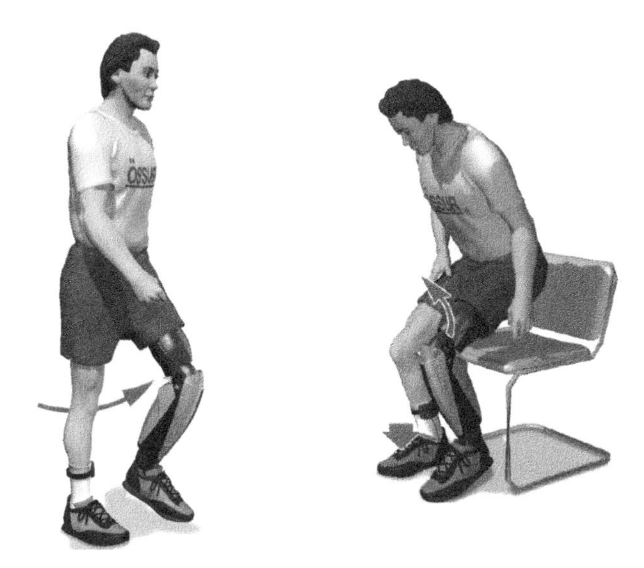

Figure 9.10 Active heel rise through the mid-swing phase (left) and strategy for assisted standing up from a chair (right)

Mid swing. The heel-rise activity which is initiated at initial swing continues during mid-swing. As the user continues to flex the hip to execute leg advancement, the intelligent powered prosthesis will increase the flexion to 60°, (see Figure 9.10 (left)), continuing to replace the activity of the BFSH and the popliteus muscle. This flexion is crucial to achieve the required ground clearance and avoid toe-stub and stumble. Following a short pause in mid-swing, the knee will than start to extend at a speed comparable to the initial swing flexion.

The transition into active swing extension that is required to assure the advancement of the lower leg is considered mid-swing. The intelligent powered prosthesis is the first and only knee joint to provide active extension of the knee during the swing phase. By means of motor activity the system ensures appropriate knee extension at the correct pace for each step and at each walking speed. Throughout the swing phase the motor replaces the function of the quadriceps and hamstring muscles to accelerate and decelerate the motion of the lower leg, thus producing a natural, smooth motion.

The anthropometric characteristics of the intelligent powered prosthesis correspond to that of the natural human leg. This means that the positioning of the CoM and the weight of the prosthetic knee joint stimulate forward motion of the user's body during active pendulum of the knee. In this way the knee allows the user to shift the bodyweight forwards instead of backwards, as in the case of amputees who kick their knee forward to ensure extension. Users comment that the fact that knee extension is automated and complete extension is guaranteed reduces the amount of attention they need to pay to the pre-positioning of the knee in preparation for initial contact. This allows the user to focus on quality of gait rather than on controlling the pendulum of the knee during the swing phase.

An added benefit of active swing extension during mid-swing becomes apparent where an obstacle interferes with the trajectory of the swing phase of the knee system, where the active swing extension will help overcome the resistance presented by the obstacle and the knee will still complete the swing trajectory.

Terminal swing. During the terminal swing phase the knee system replaces the deceleration function of a nonamputee's hamstring muscle, ensuring that the extension is controlled. As terminal swing is the preparation for initial contact and loading response, pre-positioning of the prosthetic leg is crucial for stability and shock absorption during this phase.

The passive prosthetic knee joint is able to decelerate the knee-extension motion by applying resistance at the end of the trajectory to avoid terminal impact, but as soon as initial contact occurs the user will forcefully extend the knee to gain stance stability. During this stage of the swing phase the intelligent powered prosthesis is able to control the angular position of the knee joint. This means that the prosthesis is able to maintain slight flexion in the knee during the terminal swing phase, which is a more convenient position for transition to initial contact.

Initial contact and loading response. In combination with the artificial intelligence, the knee motor can execute the required kinematic patterns corresponding to those of the human knee. The leg prosthesis does not reach full extension at terminal swing, but remains in slight flexion (5°) just like the human knee, and facilitates knee flexion and shock absorption at initial contact. As soon as the foot touches the ground and the ground reaction forces increase, the concentric and eccentric muscle activity around the knee ceases and is replaced by a spring-like action which responds according to the load applied to the knee. Depending on the fine-tuning of the software and the weight application on the knee, flexion angles from 5 to 15° can be achieved to absorb the shock. Subjective reporting by users confirms that this spring mechanism contributes to the user's feeling of stability and safety without requiring forceful extension of the prosthetic knee joint.

Objective measurements show that all amputees who use the intelligent powered prosthesis use stance flexion during initial contact to reduce the impact and smoothen the gait pattern. Of these users, most had tried passive knee designs like the ones mentioned above which allow stance flexion, but had not used them since they were considered cumbersome, difficult and instable.

The upper graph in Figure 9.11 represents the ground reaction forces applied to the prosthetic knee. The lower graph is synchronized with the upper one and represents the kinematics of the power knee. The dotted arrow indicates the period before initial contact on the prosthetic side. It shows how the knee remains in slight flexion (deviation from the baseline) during this stage, which is comparable to the human knee. The solid arrow indicates the initial contact phase (heel force increases in the upper graph) and indicates how the knee bends during this phase to provide shock absorption and to allow for natural and smooth motion.

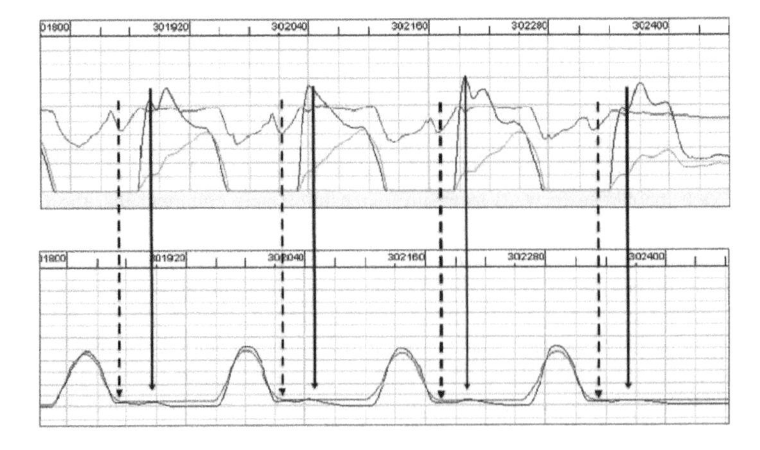

Figure 9.11 Synchronized view of the ground reaction forces acting on the prosthetic leg (upper graph) and the corresponding knee kinematics (lower graph)

9.3.2.4 Restored functional requirements

Apart from typical walking, the intelligent powered prosthesis is able to restore a variety of additional functional requirements that are not provided by the passive prosthesis. The main objective in regaining these additional functional activities is to restore postural symmetry and redistributive forces on both legs instead of applying excessive stress on the sound side ligament and muscle structures.

Standing up from a chair. When standing up from a chair, nonamputees will distribute the load over both legs. This contributes to postural symmetry and ease of standing up and reduces the strain on either of the legs.

Amputees using a passive mechanical prosthesis will mainly rely on the sound leg as it is the only active lifting power they have at their disposal. This produces an asymmetric movement of the body and puts uneven strain on the muscles used. For users whose sound leg is compromised, standing up from a chair often requires a lot of effort and is very straining.

The intelligent powered prosthesis will actively assist the action of standing up from a chair. The knee will lift the patient with the same amount of force and at the same speed as the sound leg. This enables the user to load both legs equally, maintain a symmetric position during the lift and reduce the strain on the sound limb (see Figure 9.10 (right)).

Ascending stairs. When walking upstairs a nonamputee uses an alternating strategy. First one leg and then the other is lifted up to a higher step, powered by the quadriceps muscle. In order to do so successfully the nonamputee needs to increase the knee-flexion angle and the hip flexion to bring the foot from a lower to a higher position and he/she must have sufficient quadriceps muscle power to lift the bodyweight to the next step.

When ascending stairs with a passive prosthesis, both of the above-mentioned requirements are missing. The prosthesis lacks the active power to position the prosthesis on the higher step. Once the sound leg is placed on the higher step, the user lifts him-/herself up by dragging the prosthesis to the same level as the sound leg.

The intelligent powered prosthesis has the ability to recognize the action of climbing stairs and to control actively its own motion and position to meet the requirements of the stair-climbing action. This strategy allows the amputee to climb stairs in a similar way to nonamputees (see Figure 9.12). As soon as the sound foot is placed on the first step, the knee identifies the executed function as stair climbing and provides the necessary knee flexion to allow the user to flex the hip and bring the leg up to the next step. As soon as the foot is placed on the step the knee actively extends with the same force and speed as the sound leg and lifts the subject up to the next step. During this alternating activity the user loads both legs equally and the range of postural movements is limited. In comparison with the passive prosthesis, the strain on the sound leg and lower back are reduced and the normal function is restored.

9.3.3 Conclusions

This case study demonstrates how the addition of intelligence and power to a prosthetic leg helps to satisfy more functional requirements for the above-knee amputee. Subjective user findings confirm the hypothesis that walking with an intelligent powered prosthesis requires less effort, imparts a smoother gait and facilitates negotiation of uneven terrain. Empirical clinical findings indicate that with this type of knee design the quality of the locomotion is improved, gait deviations are reduced and the functional performance of the above-knee amputee is normalized.

Among the more interesting aspects of this achievement are the ability to regenerate kinematics and control the trajectory of the knee joint in any given situation. The fact that sensory information

Figure 9.12 Alternating stair ascend

is collected at the sound leg improves the synergies between the prosthetic leg and the human leg, producing higher levels of symmetry than achieved with a passive prosthetic leg. With the addition of active lifting power, important activities of daily living can be restored, such as standing up from a chair, walking up a steep incline or climbing stairs in a symmetrical manner like a nonamputee.

A combination of continuous investigation and additional practical experience is expected to help identify and objectify these benefits, but in any case most of them are obvious from simple clinical observation. The leg prosthesis has been conceived as a symbol of individual independence, equipping the above-knee amputee with the ability to face any challenge. The concept's great strength at this stage is its freshness and novelty; with further development it should be possible to improve its efficacy.

9.4 CASE STUDY: THE CONTROL METHOD OF THE HAL (HYBRID ASSISTIVE LIMB) FOR A SWINGING MOTION

J. C. Moreno, E. Turowska and J. L. Pons

Bioengineering Group, Instituto de Automática Industrial, CSIC, Madrid, Spain

The hybrid assistive limb (HAL), has been developed at the University of Tsukuba as an assistive device for the operator's lower limb. In order to assist or enhance the human motion, the system has to generate torque as a muscle. For this, the voluntary motion of the operator is detected, gathering myoelectric information (Ferris, Sawicki and Domingo, 2005). Torque generation is required but also control of the viscoelasticity by muscle effort and adjustment of the stiffness. When the operator needs a high joint viscoelasticity, it is useful to increase the viscoelasticity of an actuator of the HAL to assist the motion. The method included in HAL-3 (see Hayashi, Kawamoto and Sankai 2005), in the case of the swinging motion of the lower leg, is presented in this case study.[1]

[1]The text in this case study is based on the paper by Hayashi, Kawamoto and Sankai (2005), reproduced by permission of IEEE,

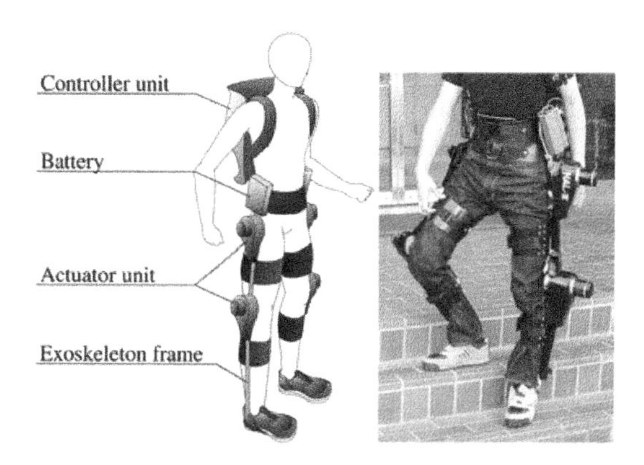

Figure 9.13 Configuration of the robot suit HAL-3. Reproduced from Control method of robot suit HAL working as operator's muscle using biological and dynamical information, Hayashi, T. Kawamoto, H. Sankai (paper appearing in Intelligent Robots and Systems (IROS 2005)) © 2005 IEEE

9.4.1 System

The HAL exoskeleton consists of a frame with actuators for knee and hip joints in each leg (see Figure 9.13). The hyperextension or hyperflexion motions are prevented with mechanical limiters at each actuator. Each joint angle is measured by potentiometers attached to the joint. Myoelectricity is detected with two sensor units attached on the skin near the flexor and the extensor driving the targeted joint. Each sensor unit is composed of two electrodes and an instrumentation amplifier. The estimation of the myoelectrical activity is defined as the envelope of the measured myoelectricity at the extensor and flexor of the human joint. For more details on motion and sensor technologies, the reader is referred to Section 6.2.

The human muscle torque is estimated by applying coefficients (obtained for each user with a least squares method during a calibration procedure) to convert from myoelectric activities to the contraction torque. A gain parameter is applied to the estimated human muscle in order to define the torque produced by the HAL.

9.4.2 Actuator control

Myoelectric activity of the driving muscles is calculated online by means of two electrodes and an instrumentation amplifier, and is used to calculate the muscle torque $\hat{\mu}$. The torque τ_μ produced by the HAL exoskeleton is given by

$$\tau_\mu = \alpha_\mu \hat{\mu} \tag{9.1}$$

where α_μ is a gain parameter.

A musculoskeletal model of the operator's lower limb equipped with the HAL is used for the estimation of the viscoelastic properties of joint muscles and control of the system. Muscles acting on a joint can be regarded as one muscle group. The model considers the muscle group around the knee joint that can produce torque towards the contracting direction but cannot produce it towards the extending direction. Thus, the muscle group needs two torque generators corresponding to the two directions. Viscoelastic properties of the muscle group can be represented as a combination of a viscous element and an elastic element. In the HAL, the two elements are defined as time-varying parameters as it is assumed that the operator can modify the viscosity and elasticity with time.

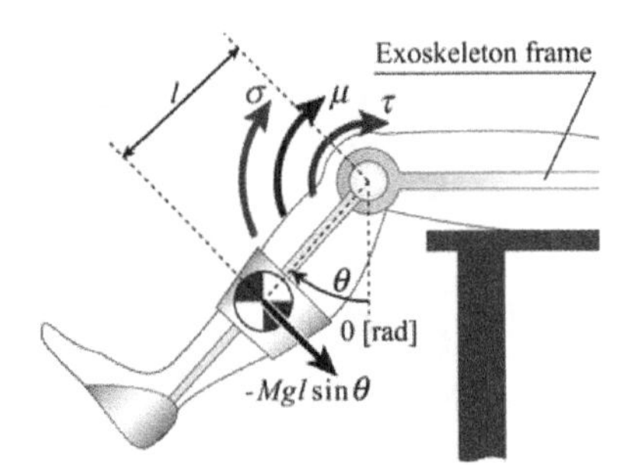

Figure 9.14 Configuration of the musculoskeletal model of operator's lower leg equipped with HAL. Reproduced from Control method of robot suit HAL working as operator's muscle using biological and dynamical information, Hayashi, T. Kawamoto, H. Sankai (paper appearing in Intelligent Robots and Systems (IROS 2005)) © 2005 IEEE

A musculoskeletal model of the operators leg, as a pendulum system, is presented in Figure 9.14. The equation of motion of the ith link of the model is expressed as

$$I_i\ddot{\theta} + (D_i + R_i)\dot{\theta} + K_i\theta_i + M_i g l_i \sin\theta_i = \tau_i + \mu_i + \sigma_i \tag{9.2}$$

where θ is the angle of joint i, I is the total inertia around the joint, D and K are the viscous and elastic coefficients of the muscle group, R is the viscous coefficient of the HAL, M is the mass of the leg, g is the gravitational coefficient, l is the distance between the joint and the centre of mass of the operator's leg and σ is the total interaction torque between adjacent links.

9.4.2.1 Control of exoskeleton viscoelastic properties

The actuator torque to control the viscoelastic properties of the HAL exoskeleton is based on impedance control methods. For more details on the underlying theory of impedance control, the reader is referred to Section 5.4. The actuator torque τ_ζ for a given joint can be expressed as

$$\tau_\zeta = \alpha_\zeta(-D_i\dot{\theta}_i - K_i\theta_i) \tag{9.3}$$

In order for the HAL to work as the muscles, the torque τ of the actuator is expressed as

$$\tau_1 = \tau_{1\zeta} + \tau_{1\mu} + \tau_{ic} \tag{9.4}$$

where τ_i is the torque τ_{ic} to compensate (partially) for the mechanical impedance. Thus, the actuator torque provided by the HAL is produced as an amplifier of both the human muscle torque and the viscoelastic properties. In consequence, the proposed method would reduce loads on the operator's muscles.

9.4.3 Performance

The proposed method has been evaluated for the swinging motion of a lower leg (see Hayashi, Kawamoto and Sankai 2005). The operator putting on the HAL sat on a chair that has enough height to prevent his foot from grounding. The operator swung his right lower leg up and down. The operator was asked not to activate other joints except the right knee.

Two experiments have been performed in order to evaluate the effectiveness of the method. The experiments consisted of comparing first the case when no assisting method applied, producing only

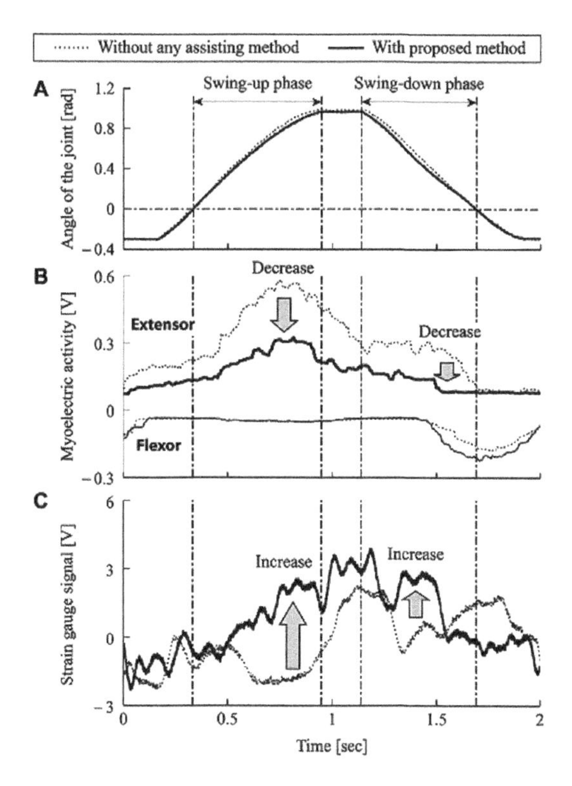

Figure 9.15 A typical cycle of the swinging motion when applying the proposed method. Reproduced from Control method of robot suit HAL working as operator's muscle using biological and dynamical information, Hayashi, T. Kawamoto, H. Sankai (paper appearing in Intelligent Robots and Systems (IROS 2005)) © 2005 IEEE

the torque τ_c compensating its mechanical impedance, and then the case when the proposed method was applied and the torque produced was according to Equation (9.4), with a manual definition of the gain parameters and a targeted frequency for the swinging motion of 0.5 Hz.

Figure 9.15 presents a typical cycle of the swinging motion with HAL assistance. The increase in the angle corresponds to the extension of the knee joint and the decrease in the angle corresponds to the flexion of the joint. For a comparison, the cycle of swinging motion without any assisting method is superposed here as dotted lines. The transition of the angle of the actuator of the knee joint is shown in Figure 9.15A. Figure 9.15B presents the myoelectric activities around the knee joint. It can be observed that the myoelectric activity of the extensor is smaller than the activity measured when not applying assistance. Also, a very reduced co-contraction of the extensor and flexor is observed for this experiment. Figure 9.15C presents the measurement of strain (for strain measurements the reader is referred to Section 6.2) on the exoskeleton frame. During the swing-up phases, the strain signal is approximately positive and, in addition, in the middle of the swing-down phase, the strain signal increased. The results indicate that the force applied to the lower leg from the HAL worked in the extending direction.

It was also found that when assistance was applied the observed myoelectric activity of the flexor increased as compared to the case without any assisting method. Thus, the operator produced the muscle torque to flex the knee joint because the HAL has restrained knee flexion more than the operator has expected. It is possible to establish a particular method to adjust the gain parameters appropriately in order to avoid such an operation.

9.5 CASE STUDY: KANAGAWA INSTITUTE OF TECHNOLOGY POWER-ASSIST SUIT

K. Yamamoto

Kanagawa Institute of Technology, Atsugi-shi, Kanagawa, Japan

In 1991 the author proposed a pneumatic power-assist suit that substantially reduced the physical burden of the caregiver wearing it (Yamamoto, Miyanishi and Imai, 1991). In 1994, a wearable powered suit was developed that was constructed with powered arms, a powered waist and powered legs (Yamamoto, Hyodo and Imai, 1996) and finally, in 2002, a standalone wearable power-assist suit was developed (Yamamoto *et al.*, 2003). This suit was equipped with rubber cuff actuators driven by micro air pumps, muscle hardness sensors and an embedded microcomputer. The muscle hardness sensor was developed to detect the muscle force driving the joints: the microcomputer was for calculation of the joint torques needed to lift heavy objects. The calculation equations were derived from body mechanics. This suit could run continuously for 20 minutes on 12 V Ni–Cd portable batteries. The latest power assist suit can generate greater assist power and is composed of a compact body, a compact embedded microcomputer and small, flat rotary sensors. It has newly designed muscle hardness sensors that are embedded in a three-dimensional mesh (Ishii, Yamamoto and Hyodo, 2005; Yamamoto *et al.*, 2004).

9.5.1 The basic design concepts

There are four basic points in the design of the power-assist suit:

1. The system must be absolutely safe, i.e. ready for all emergencies. This is assured by placing the nurse in control, i.e. the assist system is a master and slave system in one unit. In addition, when the electric power source is cut off, the air supply pumps and the exhaust valves prevent back flows from the air actuators, so that the suit continues to assist in the process of holding.
2. There are no mechanical parts on the front of the suit. The nurse's arms and chest may be in direct contact with the body of the patient carried in her arms. This produces empathy between the patient and nurse.
3. Flexible joints are implemented by a pneumatic rotary actuator using rubber cuffs. The use of these pneumatic actuators in joints make the nurse's arms, waist and legs soft to touch.
4. Assisting forces adapted to requirements for bending and stretching the joints is achieved by using the muscle hardness sensor to detect the force exerted in the muscles driving the joints. Thanks to this sensing system, smooth movements of the arms, waist and legs of the assisting suit are possible. As an additional backup and failsafe mechanism, the joint torque needed to maintain a position is calculated by means of equations derived from static body mechanics using the joint angles.

9.5.2 Power-assist suit

The exoskeleton suit (see Figure 9.16) consists of shoulders, arms, a spine, a waist, and legs. The scapula-thorax joints are put in to enable the blade bone to rotate and thus allow the hand to reach a remote point. Each joint has an angle sensor (potentiometer), and the elbow, waist and knee joints

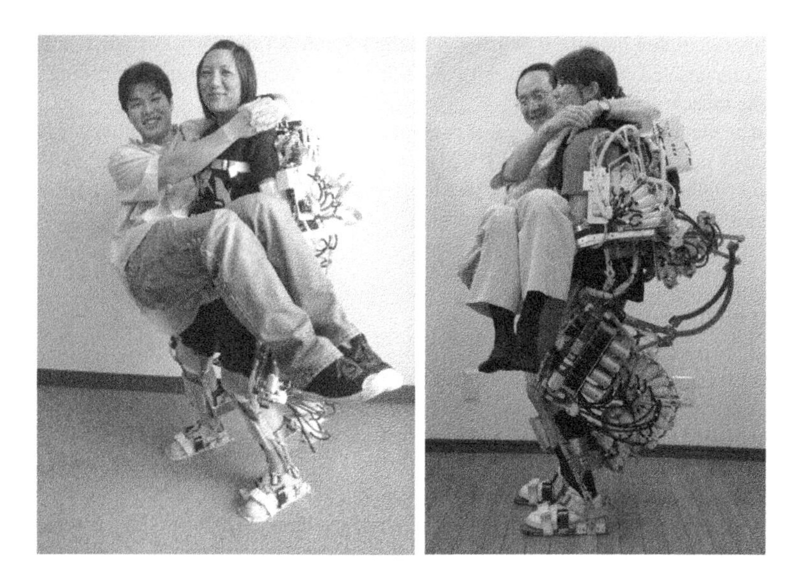

Figure 9.16 Power-assist suit

have direct-drive pneumatic actuators. The actuators consist of a rubber cuff covered with fabrics and bandages. Small DC motor-driven air pumps are directly connected to them. Each actuator generates soft assistance via air compressibility and rubber cuff elasticity, allowing users to move smoothly during normal operation. An Ni–Cd cell power source (12 V, 30 mm wide and 300 mm long) is mounted on each femur of the leg exoskeleton structure. The aluminium suit leaves the caregiver's ventral area free to allow physical contact between the caregiver and the person being assisted.

The suit control system is shown in Figure 9.17. When the nurse moves, muscle sensors on her upper arms, thighs and back detect the muscle power driving each joint. Each joint torque is calculated independently using a quasi-static physical dynamics model. This model considers the angle signal of each joint, the estimated weight of the person to be assisted, and the weights of the wearer and

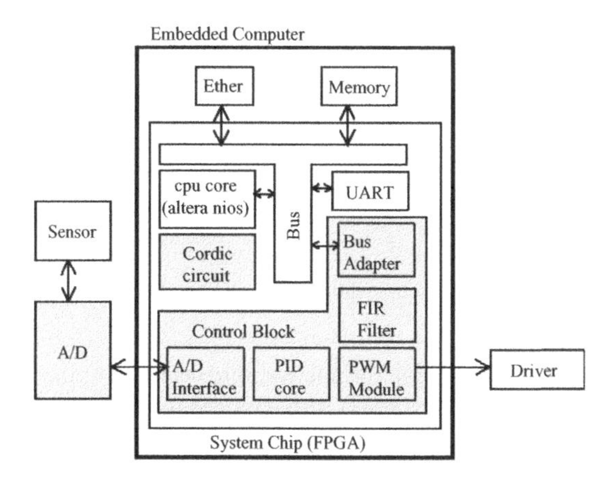

Figure 9.17 Microprocessor block diagram

the suit. Of this calculated output 50 % is used as the reference value for assistance torque. Finally, this value is compensated by a maximum of ±20 % of the reference torque according to the data provided by the muscle hardness sensor.

This is used as the final assist torque value to be generated by the actuator. That torque is calculated by the embedded microcomputer, and a PWM voltage signal is transmitted to the small air pump and the exhaust solenoid valve to control the air pressure that will provide the needed torque. The internal pressure of the actuator is PID controlled.

9.5.3 Controller

The controllers were implemented in a system on a programmable chip (SOPC) to make the suit wearable. Each controller board contains a field programmable gate array (FPGA), ADCs, an Ethernet controller, and SRAM and EEPROM. The controller core consists of a Nios 2.0 processor (32 bits) and a control block module (see Figure 9.17). The control block includes a 24-channel PWM module (18 bits), ADC interfaces, finite impulse response filters and a PID control core (16 bits). The standalone block's clock frequency is 33 MHz and the operation cycle is 20 clocks. The Nios processor is responsible for nonlinear operation and setting up of the control block parameters. This processor can be controlled remotely using the Ethernet interface. A control block design tool has been developed so that the type and number of sensors and actuators can be readily changed to support different configurations. The development environment uses a graphic user interface and the user combines required functional modules on the GUI to build the control block. The design tool automatically generates Verilog hardware description language (HDL) for the control block and the C program header file. The user can thus design the control block without having to be an expert.

9.5.4 Physical dynamics model

The role of the power-assist suit is to provide the torque required by the elbow, waist and knee when the nurse picks up or puts down the person being assisted. A muscle hardness sensor has been developed to detect the muscle power driving the joints. However, that was not enough to ensure full system reliability; to achieve that, a physical dynamics calculation model was determined in order to calculate the power needed by each joint to maintain a given posture.

Since nursing movement is slow and quasi-static, the equation of equilibrium of moment is solved using each joint angle as a variable. The outcome is the required muscle power. Muscles generating torque for the elbows, waist and knees, i.e. the biceps, back erector and thigh rectus muscles, are represented by muscle hardness sensors. The muscle forces driving the joints are estimated by computing the equations of moment balance as a function of the joint angles. These equations of moment balance are the result of gravity compensation derived from dynamics calculation using Lagrange's method (see Chapter 3).

9.5.5 Muscle hardness sensor

The muscle hardness sensor was developed for safe and sure detection of the nurse's movement and relates muscle hardness to muscle power. It was also designed to facilitate donning and doffing and stable detection. The outcome was an integral sensor embedded in the three dimensional mesh belt. It copes with the various thicknesses of subcutaneous fat and muscle at each measurement location on the nurse by adjusting the shape, height and tenseness of a silicone rubber button pressed against the muscle.

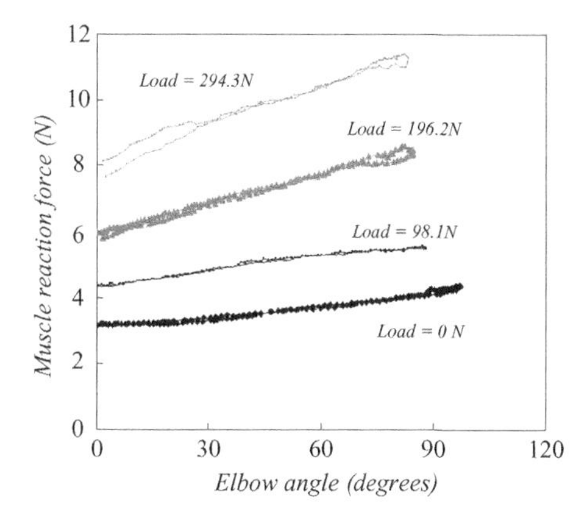

Figure 9.18 Conversion of the arm muscle hardness sensor

In experiments with male adults where each sensor was operated by a cylindrical button, bending and stretching of each joint was measured with loads (150 mm × 560 mm iron plates) placed on the forearm. Figure 9.18 shows detection with the sensor placed at the middle of the biceps. The measurement protocol included a basic task: a load on the forearm is applied while the elbow is outstretched ($\theta_e = 0°$); the user then bends it almost to a right angle and stretches it again to the initial position. Characteristics and load responsiveness are proportional, with less hysteresis and different slopes depending on the loads. Similar results were found when the sensor was placed at the lower part of the erector muscles for the waist and at the mid-point of the thigh rectus muscle for the knee. A similar protocol was used in every experiment. Trends among women were similar, but output was lower.

9.5.6 Direct drive pneumatic actuators

The actuators for the elbow and knee were structured so that aluminium plates were sequentially connected and folded in a zigzag fashion, while the actuator for the waist used one end for connection. A fabric-covered commercially available cuff (90 mm × 120 mm) for measuring blood pressure is placed in between two aluminium plates. The periphery of the plate is connected to the belt limiting the open angle. This prevents slipping of the cuff and enhanced the stability and the force transfer efficiency. For the knee actuator, an alternative combination of aluminium plates 70 mm × 150 mm and 70 mm × 170 mm is used, producing a slim, elliptical shape. The exhaust solenoid valve remains closed while power is not distributed, and the internal pressure of the cuff is maintained for safety.

A small air supply pump (30 mm ϕ, 65 mm long, 6 V DC, 1 A) is connected directly to the actuators. This has a maximum output pressure of 88 kPa, a flow rate of 8.0×10^{-5} m^3/s and a power of 1.6 W (Pa m^3/s). There is a small exhaust solenoid valve (33 mm wide, 25 mm deep, 54 mm high, 12 V DC, 0.3 A) with an effective sectional area of 2.5 mm^2.

9.5.7 Units

Because the rotary axis of a human joint is not monocentric, the exoskeleton joints were implemented using two flat gears (see Section 3.3 for a discussion on human limb biomechanics). The shoulder has two joints, one that can swing back and forth and one that can swing from side to side. The

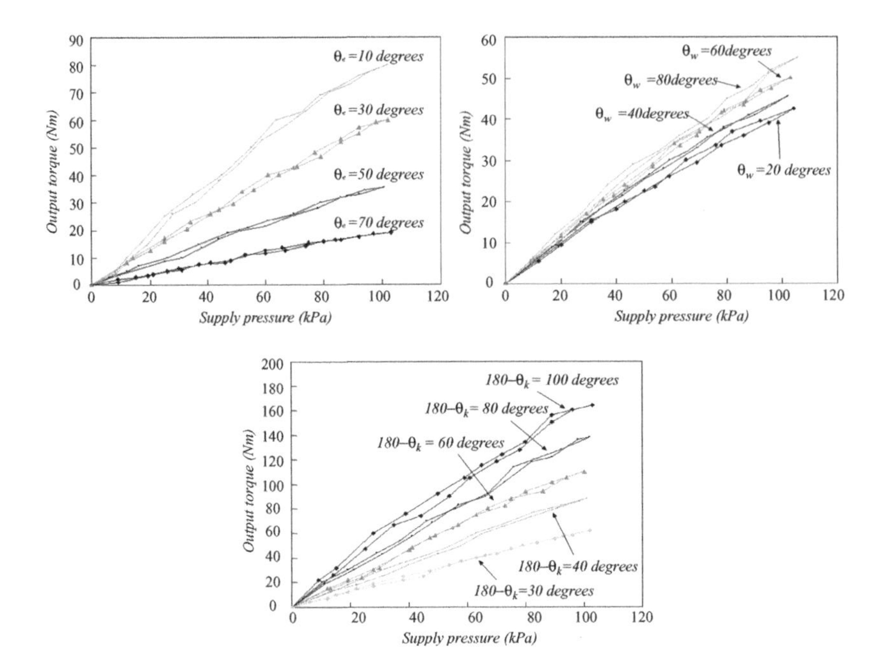

Figure 9.19 Output characteristics of the arm unit (top left) as a function of elbow angle, θ_e, output characteristics of the waist unit (top right) as a function of waist angle, θ_w, and output characteristics of the leg unit (bottom) as a function of knee angle, θ_k

rotary actuators produce extra torque to drive the elbow joints and assist in raising the arms. Output characteristics of the arm unit are shown in Figure 9.19 (top left). Linearity was good and a maximum torque of 80 N m was achieved with a supply pressure of 100 kPa.

The force produced by the rotary actuator is transmitted to the body of the suit by a four-node linkwork. Figure 9.19 (top right) shows the output characteristics of the waist unit. Again, linearity was good and a maximum torque of 55 N m was achieved, this time with a supply pressure of 108 kPa.

To make walking easy, a compliant sole was developed so that the toe could be flexed. The rotary actuators produce extra torque to assist the knee muscles. A maximum torque of 163 N m was achieved with a supply pressure of 103 kPa; see Figure 9.19 (bottom) for the output characteristics of the leg unit.

9.5.8 Operating characteristics of units

The operational characteristics of the different exoskeleton units were evaluated in experimental trials. The protocol is as follows: the wearer dons the suit and flexes the knees and waist with both arms outstretched. A 40 kg iron plate is then placed on the forearms and the waist and knees are extended while both arms lift the plate. From this position the waist and knees are flexed again until the original posture is reached. The output signals from the muscle sensor, potentiometer and actuator pressure sensor are recorded during the operation.

In order to examine the effect of the power-assist suit, the subject repeated the operation without the active assistance of the suit, but this time with a 20 kg iron plate on the forearms. Finally, both sets of data, with and without the active assistance, were compared.

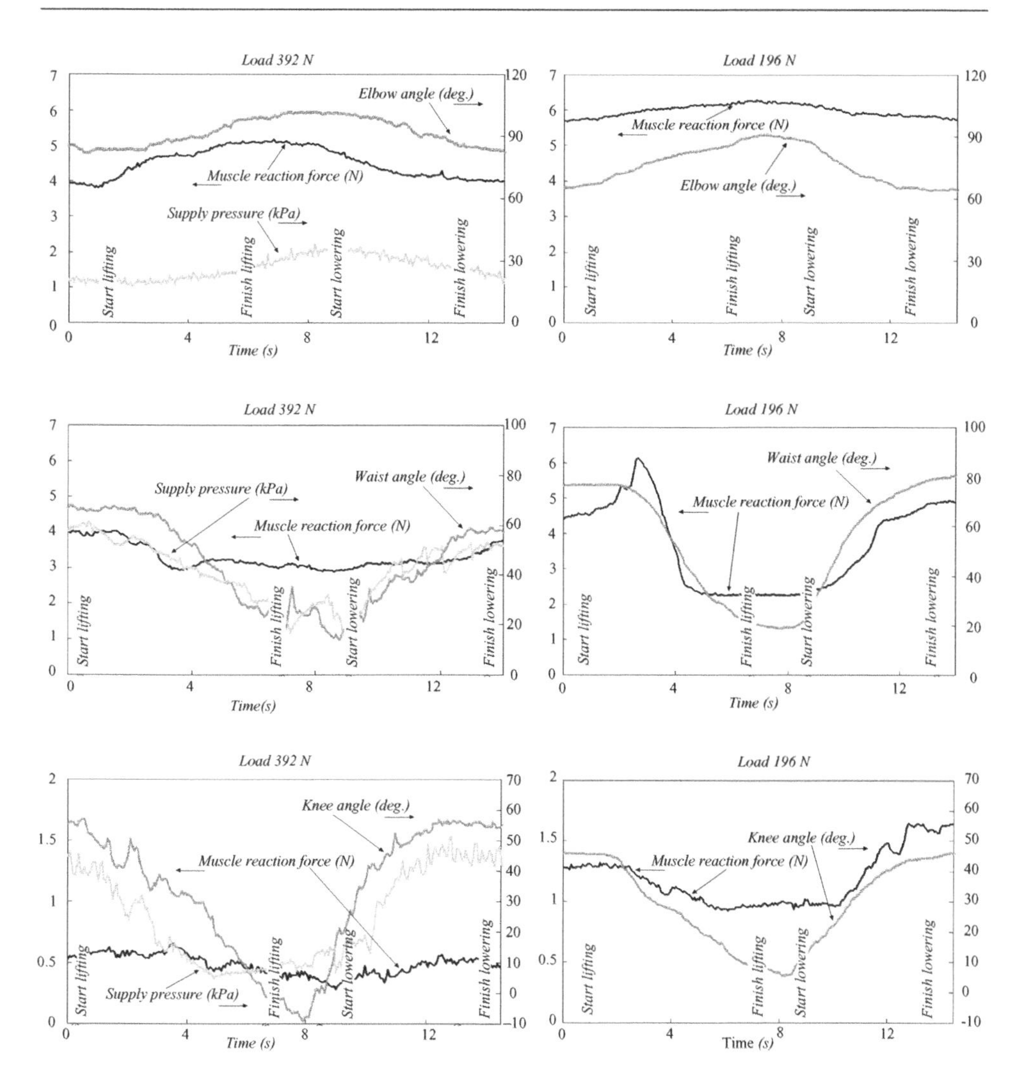

Figure 9.20 Operation of the arm unit (top left) and reaction of the arm muscle without an arm unit (top right), operation of the waist unit (middle left) and reaction of the waist muscle without the waist unit (middle right) and operation of the leg unit (bottom left) and reaction of the leg muscle without the leg unit (bottom right)

Figure 9.20 (top left) shows the operation of the arm unit. The elbow angle θ_e for stretching the arm is assumed to be 0°. The reaction of the biceps muscle, i.e, the output of the muscle hardness sensor, increases as muscle tension augments to hold the plate. This causes an increase of the actuator supply pressure to enable it to assist the elbow. The muscle tension decreases in response to reduced force in the arm while lowering the plate. Thus, the assistance becomes weaker as the plate is put down. Figure 9.20 (top right) shows the results of this process. It was found that the muscle tension was lower when the 40 kg plate was being held (i.e. when the assistance was on) than when the 20 kg plate was being held (i.e. when the assistance was off). In other words, the operation requires less muscle power when assisted by the empowering exoskeleton.

Figure 9.20 (middle left) shows the operation of the waist unit. The waist angle θ_w for stretching the waist is assumed to be 0°. The angle is minimum in the stretching posture and rises when the waist is bent again. The actuator supply pressure follows this motion, increasing or decreasing accordingly. Figure 9.20 (middle right) shows the results of exoskeleton measurement. When the waist begins to bend, the sensor output immediately rises, but afterwards the trend is repeated. Comparing the reactions of the erector muscle when the suit was worn and when it was not worn, it was found that the operation required less muscular activation even though the weight was doubled.

Figure 9.20 (bottom left) shows the operation of the leg unit. The knee angle $(180 - \theta_k)$ for stretching the knee is assumed to be 0°. Figure 9.20 (bottom right) shows the results of exoskeleton measurement. When force is applied to the leg for standing, muscle sensor output increases slightly. It decreases at the onset of standing and increases at the onset of crouching. Actuator supply pressure increases or decreases accordingly. The reaction of the muscles involved is the same as for the waist and arms units, indicating a successful reduction of the human power required to perform the task.

9.6 CASE STUDY: EEG-BASED cHRI OF A ROBOTIC WHEELCHAIR

T. F. Bastos-Filho, M. Sarcinelli-Filho, A. Ferreira, W. C. Celeste, R. L. Silva, V. R. Martins, D. C. Cavalieri, P. N. S. Filgueira and I. B. Arantes

Graduate Program on Electrical Engineering, Federal University of Espirito Santo (UFES), Brazil

Wheelchair mobility poses severe problems for disabled people in public and private environments. If users have the possibility of using their hands, they can control an electric wheelchair with a joystick and thus move with a high degree of autonomy. However, if they do not have this capability, as in the case of quadriplegics or patients with spinal cord injury, they become dependent on other people (Cassemiro and Arce, 2004).

When the impairment is caused by neuromotor disabilities – especially amyotrophic lateral sclerosis (ALS) – a new problem arises: in addition to the absence of mobility and social dependence, these people have their capacity to communicate gradually becomes chronically reduced. The patient's quality of life is then poor in general. Such people frequently experience anxiety and depression because they lack the autonomy necessary to perform ordinary tasks that they once performed independently (Borges, 2003).

Assistive technologies can help to improve the quality of life of disabled people by creating devices and ways to maximize their communication capabilities and independence. One possibility is the use of biological signals generated by the impaired individual him-herself to control the movement of his/her wheelchair or to communicate with people around him/her. The basic assumption underlying this idea is that the deformation and degeneration of muscle cells, which is a characteristic of ALS, does not destroy the patient's cognitive system (Hori, Sakano and Saitoh, 2004). Thus, the brain continues being a useful source of biological signals even when the individual suffers severe motor disabilities. In this way it is possible to recognize EEG patterns and to associate such patterns with previously defined actions in order to implement a cHRI (see Section 4.2) (Millán *et al.*, 2003; Wolpaw *et al.*, 2002).

A BCI can take the EEG of an impaired individual as the input of the assisting system, recognize a short set of readily generated voluntary brain patterns and associate them with a group of previously defined tasks (e.g. to control the movement of the wheelchair). Such an interface provides a communication channel between the impaired individual and the world around him/her and between the impaired individual and a robotic wheelchair. Such an interface has been used in other applications (Ferreira *et al.*, 2006; Frizera-Neto *et al.*, 2006) and has been recently used to allow an impaired individual to control his wheelchair, as shown in the experiments reported here. In this specific case, as the same experiments show, a cHRI of this kind was found to guarantee a satisfactory level of user mobility.

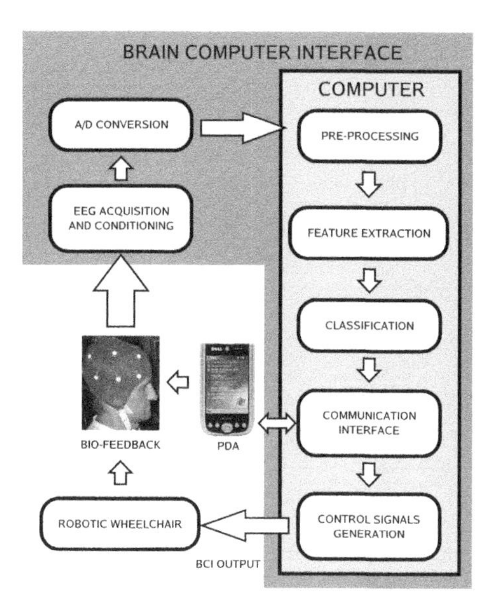

Figure 9.21 The structure of the proposed cHRI

The structure of the cHRI and the way it interacts with the impaired individual and the wheelchair are shown in Figure 9.21. The EEG signals are processed and digitized through a high-resolution ADC. The signal is then filtered through a passband digital filter in the frequency range of 8–13 Hz (the α-band). From this, the signal variance is derived and passed on to a classifier. On the basis of a predefined threshold, the classifier identifies whether or not the user wishes to select a symbol shown on the screen of the PDA. If he/she so wishes, the communication interface asks the PDA for the information necessary and sends it to the next module, which is responsible for generating the control actions needed to make the robotic wheelchair execute the task chosen by the user. The feedback loop is closed through the operator him-herself (biofeedback, an instance of cHRI).

The PDA is installed on the wheelchair in such a way that it is always visible to the user. It provides a graphic interface containing the options available to the operator, including pre-programmed movements of the wheelchair, a virtual keyboard for text edition and symbols to express some basic needs such as sleep, drink and eat. For all these cases, a specific option is selected by means of a procedure that scans the rows and columns in which the icons are distributed in the PDA screen. A voice player confirms the selected option, providing feedback to the user and a means of communication with the people around him/her.

9.6.1 EEG acquisition and processing

The EEG acquisition system includes a signal conditioning board and a quantization board. The latter is responsible for signal quantization and filtering. The signal conditioning board has two acquisition channels that can be connected in a bipolar or a unipolar configuration. A third electrode is used as the reference for the amplifier and is connected to the operator's right ear.

A highpass filter with a cutoff frequency of 0.1 Hz prevents saturation of the amplifiers from continuous voltage caused by the coupling between the electrode and the skin. A fourth-order lowpass Butterworth filter with a cutoff frequency of 32 Hz limits the spectrum of the acquired signal to

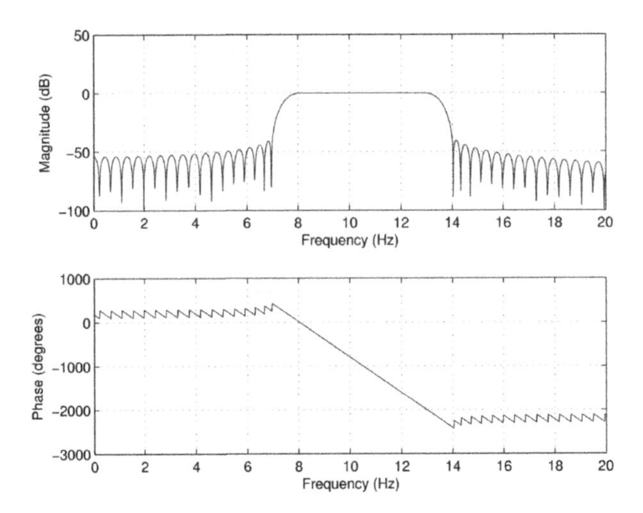

Figure 9.22 A α-band (8-13 Hz) filter response

the EEG band and attenuates 50–60 Hz artefacts such as contaminating noise and disturbances generated by muscle movements, electrode displacement or electromagnetic induction (Webster, 1998). Requirements of high input impedance, high common-mode rejection ratio and low noise are also satisfied by the signal conditioning board. The gain of this circuit is adjustable and distributed in two amplification stages, so that the output signal can be adjusted for different users.

The second part of the acquisition system is analog-to-digital conversion, which is based on a 22-bit AD7716 ADC. A lowpass digital filter is also included in the chip, with selectable cutoff frequencies of 36.5, 73, 146, 292 and 584 Hz. The sampling rate is 140 Hz, so the cutoff frequency of the lowpass filter is set to 36.5 Hz. After being quantized and filtered as shown in Figure 9.22, the signal is sent to the mini-ITX computer through the parallel port.

The type of EEG pattern determines the signal processing methods to be used (Lehtonen, 2003). Event-related synchronization and desynchronization (ERS and ERD respectively) patterns are used in this work (see Section 4.2.2). These are characterized by meaningful changes in the signal energy in specific frequency bands. An energy increase is associated with an ERS, while an energy decrease is associated with an ERD (Pfurtscheller and Silva, 1999). The α frequency band (8–13 Hz) is used to detect these patterns, and the signals are acquired in the occipital region of the head through electrodes placed at positions O_1 and O_2 (see Section 4.2).

The alpha rhythm of an operator who has his/her eyes open (or is receiving visual stimulus or is concentrated) remains at a low energy level. When his/her eyes are closed (or he/she is receiving no visual stimulus or is relaxed), there is a considerable energy increase, characterizing an ERS. These energy changes can be detected through the variance of the filtered EEG, as shown in Figure 9.23.

The variance of a dataset is given by

$$s = \frac{1}{N}\sum_{k=1}^{N}(x_k - \mu)^2 \tag{9.5}$$

where N is the number of samples and

$$\mu = \frac{1}{N}\sum_{k=1}^{N} x_k \tag{9.6}$$

is the mean of the data set. The value of 280 samples used here for N was determined empirically.

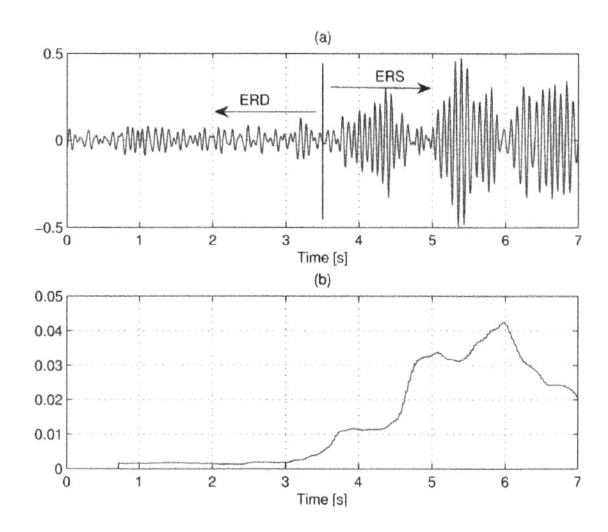

Figure 9.23 (a) Filtered EEG signal with ERD and ERS and (b) variance increase during an ERS

The variance is the input from a threshold-based classifier, and it is therefore possible to detect whether or not the operator wishes to select a symbol presented on the PDA screen. The embedded mini-ITX calculates the control actions needed to accomplish the chosen task and sends reference commands to the low-level control module (MSP430).

9.6.2 The PDA-based graphic interface

The PDA used in this work is a DELL Axim X50, running Windows Mobile 2003. It has a 3.5 inch display, a 520 MHz processor and 64 MB of memory. The PDA communicates with the mini-ITX through a serial line and the integrated circuit MAX232 handles the voltage-level conversion between TTL and RS-232.

The software application running in the PDA is called robotic wheelchair control (RWCC) and was developed using *eMbedded Visual C++ 3.0*. The programmed interface is a simple one, which any user can work with to make his/her choices; this reinforces the user's self-confidence in controlling the wheelchair.

One of the functions of RWCC is to control the wheelchair. This is done through the MOVEMENT screen (Fig. 9.24), which presents to the operator a set of symbols corresponding to movements of the robotic wheelchair. The first command sets the wheelchair in motion, and the next – regardless of where the automatic scan is – halts it. For safety's sake only short rearward displacements are allowed because of the lack of visibility in that direction.

9.6.3 Experiments

Figure 9.24 shows a user testing the complete system. Before operating the system, it is essential to clean thoroughly the points on the user where the electrodes are to be connected (positions O_1 and O_2 on the head and the right earlobe). A special gel is then applied to the electrode and to the user's skin to reduce electrical impedance. All the users involved in the experiments were able to command the wheelchair. Note that the kind and volume of the user's hair does not interfere with the quality of the acquired signal.

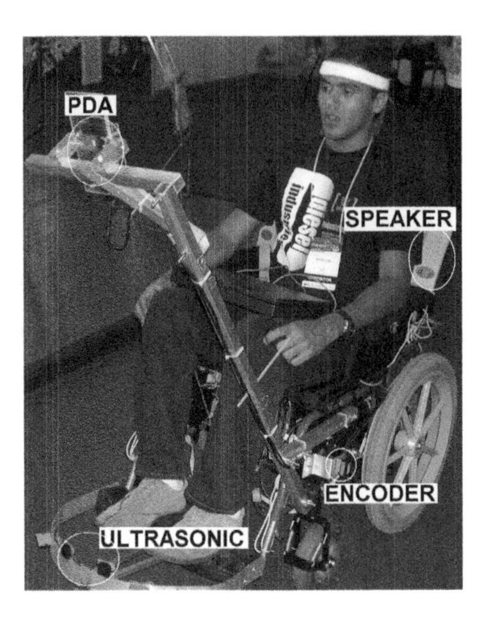

Figure 9.24 Testing the system prototype

The analysis of the signal in the α-band and its variance shows very clearly when the user closes his/her eyes (high variance), generating an ERS. This indicates that the user wishes to select the command currently highlighted on the PDA screen. A threshold-based classifier is used to detect the state of the user, which is either eyes closed (ERS) or eyes open (ERD). An adjustable hysteresis zone is included in the classifier to prevent small variance changes from generating unwanted selections.

Artefacts related to blinking, whose main frequency is around 5 Hz, are eliminated from the filtered signal. This was effectively verified during the experiments. The system is not affected by muscle activity. In fact, when the user closes his/her eyes to generate the selection signal, EMG signals are generated; however, they do not register as they are removed by the passband filter that delimits the signal to the α-band. In this way the filter guarantees that only EEG signals in the alpha band are taken into account.

9.6.4 Results and concluding remarks

The system as developed was tested by healthy users in indoor and outdoor environments. In both situations the results were satisfactory according to the subjective evaluation of the users. The acquisition system was found to be efficient when choosing commands to control the wheelchair through EEG signals. Only minimal knowledge of the interface and a very short training period is required to operate the system.

The chief advantages of the system here developed, from the point of view of prospective users of this assistive technology, are that the electrodes are easy to fit, the graphical interface is simple and the system is easy to adapt to commercial electrical wheelchairs. Several improvements to the current version of the autonomous robotic wheelchair are currently being included:

- Incorporation of a controller that can safely guide the wheelchair from its current location to another predefined location. The operator will select not individual movements but an icon representing the desired goal (bedroom, kitchen, bathroom, etc.).

- Connection of all sensors to the microcontroller-based board and implementation of communication between the wheelchair and other sensors placed about the user's habitat, thus configuring an *intelligent space* for wheelchair navigation.
- Incorporation of a video camera system to capture eyeball movement (videooculography).
- Tests with wavelet and neural network classifiers applied to signals from the motor cortex (positions C_3, C_4 and C_z) in such a way that mental states can be recognized, thus implementing a more intuitive way to command the wheelchair without having to close one's eyes or relax.

Acknowledgements

The authors wish to thank CAPES, a foundation of the Brazilian Ministry of Education, FAPES, a foundation of the Secretary of Science and Technology of the State of Espirito Santo (Process 30897440/2005), and FACITEC/PMV, a fund of Vitoria City Council supporting Scientific and Technological Development, for providing financial support for this research.

REFERENCES

Aaron, S.L., Stein, R.B., 1976, 'Comparison of an EMG-controlled prosthesis and the normal human biceps brachii muscle', *American Journal of Physical Medicine* **55**: 1–14.

Behrman, A.L., Harkema, S.J., 2000, 'Locomotor training after human spinal cord injury: a series of case studies', *Physical Therapy* **80**: 688–700.

Blaya, J., Herr, H., 2004, 'Adaptive control of a variable-impedance ankle–foot orthosis to assist drop foot gait', *IEEE Transactions on Neural Systems Rehabilitation Engineering* **12**: 24–31.

Borges, C.F., 2003, 'Dependency and death of the mother of family: the solidarity of the family and the community in taking care of the patient suffering from amyotrophic lateral sclerosis', *Psychology in Study* **8**: 21–29.

Browning, R.C., Modica, J.R., Kram, R., Goswami, A., 2007, 'The effects of adding mass to the legs on the energetics and biomechanics of walking', *Medicine and Science in Sports and Exercise* **39**: 515–525.

Burke, M.J., Roman, V., Wright, V., 1978, 'Bone changes in lower limb amputees', *Annals of the Rheumatic Diseases* **37**(3), 252–254.

Colombo, G., Wirz, M., Dietz, V., 2000, 'Driven gait orthosis for improvement of locomotor training in paraplegic patients', *Spinal Cord* **39**: 252–255.

Colombo, G., Joerg, M., Schreier, R., Dietz, V., 2001, 'Treadmill training of paraplegic patients using a robotic orthosis', *Journal of Rehabilitation Research and Development* **37**: 130–134.

Cassemiro, C.R., Arce, C.G., 2004, 'Visual communication through a computer in amyotrophic lateral sclerosis', *Brazilian Archives of Oftalmology* **67**: 295–300.

Davis, S., Tsagarakis, N., Canderle, J., Caldwell, D.G., 2003, 'Enhanced modelling and performance in braided pneumatic muscle actuators', *International Journal of Robotics Research* **22**: 213–227.

Dietz, V., Wirz, M., Curt, A., Colombo, G., 1998, 'Locomotor pattern in paraplegic patients: training effects and recovery of spinal cord function', *Spinal Cord* **36**: 380–390.

Eng, J.J., Winter, D.A., 1995, 'Kinetic analysis of the lower limbs during walking: what information can be gained from a three-dimensional model?', *Journal of Biomechanics* **28**: 753–758.

Esquenazi, A., DiGiacomo, R., 2001, 'Rehabilitation after amputation', *Journal of American Podiatric Medical Association* **91**(1): 13–22.

Ferreira, A., Bastos-Filho, T.F., Sarcinelli-Filho, M., Auat, F.C., Postigo, J.F., Carelli, R., 2006, 'Teleoperation of an industrial manipulator through a TCP/IP channel using EEG signals', in *Proceedings of the International Symposium on Industrial Electronics (ISIE 2006)*, pp. 3066–3071.

Ferris, D.P., Sawicki, G.S., Domingo, A., 2005, 'Powered lower limb orthoses for gait rehabilitation', *Topics in Spinal Cord Injury Rehabilitation* **11**: 34–49.

Ferris, D.P., Gordon, K.E., Sawicki, G.S., Peethambaran, A., 2006, 'An improved powered ankle–foot orthosis using proportional myoelectric control', *Gait and Posture* **23**, 425–428.

Frizera-Neto, A., Celeste, W.C., Martins, V.R., Bastos-Filho, T.F., Sarcinelli-Filho, M., 2006, 'Human–machine interface based on electro-biological signals for mobile vehicles', in *Proceedings of the International Symposium on Industrial Electronics (ISIE 2006)*, pp. 2954–2959.

Gordon, K.E., Ferris, D.P., 2007, 'Learning to walk with a robotic ankle exoskeleton', *Journal of Biomechanics* **40**: 2636–2644.

Gordon, K.E., Sawicki, G.S., Ferris, D.P., 2006, 'Mechanical performance of artificial pneumatic muscles to power an ankle-foot orthosis', *Journal of Biomechanics* **39**: 1832–1841.

Gharooni, S., Heller, B., Tokhi, M., 2000, 'A new hybrid spring brake orthosis for controlling hip and knee flexion in the swing phase', *IEEE Transactions on Neural Systems Rehabilitation Engineering* **9**: 106–107.

Harkema, S.J., 2001, 'Neural plasticity after human spinal cord injury: application of locomotor training to the rehabilitation of walking', *Neuroscientist* **7**: 455–468.

Hay, J.G., 1985, *The Biomechanics of Sports Techniques*, 3rd edition. Prentice-Hall Inc, Englewood Cliffs, Lew Jersey.

Hayashi, T., Kawamoto, H., Sankai, Y., 2005, 'Control method of robot suit HAL working as operator's muscle using biological and dynamical information', in *IEEE/RSJ International Conference on Intelligent Robots and Systems (IROS2005)*, pp. 3036–3038.

Hesse, S., Bertelt, C., Jahnke, M.T., Schaffrin, A., Baake, P., Malezic, M., Mauritz, K.H., 1995, 'Treadmill training with partial body weight support compared with physiotherapy in nonambulatory hemiparetic patients', *Stroke* **26**: 976–981.

Hori, J., Sakano, K., Saitoh, Y., 2004, 'Development of communication supporting device controlled by eye movements and voluntary eye blink', in *Proceedings of the 26th Annual International Conference of the IEEE EMBS*.

Irby, S.E., Kaufman, K.R., Wirta, R.W., Sutherland, D.H., 1999, 'Optimization and application of a wrap-spring clutch to a dynamic knee–ankle–foot orthosis', *IEEE Transactions on Rehabilitation Engineering* **7**: 130–134.

Ishii, M., Yamamoto, K., Hyodo, K., 2005, 'A stand-alone wearable power assist suit – development and availability', *Journal of Robotics and Mechatronics* **17**(5): 575–583.

Jezernik, S., Scharer, R., Colombo, G., Morari, M., 2003, 'Adaptive robotic rehabilitation of locomotion: a clinical study in spinally injured individuals', *Spinal Cord* **41**: 657–666.

Kepple, T.M., Siegel, K.L., Stanhope, S.J., 1997, 'Relative contributions of the lower extremity joint moments to forward progression and support during gait', *Gait and Posture* **6**: 1–8.

Klute, G.K., Czerniecki, J.M., Hannaford, B., 1999, 'Mckibben artificial muscles: pneumatic actuators with biomechanical intelligence', in *IEEE/ASME International Conference on Advanced Intelligent Mechatronics*. Atlanta, Georgia, IEEE, Piscataway, New Jersey.

Klute, G.K., Hannaford, B., 1998, 'Fatigue characteristics of mckibben artificial muscle actuators', in *IEEE/RSJ International Conference on Intelligent Robots and Systems*, Victoria, British Columbia, Canada, IEEE, New York.

Klute, G.K., Hannaford, B., 2000, 'Accounting for elastic energy storage in mckibben artificial muscle actuators', *Journal of Dynamic Systems, Measurement and Control* **122**: 386–388.

Kulkarni, J., Gaine, W.J., Buckley, J.G., Rankine, J.J., Adams, J., 2005, 'Chronic low back pain in traumatic lower limb amputees', *Clinical Rehabilitation* **19**(1): 81–86.

Lehtonen, J., 2003, 'EEG-based brain computer interfaces', Master Thesis, Helsinki University of Technology, Espoo, Finland.

Millán, J.R., Renkens, F., Mouriño, J., Gerstner, W., 2003, 'Non-invasive brain-actuated control of a mobile robot', in *Proceedings of the 18th International Joint Conference on Artificial Intelligence*.

Moreno, J.C., Brunetti, F., Pons, J.L., Baydal, J., Barberà, R., 2004, 'An autonomous control and monitoring system for a lower limb orthosis', in *Conference Proceedings of the 26th Annual International Conference of the IEEE Engineering in Medicine and Biology Society*, vol.3, pp. 2125–2128.

Moreno, J.C., Brunetti, F., Pons, J.L., Baydal, J., Barberà, R., 2005, 'Rationale for multiple compensation of muscle weakness walking with a wearable robotic orthosis', in *Proceedings of the IEEE International Conference on Robotics and Automation ICRA*, pp. 1914–1919.

Moreno, J.C., Rocon, E., Ruiz, A., Brunetti, F., Pons, J.L., 2006, 'Design and implementation of an inertial measurement unit for control of artificial limbs: application on leg orthoses', *Sensors and Actuators B* **118**: 333–337.

Neptune, R.R., Kautz, S.A., Zajac, F.E., 2001, 'Contributions of the individual ankle plantar flexors to support, forward progression and swing initiation during walking', *Journal of Biomechanics* **34**: 1387–1398.

Norvell, D.C., Czerniecki, J.M., Reiber, G.E., Maynard, C., Pecoraro, J.A., Weiss, N.S., 2005, 'The prevalence of knee pain and symptomatic knee osteoarthritis among veteran traumatic amputees and non-amputees', *Archives of Physical Medicine and Rehabilitation* **86**(3): 487–493.

Perry, J., 1992, *Gait Analysis: Normal and Pathological Function*, Slack Inc, Englewood Cliffs, New Jersey.

Pfurtscheller, G., Silva, F.H.L., 1999, 'Event-related EEG/MEG synchronization and desynchronization: basic principles', *Clinical Neurophysiology* **110**: 1842–1857.

Reinkensmeyer, D.J., Aoyagi, D., Emken, J.L., Galvez, J.A., Ichinose, W., Kerdanyan, G., Maneekobkunwong, S., Minakata, K., Nessler, J.A., Weber, R., Roy, R.R., De Leon, R., Bobrow, J.E., Harkema, S.J., Edgerton, V.R., 2006, 'Tools for understanding and optimizing robotic gait training', *Journal of Rehabilitation Research and Development* **43**: 657–670.

Reynolds, D.B., Repperger, D.W., Phillips, C.A., Bandry, G., 2003, 'Modeling the dynamic characteristics of pneumatic muscle', *Annals of Biomedical Engineering* **31**: 310–317.

Sawicki, G.S., Domingo, A., Ferris, D.P., 2006, 'The effects of powered ankle–foot orthoses on joint kinematics and muscle activation during walking in individuals with incomplete spinal cord injury', *Journal of Neuroengineering and Rehabilitation* **3**: 3.

Suzuki, K.M., Takahama, M., Mizutani, Y., Arai, M., Iwai, A., 1983, 'Locomotive mechanics of normal adults and amputees', in *Biomechanics VIII-A* (eds. H. Matsui and K. Kobayashik) Champaign, Illinois. Human Kinetics Publishers, pp. 380-385.

Tondu, B., López, P., 2000, 'Modeling and control of mckibben artificial muscle robot actuators', *IEEE Control Systems Magazine* **20**: 15–38.

Webster, J.G., 1998, *Medical Instrumentation. Application and Design*, John Wiley & Sons, Inc.

Winter, D.A., 1983, 'Energy generation and absorption at the ankle and knee during fast, natural, and slow cadences', *Clinical Orthopaedics*, 147–154.

Wolpaw, J.R., Birbaumer, N., McFarland, D.J., Pfurtscheller, G., Vaughan, T.M., 2002, 'Brain–computer interfaces for communication and control', *Clinical Neurophysiology* **113**: 767–791.

Yamamoto, K., Hyodo, K., Imai, M., 1996, 'Development of powered suit for assisting nurse labor', *Research Reports of Kanagawa Institute of Technology, Part B* **20**: 29–43.

Yamamoto, K., Miyanishi, H., Imai, M., 1991, 'Development of pneumatic actuator for powered arm', in *Proceedings of the JHPS Autumn Meeting* (in Japanese), pp. 85–88.

Yamamoto, K., Ishii, M., Hyodo, K., Yoshimitsu, T., Matsuo, T., 2003, 'Development of power assisting suit for assisting nurse labor – miniaturization of supply system to realize wearable suit', *JSME International Journal, Series C*, **46**(3): 923–930.

Yamamoto, K., Ishii, M., Noborisaka, H., Hyodo, K., 2004, 'Stand alone wearable power assisting suit – sensing and control systems', in *Proceedings of the 2004 IEEE International Workshop on Robot and Human Interactive Communication*, pp. 661–666.

10

Summary, conclusions and outlook

J. L. Pons, R. Ceres and L. Calderón

Bioengineering Group, Instituto de Automática Industrial, CSIC, Madrid, Spain

This book is devoted to a detailed analysis of the field of *wearable robotics*. The book begins with an introduction of the distinctive features of wearable robots as a species of personal robots in close cooperation with the human (the wearer) through a dual interaction, both cognitive and physical. The intrinsically biomechatronic principle underlying wearable robots is highlighted from the very first pages of the book.

The entire structure and layout of the book is designed to stress this dual interaction and biomechatronic approach to wearable robotics. Since bioinspiration is a fundamental part of biomechatronics, the first thematic chapter is devoted to analysing how nature, in particular human biological and functional structures, may influence the design of wearable robots. Human mechanics is then introduced in Chapter 3 as a source of information for designing wearable robots.

The following two Chapters 4 and 5, are devoted to a study of human–robot interaction, first from a cognitive standpoint and then from a physical perspective. Here the opportunity is taken to explain how the two systems interact and influence each other when coupled to the counterpart.

Wearable robot technologies are introduced in Chapters 6 and 7. The former describes sensors, actuators and portable energy storage technologies while the latter focuses on communication networks for wearable robots.

Although all chapters include illustrative examples of current or recent research projects worldwide, the book devotes two specific chapters 8 and 9, to the introduction of outstanding research activities in the field of upper limb wearable robots and lower limb/full-body wearable robots respectively.

The following sections summarize the most important aspects addressed throughout the book and point to likely trends in various aspects of wearable robotics.

10.1 SUMMARY

In Chapter 1, wearable robots are introduced as person-oriented robots. Of all the kinds of wearable robots, exoskeletons arouse the most interest in that their mechanical structure maps on to the anatomy of the human limb to produce a close cognitive and physical H–R interaction.

Wearable Robots: Biomechatronic Exoskeletons Edited by José L. Pons

Briefly then, there is a cognitive human–robot interaction, cHRI, in which a flow of information can be established in either or both directions. This cognitive interaction may be either conscious – at various different levels – or involuntary, or a combination of the two. This cognitive interaction is supported by an *ad hoc* cognitive interface, cHRi, but it can also be triggered by a physical interaction, pHRI, making use of the physical interface, pHRi. The cognitive processes may be confined to one side of the human–robot interface, but in some cases they are coupled to the counterpart by this intrinsic dual interaction between human and robot.

The first chapter includes a brief historical note to illustrate how the idea of artificially extending the physical capability of humans by means of assistants (robots) dates back to the Greek philosophers. Figure 10.1 presents milestones in the evolution of this idea up until the last century, when the field of robotics started to develop, and on to the recent emergence of wearable robotics as a very particular instance of the former.

Chapter 1 also addresses the biomechatronic concept of wearable robots. There are three aspects of wearable robots, which tie in with a biomechatronic approach: firstly, wearable robots are mechatronic systems in close interaction with a biological system (the human wearer); secondly, the design of wearable robots and their components is bioinspired; and, finally, most of the design procedures adopted in the field of wearable robotics are biologically inspired. It is therefore fair to say that wearable robotics is one of the scientific disciplines where the three distinctive aspects of biomechatronics are most clearly appreciable.

10.1.1 Bioinspiration in designing wearable robots

The basis for bioinspiration in wearable robotics is addressed in depth in Chapter 2. Western philosophy recognizes nature as a model to follow, and thus the understanding of biological systems is a first step in bioinspiration. The chapter presents the general principles whereby functional aspects of living creatures are optimized through a biological evolutionary process. One of the most important functions to be optimized by living creatures is energy consumption, but minimization of damage and a compromise between power and precision of movements are also vital. This process of functional optimization has led to very efficient motor control mechanisms in living creatures.

The process of optimization through evolution may itself be the basis for bioinspired evolutionary design algorithms, as in the case of optimization through genetic algorithms, where the steps of reproduction, crossover, mutation and elitism are included in the optimization of systems – e.g. a wearable robot or one of its components as in Case Study 3.5.

Bioinspired designs require an understanding of living creatures in terms of functions and function-supporting structures, and so Chapter 2 devotes several pages to analysing biological models. In particular, neuromotor control structures are studied in detail. This study includes a brief analysis of the human nervous system and the motor control mechanism, e.g. internal models, central pattern generators and sensorimotor reflexes.

Muscles are the actuators in the human motor system. The musculoskeletal system is partially addressed in this chapter, but it is also analysed in Chapter 3, in particular with regard to the mechanics of human limbs. Here in Chapter 2 the focus is on muscular physiology as a model in designing wearable robots. This is complemented by the study of sensorimotor mechanisms as low-level motor control mechanisms in the hierarchical human motor control structure. The discussion on neuromotor control structures and mechanisms as a source of inspiration ends with a note on how these bioinspired designs might be used in a recursive interaction to explain partially understood biological systems.

Levels of biological inspiration in engineering design depend on the level of understanding of biological systems. Thus, biomimetism is defined as a replication of observable behaviour and structures in living creatures, while bioimitation is defined as a replication of the dynamics of these control

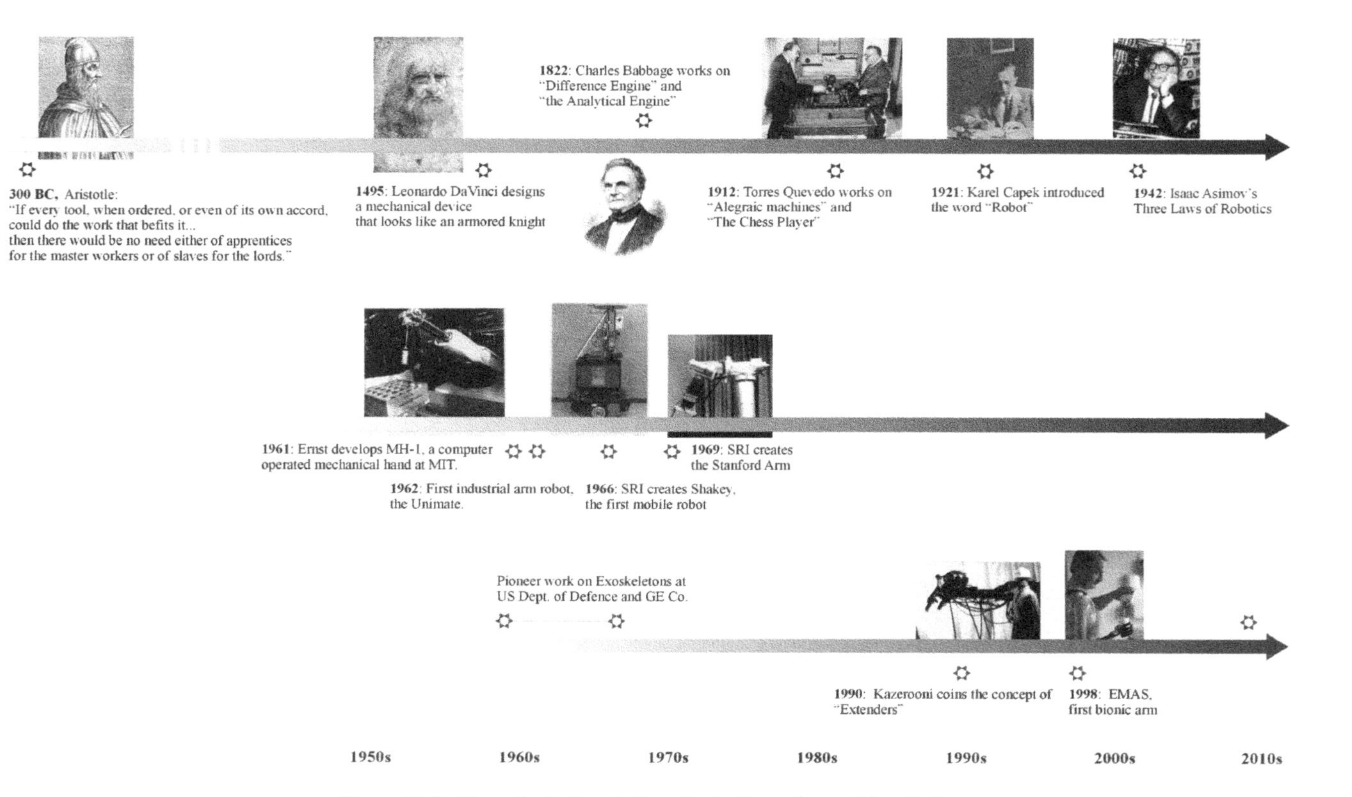

Figure 10.1 Chronological evolution of robotics and wearable robotics

structures. The former does not entail a full understanding of the mechanisms involved in the function of a biological system, but rather a replication of behaviour. The latter implies a full understanding of the function of living creatures and a modelling of function dynamics.

Chapter 2 ends with three case studies. Case Study 2.5 shows how human gait is analysed to drive the design of efficient walking robots, with clear spinoff applications in the control of lower limb robotic exoskeletons. Case Study 2.6 shows the bioinspired design of the hierarchical control structure and mechanisms for a wearable hand robotic prosthesis. Finally, Case Study 2.7 illustrates how internal models, central pattern generators and reflexes can be used to stabilize limit-cycle control of lower limb exoskeletons.

10.1.2 Mechanics of wearable robots

Chapter 3 analyses the mechanics – i.e. the kinematics and dynamics – of both actors, the wearable robot and the human wearer. In so doing, a common framework, the Denavit–Hartenberg formulation, is adopted for modelling the mechanics. In this approach, certain reasonable assumptions are adopted in order to be able to use the methods commonly applied in robotics for the analysis of human kinematics and dynamics, e.g. modelling the human body as a chain of rigid links where each segment has certain properties, like length or inertia, which approximate those of humans; these segments are linked by joints that imitate human ones in terms of degrees of freedom (DoFs) and ranges of motion (RoMs).

The Denavit–Hartenberg formulation is first introduced in a section devoted to robot mechanics. Here, forward and inverse kinematic and dynamic problems are formulated. The equations of motion for the wearable robot are arrived at by following the Lagrange formulation. This section devotes special attention to concepts like mobility, redundancy, workspace and singular configurations of the robotic kinematic chain, as they are of particular interest later in the chapter when analysing the kinematic compliance between human and robot, and in particular the problem of kinematic redundancy in human–exoskeleton systems.

Biomechanics, the application of methods and techniques from Mechanical Science (Physics and Engineering) to the analysis and understanding of biological systems, is addressed in this book with the Denavit–Hartenberg formulation. The assumption of rigid links, which was reasonable when studying the kinematics and dynamics of human motion, is no longer valid when analysing the physical interaction between human and robot. Thus, while in this chapter this assumption was adopted to analyse human mechanics, in Chapter 5 the transmission of forces through soft tissues is discussed from the standpoint of robot control, taking into account models of soft tissue deformation.

Chapter 3 introduces the medical description of human movements for the first time. It then goes on to deal with the kinematics of the upper and the lower human limbs in terms of human movements, DoFs, RoMs, the musculoskeletal system responsible for movements at each level of articulation and the characteristics of these movements. Following this analysis, the Denavit–Hartenberg formulation is imported to derive kinematic models of the human limbs, which are then extended as dynamic models.

This chapter also includes three case studies. Case Study 3.5 illustrates aspects from this chapter and from Chapter 2. On the one hand, it illustrates the use of genetic algorithms as a tool to optimize the bioinspired design of a knee joint for a wearable lower limb exoskeleton and, on the other hand, it illustrates how the kinematics of the anatomical joint of the knee is modelled and how the exoskeleton knee fits the anatomical joint kinematically. This has been shown to lead to reduced involuntary interaction forces in the exoskeleton–human support. Case Study 3.6 shows a redundant pronation–supination joint for an upper limb exoskeleton so designed that it is kinematically compliant with the kinematics of the human limb. This particular instance of pronation–supination movement is further analysed in Case Study 8.5. Finally, Case Study 3.7 is included to illustrate how a dynamic model of the human–exoskeleton combination can be used to work out dynamic characteristics of human movements; in particular, in this case study, the model is used to derive power and torque characteristics of tremor at different joints in the upper limb.

10.1.3 Cognitive and physical human–robot interaction

Humans and wearable robots interact both cognitively and physically. The cognitive interaction between the two actors is analysed in detail in Chapter 4. Cognitive processes occur both on the human and on the robot side. The cognitive interaction is a result of the flow of information between these two cognitive processes.

Since the cognitive process in the human takes place at different levels, thus involving different neuromotor structures, the cHRI in the direction from the wearer to the robot can be established from data collected at various levels. The data typically used as a basis for establishing a cHRI are either bioelectrical, e.g. EEG, ENG, EMG and EOG, or biomechanical. This chapter discusses cHRIs based on a selection of these data sources, specifically EEG, EMG and biomechanical data. The chapter therefore deals with data in descending order of abstraction within the cognitive process in the human.

Some of these cHRIs are intrinsically unidirectional, e.g. in the case of EMG- and EEG-based interaction. This means that only information from human cognitive processes can be used to command the wearable robot. If a flow of information is desired in the other direction, e.g. to provide the human with feedback information on processes taking place on the human side, other modes of interaction are required, for instance those based on biomechanical interaction (tactile or haptic), on ENG or on other natural modes (visual, auditory).

The chapter describes the physiology of the bioelectrical activity (EEG and EMG) used for cHRI. It then goes on to describe models for both bioelectrical processes on the basis of this description, along with processing techniques and algorithms used for feature extraction, classification and recognition of cognitive processes. The case of cHRI based on biomechanical data is slightly different in that it is not as easily formalized as in the case of EEG and EMG. Here biomechanical models are described according to the type and application of the wearable robot. Thus, for lower limb applications gait activity, such as the paradigmatic activity to be supported by means of wearable robots, is modelled as a cyclic process. Upper limb function is much more diverse and therefore cannot be classified in this way.

The chapter devoted to cHRI includes four case studies. The first two illustrate the development of a cHRI based on biomechanical data in the particular instances respectively of lower limb exoskeletons for supporting gait and upper limb wearable robots for suppressing tremor. In the first study, a fuzzy set of rules is used to detect transition events during normal gait activities. In the second, a combination of Benedict–Bordner and weighted-frequency Fourier linear combiner filters are applied to identify and track the tremorous component of human upper limb movement as an input to the tremor suppression exoskeleton. The third case study in Chapter 4 illustrates the use of cortical activity to drive a robot. It is not strictly a wearable robot application but describes a control method that will foreseeably be present in the next generation of wearable robots. The last case study in this chapter illustrates concepts from Chapter 4, describing the implementation of a cHRI based on hand postures and gestures, and from Chapter 7 since the cHRI system is based on a wireless sensor network (WSN).

Chapter 5 deals with the human–robot physical interaction. Safety and dependability issues in wearable robotics are directly affected by how this physical interaction is controlled and implemented. The exertion of force and pressure on the human actor is a delicate issue. This force is exerted through soft tissues comprising muscle, fat, blood vessels and nerve tissue, and incorrect application of these forces may cause damage. Therefore the chapter starts by briefly describing the physiological factors affecting such interaction. This short description of the human sensory system complements the analysis of neuromotor structures and mechanisms already introduced in Chapter 2.

Chapter 5 then studies the various different factors affecting wearable robot design in terms of the pHRI. A first section is devoted to analysing the physical interaction resulting from incompatible kinematic design of wearable robots. Human limbs are prone to intersubject variability (both at the level of kinematic parameters in a Denavit–Hartenberg model, e.g. length of bones, and at the level of dynamic parameters of the limbs, e.g. mass and volume). In addition, kinematics of human limbs are also subject to variability within an individual; for instance, the ICR for a particular joint moves with joint excursion, producing kinematic incompatibility between wearable robots and human limbs. This incompatibility is then classified into incompatibility resulting from macromisalignments, e.g. due to oversimplified robot joints, and incompatibility from micromisalignments resulting from intersubject variability. The section then discusses possible criteria when designing wearable robots to overcome this kinematic incompatibility and thus reduce undesired interaction forces.

Since application of forces is typically achieved through soft tissues, the chapter analyses human tolerance of pressure and describes some of the models used in the literature to describe the behaviour of soft tissues under controlled forces. The analyses of tolerance of force and tissue models originate recommendations for the design of mechanical supports between the two actors.

This chapter also describes control of the physical interaction. Firstly, a model for the mechanical behaviour of human limbs under external load is introduced. Then the robot is described in terms of a controllable impedance, although additional control schemes are also discussed. Following the presentation of the mechanical behaviour of the two actors, a section is devoted to the human and the robot in a closed-loop control scheme, which is briefly discussed from the special perspective of various application scenarios: empowering, telemanipulation, rehabilitation and functional compensation.

Since both actors in this physical interaction 'run' independent motor control mechanisms, the physical interaction most often results in physically triggered cognitive interactions. This is described in this chapter through the example of a wearable upper limb robot for tremor suppression, where these cognitive interaction are apparent. The notion of having two independent control systems interacting directly raises questions of stability. This is discussed in the last section of the chapter.

Chapter 5 includes four case studies. The first analyses and quantifies forces resulting from nonergonomic and kinematically incompatible wearable robot designs. In so doing, first the theoretical constraint displacements are worked out and then the interaction force is experimentally quantified. The second case study is an analysis of human tolerance of pressure on both the upper and the lower limbs. Force was applied with an indentor in several areas of the human limbs and the pressure thresholds were recorded, producing a map of sensitive areas and a set of recommendations when designing mechanical supports for wearable robots. The last two case studies discuss the control of joint human–robot mechanical impedance in upper and lower limb WRs respectively.

10.1.4 Technologies for wearable robots

As was mentioned in Chapter 1 and discussed at length in Chapters 6 and 7, one of the limiting factors for the deployment of wearable robots is fundamentally technological. Sensors, actuators and power storage technologies in a networked environment are all required for a truly wearable solution. This book devotes two chapters to technologies for wearable robots. Chapter 6 addresses sensor, actuator and power storage technologies, while Chapter 7 deals with communication networks for wearable robots and other wearable applications.

Sensor technologies are classified into sensors for measuring biomechanical variables and sensors for measuring bioelectrical variables, thus complementing the analyses of the cHRI based on biomechanical and bioelectrical information in Chapter 4. In addition, sensors for measuring microclimate at the human–robot interface are also considered. As to actuator technologies, the book presents the requirements in terms of power, force and bandwidth for actuator technologies and describes the most salient implementations in the literature. In addition, some novel alternative actuator technologies (ERF-MRF, SMAs, EAPs) are introduced, although the reader is referred to specialized books for in-depth analysis of these technologies. Communication networks for wearable robots are addressed in Chapter 7.

10.1.5 Outstanding research projects on wearable robots

The different topics analysed throughout the book are illustrated in Chapters 8 and 9. The former includes outstanding research projects in the area of upper limb wearable robots. The latter address lower limb and full-body wearable robots.

Case Study 8.1 introduces the robotic exoskeleton called WOTAS (wearable orthosis for tremor assessment and suppression), which provides a means of testing and validating nongrounded control strategies for orthotic tremor suppression. This case study describes the general concept of WOTAS in detail, outlining the special features of the design and the selection of system components. It also describes the implementation of the two control strategies developed for tremor suppression with exoskeletons. These control strategies rely on a cHRI based on the detection of tremor information, e.g. tremor onset and dynamic parameters, from biomechanical data. The two strategies are based on biomechanical loading (described in Case Study 5.7) and notch filtering of the tremor through the application of internal forces (described in Section 5.4). Results from experiments using these two strategies on patients with tremor are summarized. Finally, results from clinical trials are presented, indicating the feasibility of ambulatory mechanical suppression of tremor.

Section 8.2 presents the CyberHand upper limb robotic prosthesis. The most outstanding characteristic of the device is that it recreates the natural link that exists between the hand and the central nervous system by exploiting the potentialities of implantable interfaces with the peripheral nervous system.

Case Study 8.3 presents a novel human arm exoskeleton called EXARM. The EXARM is a human–machine interface for master–slave robotic teleoperation with force feedback. The purpose of this exoskeleton is to assist International Space Station (ISS) crew in different tasks such as assembly, inspection and transportation. These activities are carried out by EUROBOT, a humanoid space robot on the outside of the ISS, which is remote-controlled by astronauts. There are two different modes of operation: autonomous, where the robot movement is planned and programmed offline, and manual, where the robot reproduces the user's movements.

Case Study 8.4 introduces the NEUROBOTICS exoskeleton (NEUROExos), which is a platform intended to fuse neuroscience and robotics. The study illustrates how the different aspects addressed in the book influence the design of an exoskeleton. The research on this exoskeleton focuses on the development of a natural control interface based on natural user behaviour in order to develop novel control strategies for proper control of the pHRI. The human control strategy is investigated and replicated in the external actuator system; in this way the robot impedance can be continuously matched to the human arm impedance by means of agonist/antagonist control of the NEUROExos joints. Special attention is also paid to the development of kinematic coupling of the robot mechanical structure with the human arm, and to the actuation system that will generate the required torques at the joints.

Section 8.6 introduces a soft-actuated arm exoskeleton for use in physiotherapy and training. The chief characteristics of this exoskeleton are that it is activated by pneumatic muscle actuators (pMAs) and has 3 DoFs in the shoulder, 2 DoFs in the elbow and 2 DoFs in the wrist. This section describes the different functions and the structure of the device, and also presents a method for torque-position control and the dynamics of the arm–exoskeleton. Finally, the authors of the case study present some experiments performed for system validation.

Case Study 9.1 presents the biomechanical aspects considered in the design of the GAIT and ESBiRRo exoskeletons at IAI-CSIC, providing details of the integrated systems and the results of their application to patients. The GAIT exoskeleton is presented as an example of a controllable

KAFO, and the ESBiRRo exoskeleton is introduced in the form of a bilateral HKAFO incorporating a limit-cycle walking strategy. Particular aspects of the GAIT exoskeleton are described in Case Studies 3.5 (the compliant orthotic knee joint), 5.8 (stance stabilization during gait), 4.5 (gait control based on learned patterns) and 6.7 (knee actuator design), and aspects of the ESBiRRo exoskeleton are described in Case Study 2.5.

Case Study 9.2 presents a powered ankle–foot orthosis to provide plantar flexion assistance during walking. The system is suitable for gait rehabilitation after neurological injury, and also as a tool to investigate the observed adaptations of the human nervous system. The control system drives artificial dorsiflexor and plantarflexor muscles (pneumatic actuation) based on EMG signals. The system was evaluated with healthy subjects and the orthoses demonstrated substantial assistance of ankle motion after training, producing around 50 % of normal ankle plantar flexor torque in stance.

Intelligent and powered leg prostheses are presented and discussed in Case Study 9.3. The study provides details of the rationale for integration of a multilevel human/system interface through advanced sensory control, artificial intelligence and actuation techniques. It also presents the detailed analysis required to define the partial objectives of a leg prosthesis during human walking. The authors describe the role played by artificial intelligence in the operation and observation of human–system interfacing. Advanced motorized knee units included in the most advanced leg prosthetics are presented, together with considerations on how to achieve natural performance in different scenarios.

The hybrid assistive limb developed at the University of Tsukuba is presented in Case Study 9.4. The study introduces the system as an assistive device for the operator's lower limb. It presents the method included in version HAL-3 for the case of a swinging motion of the lower leg, describing how the voluntary motion of the operator is detected by gathering myoelectric information and how the HAL generates torque and controls the viscoelasticity of human joints.

Case Study 9.5 presents a pneumatic full-body exoskeleton which was developed at Kanagawa Institute of Technology. This is a classic example of exoskeleton control based on purely physical interaction. To detect the user's movement surely and safely, the project has developed a muscle hardness sensor, which basically consists of a load cell that measures the force exerted by the muscles. Direct-drive pneumatic actuators were designed for the elbows, knees and waist using generic cuffs for blood pressure measurement. The exoskeleton does not include any structure in front, as it is intended to help nurses carry patients bodily.

Cognitive control of a robotic wheelchair is presented in Case Study 9.6. A robotic wheelchair cannot strictly speaking be considered a wearable robot; however, the system introduced in this study is a good illustration of the concept of an EEG-based cHRI. This section deals with the control architecture of a robotic wheelchair and the acquisition, processing and classification of EEG patterns as a command input to drive the wheelchair; it also looks briefly at the experimental validation of the system.

10.2 CONCLUSIONS AND OUTLOOK

Wearable robotics is the next logical step after service robots and personal robots, the difference being the closer cognitive and physical interaction between wearable robots and humans. Given the intrinsic combined operation of humans and robots in wearable robotics, a systemic biomechatronic approach is required in order to develop this scientific area fully.

The direction of modern biomechatronics, in particular in the area of wearable robot design, must exploit four research avenues:

- Increasing miniaturization, chiefly in component design, so that more compact sensor, actuator and energy storage technologies can be adopted. Miniaturization will pave the way for lower energy consumption by these technologies. This in turn will make it possible to establish wireless communication networks from which wearable robotics can benefit.

- Increasing intelligence, mainly in the areas of intelligent cognitive and physical human–robot interaction. Cognitive interaction should naturally detect user intention as an input to control the robot. It needs to be two-way in order to allow proprioceptive feedback. Safety and dependability of physical interaction should be a priority in cooperative human–robot systems.
- More compact solutions based on multifunctional components are to be developed. This falls into the first avenue of research, since sharing several functions, e.g. actuation, sensing and control, in the same component also contributes to miniaturization of designs.
- Integration of hybrid (artificial and biological) systems. This will foreseeably result in systems where the borderline between artificial and biological components eventually disappears and the biological and artificial components are closely interfaced.

As stressed throughout the book, the cognitive interaction between wearer and robot is of paramount importance. As to cHRI schemes based on EMG, it is fair to say that the chief drawback of sEMG lies intrinsically in the sensing technology. The measurements are strongly dependent on several factors and may change in response to movements, sweating and skin conditions. Crosstalk is another major problem with sEMG, especially when many commands or classes are needed to control the WR. The best way to reduce crosstalk is to use distance muscles, but this yields more complex and unnatural contraction patterns. The other option for reducing crosstalk is to use signal processing techniques. Again, this will be strongly dependent on time-dependent conditions and random factors. In addition, the signal processing methods used to eliminate the crosstalk introduce undesired delays in obtaining the final output and are therefore not suitable.

The different techniques that have been proposed for feature extraction generally yield good results, which can be used to control WRs.

Alternative means of monitoring muscle activity are currently being studied (Farrell and Weir, 2005). In the early 1980s, the use of PNS signals for HMi was already envisaged, reinforced by new technologies such as nerve cuff electrodes. The most widely accepted way is to use PNS signals to decode natural control commands for WR control, for instance in hand prostheses (see Case Study 8.2 and Dario *et al.* 2005). The future of these technologies is promising, and the bionic man no longer looks entirely like a dream.

The trends in new-generation brain-based cHRI are moving in two main directions: the use of new algorithms to improve feature extraction and classification and the use of electrodes implanted in the cortex. The use of new algorithms is related to the application of novel signal processing techniques or to the application of conventional techniques on new bands of the EEG signal.

One of the main directions of research is to make cHRI more flexible and adaptable. EEG signals are dynamic, and therefore the control system must be able to adapt in such a way that it can recognize the correctness of the user's choices and adjust to the changes in the user EEG. Adaptation to the user can be accomplished by evaluating his/her response to the outputs of the interface system. Recent studies on the identification and use of error potential in BCI systems point the way to better integration of the interface system. Such integration can be used to implement online training of the system just as the user trains her-himself to use the BCI (Buttfield, Ferrez and Millán, 2006).

Another approach is to use the time evolution of features. Brain processes are not discrete; they evolve from one state to another over time. Thus, evaluation of the evolution of characteristics can be a help in assessing mental states (Bashashati, Ward and Birch, 2005).

The use of implanted electrodes has recently been considered as an approach to cHRI, but studies on humans have been very limited owing to safety and ethical issues. Most of the research in this area is conducted on animals (rats, cats and monkeys) (Hochberg *et al.*, 2006; Lebedev and Nicolelis, 2006). For more information on this topic, the reader is referred to Case Study 4.7, where the application of cortical brain activity for BCI is discussed.

Safety and dependability of wearable robots is a major concern when managing the physical human–robot interaction. Safety and dependability are both dependent on issues like the controlled application of load between the two actors, the kinematic compatibility of the two kinematic chains, the human limb and the wearable robot, ergonomics and comfort.

There are two main aspects to be borne in mind when designing support systems for wearable robots: the anatomical areas and structures that can support effective loads and the maximum levels of pressure that these structures can handle without compromising safety and comfort. Also, joint movement areas, bony prominences, surface tendons, surface vessels or nerves and highly irrigated areas must be avoided in the design of systems for load transmission.

The study in Section 5.6 also indicates that, on the one hand, tolerance of pressure is uniform over the entire forearm but is higher in the hand. On the other hand, the areas of least tolerance of pressure in the lower limb are over the tarsal bones and the inner face of the leg.

Kinematic incompatibility between actors, i.e. offsets between their axes of rotation, is a source of constraint on displacements and forces between an exoskeleton and a human limb. Case Study 5.5 shows that ergonomic design of an exoskeleton is important. If an exoskeleton is nonergonomic, shear forces between the operator's limbs and the exoskeleton can cause discomfort to users. Such shear forces are produced by kinematic mismatch alone and have nothing to do with actuation. To eliminate these interaction forces, passive joints can be incorporated in the mechanical structure of the exoskeleton, as proposed in Section 5.2.

An alternative way of eliminating undesired interaction forces is to look closely at human biomechanics for insights that can be used for compliant kinematic designs. This is the approach discussed for instance in Section 3.5, where the bioinspired design of a knee joint for a lower limb wearable robot is presented.

Another possible source of discomfort is the microenvironmental condition at the interface between wearable device and human skin. Microclimate is one of the most important issues as far as comfort is concerned. Wearable devices modify dry heat exchange by convection, conduction and radiation and the transfer of damp by evaporation. Such modifications can increase sweating through heat accumulation in the body parts covered by a wearable device; sweat can accumulate between the body and the wearable device and may cause discomfort and maceration of the epidermis. Therefore, thermal comfort needs to be evaluated by means of microclimatic sensing in order to select the most suitable material and design for improved comfort.

Wearable robots were first proposed in the military arena. Currently they are being successfully proposed in rehabilitation, functional compensation of physical impairment, empowering (in general assistance), telemanipulation and space. Applications will foreseeably be extended and the coming years will probably see closer-knit and hybrid human–robot systems. If this development is to be successful, special attention must be paid to safety, dependability and ethics.

REFERENCES

Bashashati, A., Ward, R.K., Birch, G.E., 2005, 'A new design of the asynchronous brain computer interface using the knowledge of the path of features', in *Proceedings of the Second International IEEE Engineering in Medicine and Biology Society Conference on Neural Engineering*, pp. 101–104.

Buttfield, A., Ferrez, P.W., del Millán, J.R., 2006, 'Towards a robust BCI: error potentials and online learning', *IEEE Transactions on Neural Systems and Rehabilitation Engineering* **14**(2): 167–168.

Dario, P., Carrozza, M., Guglielmelli, E., Laschi, C., Menciassi, A., Micera, S., Vecchi, F., 2005, 'Robotics as a future and emerging technology: biomimetics, cybernetics, and neuro-robotics in European projects', *IEEE Robotics and Automation Magazine* **12**(2): 29–45.

Farrell, T., Weir, R., 2005, 'Pilot comparison of surface vs. implanted EMG for multifunctional prosthesis control', *Proceedings of the 9th International Conference on Rehabilitation Robotics*, pp. 277–280.

Hochberg, L.R., Serruya, M.D., Friehs, G.M., Mukand, J.A., Saleh, M., Caplan, A.H., Branner, A., Chen, D., Penn, R.D., Donoghue, J.P., 2006, 'Neuronal ensemble control of prosthetic devices by a human with tetraplegia', *Nature* **442**: 164–171.

Lebedev, M.A., Nicolelis, M.A.L., 2006, 'Brain–machine interfaces: past, present and future', *Trends in Neurosciences* **29**(9): 539–546.

Index

Wearable Robots: Biomechatronic Exoskeletons Edited by José L. Pons

Printed and bound in the UK by
CPI Antony Rowe, Eastbourne

Printed and bound by CPI Group (UK) Ltd, Croydon, CR0 4YY
16/04/2025
14658391-0001